Advances in Geographical and Environmental Sciences

Advances in Geographical and Environmental Sciences synthesizes series diagnostigation and prognostication of earth environment, incorporating challenging interactive areas within ecological envelope of geosphere, biosphere, hydrosphere, atmosphere and cryosphere. It deals with land use land cover change (LUCC), urbanization, energy flux, land-ocean fluxes, climate, food security, ecohydrology, biodiversity, natural hazards and disasters, human health and their mutual interaction and feedback mechanism in order to contribute towards sustainable future. The geosciences methods range from traditional field techniques and conventional data collection, use of remote sensing and geographical information system, computer aided technique to advance geostatistical and dynamic modeling.

The series integrate past, present and future of geospheric attributes incorporating biophysical and human dimensions in spatio-temporal perspectives. The geosciences, encompassing land-ocean-atmosphere interaction is considered as a vital component in the context of environmental issues, especially in observation and prediction of air and water pollution, global warming and urban heat islands. It is important to communicate the advances in geosciences to increase resilience of society through capacity building for mitigating the impact of natural hazards and disasters. Sustainability of human society depends strongly on the earth environment, and thus the development of geosciences is critical for a better understanding of our living environment, and its sustainable development.

Geoscience also has the responsibility to not confine itself to addressing current problems but it is also developing a framework to address future issues. In order to build a 'Future Earth Model' for understanding and predicting the functioning of the whole climatic system, collaboration of experts in the traditional earth disciplines as well as in ecology, information technology, instrumentation and complex system is essential, through initiatives from human geoscientists. Thus human geosceince is emerging as key policy science for contributing towards sustainability/survivality science together with future earth initiative.

Advances in Geographical and Environmental Sciences series publishes books that contain novel approaches in tackling issues of human geoscience in its broadest sense — books in the series should focus on true progress in a particular area or region. The series includes monographs and edited volumes without any limitations in the page numbers.

More information about this series at https://link.springer.com/bookseries/13113

Narayan Chandra Jana · Anju Singh · R. B. Singh
Editors

Livelihood Enhancement Through Agriculture, Tourism and Health

Editors
Narayan Chandra Jana
Department of Geography
The University of Burdwan
Bardhaman, West Bengal, India

R. B. Singh
Department of Geography, Delhi School of Economics
University of Delhi
Delhi, India

Anju Singh
Department of Geography, Aditi Mahavidyalaya
University of Delhi
Delhi, India

ISSN 2198-3542 ISSN 2198-3550 (electronic)
Advances in Geographical and Environmental Sciences
ISBN 978-981-16-7309-2 ISBN 978-981-16-7310-8 (eBook)
https://doi.org/10.1007/978-981-16-7310-8

This Springer imprint is published by the registered company Springer Nature Singapore Pte Ltd.
The registered company address is: 152 Beach Road, #21-01/04 Gateway East, Singapore 189721, Singapore

Message

The world is currently facing two enormous crises in the form of the COVID-19 pandemic and the climate emergency. As I write this short piece, the headlines are all about two key challenges facing humanity. While the novel coronavirus continues to wreak havoc in most countries, extreme weather events are becoming ever more frequent and damaging. Even in developed countries, where substantial proportions of the adult population have been vaccinated, coronavirus infection rates and deaths are still rising. When the health care and economic stresses brought about by the pandemic are combined with the reality of rapid anthropogenic climate change, the need for actions towards a more sustainable future becomes all the more obvious. Accordingly, the appearance of a volume that comprehensively explores three themes central to improving human livelihoods is both pertinent and timely. *Agriculture* is seriously threatened by climate change and, given the negative impact of the coronavirus on the global economy (especially on the rural and urban poor and not least in South Asia), it is more important than ever to find ways to improve both its efficiency and sustainability. *Tourism* can make significant contributions to local and regional economies, but the industry has been brought to its knees due to imposed travel restrictions and lockdowns. *Health* issues have been at the very forefront of our minds in the last 2 years (sadly, we all know people who have succumbed to COVID-19), and the need to improve our understanding of the spread of disease and access to adequate healthcare facilities must be considered a priority. The chapters in this volume testify to the role that geographers can play in addressing the multiple challenges that society currently faces and I congratulate the editors on bringing together such a comprehensive range of studies in the quest for a more sustainable future.

Prof. Michael E. Meadows
B.Sc. (Hons), Sussex, Ph.D. Cantab.
FSSAG, FRGS, FRSSAf, FAAS,
President: International Geographical Union
University of Cape Town
Cape Town, South Africa

Nanjing University
Nanjing, China

Foreword

Geography has been dealing with a combination of issues. It is known for integrating themes in a spatial form. This discipline attempts to analyse in totality and hence creates a better ecosystem for understanding and unfolding the complicated patterns. This volume on *Livelihood Enhancement through Agriculture, Tourism and Health* is a ready example of the contributions made by the eminent geographers in this direction. Narayan Chandra Jana, Anju Singh and R. B. Singh have joined hands in bringing out this edited volume. Needless to say, I am delighted to write the Foreword acknowledging their contributions in promoting Geography.

The changes are continuously happening which may be slow and sometimes not even conspicuous. However, the spatial interrelations are endlessly altering. The associated issues change in different ways. Catching up with such changes is really a challenge. Due to global warming, climate change and population increase, there has been acute livelihood crisis all over the world, especially in the underdeveloped regions. In fact, Asian countries are the largest contributor to world population where lack in the source of livelihood is a regular phenomenon. It is also one of the priority areas of the government while formulating planning strategies for regional development. In all these countries, water shortage, crop failure, low productivity, deforestation and the like exert a serious blow to the means of livelihood, which present a million dollars question mark on the quality of life as well. Not all but some of the pressing issues on agriculture, tourism and health for enhancing livelihood have been discussed in this publication. Examples from aboard have also been included. Considering some indications of possible reductions of population is some areas of the world in future, the analysis and strategies have become more interesting and complicated.

The present volume is a collection and compilation of 26 research papers of diverse interests on agriculture, tourism and health towards livelihood enhancement. The authors contributed to this book are mainly from India, Bangladesh, Sri Lanka, Brazil, Canada and Hungary. The range of issues focused on national, regional and local dimensions are not only confined to the discipline of geography but also to the interest areas of allied fields as well.

I would like to appreciate the editors for their arduous task in bringing out this precious volume of contemporary relevance. This book would be an addition to the existing literature and may attract the attention of geo-scientists, researchers, development planners and decision-makers engaged in dealing with the multifarious aspects of livelihood, agriculture, land use studies, tourism management, health care, tribal development and so on in the South Asian countries and beyond. I am confident that this volume will be well received in the academic platform and intellectual arena.

The name of R. B. Singh has been associated with this volume who had made important contributions to Indian and global geography initiatives. This volume should be considered as a precious remembrance of his contributions as well.

August 2021

Dr. Prithvish Nag
Former Vice Chancellor: MGKVP
Varanasi, India

UPRTOU, Prayagraj (Allahabad)
SSVV, Varanasi, India

DDU Gorakhpur University
Former Surveyor General of India
Former Director, NATMO Kolkata
West Bengal, India

Preface

In the present juncture of socio-economic and environmental crisis, there are million dollars questions on the access to sources and means of livelihoods. People of the developing and underdeveloped countries suffer a lot because of the large population pressure in comparison to their existing resource base. Poverty and malnutrition is the common phenomena. Therefore, access to natural resources by the poor is essential for sustainable poverty reduction. The livelihoods of rural people without access or with very limited access to natural resources are vulnerable because they will face difficulty in obtaining food, shelter and other assets. Livelihoods of the poor can be enhanced through economic diversification, i.e. agriculture, tourism, health and other activities.

Agriculture is the backbone of our economic system. It not only provides food and raw material but also employment opportunities to a very large number of populations. Higher temperatures will have an impact on yields, while changes in rainfall could affect both crop quality and quantity. Climate change could increase the prices of major crops in some regions. For the most vulnerable people, lower agricultural output means lower incomes. Climate Change is expected to increase the risk of illnesses and death from extreme heat and poor air quality. The recent evidence is COVID-19 pandemic. Climate change will also affect the occurrence of infectious diseases. A number of well-known diseases are climate-sensitive, such as malaria, dengue fever, cholera, etc. Tourism is considered to be an industry and alternative source of economy. It can generate employment opportunities and boost up the national economy. So focusing on the issues of agriculture, tourism and health for livelihood enhancement is relevant in the present juncture of crisis.

This book covers 26 research papers of diverse interests related to Livelihood, Agriculture and Land Use, Tourism, Urban Issues, Disease and Healthcare Issues, Tribal Development, Development and other Technological issues in the South Asian Countries and beyond. The issues covered at different levels are: livelihood transformations and sustainability, mobility, migration and livelihood of informal migrant workers during COVID-19 pandemic, livelihood change and water scarcity, increase of sediment delivery due to tobacco cropping, irrigation management planning, small holder tea farming, transition of traditional agriculture, land use–land cover dynamics in micro-watershed, spatio-temporal changes in crop combination and development of sericulture, opportunities and the challenges of tourism industry, tourism development in Himalayas and tourism potentials of fossil parks, quality of life in million plus urban agglomeration, urban rejuvenation and social sustainability, opportunities and challenges of urban wetlands, spatial inequalities of amenities in a city, land suitability analysis of settlement concentration using GIS-based multi-criteria decision-making technique, GIS-based healthcare accessibility analysis, critical analysis of dengue outbreak, ground reality of tuberculosis patients and road connectivity and nodal accessibility of maternal healthcare service centres, socio-ecological perspective of tribal development, development of tribal livelihood, fourth paradigm in geographical sciences and establishing relationships of cellular communication coverage.

It is important to note that Geography is a subject of inter-disciplinary nature. The range of issues on national, regional and local dimensions discussed here are not merely the sole areas of geography but also the interest areas of other disciplines as well. The diverse issues dealt in detail in this proposed volume may attract the attention of geo-scientists and researchers of allied fields like Livelihood, Agriculture, Land Use Studies, Tourism management, Health care, Tribal Studies and so on. This book may be of immense help to the researchers, scientists, planners and decision-makers engaged in dealing with problems of livelihood, agriculture, tourism, tribes and healthcare aspects in the developing countries and beyond.

India is predominantly a rural country, which is witnessing several transformations in all its major domains. The rural settlements and their livelihoods are also no exception and observing changes in new economic order. The trends suggest that there is declining share of agriculture in the national economy, whereas urban population is increasing at a faster rate, which threatens agricultural environs and influence their economic activities significantly. It also adds complications to rural livelihood sustainability. Chapter 1 entitled *Livelihood Transformations and Sustainability in India* by Shahab Fazal, Deepika Vashishtha and Salma Sultana attempts to evaluate transforming status of rural livelihood sustainability in the states of India. The present study is primarily based on the secondary sources of data, collected from various governmental and ministerial publications. UNDP's normalization method was incorporated to standardise indicators and a modified form of IPCC's Vulnerability Index was used to develop 'Sustainable Livelihood Index' (SLI). This index is taken as a base for formulating 'Livelihood Ladder', adapted from the Oxfam Report. Major findings of this paper reveal that there are large-scale inter-state disparities for different assets, where central and eastern states of India are found to be poor on livelihood sustainability due to their lower human, social and financial capital and thus more vulnerable to present-day shocks and stresses, while southern and northern states are better placed in terms of livelihood sustainability.

The present study in Chap. 2: *The Exodus in Times of Pandemic: Mobility, Migration and Livelihood of Informal Migrant Workers During COVID-19 Crisis in India* by Twisha Singh and Anuradha Banerjee focuses on an assessment of the impact of COVID-19 on mobility and livelihood of informal migrant workers in India. The paper is based on a theoretical background built upon the arguments of development economists and scholars of migration. Case studies based on media reporting and newspaper clippings, presentation by scholars have been used to highlight the context of the discussion. The analysis also rests on data collected from Government documents, NSSO and the Census of India, 2011. The focus is on the lower end of the diverse labour market in India, that on one hand constitutes insecure tail end jobs, unequal wages and poor access to basic amenities and infrastructure, deprivation of social security on one hand and on the other constitutes a major part of Government's nativist project, as it is an essential intra-national economic tool for development, however categorized by unevenness at its core. The paper places an exploratory effort to identify the institutionalized inequalities and mounting vulnerabilities during the 'exodus' of the informal migrant workers, amidst the nationwide lockdown. The research finds that there is a necessity to rethink mobility and migration that has become apparent due to the COVID-19 crisis, and the fallout out of which has been extremely detrimental to the poorest section of the Indian workforce, i.e. the informal migrant workers, of which a vast section constitute the seasonal labourers. This also raises the issue of redistributive conflict and locating the problematic in the allocation of resources on one hand and on the efficacy of the government schemes in reaching the defenseless 'poorest of the poor' on the other. The question here is also indicative of the inherent and persistent problems of data and reliable estimates involving this vast segment of the informal labour, required for enforcement of social safety nets.

Water scarcity is a main constraint and has become a serious threat in meeting domestic purposes as well as the livelihood as a whole of the resettled people in Kilinochchi district, Sri Lanka. Keeping in view the current scenarios, the paper in Chap. 3 entitled *Challenges in Livelihood of Residents in Kilinochchi District, Sri Lanka Due to Water Scarcity* by P. Kirishanthan, call for an investigation on challenges in livelihood due to water scarcity and its impacts on sustainable development goals. This research is carried out in Karachchi Divisional Secretariat Division that belongs to Kilinochchi district. The study manipulates a combined approach of both qualitative and quantitative methods. It relies mainly on primary data that were collected through household survey, structured interviews and observation, while secondary data from certain sources were also used. The questionnaire survey was conducted with randomly selected 214 households from the purposively selected 5 Grama Niladhari divisions. Qualitative data and quantitative data of the study were analysed using content analysis, and descriptive statistics, respectively. Frequency distribution tables, pie charts and graphs were used to display the findings of the data analysis. Findings of the study include various livelihood challenges: monthly income, cost of living, agriculture, home gardens, livestock, inland fishing, unemployment and small business which are negatively impacted by water scarcity in the study area. Further, the study indicates that the sustainable development goals no.: 1- No Poverty, 3- Good Health and Well-being, 5- Gender Equality, 8- Decent Work and Economic Growth and 9- Industry, Innovation and Infrastructure are highly constrained to achieve due to the negative impacts on livelihoods caused by water scarcity, and that SDG 6- clean water and sanitation and SDG 2-Zero Hunger are extremely threatened in the study area due to water scarcity. Therefore, relevant stakeholders should take urgent actions to ensure the availability and accessibility of water to address the challenges faced by the households in Karachchi, Kilinochchi district, in terms of livelihoods and sustainable development.

In Chap. 4 entitled *Tobacco Cropping Increases Sediment Delivery in a Subtropical Agricultural Catchment in Southern Brazil* by Edivaldo Lopes Thomaz, Fátima Furmanowicz Brandalize, Valdemir Antoneli and João Anésio Bednarz; the study shows how the agricultural calendar affects the concentrations of suspended sediments both spatially and temporally in an agricultural catchment. A nested monitoring design was deployed with a group of small headwaters within a second-order catchment. Suspended sediment concentration (SSC) was performed at each monitoring site with a set of rising-stage sampling collectors. Analysis of the land use dynamics showed a clear intra-annual sediment transfer into the aquatic system. The SSC in the catchment differed in summer under tobacco cropping (*Nicotiana tabacum* L) compared to that in winter under oat cropping (*Avena sativa* L). The accumulated sediment in summer was 67% higher than in winter. We found that the structural and functional hydro-geomorphic connectivity in an area with tobacco crops in the channel expansion zone caused significant hillslope-channel sediment transfer. In addition, the tobacco areas have plenty of connectors causing disruptions and enhancing the sediment delivery into the stream.

Digital Elevation Model (DEM) is thought to be an important aspect in the irrigation management. In order to ensure food security in a land-hungry country with high population density, the Government of Bangladesh has decided to use every inch of agricultural land for cropping. Therefore, DEM is essential for proper irrigation management in Bangladesh. The main objective of the study of the present paper in Chap. 5 entitled *Digital Elevation Model and Irrigation Management Planning in Bangladesh* by M Manzurul Hassan and Md.Ashraf Ali is to prepare DEM to explore the areas suitable for irrigation in Kishoreganj upazila/subdistrict (205 square kilometres in area) under Nilphamari district with water supply from surface water source. Printed topographic maps prepared between 1960 and 1964 with contour lines (with 1:15,840 for 1 foot contour maps) from Bangladesh Water Development Board (BWDB) were used for preparing the DEM. The collected topographic maps for the study site were digitized with ArcGIS (version 10x) format. Field investigations were carried out using Global Positioning System (GPS, Model: Garmin eTrex 30) for relevant positional information, levelling devices (G2-32X) for exploring spot-height within a selected Location of Interest (LOI) and Google Earth images for identification of physical features. The study site is mainly a gentle slope with Reduced Level (RL) between 30.0 and 48.0 metres from the Mean Sea Level (MSL). Soil property for agriculture and agricultural productivity is very suitable in the study site. The DEM Map, Area-Elevation-Discharge Curve and GPS data show different land features in the study area for irrigation. Our study shows that about 11460 hectares (55.9%) land are available for irrigation subject to provide sufficient water during the dry season with developing new canals from the nearby perennial Teesta River. The irrigation suitable area covers very gently slope within the RL of 37.0–44.0 metre from MSL. The prepared DEM shows different physiographic characteristics and suitable irrigation areas in the study site. Suitable land for irrigation covers very gentle slope and there is high opportunity to increase suitable land area with developing new canals from nearby Perennial River. The prepared DEM will be helpful for future irrigation and agricultural development planning. The delineated small, narrow and terrain features in the study site can be utilized for potential irrigation.

In Chap. 6, Chinmoyee Mallik focuses on *Smallholder Tea Farming in West Bengal, India: An Exploratory Insight*. India, like most of the developing countries, is dominated by small holder farmers. While these small farms were typically of subsistence type, recently a considerable proportion of them have massively shifted in favour of cash crops in many parts of the country. This is very intriguing because escalating economic vulnerability of the small farmers due to erosion of state support from the farm sector in the neo-liberal policy context and concomitant monetization of small holder economy are self-contradictory. Although South-East Asian countries have already experienced such a phenomenon few decades back with respect to the rubber production, the Indian tea production, particularly in Assam and West Bengal, is following a similar trajectory. This paper is mainly based on Agricultural Census of India, National Sample Survey unit level data and an exploratory field work in the tea producing district of Jalpaiguri in West Bengal undertaken in 2019. The field work consists of an exploratory quantitative survey as well as in-depth interviews of few small tea growers to understand the recent trends and patterns of restructuring of the pre-existing agricultural system in the region. This paper seeks to draw insights from cropping pattern shift away from food crop towards cash crops and the socio-political and economic environment associated with this recent phenomenon. It emerges that the small farmers have shifted cropping away from food crops to cash crops and that the small farmers who have adopted tea farming have mostly replaced paddy cultivation.

Chapter 7 deals with *The Transition of Traditional Agriculture in Nagaland, India: A Case Study of Shifting (Jhum) Cultivation* by Devpriya Sarkar. *Jhum* cultivation is a practice of clearing the vast forest land for cultivating crops, where the land is left fallow after one or two growing cycles. It is the dominant form of traditional agricultural practice and continues to be the significant component of the livelihood of the state Nagaland's village communities. However, this has undergone many changes over time through various evolving practices and policy interventions. Further, the field survey found out that these self-sufficient communities are also looking towards modern economic development and activities and aim to earn more for a 'better' life. Such a changing attitude is showing reflection in their everyday lives, while state policies and innovations find ways to penetrate these traditional systems.

In Chap. 8, Raj Kumar Samanta and Narayan Chandra Jana focuses on the *Land Use-Land Cover Dynamics in Baku Micro-watershed Area of Ausgram Block—II, Purba Barddhaman District, West Bengal, India*. Baku Micro-watershed project has a positive impact on land use land cover aspects like increase of double-crop area, plantation activities, etc. It indicates a good mark of watershed development. In this paper, such kinds of spatio-temporal changes of land use land cover patterns have been analysed. Relation of physiography and soil with land use has been identified. Here, low lands cover fertile soil with more prosperous agriculture and uplands are the areas of laterites with forest cover. The maximum portion of soils is sandy and acidic. Low lands indicate good quality land with a gentle slope and good quality soil. Land assessment and planning has been carried out for further better land utilization. The paper is based on extensive field works. Simple statistical and cartographic techniques have been applied to show various results.

Chapter 9 deals with *Spatio-temporal Changes of Crop Combination in Selected C.D. Blocks of Purba Bardhaman District, West Bengal, India* written by Chanchal Kumar Dey and Tapas Mistri. Undivided Burdwan district was forever known as the *Granary of Bengal* but now, Purba Bardhaman district has occupied the greater parts of the agriculturally advanced area. To find out the agricultural zones of the district, crop combination method has been used to understand in which areas how many crops are cultivated and to what extent. So the present study is concentrated to find out the spatio-temporal changes of crop combination and their comparative analyses in selected blocks located in both sides of the Damodar River in Purba Bardhaman district from 2000–01 to 2015–16. The present study is based on mainly secondary data and perception studies of the farmers. QGIS tools and Coppock's crop combination method have been used. Finally, changing trends have analysed that shows two blocks having unchanged status, while four blocks are gradually decreasing and two blocks are showing the rising trend.

Chapter 10 is devoted to discuss the *Development of Sericulture in Murshidabad with Special Reference to Women's Participation* by Abhirupa Chatterjee. Sericulture, being an agro-based labour intensive industry, includes both agricultural and industrial aspect and thus refers to the activities from cultivation of silkworm food plants, rearing of silkworms, obtaining silk up to weaving. As this industry mainly depends on human power, it helps to provide an ample employment opportunities to the developing counties, likewise in India and considered as a remunerative cash crop, whereas being retreated from the developed countries because of the increasing labour cost. Silk known as "Queen of Textiles" is an inseparable part of Indian ritual. India has secured second position in raw silk production with more than 18% of the world's total production. Women play a vital role in this industry as 60% of the work has been done by them and simultaneously 80% of silk is consumed by them. In West Bengal, sericulture plays as an important role in rural avocation by creating family employment round the year. The Murshidabad district of West Bengal is well equipped in the production as well as weaving of silk and so, as a matter of fact the silk industry of West Bengal which is mainly confined around this state sometimes goes by the name of 'Murshidabad Silk'. This paper intends to analyse the active participation of women in the development and also the current status of sericulture as well as the silk industry of Murshidabad.

In Chap. 11, K. M. Rezaul Karim tries to explore the *Opportunities and the Challenges of Tourism Industry in Bangladesh.* Tourism is a growing industry all around the world. Though Bangladesh has a huge prospect to improve tourism due to its natural scenery and enriched heritage, the industry flops to the extent of its end because of its different challenges. The *Sundarban* mangrove forest, *Shatgombuj* Mosque, *Paharpur* Buddhist Vihara are the three world heritage sites in Bangladesh. The total contribution of tourism to GDP is 4.4%, but the global contribution is

10.4% in 2018. Attracted by the natural beauty of the country, a significant number of domestic and overseas tourists visit its different tourist sites. Despite its immense potentials, the sector is facing different challenges like inadequate infrastructure and backward communication system, deficiency of accommodation facilities, lack of safety and security, scarcity of professionalism, lengthy visa processing, political instability, which are discouraged both international and internal tourists to visit the attractive places in the country. Tourism also brings socio-economic and environmental benefits for the country, albeit mass tourism is also associated with negative effects on the social environment. The opportunities for tourism of the country are religious tourism, sports tourism, eco-tourism, educational tourism, spa tourism, rural tourism, cultural tourism, etc. But there is a lack of research and plan to explore the development of tourism industry of Bangladesh. The paper tries to focus on the important and attractive tourist spots and the impact of tourism on the economy of the country. It also explores the challenges and opportunities of the tourism sector of Bangladesh.

Chapter 12 focuses on the *Socio-economic Development Through Tourism: An Investigative Study for the Himalayan State Sikkim, India* authored by Debasish Batabyal and Dillip Kumar Das. Having been accorded the statues of the largest service industry in India, tourism is an instrument for economic development and employment generation, particularly in remote and backward areas. Tourism in the Indian Himalayan Region has shown a perpetual and increasing trend over many decades despite several disasters and crises. Sikkim is one of the peaceful, hospitable small Indian states boasts of rich ecological and cultural diversities. Tourism in almost all the alpine region of the Indian sub-continent is conceptualized to have been a mean of spending from disposable and discretionary income, mostly for non-essential activities. This old and stagnant idea has been changing drastically as tourism either energies through a total long-lasting experience or an essential mean of the present time. On the supply side aspect, it is imperative to provide new avenues for income and employment in the destination. This article has dealt with the modern socio-economic environment of tourism in the backdrop of its essential sustainable development indicators. More specifically, this article has shown how tourism phenomenon is influencing the ecology and community benefits with important tourism marketing and destination supply trends. The study is based on primary data collection of tourists and local community in Sikkim and the statistical tools used herein is Kendall's Coefficient of Concordance.

In Chap. 13, an attempt has been made by Rahul Mandal and Premangshu Chakrabarty to explore *Tourism Potentials of Fossil Parks as Geoheritage Sites: A Study in Western and South Western Region of West Bengal, India.* Fossils are paleontological treasures manifesting preserved remains, impression or traces of organisms that existed in past geological ages. Fossil parks are one of the major geotourism attractions especially when they attain the status like world heritage sites. From geotourism promotion perspective, the fossil parks have exceptional heritage and scientific values as admired by UNESCO. Geosite and geomorphosite tourism is still at its juvenile stage in India despite of the positive efforts of Geological Survey of India who recognized 26 geological sites as National Geological Monuments including the seven fossil parks. With increasing interest on fossils as geoheritage, geopark network of the country has been strengthened with inclusion of new sites. In the year 2006, angiosperm wood fossils have been discovered from Illambazar forest of Birbhum District in and around a tribal village named Amkhoi. A fossil park is inaugurated in the year 2018 and its success encourages its extension and further planning for geotourism. This paper is an attempt to apply SWOT-AHP analysis on Amkhoi Fossil Park, the first fossil park of West Bengal which is youngest among the fossil parks of the country in order to evaluate its sustainability aspects.

Chapter 14 attempted for *Classifying the Million-Plus Urban Agglomerations of India—Geographical Types and Quality of Life* by habil. Zoltán Wilhelm, Róbert Kuszinger and Nándor Zagyi. India is one the fastest growing and developing economies, and also societies of the world. An evident consequence of this is urbanisation, which poses a huge challenge for the population and the political decision-makers of the country and is also one of the most topical research trends of the social geographical researches concerning India. The paper first introduces the general urbanization trends experienced in the sovereign India in the 1951–2011 period, in the framework of an analysis of statistical data recorded in censuses, indicating the volume and trends of urbanization. This is followed by the demonstration of the structural features and diverse development paths of the million-plus agglomerations (i.e. agglomerations with at least a million inhabitants), connected to one of its main characteristics depicted by this introductory summary: metropolization. Using the quantitative categories defined during their analysis, the authors classify the metropolises of India in urbanization types, with the method of cluster analysis. In what follows, we sought to answer whether any correlation could be justified between these urbanization types and the complex quality of life indicators we generated for the central settlements of the agglomerations.

Chapter 15 is devoted to discuss *Urban Rejuvenation and Social Sustainability in Smart City: An Empirical Study of Community Aspirations* written by Virendra Nagarale and Piyush Telang. Sustainable environment is the need of any society. Urban renewal in Smart Cities is the major concern for incorporation of sustainable environment in cities. Significant increase in population and lack of proper planning strategy has led to the series of problems of urban decay in metropolitan areas intimidating community well-being and security. To seize urban decay, Rejuvenation is normally an adopted move towards regeneration of rundown areas. Rejuvenation often results negatively and may lead to bother existing social networks. The success of the renewal practices mostly depend on active participation of residents. The approach of sustainable development in urban rejuvenation should balance the interests of stakeholders in different socio-economic and demographic class. According to Ease of Living Index 2018, Pune ranks highest among 111 cities in India while in Smart city ranking by Ministry of Housing and Urban Affairs (MoHUA) in 2019 Pune stood 11th with 213.50 marks. For ranking purpose, MoHUA considers a variety of factors like performance of civic institution, spent expenditure and implementation of different projects in 5 years. On the one hand, Pune is considered as the most liveable city in contrast its ranking being a smart city shows negative effect. Pune

City is selected for the present study that is governed by Pune Municipal Corporation (PMC). PMC governs 331.26 sq. km area comprising 15 administrative and 144 electoral wards. The total population of PMC is 3371626, where 452240 are SCs and 37630 are STs. The present study explores the preferences and aspirations of citizens in regard to urban renewal through smart city mission. The questionnaire survey is been used to collect the responses of citizen from Pune Municipal Corporation. This paper aims to assess how residents perceive the urban renewal strategy through smart city mission and to identify the responses in view of individual's socio-economic and demographic structure. The results show that how age, educational level, employment status, etc., changes the perspective of responses.

In Chap. 16, Prasanna Priyadarsini and Ashis Chandra Pathy focus on the *Urban Wetlands: Opportunities and Challenges in Indian Cities—A Case of Bhubaneswar City, Odisha.* Urban wetlands functions differently than those in natural areas. In natural wetland, the water level is not changing rapidly until and unless natural events occur in urban context, the water level of wetlands can fluctuate more rapidly due to anthropogenic activities which affect the ecosystem services of the wetland. The landscape of urban wetlands does not only have the role of carbon sink, water accumulation, cleaning and drainage but also it binds the nature with city dwellers. Yet, as built up spaces within urban areas have increased, these treasures have undergone a drastic decline. Analysis of published land use and land cover data from 22 cities by Wetlands International South Asia team, indicates that during 1970–2014, every one square kilometre increase in built-up area matched up with a loss of 25 ha wetlands. The main thrust of the study is to analyse the major causes behind wetland loss in the capital city and also to assess the wetland ecosystem services for the existing wetlands, its conservation and management.

In Chap. 17, an attempt has been made for *Analyzing Spatial Inequalities of Amenities in Jammu City Using Geo-Informatics* by Rajender Singh, Kavleen Kaur, Sarfaraz Asgher, Davinder Singh and Sandeep Singh. Since the partition of India and Pakistan, the City of Jammu has emerged as a most favorable and suitable destination to settle down, for the people coming from other side of the border during 1947 and 1965 war between India and Pakistan and also people from inside the country thereafter. During this period, the city of Jammu has shown unprecedented growth of population from 157708prs in 1971 to 576195prs in 2011. Keeping in view the flow of migrants and the widespread growth of population in Jammu city, current research is an attempt to analyse the availability and disparities in the spatial distribution of basic amenities using simple geospatial technique among the different wards of Jammu City. The outcome of the research shows that basics amenities among the different wards of Jammu City are not uniformly distributed. With a composite score of more than 43.47, about 26% of the total wards show high levels of overall development. There are only seven wards falling in the medium line. On the contrary, a highly drastic picture has been observed, as more than 63% of the total wards (i.e. 45 numbers of wards) show low to very low levels of overall development. Broadly speaking, old city wards are found to have adequate urban amenities when compared to peripheral wards. The findings of the current research are extremely useful to decision-makers for the comprehensive formulation of urban policy.

Chapter 18 is devoted to *Land Suitability Analysis for Settlement Concentration in Fringe Area of Siliguri Town, West Bengal (India)—A GIS-Based Multi-Criteria Decision-Making Approach* by Sanu Dolui and Sumana Sarkar. One of the crucial questions among the urban planner is to determine suitable locations for future urban expansion, especially in areas adjacent to large cityscapes. Plain land, fertile soil, an opportunity for a livelihood and a decent transportation system has always encouraged human habitation; however, adverse physical environments and inadequate livelihood opportunities have always constrained urban expansion. The present study area, Siliguri town, stretched over Darjeeling and Jalpaiguri districts which is third largest urban agglomeration in the Indian state of West Bengal and continues to grow at a rapid pace. After considering various socio-economic, environmental and physical factors, the final suitability map of settlement concentration was prepared using remote sensing technique, AHP and FAHP method. Among the selected factors, seven are found as favourable factors, viz. elevation, slope, distance from the river and road, distance from main settlement patches, changes in an existing built-up area, night-time light images; and three are discouraging factors, viz. dense forest cover, river flood-prone area, distance from tea garden and protected landscape. The generated thematic maps of these criteria were standardized and given weights according to their importance to each other using a pairwise comparison matrix applying AHP and FAHP methods. The final suitable map was classified into four suitable zones; in the highly suitable zone, 92.66% of pixels are matched with both the weightage method. This study revealed that FAHP was marginally more useful than AHP in detecting future urban expansion. This research may be useful for optimizing land use planning and help urban planners in the decision-making process.

Chapter 19 focuses on the *GIS-Based Healthcare Accessibility Analysis—A Case Study of Selected Municipalities of Hyderabad* by Srikanth Katta and B. Srinagesh. The spatial access and dynamics of a changing population in urban areas with changing healthcare needs require frequent and logical methods to evaluate and assist in primary healthcare access and planning. Spatial or geographical access is an important aspect in the planning process. Healthcare accessibility analysis based on GIS is a logical method which can be applied to test the degree to which equitable access is obtained. In reality, a person will always go to their closest facility; GIS analysis is however based on this assumption of this rational choice. Inputs to the analysis are supply in the form of healthcare facilities and demand estimates in the form of people who are actually seeking the healthcare service. Hyderabad healthcare system is a dual system made up of private and public healthcare facilities. Private healthcare system is expensive and only affordable to rich class. In the present study, GIS analysis is applied to determine three distinct demand scenarios based on a combination of three variables: (a) Household income groups, (b) Age criteria, (c) Chronic diseases and Healthcare emergencies. GIS is used to determine catchment or buffer areas for each healthcare facility, allocating demand to its closest healthcare facility limiting access based on facility capacity and accessibility through a road network. The catchment or buffer area analysis results from each of the three demand scenarios are compared with actual situations in the form of nearest facilities and mapped origins of number of users at each facility. The major objective of the study is to show the use of GIS to quantify and improve the access to healthcare resources in terms of availability (supply of services which meets the population needs) and Accessibility (physical access along with travel time and cost) in Circle No. 9 of GHMC, Hyderabad.

Chapter 20 is devoted to discuss the *Dynamics of Disease Diffusion: A Critical Analysis of Dengue Outbreak in Kolkata and Adjacent Areas* by Teesta Dey. While medical advancement has conquered most infectious diseases in the twenty-first century, the majority of the tropical and sub-tropical developing countries are still fighting against various vector-borne diseases as an inevitable consequence of the climate change. The predominance of vector-borne diseases in Kolkata has become a continuous threat to human health. Rapid increase of disease incidence, proliferation and fatality rates among dengue patients creates multidimensional effects on the socio-political scenario of urban daily lives. In the last 10 years, the city has experienced a gradual transformation of Dengue Ecology among the city dwellers. Over time, the virus has changed its temporal Disease–Population Dynamics including its transmissibility, rate of replication, infectivity and virulence. Major dengue outbreaks in this city exhibit the ineffective surveillance, improper urban environmental planning, vulnerable living status, insufficient political strategies, poor perception quality and ignorance. High rate of dengue-induced death poses an important question regarding the effectivity of health planning in a metro city like Kolkata. In addition to that, discrepancies in dengue-related reports and data have become associated with disease ecology and the health politics of Kolkata. In this context, an attempt has been made to focus on the spatio-temporal transformation of the dengue virus, to analyse the severity pattern among the patients, to generate dengue-prone area mapping based on spatial autocorrelation method, to correlate the spatiality factor of dengue pockets with local socio-economic conditions and finally to study the changing dengue Dynamics in this city from a geographical perspective.

In Chap. 21, Tapan Pramanick, Deb Kumar Maity and Narayan Chandra Jana have made an attempt to analyse the *Tuberculosis Patients in Malda District of West Bengal, Eastern India: Exploring the Ground Reality*. Tuberculosis is one of the leading causes of mortality in the world which is caused by mycobacterium tuberculosis and spread through the air. This study tries to analyse the socio-demographic profile of tuberculosis patients and their behavioural conditions. The present study was conducted among the all tuberculosis unit (TU) of Malda District during 2019. Malda District is divided into 20 tuberculosis unit and each TU comprises more than two lakh population. A total number of 167 sample survey was carried out during a primary survey using simple random sampling technique, where 89 male patients and 78 female patients were surveyed using a structured questionnaire. Total sampling size is selected from four categories patients, i.e. 91 from category-I patients, 28 from Category-II patients, 43 from category-IV (MDR) patients and 5 from Category-V (XDR) patients. The most common age group found from the survey is 21-30 years were 58 (34.73%). The least common age group is<60 years 12(7.18%). The Category-I patients were 91(54.49%) and MDR 43(25.75%). 127 (74.06%) patients cannot use a mask. Among the total patients 78 (46.70%) were illiterate and 5 (2.99%) graduate. 51 (30.54%) were SC category, 101 (60.47%) were income level <5000.

Data gathered from this survey found that 97 (58.08%) population are daily labour and 39 (23.35%) are *bidi* workers. From the demographic profile of the surveyed population it is found that mostly illiterate and unconscious people are affected by this disease, not only educationally backward but socially backward peoples are also affected by this disease. People having malnutrition, drug addiction and smoking represent the main section of tuberculosis patients, which is a concern for health planners. Findings of this study may be useful for identifying the sections of people who are being affected by this disease and the planners and medical practitioners may plan to minimize the rate affected people through understanding the situation of the study area based on this study.

Chapter 22 devoted to *An Assessment Study on Hierarchical Integrity of Road Connectivity and Nodal Accessibility of Maternal Health Care Service Centres in Itahar Block, Uttar Dinajpur District, West Bengal* by Madhurima Sarkar, Tamal Basu Roy and Ranjan Roy. Access and reachability to healthcare centre is an important issue for lucrative delivery of health services to its recipient. The basic essence of emergency service facility is to assess the degree of reachability through its connectedness and accessibility. Better connectivity and accessibility would provide nodal services with greater extent. The first and foremost objective of this study is to recognize the spatial location of different healthcare service centres as a nodal service point with its linkage perspective in Itahar block, Uttar Dinajpur District. The study has given its emphasis on to recognize the relative location of maternal healthcare service centres, its network alignment, connectivity and accessibility. The study put its effort to highlight the fact that only maternal healthcare services equipped with better quality is not enough to give its optimum until and unless the better accessibility is achieved through the said services to its recipient. The entire study involves in acquiring, and analysis of the spatial data such as discrete location of the maternal healthcare centres, its weathered road connectivity, degree of reachability and to recognize the spatial extent of its services for each and individual healthcare service centre. In this regard, the road network connectivity to each healthcare centre has been taken into consideration. The entire analysis has been carried out through the Geospatial analysis techniques. A matrix algebra technique, different algorithm regarding network analysis has been carried out to assess and evaluate the connectivity and accessibility of maternal healthcare service centres in Itahar block.

Chapter 23 focuses on the *Strategies for Sustainable Tribal Development in Purulia District, West Bengal, India: A Socio-ecological Perspective* by Somnath Mukherjee. Social ecology is simply an approach to understand the human world through some interconnected elements like ecology, social structure, culture, economy and polity. The social ecology can interpret the societal condition and many plans and policies for sustainable development of both man and environment can also be made by the study of social ecology. Social ecology discerns that the outrage of the development and growth have destroyed the ecological settings of the tribes. The tribes have been uprooted from their pristine land and compelled to live with mixed culture. The said actions affect both the tribal economy and the tribal identity. The Kharia Sabar of the eastern part of the Chotanagpur plateau is the most affected tribal community who lost their traditional habitat and has compelled to live with the stigma of criminality followed by extreme poverty and social exclusion. The present study explores some strategies and suggestions for the sustainable tribal development by understanding the life and livelihood of the tribe under socio-ecological lines.

In Chap. 24, Sumanta Kumar Baskey and Narayan Chandra Jana have analysed the *Development of Tribal Livelihood in Manbazar-II Block of Purulia District, West Bengal, India.* Development is a multi-dimensional phenomenon. It includes some variables like the level of economic growth, level of education, level of health services, degree of modernization, status of women, level of nutrition, quality of housing, distribution of goods and services, access to communication, etc. Development of tribal livelihood is not uniform among the five tribes in Manbazar-II block because they have primitive traits, geographical isolation and distinct culture, economic backwardness and as well as of their limited engagement on different functional activities. It has been examined with mentioned variables regarding some selected developmental indices. Disparity has been found within them. Overall development will be increased through improvement of common minimum needs and their awareness. And also true development requires Government action especially for women. Role of social development such as literacy (particularly female literacy) in promoting basic capabilities emerges as the prerequisite to overall development.

Chapter 25 has discussed about *The Fourth Paradigm in Geographical Sciences* by Sandeep Kundu. Science has evolved from being empirical (1000 years ago) and theoretical (100 years ago) to computational (decades ago) and now, in the twenty-first century, it has entered into a new paradigm. This new paradigm, referred to as the 'Fourth Paradigm' is based on data-driven science and is changing the way we derive scientific insights. High volumes of data generated at varying speeds, spread over different geographies and stored digitally in a variety of formats constitute 'Big Data' which on which scientific analytics are being targeted. Artificial intelligence and machine learning algorithms constitute the core of present-day analytics providing new insights different from the traditional approaches like empirical, theoretical and computational. Industry and businesses are leading the way forward in leveraging onto this fourth paradigm which has huge potential in Geographical Sciences. Burning issues in health, population, migration, public policy, society, sales & marketing, climate and environment or energy & sustainability are now being addressed through this new paradigm. This article discusses the fundamentals of data science with linkages to geographical science and summarizes few key applications of 'The Fourth Paradigm' in Geographical sciences.

Chapter 26 has focused on *Establishing Relationships of Cellular Communication Coverage Provided by Governmental and Non-governmental Companies as a Function of Digital Elevation, Population Density, and Transport Infrastructure in Jodhpur District, Rajasthan* by Aswathy Puthukkulam, Sanjay Gaur, T. R. Vinod and Anand Plappally. With recent unplanned development and steep human population density increase in Jodhpur District during 2010–2019, spectral congestion can be an impending problem. The article emphasis lies in the analysis of coverage calculated based on Okamura–Hata model with respect to several distinct parameters. Jodhpur city, lying on the Vindhyan porous plateau (with high water management potential) has the highest density of mobile towers in the district. Mobile towers installed by government communication companies are mostly across the rustic campaign. These are distributed with relatively uniform density. Most of the district is characterized by negative normalized difference vegetation indices and low slopes confirming desert (Thar Desert) climate. Therefore, population density, railway routes, road infrastructures and townships are major parameters which define cellular tower installations. Governmental company-based tower installations account for less than 20% of the non-governmental cellular tower installations. Coverage of the governmental towers is much higher than those of non-government cellular tower installations in terms of land surface area. WebApp Builder in ArcGIS is utilized to present this scenario which here at provides an opportunity to perform design, development and planning of future cellular tower installations.

The editors are grateful to the authors of 26 chapters of this present book hail from India, Bangladesh, Sri Lanka, Hungary, Brazil, Canada and Singapore for their valuable contributions on the livelihood enhancement through agriculture, tourism, health, etc. We are thankful to Dr. Sujay Bandyopadhyay and Mr. Sasanka Ghosh of Kazi Nazrul University; Tapan Pramanick, Buddhadev Hembram and Ratan Pal of Burdwan University; and Sri Kalikinkar Das of Gour Banga University, West Bengal, India, for active cooperation in the preparation of this manuscript. We are also grateful to Springer Nature, Singapore, for accepting this volume for publication.

Bardhaman, India — Narayan Chandra Jana
Delhi, India — Anju Singh
Delhi, India — R. B. Singh

Contents

Editors and Contributors

About the Editors

Dr. Narayan Chandra Jana is an Applied Geographer with a postgraduate and doctoral degrees in Geography, postgraduate degree in Disaster Mitigation, PG diploma in Sustainable Rural Development and diploma in Tourism Studies. He has contributed more than 100 research papers published in various national and international journals and edited volumes. He has *authored three books* entitled (i) The Land: Multifaceted Appraisal and Management (with Prof. N. K. De), (ii) Transformation of Land: Physical Properties and Development Initiatives, (iii) Tsunami in India: Impact Assessment and Mitigation Strategies; *jointly edited five books* entitled (i) Disaster Management and Sustainable Development: Emerging Issues and Concerns (with Prof. Rajesh Anand and Dr. Sudhir Singh), (ii) Human Resources (with Prof. Sudesh Nangia and Prof. R. B. Bhagat), (iii) West Bengal: Geo-Spatial Issues, (iv) Resources and Development: Issues and Concerns (with L. Sivaramakrishnan and others) and (v) Population Dynamics in Contemporary South Asia: Health, Education and Migration (with Prof. Anuradha Banerjee and Dr. Vinod Kumar Mishra). He was also actively engaged in postdoctoral research and teaching in the Centre for the Study of Regional Development, Jawaharlal Nehru University, New Delhi, for 5 years. He was the Coordinator (Eastern India) of the *International Geographical Union: Commission on Geography of Commercial Activities* (1992–96). He is the Life Member of 24 academic societies of repute. He was the Vice-President of *National Association of Geographers, India* (NAGI), Delhi (2011–12, 2012–13, 2013–14 and 2019–20) and was the Convener of 33rd Indian Geography Congress, 2011; 35th Indian Geographers' Meet, 2013 and XIV IGU-INDIA International Conference, 2020. He was the Deputy Coordinator of UGC-SAP-DRS Programme (2012–17) and the Coordinator of DST-FIST Programme (2012–13). He is the Member of the Editorial Board of *Indian Journal of Landscape Systems and Ecological Studies*, Kolkata, and Member of the Advisory Board of the journal *Earth Surface Review*, Gorakhpur. He was the Secretary of *Institute of Indian Geographers* (2016–19) and Founder Secretary of *Association of Bengal Geographers*. He is also the Editor of newly launched journal *Contemporary Geographer*. He is the Member of PG Board of Studies in Geography in Kazi Nazrul University, Asansol; Bankura University and Coochbehar Panchanan Barma University, as well as BRS Member in Geography in the University of Gour Banga, Malda and Bankura University, West Bengal. He visited *Nepal (1994), Sri Lanka (2012), Bangladesh (2013), Thailand (2013), Russia (2014), China (2016), Japan (2016)* and *Thailand (2017)* for various academic purposes. He has delivered about 100 lectures in Academic Staff College, UGC-sponsored national seminars and various academic departments of different universities. He has successfully guided 11 M. Phil and 11 doctoral dissertations. He has completed one major research project entitled Tribal Livelihood and Sustainable Development in Mayurbhanj, Orissa, sponsored by ICSSR and one small research project on Wasteland sponsored by NRDMS of DST. He has also conducted one research methodology course sponsored by ICSSR. His areas of research interest cover Applied Geomorphology, Hazards & Disasters, Environmental Issues, Land Use and Rural Development. He has been nominated as Steering Committee Member in *IGU Commission on Research Methods in Geography*. Currently, He is a Professor (Former Head) in the Department of Geography & Coordinator, M. Sc. in Geospatial Science, University of Burdwan, West Bengal, India.

Dr. Anju Singh is now serving as an Associate Professor in the Department of Geography, Aditi Mahavidyalaya, University of Delhi. She specializes in the field of climatology, land use and land cover study, urban geography, water resource management and coastal ecosystems. She has received international travel grants to present research findings at international forums in Japan. She has published one book and more than ten chapters in edited books and in journals of national and international repute. She is currently actively associated with the National Association of Geographers, India (NAGI), and the Association of Geographical Studies (AGS). She is also a Member of the prestigious BRICS Countries project on Satellite Validation sponsored by the Department of Science and Technology, Government of India. She is also one of the editors of the Springer volume on Water Resources Management and Sustainability. She has been invited by the National Institute of Open Schooling (NIOS) and the Indira Gandhi National Open University (IGNOU) for preparing reading material.

Dr. R. B. Singh (Date of Birth: 3 February, 1955), is an Ex. Professor since 1996, Coordinator UGC-SAP-DRS III (2014–2019) and the Ex. Head in the Department of Geography, Delhi School of Economics, University of Delhi, Delhi-7 (2013–2016 and 2019–20). He earlier served as UGC Research Scientist-B/Reader (1988–1996), Lecturer (1985–1988) and CSIR Pool Officer (1983–1985).

Recently, he got elected as the first Indian and the second Asian Secretary General and Treasurer of the International Geographical Union for 2018–22. He is formerly Chair, Research Council, CSIR—Central Food Technological Research Institute, Mysore; Ex-Member, Research Council, CSIR—Central Institute of Medicinal and Aromatic Plants, Lucknow; Member of International Science Council (earlier ICSU) Prestigious Scientific Committee-Health and Wellbeing in Changing Urban Environment-System Analysis Approach since 2016. He was the Vice-President in International Geographical Union (IGU) since 2012 and is elected again for the second consecutive term (2016–18) of the highest world geographical body. He is invited by IAP-Global Network of Science Academies to join the Working Group for statement on Science and Technology for

Disaster Risk Reduction. He was unanimously Elected President of the Earth System Science Section of the Indian Science Congress Association for 2019–20. NITI Aayog, Government of India, invited him as a Member of the prestigious committee for preparing Vision India—2035. Chair, Task Force, Landslides Awareness, National Disaster Management Authority, Government of India, 2018–2019. He is also International Science Council GeoUnions Standing Committee on Risk and Disaster Management. He is also associated with the prestigious programme such as Co-Chair ISC-CODATA-PASTED, Member Earth System Governance and IAP–Global Network of Science Academies representative on Disaster Risk Reduction. He is one of the three founders of the Centre for Himalayan Studies, University of Delhi, and now Advisory Committee Member.

He has to his credit 15 books, 39 edited research volumes and more than 247 research papers including 123 published in national and international journals (i.e. Remote Sensing, Climate Dynamics, Current Science, Singapore Jl. Of Tropical Geography, Energies, Sustainability, Theoretical and Applied Climatology, Environmental Science and Policy, Physical Geography, Advances in Meteorology, Physics and Chemistry of the Earth, Agriculture, Ecosystem and Environment, Hydrological Processes, Mountain Research and Development, Journal of Mountain Science, Climate, Frontiers in Environmental Science, Advances in Earth Science, Advances in Limnology, European Jl. Of Geography, Asian Geographer, Environmental Economics, Cities and Health, Tourism Recreation Research). He was the Special Series Editor of prestigious journals like Sustainability, Advances in Meteorology, Physics and Chemistry of the Earth, NAM Today. He is the Editorial Committee Member of Jl. of Mountain Science. In 1988, the UNESCO/ISSC (Paris) awarded him Research and Study Grants Award in Social and Human Sciences. He was also associated with prestigious international collaborative research programs such as ICSSR-IDPAD, CIDA-SICI, DFID, Finland Academy of Sciences, UGC and Ministry of Agriculture. He has supervised 38 Ph.D. and 81 M.Phil students. He is a Springer Series Editor—Advances in Geographical and Environmental Sciences; Sustainable Development Goals.

He was awarded prestigious Japan Society for Promotion of Science (JSPS) Research Fellowship at Hiroshima in 2013 and Several Travel Fellowships/Support from UNEP, UNITAR, UNISDR, IAP, UNU, UNCRD, WCRP, IAHS, IGU, NASDA, INSA, UGC, SICI, MAIRS and Univ. of Delhi, etc., for participating and presenting papers, Chairing session and discussing research projects in about 45 countries. He was also associated with Nordic Inst. of Asian Studies, Copenhagen (Denmark) in 1998 and Visiting Professor for delivering invited Lectures at the University of Turku (Finland). He was one of the contributors in the famous-The World Atlas-Earth Concise, Millennium House Ltd., Australia. He was invited by UGC for Preparing National Level CBCS Syllabus for Undergraduate Geography in 2015. He is also the Chair of the UGC prestigious committee for preparing Learning Outcome-based Curriculum Framework since July 2018. Recently, UGC-Consortium for Educational Communication invited him as Academic Advisory Council for CEC MOOCS on SWAYAM. He has been the Chairman-Governing Body of the two prestigious Delhi University Colleges, i.e. Kamla Nehru College and Shaheed Bhagat Singh College. He has been expert in the prestigious Committees of the Government of India—Ministry of Environment and Forests, Department of Science and Technology, National Disaster Management Authority (NDMA), ICSSR, CSIR, UGC, UGC-CEC, NCERT, UPSC NIOS, etc. He is the Fellow of NIE.

Contributors

Md. Ashraf Ali Design Engineer, Water and Environment, Mott MacDonald Ltd., Dhaka, Bangladesh

Valdemir Antoneli Department of Geography, Universidade Estadual do Centro-Oeste, UNICENTRO, Irati, Paraná, Brazil

Sarfaraz Asgher Department of Geography, University of Jammu, Jammu, India

Anuradha Banerjee Centre for the Study of Regional Development, School of Social Sciences, Jawaharlal Nehru University, New Delhi, India

Sumanta Kumar Baskey Department of Geography, Vivekananda College, East Udayrajpur, Kolkata, West Bengal, Madhyamgram, India

Debasish Batabyal Amity University, Kolkata, India

João Anésio Bednarz Department of Geography, Universidade Estadual do Centro-Oeste, UNICENTRO, Irati, Paraná, Brazil

Fátima Furmanowicz Brandalize Escola Pública Municipal, Escola Esperança C. Chuilki, Irati, Paraná, Brazil

Premangshu Chakrabarty Department of Geography, Visva-Bharati University, Santiniketan, West Bengal, India

Abhirupa Chatterjee Nagaland University, Lumami, Nagaland, India

Dillip Kumar Das Department of Tourism Management, The University of Burdwan, Burdwan, West Bengal, India

Chanchal Kumar Dey Ph.D. Research Scholar, Department of Geography, The University of Burdwan, Purba Bardhaman, West Bengal, India

Teesta Dey Department of Geography, Kidderpore College, Kolkata, India

Sanu Dolui Department of Geography, The University of Burdwan, Burdwan, WB, India

Shahab Fazal Department of Geography, Aligarh Muslim University, Aligarh, Uttar Pradesh, India

Sanjay Gaur Jodhpur Institute of Engineering and Technology, Rajasthan Technical University, Kota, Rajasthan, India

M. Manzurul Hassan Department of Geography and Environment, Jahangirnagar University, Savar, Dhaka, Bangladesh

Narayan Chandra Jana Department of Geography, The University of Burdwan, Bardhaman, West Bengal, India;
Department of Geography, The University of Burdwan, Golapbag, Burdwan, India

K. M. Rezaul Karim Department of Sociology, Government M. M. College, Jashore, Bangladesh

Srikanth Katta Department of Geography, University College of Science, Osmania University, Hyderabad, India

Kavleen Kaur Department of Geography, University of Jammu, Jammu, India

Sandeep Kundu National University of Singapore, Singapore, Singapore

Róbert Kuszinger Doctoral School of Earth Sciences, University of Pécs, Pécs, Hungary

Deb Kumar Maity Department of Geography, West Bengal State University, Barasat, Kolkata, India

Chinmoyee Mallik Department of Rural Studies, West Bengal State University, Barasat, Kolkata, India

Rahul Mandal Department of Geography, Visva-Bharati University, Santiniketan, West Bengal, India

Tapas Mistri Assistant Professor, Department of Geography, The University of Burdwan, Purba Bardhaman, West Bengal, India

Somnath Mukherjee Department of Geography, Bankura Christian College, Bankura, West Bengal, India

Virendra Nagarale Department of Geography, SNDT Women's University, Pune Campus, Pune, India

Ashis Chandra Pathy Dept. of Geography, Utkal University, Bhubaneswar, India

Anand Plappally IIT Jodhpur, Jodhpur, Rajasthan, India

Tapan Pramanick Department of Geography, The University of Burdwan, Bardhaman, West Bengal, India

Prashna Priyadarsini Dept. of Planning, College of Engineering and Technology, Bhubaneswar, India

Kirishanthan Punniyarajah Department of Geography, University of Colombo, Colombo, Sri Lanka

Aswathy Puthukkulam Jodhpur Institute of Engineering and Technology, Rajasthan Technical University, Kota, Rajasthan, India

Ranjan Roy Department of Geography and Applied Geography, North Bengal University, Siliguri, West Bengal, India

Tamal Basu Roy Department of Geography, Raiganj University, Raiganj, West Bengal, India

Raj Kumar Samanta Department of Geography, The University of Burdwan, Burdwan, West Bengal, India

Devpriya Sarkar Jawaharlal Nehru University, New Delhi, India

Madhurima Sarkar Department of Geography, Raiganj University, Raiganj, West Bengal, India

Sumana Sarkar Department of Geography, The University of Burdwan, Burdwan, WB, India

Davinder Singh Department of Geography, University of Jammu, Jammu, India

Rajender Singh Department of Geography, University of Jammu, Jammu, India

Sandeep Singh Lecturer in Geography, School Education Department, Jammu, India

Twisha Singh Department of History and Classical Studies, McGill University, Montreal, Canada

B. Srinagesh Department of Geography, University College of Science, Osmania University, Hyderabad, India

Salma Sultana Department of Geography, Aligarh Muslim University, Aligarh, Uttar Pradesh, India

Piyush Telang ICSSR Project, Department of Geography, SNDT Women's University, Pune Campus, Pune, India

Edivaldo Lopes Thomaz Department of Geography, Soil Erosion Laboratory Universidade Estadual do Centro-Oeste, UNICENTRO, Guarapuava, Paraná, Brazil

Deepika Vashishtha Department of Geography, Aligarh Muslim University, Aligarh, Uttar Pradesh, India

T. R. Vinod Center for Environment and Development, Thiruvanathapuram, Kerala, India

Habil Zoltán Wilhelm Faculty of Sciences, University of Pécs, Pécs, Hungary

Nándor Zagyi Szentágothai Research Centre, University of Pécs, Pécs, Hungary

Chapter 1
Livelihood Transformations and Sustainability in India

Shahab Fazal, Deepika Vashishtha, and Salma Sultana

Abstract India is predominantly a rural country. It is witnessing several transformations in all its major domains. The rural settlements and their livelihoods are also no exception and observing changes in new economic order. The trends suggest that there is declining share of agriculture in the national economy whereas urban population is increasing at a faster rate, which threatens agricultural environs and influence their economic activities significantly. It also adds complications to rural livelihood sustainability. This study is primarily based on the secondary sources of data, collected from various governmental and ministerial publications. It attempts to evaluate transforming status of rural livelihood sustainability in the states of India. UNDP's normalization method was incorporated to standardize indicators and a modified form of IPCC's vulnerability index was used to develop "Sustainable Livelihood Index" (SLI). This index is taken as a base for formulating "Livelihood Ladder", adapted from the Oxfam Report. Main findings of this paper reveal that there are large-scale inter-state disparities for different assets, where central and eastern states of India are found to be poor on livelihood sustainability due to their lower human, social, and financial capital and thus more vulnerable to present-day shocks and stresses while southern and northern states are better placed in terms of livelihood sustainability.

Keywords Rural · Livelihoods · Sustainability · Vulnerability · Agriculture · Transformations · India

1.1 Introduction

India is a country of villages, as per Census 2011, 68.84% population is rural with 893 million people, making it the largest rural country of the world. Majority of this population depend directly or indirectly on agriculture for their livelihoods. It is a source of living for an estimated 86% of rural people across the globe (World Bank

S. Fazal (✉) · D. Vashishtha · S. Sultana
Department of Geography, Aligarh Muslim University, Aligarh, Uttar Pradesh, India

N. C. Jana et al. (eds.), *Livelihood Enhancement Through Agriculture, Tourism and Health*, Advances in Geographical and Environmental Sciences,
https://doi.org/10.1007/978-981-16-7310-8_1

2008). The global rural population is now close to 3.4 billion and Africa and Asia are home to nearly 90% of the world's rural population (United Nations 2018). However, the recent trends suggest a decline in rural population of the world from 60% in 1980 to 45% in 2017 (World Bank 2018). While urban areas are experiencing a constant and steady growth of population (WHO 2018). The similar patterns are also observed in India, where urban population is increasing persistently (Census 2011). However, the magnitude of change for both rural and urban population does not match to the same pace as it is still inclined and dominated by rural population (Sudhira and Gururaja 2012). Despite the rise of urbanization, more than half of India's population is projected to be rural by 2050 (United Nations 2012; NITI Ayog 2017). Moreover, the absolute size of rural population is still too critical and enormous as it is two and half times bigger than the total population of USA (Sekhar and Padmaja 2013; The Hindu 2011).

The rural areas in India are experiencing social, economic, political, demographic, cultural, and ecological transformations mainly due to increasing influence from urban areas, bringing features of urban environment into rural settings, shifts in rural ecology, changes to systems, and processes that impact rural people's way of living and livelihoods encompassing agricultural as well as non-agricultural sectors (Ramphul Ohlan 2016; Patil and Dhere 2012).

However, in the process of economic transformation, agriculture sector loses its importance due to its eroding contribution in national income (Webb and Block 2012). Economic policy reforms of 1991 and the formation of WTO in 1995 have brought structural transformations in the Indian agricultural sector (Kamata et al. 2007). The Central Statistical Organization (CSO) reveals that in 1950–51, the share of agriculture in GDP was around 55% which has declined to less than 17% in 2016–17 (Patnaik and Prasad 2014). But still, a very high proportion of labor force (nearly 60%) continues to depend on agricultural sector (Dev 2018). This negative trend has major repercussions from the viewpoints of rural poverty and inequality (Datt and Mahajan 2016; Ramchandani and Karmarkar 2014; Barah 2007).

Since last few years, it has been observed that these transformations in rural areas are exacerbating the sustainability of rural livelihoods (Gupta 2015; Moran et al 2007). One of the key challenges in this regard owes to increasing urbanization process which results in the rapid conversion of fertile agricultural land to urban land use (Fazal 2000). As a result, land-based livelihoods of small and marginal farmers are increasingly becoming unsustainable. As their land has failed to support their families' requirements, they are forced to look at alternative means for supplementing their livelihoods (Banu and Fazal 2017). It is increasingly the case with rural workers, who are more foot loose than before and there is considerable seasonal migration from rural to urban areas for short- and medium-term employment under a variety of arrangements (Agrawal 2008; Petersen and Pedersen 2010). This distress migration badly affects different aspects of their lives. It clearly reflects the poorly productive rural livelihoods in India (Fazal 2014).

Although rural livelihoods are in crisis state, nevertheless, sustainability of rural livelihoods is increasingly attaining central position to the debate about rural development, poverty reduction, and environmental management (Scoones 1998, 2009).

It has been embraced by a number of development agencies, with UNDP and the Department for International Development (DFID).

Keeping all these aspects in view, this study attempts to evaluate the present status of rural livelihood and its sustainability status in the states of India.

1.2 Study Area

This study is carried out for the 28 states of India, which is the seventh largest country, marked by physical and economic diversity with an area of 3.288 million sq. km. India is the world's second most populous country with a population of 1.27 billion (FAO 2018; World Bank Data 2018; Bloom 2011; Nagdeve 2009) (Fig. 1.1). Agriculture, with its allied sectors, is the largest source of livelihoods in India. 70% of its rural households still depend primarily on agriculture for their livelihood, with 82% of farmers being small and marginal (FAO 2018; Bairwa et al. 2014).

Agricultural sector is not merely a source of livelihood but a way of life. It is the main source of food, fodder, and fuel and is the basic foundation of economic development (Mehta 2014; FAO 2012). However, this sector is undergoing various transformations, influencing its sustainability. Moreover, policy reforms of 1991, historical background of states, varying degree of availability and access to assets, capabilities, and their abilities to cope with adverse circumstances largely influence the sustainability of rural livelihoods.

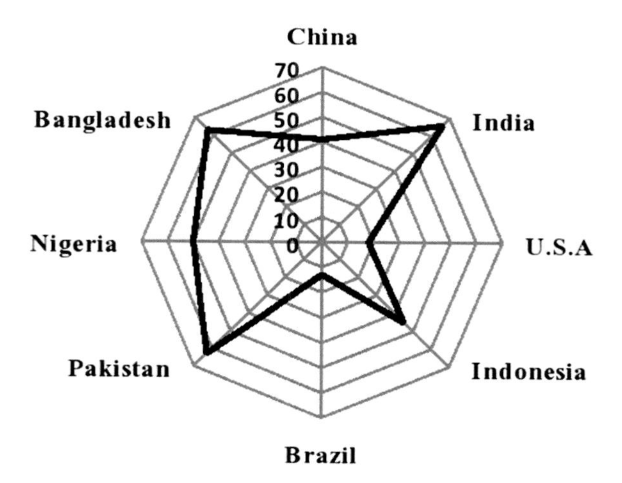

Fig. 1.1 Percentage of rural population in top most populous countries of the world. *Source* World Bank Data, 2018

1.3 Livelihood Sustainability: Meaning and Concept

Livelihoods are "means of making a living", the various activities and resources that allow people to live (FAO 2007). Chambers and Conway (1992) elaborated "A livelihood comprises the capabilities, assets and activities required for a means of living; a livelihood is sustainable which can cope with and recover from stress and shocks, maintain or enhance its capabilities and assets, and provide sustainable livelihood opportunities for the next generation; and which contributes net benefits to other livelihoods at the local and global levels and in the short and long-term". It influences people's lives and well-being, particularly those of the poor in the developing world (Bebbington 1999; Carney 1998; Davies 1996; Rennie and Singh 1996; Bernstein et al. 1992).

Sustainability has many dimensions, e.g., economic sustainability, social sustainability, environmental sustainability, etc. Chambers and Conway used the term "sustainability" for "safeguarding and long term livelihoods". It refers to the maintenance and enhancement of capabilities both now and in the future (Juri Lienert and Paul Burger 2015). Sustainability also aims toward achieving a balance of economic, social, and environmental goals (Greenroads 2008). It is equity and harmony extended into the future (Mega and Pedersen 1998).

Conceptually sustainable livelihoods focus on changing views of poverty, recognizing the diversity of aspirations, the importance of assets and communities, and the constraints and opportunities provided by institutional structures and processes (Ashley and Carney 1999; Sen 1981; Swift 1989; Chambers and Conway 1992; Moser 1998). It is a way of putting people at the center of development, thereby increasing the effectiveness of development assistance (Carney 1998; DFID 2000; Alterelli and Carloni 2000). The Brundtland Commission Report of 1987 offered the first appearance in policy debate of what was conceptualized later as sustainable livelihoods. Furthermore, it was through the 1992 IDS paper that the concept gained widespread acceptance.

1.4 Materials and Methods

1.4.1 Database

This study employs secondary data which were taken from different sources like Census of India, Office of Registrar of India, 2011; Rural Development Statistics, Government of India (GOI), 2011–12; All India Survey on Higher Education, MHRD, Department of Higher Education, GOI, 2012–13; Bulletin of Rural Health Statistics, India, 2014; District Level Household and Facility Survey, 2012, Ministry of Health and Family Welfare, GOI; Sample Registration System, Office of Registrar of India, 2013; National Sample Survey Organisation (66th Round,

2009–10 and 68th Round, 2012), Ministry of Labour and Employment, GOI, 2011–12; Economic Survey, GOI, 2014–15; Sample Registration System (SRS) Bulletin, Ministry of Home Affairs, 2013; Ministry of Road Transport and Highways, 2012; Reserve Bank of India, 2013; Open Government Data (OGD) Platform; Department of Statistics and Information Management, RBI, 2013; India State of Forest Report, 2017; Central Water Commission, 2011; Planning Commission, GOI, 2012; Directorate of Economics and Statistics, 2013–14, Department of Agriculture, Cooperation and Family Welfare, 2015–16. The details of all these data sources are explicitly mentioned in Table 1.1.

1.4.2 Selection of Components and Indicators for Sustainable Livelihood Index (SLI)

An effort has been made to identify appropriate indicators at state level for assessing livelihood sustainability in India. The identification and selection of variables for analyzing well-being, livelihood, quality of life, etc. have always been a contentious issue (Haan 2012; Patidar 2018). The Sustainable livelihood Index (SLI) used in this research is composed of three interacting components of Asset Index, Economic Vulnerability Index, and Sensitivity Index.

(1) **Assets Index**: It is considered as the composite of people, their activities, and accessibility to tangible and intangible resources to develop their livelihoods. Greater access to assets is positively associated with sustainable livelihoods. In this present SLI, asset index comprised of five minor components which are human, physical, social, natural, and financial capital. These were further sub-classified into 31 indicators (Table 1.1) to portray the status of resources in all the 28 states of rural India.

(2) **Economic Vulnerability Index**: It represents economic vulnerability and was employed to measure the magnitude to which a state is susceptible to varying levels of stresses and shocks. Vulnerability has two aspects: external and internal. Higher the economic vulnerability score, greater would be the hindrances in the way of achieving sustainable livelihoods.

(3) **Sensitivity Index**: It refers to the degree to which a system is affected either adversely or beneficially. Here, sensitivity is viewed as complementary to asset index. Both helps in measuring the preparedness levels of state to combat economic vulnerability. The basic difference between these two indices lies in the fact that the asset index assesses the status of livelihood sustainability at macro-level while the latter one emphasizes on micro-level, i.e., household level. The score of sensitivity is positively associated with livelihood sustainability.

Table 1.1 Development of rural livelihood sustainability index (RLSI) comprised of major components, minor components and indicators for the states of India

				Maximum and minimum values determined for each indicator		
Major components	Minor components	Sr. No.	Indicators	Max. value	Min. value	Source of data
Assets	Human capital	1.	Work participation rate	100%	0%	Census of India, 2011
		2.	Total literacy rate	100%	0%	Census of India, 2011
		3.	Female literacy rate	100%	0%	Census of India, 2011
		4.	Sex ratio	1100	600	Census of India, 2011
	Physical capital					
	(a) Education facility	5.	Gross enrolment ratio upto 17 years of age	100%	0%	Rural Development Statistics, GOI, 2011–12
		6.	Gross enrolment ratio in higher education	100%	0%	AISHE, 2012–13, MHRD, Dept. of Higher Education, GOI
		7.	Gender parity in higher education	150%	0%	AISHE, 2012–13, MHRD, Dept. of Higher Education, GOI
	(b) Health facility	8.	No. of health centres including sub centres, PHCs and CHCs	15 (rounded)	1	Bulletin of Rural Health Statistics in India, 2014
		9.	Institutional delivery	100%	0%	DLHS 4, 2012–13, Ministry of Health and Family Welfare, GOI

(continued)

Table 1.1 (continued)

Major components	Minor components	Sr. No.	Indicators	Maximum and minimum values determined for each indicator: Max. value	Min. value	Source of data
		10.	Children vaccination	100%	0%	DLHS 4, 2012–13, Ministry of Health and Family Welfare, GOI
		11.	Infant surviving ratio	1000	900	SRS, Office of Registrar of India, Ministry of Home Affairs, 2013
	(c) Food security	12.	No. of meals per day	3 (ideal)	1 (at least)	NSSO: 66th Round, (2009–10), GOI (2012)
		13.	Calorie intake per day	2700	1400	NSSO: 66th Round, (2009–10), GOI (2012)
		14.	Per capita monthly consumer expenditure for goods and services	3000	700	NSSO: 66th Round, (2009–10), GOI (2012)
	(d) Transport and communication facility	15.	Road density	6000	100	Ministry of Road Transport and Highways, 2012
		16.	Metalled roads	100%	0%	Ministry of Road Transport and Highways, 2012

(continued)

The specific indices of these three major components were selected according to the factors that affect livelihood sustainability of rural population in India (Table 1.1). At the next level, these three major components were classified into five minor components, which were further sub-divided into 44 indicators as proxies to calculate the SLI.

Table 1.1 (continued)

				Maximum and minimum values determined for each indicator		
Major components	Minor components	Sr. No.	Indicators	Max. value	Min. value	Source of data
		17.	Telecommunication density	100%	0%	Rural Development Statistics, GOI, 2011–12
	Social capital	18.	Kisan credit cards	100%	0%	Reserve Bank of India, 2013
		19.	Commercial banks	6000 (based on population size)	50	Dept. of Statistics and Information Management, RBI, 2013
		20.	Women employment rate	100%	0%	Reserve Bank of India, 2013
		21.	Female headed households	300	20	NSSO Report No. 554, 2011–12
	Natural capital	22.	Percentage of forest area	100%	0%	India State of Forest Report, 2017
		23.	Capacity utilization of soil testing laboratories	150%	0%	Dept. of Agriculture, Cooperation and Family Welfare, 2015–16
		24.	Groundwater	200%	0%	Central water commission, 2011
		25.	Cropping intensity	200%	0%	Directorate of Economics and Statistics, DAC & FW, 2013–14

(continued)

Table 1.1 (continued)

				Maximum and minimum values determined for each indicator		
Major components	Minor components	Sr. No.	Indicators	Max. value	Min. value	Source of data
	Financial capital	26.	Monthly per capita expenditure	2700	800	NSSO Report, 68th Round, 2011–12
		27.	Population above poverty line	100%	0%	Planning Commission, GOI, 2012
		28.	Main workers	100%	0%	Census of India, 2011
		29.	Cultivators	100%	0%	Census of India, 2011
		30.	Other workers	100%	0%	Census of India, 2011
Economic vulnerability	Exposure	31.	Unemployment rate	100%	0%	NSSO, 68th Round, 2012
		32.	Population below poverty line	100%	0%	Planning Commission, GOI, 2012
		33.	Agricultural labourers	100%	0%	Census of India, 2011
		34.	Marginal workers	100%	0%	Census of India, 2011
Sensitivity	Housing condition	35.	Good condition houses	100%	0%	Census of India, 2011
		36.	2+ Rooms dwellings	100%	0%	Census of India, 2011
		37.	Electrified households	100%	0%	Census of India, 2011
	Sanitation facility	38.	Tap water from treated source	100%	0%	Census of India, 2011
		39.	Water within premises	100%	0%	Census of India, 2011
		40.	Safe drinking water	100%	0%	Census of India, 2011

(continued)

Table 1.1 (continued)

				Maximum and minimum values determined for each indicator		
Major components	Minor components	Sr. No.	Indicators	Max. value	Min. value	Source of data
		41.	Toilet facility	100%	0%	Census of India, 2011
		42.	Waste water outlets connection	100%	0%	Census of India, 2011
		43.	Banking facility	100%	0%	Census of India, 2011
		44.	Access to specified assets like T.V., mobile, bicycle, scooter and car	100%	0%	Census of India, 2011

Notes The main purpose of selecting these 44 indicators is to depict a comprehensive picture pertaining to the status of rural livelihood sustainability in the 28 states of India. As Telangana, the 29th state, was carved out in the year 2014, it has not been included. Since data for various indicators like those illustrating social and natural capital was unavailable for the Union Territories, they were also excluded to avoid any extremity in data analysis. Maximum and minimum values are set in order to standardize all the indicators into indices, the value of which range from 0 to 1

1.4.3 Formulation of Sustainable Livelihood Index (SLI)

The Sustainable Livelihood Index (SLI) was used to assess variety of livelihood elements of rural population. This index was calculated by the method of equalizing the contribution of each indicator through an average weighted approach. The method used to standardize each indicator was adapted from the method used in calculating Human Development Indices (UNDP 2015). As the indicators selected under major and minor components for calculating sustainable livelihood index differ in their nature and scale, each indicator has been standardized by Formula (1.1) and thus indices for minor components were formed. Eventually these indices were averaged to develop three major indices, i.e., assets index, economic vulnerability index, and sensitivity index.

$$\text{Index } A_i = \frac{A_i - A_{\min}}{A_{\max} - A_{\min}} \quad (1.1)$$

where A_i is the actual value of an indicator of the component and $A_{\max}$ and $A_{\min}$ are the maximum and minimum values of the indicator from the entire dataset which have been pre-determined. After standardization, the value of indices ranges from 0 to 1 to represent extreme unsustainability to high sustainability, respectively, and these indices are free from any measurement units. When each of these indicators was standardized, the value for the components having more than one indicator was derived by averaging the sub-component values using the following Formula (1.2):

$$C_i = \frac{\sum_{i=1}^{n} \text{Index } A_i}{n} \quad (1.2)$$

Source Gautam and Andersen (2016).

where C_i is one of the three major components, i.e., assets, economic vulnerability, and sensitivity. Index A_i is the minor component (s) that makes up the major component and n is the number of minor components in each major component. Here, each index ranges between 0 and 1. Higher scores for asset and sensitivity index approaching to 1 indicate sustainable conditions while high scores for economic vulnerability index reflect poorly sustainable situation in the state. Mean and standard deviation method was used to classify the data in to five categories to get representative status of livelihood sustainability among Indian states.

In this study, an alternative method of measuring SLI was also calculated. It was adapted and modified from the IPCC vulnerability index (IPCC 2001). The degree of sustainability is measured using these three calculated indices of assets, sensitivity, and economic vulnerability. It is a well-recognized fact that the areas having higher value of assets and sensitivity index but lower economic vulnerability are more sustainable and immune to cope up with varying stresses and shocking events (Hahn et al. 2009). Thus, this Sustainable Livelihood Index (SLI) was formulated by adding up the scores of assets and sensitivity index and then their composite scores were divided by the economic vulnerability index. In this way, finally, SLI was formulated and divided into four categories called "Livelihoods Ladder". This concept is originally taken from the "Oxfam Report" and was modified according to the suitability of the present study. The main idea to utilize this ladder was to identify the appropriate benchmarks that help in determining transitions between different rungs on the ladder and thus make it easier to categorize states of India depending on their levels of livelihood sustainability.

After calculating the modified equation of IPCC vulnerability index, the scale for this ladder was derived, which ranges between 0 and 15. The quartile method was used to classify the ladder into accumulating, adapting, coping, and surviving categories. Here, the lowest ladder of sustainability was named as surviving while the most sustainable one was designated as accumulating (Table 1.2).

1.5 Results and Discussions

(A) Status of Selected Assets in Rural India

Assets are the means for people to make their living. It also helps in achieving their aspirations. The sustainable livelihood framework identifies five core asset categories or types of capital upon which livelihoods are built (DFID 1999).

Table 1.2 Details about the transitional stages of livelihood ladder

Livelihood Ladder	Characteristics	Range from calculated scale	Ladder' feature
Accumulating	Life is going well. States own and control an increasing range of assets and can cope with a range of stress and shocks	Greater than 6.8	+ Livelihood + Sustainability
Adapting	Life is tolerable. States own and control some assets, especially financial. However, it is not accumulating and has potential vulnerability to shocks, e.g. increasing rate of unemployment	4.6–6.8	+ Livelihood – Sustainability
Coping	Life is just getting by. States can cope with minor shocks but cannot endure major ones. Decreasing range of assets and poor sensitivity make them economically vulnerable	3–4.5	– Livelihood + Sustainability
Surviving	Life is a constant battle. States are extremely vulnerable to both minor and major eternal shocks with poor social and human assets, e.g. debt from the banks, illiteracy etc	Less than 3	– Livelihood – Sustainability

Notes + denotes positive feature of livelihood ladder while – denote negative feature of livelihood ladder
Adapted from "Oxfam Report" (2006)

(a) **Human Capital**: Human capital represents persons' ability to pursue different activities. Here emphasis was given to individual's skills, knowledge, ability, and opportunity to work (Ellis 2000; Ellis and Biggs 2001). Human capital enables to orient activities and achieve their livelihood objectives. In this paper, human capital was taken as a compendium of four indicators. These were (1) literacy rate, (2) female literacy rate, (3) work participation rate, and (4) sex ratio (Table 1.1).

Literacy is one of the basic indicators for rural development and in some ways person's ability and skills to attain livelihood. Since independence, the level of literacy, especially in rural areas, has improved in India (James and Goli 2017). The rural literacy rate in India is 67.8% (Census 2011), but it has great regional variations. Southern states like Kerala (92.98%), Goa (86.65%), and Tamil Nadu (73.54%) have high literacy rate but in central Indian, populous states like Bihar (59.78%), Jharkhand (61.11%), and Uttar Pradesh (65.46%) have poor literacy rate.

The female literacy is also a critical indicator in this regard, where again despite significant improvements, large-scale disparity exists. However, the trend is more or less similar to general literacy rate among the states (World Economic Forum 2018). But it reflects the varied access and opportunity of livelihood contribution of the individuals. Kerala (90.81%) held the first rank. On the contrary, Rajasthan (45.80%) was at the bottom, while Jharkhand (48.91%), Bihar (49%), and Andhra Pradesh (51.54%) also showed poor literacy rate (Census of India 2011).

Another indicator for measuring human capital was work participation rate, which is the percentage of total workers to the total population (James 2011). India as a whole has the work participation rate of 39.79%, whereas rural work participation rate is 41.8% (Census 2011). Here again Nagaland with (54%), Tamil Nadu (50.7%), Maharashtra (49.8%), and Kerala (36.3%) had higher participation rate, while Uttar Pradesh (33.4%) and Bihar (34%) reported lower work participation rate.

Sex ratio is critical for the society and its economic performance. It is the number of females per thousand males. India has 943 females per thousand males (Census 2011). Interestingly, it is higher for rural areas (947) than the urban areas (926). However, Haryana (882), Jammu and Kashmir (908), and Uttar Pradesh (918) had poor sex ratio with less than 920 females per 1000 males, while Kerala (1078) and Goa (1003) recorded favorable female sex ratio with more than 1000 females per 1000 males in rural areas (Census 2011; GOI 2010).

The analysis revealed that the status of human capital among states of India is not very satisfactory. The human capital score for India was 0.315. Nineteen states had their scores higher than the national average. Prominent among them were Kerala (0.785), Himachal Pradesh (0.700), and Goa (0.625), while nine states had human capital score lower than the national average like Bihar (0.075), Jammu and Kashmir (0.101), and Uttar Pradesh (0.132). Most of the populous states, e.g., West Bengal (0.357), Assam (0.331), and Andhra Pradesh (0.402) reported human capital score near national average (Table 1.3 and Fig. 1.2).

(b) **Physical Capital**: It refers to the basic strength and ability of people to make a living (Solesbury 2003; United Nations 2000). This includes individual's capacity as well as available resources and also access to basic infrastructure required. This paper attempted to assess the status of physical capital among states of India using four indicators. These were education, health, food security, and transport and communication (Table 1.1).

India has made significant progress in physical capital over the last few decades. A lot of emphasis was on education sector to improve literacy and education among females. Government initiatives such as Sarva Siksha Abhiyan and The Right to Education Act have improved the status but quality of education, vocational training, skill development, and low investments in higher education are still the critical gaps. The second indicator is health. There have been enhancements in general health facilities resulting in improved life expectancy (Bhagat 2014). Programs like National Rural Health Mission (NRHM) and Ayushman Bharat, etc. are paving way for secured health care. Nonetheless, current state of our health sector requires actions across multiple dimensions.

Table 1.3 Capital indices and composite capital index in states of India

States	Human capital index	Physical capital index	Social capital index	Natural capital index	Financial capital index	Composite capital index
Andhra Pradesh	0.402	0.481	0.401	0.387	0.515	0.437
Arunachal Pradesh	0.256	0.353	0.286	0.466	0.508	0.374
Assam	0.331	0.292	0.181	0.404	0.381	0.318
Bihar	0.075	0.242	0.155	0.441	0.197	0.222
Chhattisgarh	0.467	0.262	0.313	0.490	0.178	0.342
Goa	0.625	0.655	0.805	0.340	0.728	0.630
Gujarat	0.401	0.354	0.310	0.472	0.448	0.397
Haryana	0.203	0.559	0.172	0.483	0.618	0.407
Himachal Pradesh	0.700	0.628	0.596	0.725	0.584	0.646
Jammu and Kashmir	0.101	0.432	0.215	0.553	0.485	0.357
Jharkhand	0.245	0.224	0.330	0.225	0.128	0.230
Karnataka	0.463	0.467	0.374	0.310	0.471	0.417
Kerala	0.785	0.808	0.340	0.387	0.733	0.611
Madhya Pradesh	0.302	0.294	0.244	0.515	0.262	0.323
Maharashtra	0.544	0.497	0.323	0.323	0.490	0.435
Manipur	0.491	0.386	0.340	0.238	0.425	0.376
Meghalaya	0.427	0.258	0.376	0.386	0.546	0.398
Mizoram	0.639	0.459	0.322	0.288	0.537	0.449
Nagaland	0.584	0.323	0.349	0.326	0.610	0.438
Odisha	0.417	0.326	0.204	0.402	0.199	0.310
Punjab	0.257	0.638	0.261	0.601	0.697	0.491
Rajasthan	0.246	0.414	0.323	0.343	0.500	0.365
Sikkim	0.534	0.527	0.425	0.943	0.583	0.602
Tamil Nadu	0.562	0.622	0.392	0.412	0.506	0.499
Tripura	0.563	0.420	0.218	0.437	0.448	0.417
Uttar Pradesh	0.132	0.256	0.194	0.433	0.302	0.263
Uttarakhand	0.480	0.466	0.437	0.679	0.571	0.527
West Bengal	0.357	0.379	0.178	0.438	0.362	0.343
India	0.315	0.344	0.274	0.438	0.385	0.351

Source Computed from secondary data (sources are given in Table 1.1)

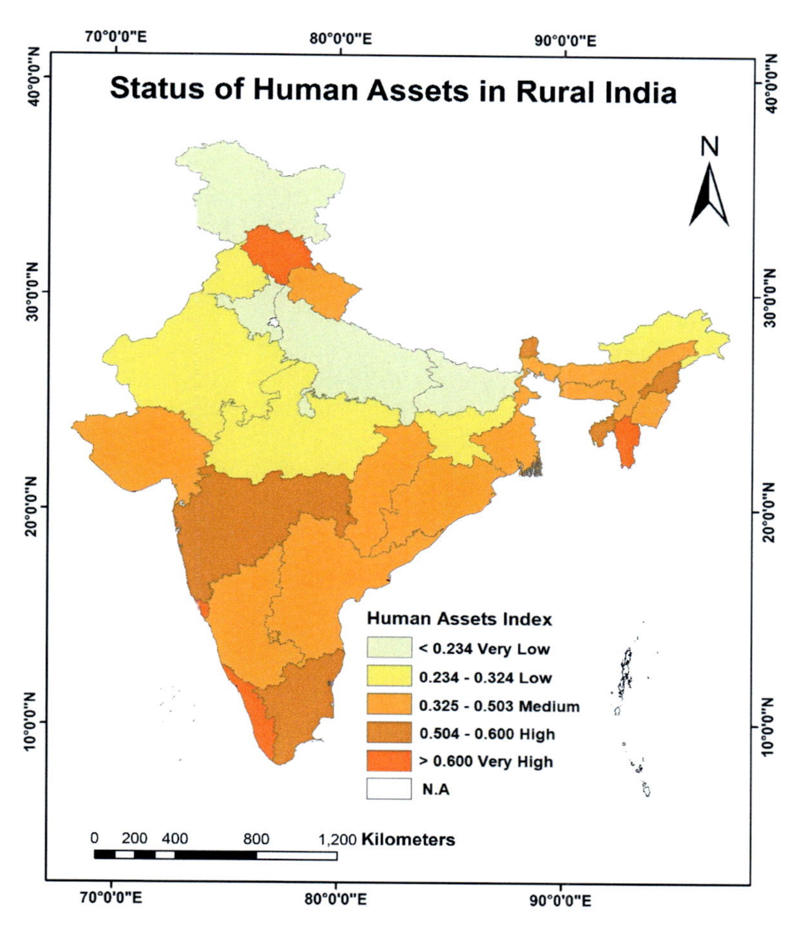

Fig. 1.2 Status of human assets in rural India

Similarly, India has been successful in achieving self-sufficiency in food grains especially after the green revolution. National Food Security Bill and Public Distribution System are the most important mediums to ensure food security (GOI 2013). However, despite such large increase in production of food grains, access to food continues to be a challenging issue. The fourth indicator is transport and communication. It is also called as the lifeline of a country and its economy (Singh and Hiremath 2010). Both are playing significant roles in improving people's mobility and livelihood opportunities. Nonetheless, the focus is on its extension to remote areas, particularly in rural areas. The state-wise analysis of all these indicators is as follows.

(1) **Education**: Education is one of the powerful instruments to improve the general status of people. It helps in getting better livelihood options that result in

reducing poverty and inequality, as well as enhances nation's competitiveness in the global economy. Three indicators were used to measure educational status in India. They were (i) gross enrollment ratio at primary and secondary level, (ii) gross enrollment ratio in higher education, and (iii) gender parity especially in higher education.

The first indicator is primary- and secondary-level Gross Enrollment Ratio[1] (GER) from 7 up to 15 years of age. Kerala (96.89%) occupied first rank in India for this indicator. Nine other states recorded their GER above 90%, while Bihar was on the other extreme with (77.57%), followed by Uttar Pradesh (82.53) and Rajasthan (83.18%) (Rural Development Statistics 2012).

Education is the essence of human resource development. A higher gross enrolment ratio in higher education is closely associated with advancements in research, science, and technology. It was recorded highest in Tamil Nadu (42%), while Odisha (10.1%) had the lowest gross enrollment ratio in higher education followed by Assam (11.2%) and Nagaland (11.8%). Analysis also revealed that enrollment status of 22 states in higher education was below the national average (0.483) (AISHE 2012–13).

The third indicator was Gender Parity[2] in Higher Education. It was reported highest in Kerala (0.988), followed by Sikkim (0.727) and Goa (0.714). However, Madhya Pradesh, Nagaland, and Tripura recorded lower score for gender parity[1] (AISHE 2012–13).

The analysis suggested that national average was 0.315 for the composite index of education with Kerala (0.800) followed by Tamil Nadu (0.713) and Goa (0.652) had better scores. Madhya Pradesh (0.274) and Tripura (0.304) were close to national average, while Bihar (0.076), Nagaland (0.142), and Arunachal Pradesh (0.223) showed poor status. There were totally nine states which recorded their scores lesser than the national average. Majority of them were from central, northeastern, and eastern states of India.

(2) **Health Facility**: Health facility is also crucial to the society, its availability and access are critical for human development (WHO and UNICEF 2013). Four indicators were used to measure status of health facility among states of India. They were (i) institutional delivery, (ii) infant surviving ratio, (iii) children's vaccination, and (iv) number of health centers including sub-centres, public health centers and community health centers (Table 1.1).

Institutional deliveries reduce the risk of maternal and child deaths and enhance their survival chances. In spite of different efforts by the government, still, it is very low in rural areas (47%). However, some states like Kerala (99.6%), Sikkim (98.9%), and Goa (97.1%) had achieved remarkable status in this field, while some other states

---

[1] Gross enrollment ratio (GER): It is defined as the number of students enrolled in a given level of education, regardless of age, expressed as a percentage of the official school-age population corresponding to the same level of education.

[2] Gender Parity Index (GPI) in enrollment at any level of education is the ratio of the number of female students enrolled to the number of male students enrolled at that level.

like Chhattisgarh (39.5%), Jharkhand (46.2%), and Meghalaya (47.3%) revealed a sad reality (Ministry of Health and Family Welfare 2012–13).

The second indicator is Infant Surviving Ratio. Manipur and Goa enjoyed the first place with 989 children per 1000, while Madhya Pradesh (941), Odisha, Uttar Pradesh (943), and Assam (945) were at the bottom of the pile (Sample Registration System 2013).

Another basic indicator was children's vaccination. Goa was well placed with first rank (89%) followed by Sikkim (85.2%) and Kerala (82.5%), while the situation was found worse in most of the northeastern states of India. Nagaland ranked lowest with (35.6%) (DLHS 4, 2012–13). It is noteworthy to mention that Bihar, Jharkhand, Odisha, and Chhattisgarh's performance was relatively better in this regard as the percentage of vaccination for all these states was more than 68.

The availability and access to health centers, including sub-centers, primary health centers (PHCs), and community health centers (CHCs), are crucial for the improved health status. On an average, there were 2.19 health centers (sub-centers, PHCs, and CHCs) per 10,000 rural populations with the highest in Mizoram (0.998) (Rural Health Statistics 2014). The national score for availability of health centers was 0.228. Although six states, namely, Bihar (0.010), Uttar Pradesh (0.047), Jharkhand (0.074), West Bengal (0.085), Haryana (0.087), and Madhya Pradesh (0.097) recorded score even less than (0.1).

The composite index for health in India is not very encouraging. The national average for health was 0.228 and only four states—Goa (0.844), Kerala (0.783), Sikkim (0.740), and Mizoram (0.683)—could make a place in very high category, while 11 states belonged to low and very low health facility with the lowest in Meghalaya (0.169) followed by Uttar Pradesh (0.174).

(3) **Food Security**: India is a large and populous country and access and availability of food is crucial for the population (IFPRI 2019). The access and quality of food were measured using three indicators, namely, (i) number of meals per day, (ii) daily calorie intake, and (iii) per capita monthly consumer expenditure (Table 1.1).

The first indicator was number of meals per day. On an average, 74.25 meals were taken in a month in rural areas, while ideally, people should have at least three meals a day (NSSO[3] 2011). This number was recorded highest in Tamil Nadu with (89.19 meals), followed by Tripura (88.7) and Kerala (88.4), while lowest was reported in Meghalaya (64.58 meals) (NSSO: 66th Round 2010).

The second indicator was calorie intake per day. Planning Commission has suggested 2400 kilocalories per day in rural areas. Jammu and Kashmir was ranked first among the states with (2655) per consumer unit per day calorie, while lowest was recorded in Jharkhand (1769), followed by Chhattisgarh (1847 kilocalories per day) (NSSO: 66th Round 2010).

The average of monthly per capita consumer expenditure on food and non-food items for rural India was Rs. 1278 (NSSO 2014). Ten states recorded their scores less

[3] NSSO stands for National Sample Survey Organization, which comes under the Ministry of Statistics, Government of India and conducts socio-economic surveys on various subjects like employment, consumer expenditure, health services, etc.

than the national score (0.245). The lowest among them was Odisha (Rs. 880.24). In contrary to it, Kerala occupied first position with Rs. 2509.92, followed by Goa (Rs. 2458.24) (NSSO: 66th Round 2010).

The composite food security index reported that more than nine states recorded their scores less than the national average (0.369) with Jharkhand (0.081) being at the bottom. Most of these states are located in central and northeastern parts of India. However, Kerala enjoyed the topmost position with (0.836) followed by Punjab (0.832).

(4) **Transport and Telecommunication Facilities**: Efficient and affordable transportation is an important driver for economic growth and livelihood status in rural areas. Three indicators were chosen to measure their status. These were (i) road density, (ii) metalled roads, and (iii) telecommunication density.

The average road density in rural India is 1580 km. per 1000 sq. km. However, there are substantial disparities among the states and also for the urban and rural areas. Seventeen states of India lied below the national average (0.463). Jammu and Kashmir with 163.58 km per 1000 sq. km recorded the lowest road density in rural areas, while Kerala with a road network of 5543.52 km per 1000 sq. km surpassed all other states (Ministry of Road Transport and Highways 2012).

The second indicator was metalled roads. These are the roads that have a gravel surface, chip seal, tarmac, or concrete. Haryana (90.71%) held the topmost position for this indicator, followed by Gujarat (89.84) and Punjab (89.18%) (Ministry of Road Transport and Highways 2012). The national score for surfaced roads is 0.664 (Ministry of Road Transport and Highways 2012).

Telecommunication density plays a very significant role in improving links between rural and urban areas and also helps in minimizing distance across borders. In India, this density was recorded highest in Himachal Pradesh (0.987), followed by Punjab (0.801) and Kerala (0.758), while it was reported lowest in Jharkhand (0.102), Chhattisgarh (0.121), and Arunachal Pradesh (0.125). More than ten states had their scores lesser than the national average (0.462) (Rural Development Statistics 2012).

The composite physical index (Table 1.3) reveals that Kerala (0.808) stood first among all its indicators' performance, followed by Goa (0.655) and Punjab (0.638). The national average for physical assets in rural India is 0.344. Analysis reported that states of central and eastern India like Odisha (0.326), Madhya Pradesh (0.294), Bihar (0.242), Jharkhand (0.224), and Uttar Pradesh (0.256) fell under the low and very low categories. There is one common thread among these states except Odisha and Madhya Pradesh that they all are densely populated owing to large-scale agriculturally fertile plains. While southern and northern states, particularly Punjab, Haryana, and Himachal Pradesh, were better placed (Fig. 1.3). Moreover, there were 19 states in total with their scores greater than the national score (0.344).

(c) **Social Capital**: The Indian rural society is diverse in social and economic terms. In traditional rural society, there was significant mutual support and cooperation but with time, it diminished. Thus, to assess the present status of external and integral support in rural areas, four indicators were evaluated, namely, (i) Kisan Credit Cards (KCC), (ii) Existence of Commercial Banks in rural areas, (iii) Female Workforce Participation Rate, and (iv) Female Headed Households (Table 1.1).

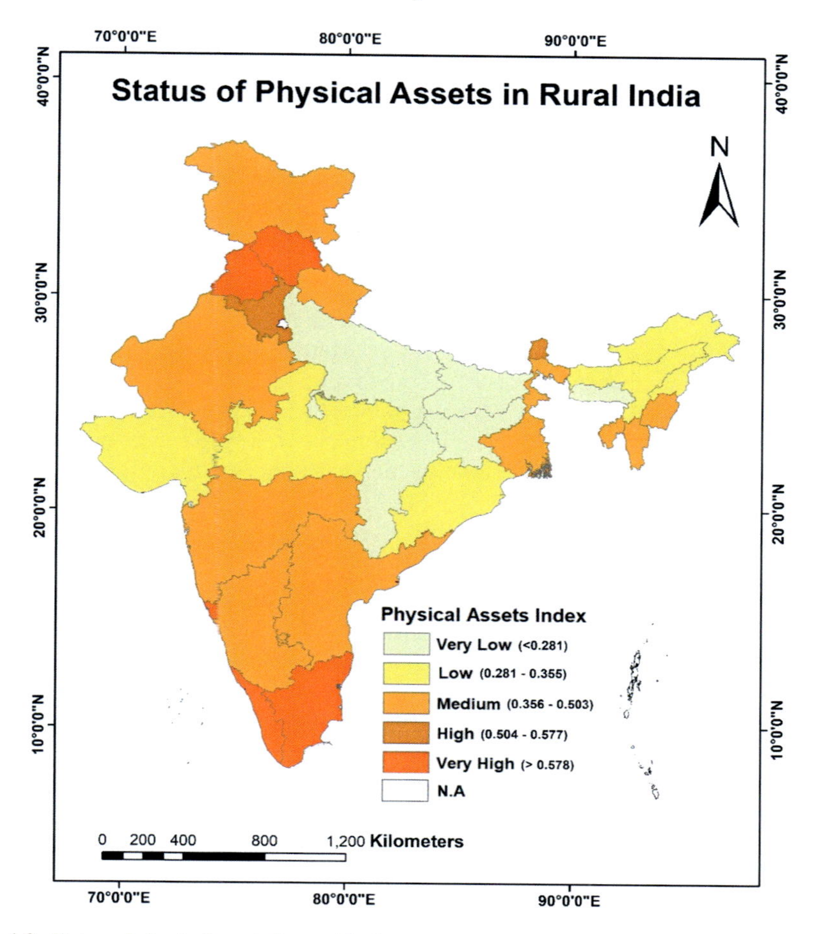

Fig. 1.3 Status of physical assets in rural India

Kisan Credit Card (KCC) scheme was started by the government of India in the year 1998–99 to provide adequate and timely support from the banking system to the farmers in a flexible and cost-effective manner. There were ten states which fell below the national average (0.209) and major states among them which had very low percentage of KCC were Odisha (0.020), followed by Chhattisgarh (0.028), Rajasthan (0.121), and Madhya Pradesh (0.152). On the contrary, Goa had first rank with (0.980), followed by Punjab (0.477) and Gujarat (0.469) recording higher percentage of KCC (Reserve Bank of India 2013; Talreja 2014).

The second indicator was the existence of Commercial Banks (CBs) in rural areas. The analysis presented that majority of the central and the northeastern states reported their scores less than the national average (0.120) with Bihar (0.021), Manipur (0.023), Madhya Pradesh (0.039), and Uttar Pradesh (0.048). However, there are glaring gaps existing in rural India regarding these commercial banks as Goa with 0.997 occupied the topmost position proceeded by Himachal Pradesh (0.334) and Punjab (0.251) (Reserve Bank of India 2013).

Female Workforce Participation Rate (FWPR) is considered as a driver of livelihood transformation and economic growth. However, it has marginally decreased from 25.7% in 2001 to 25.51% in 2011 (GOI 2011).

Eleven states had their scores lower than the national average (0.414), where Uttar Pradesh (0.105), West Bengal (0.134), and Bihar (0.155) reported poor FWPR, while Himachal Pradesh (0.871), Andhra Pradesh (0.805), and Sikkim (0.797) had encouraging FWPR (Ministry of Statistics and Programme Implementation 2014).

The fourth indicator was Female Headed Households (FHHs). They are also associated with some form of household- and society-level support to carry forward livelihood activities and women empowerment (Planning Commission 2014). Analysis revealed that there is a positive relationship between FHHs and sex ratio. States reported favorable female sex ratio had higher number of FHHs and vice versa. Haryana (0.115), Madhya Pradesh (0.127), Jammu and Kashmir (0.148), and Bihar (0.160) recorded lower percentage of FHHs. Opposite to them, Goa (0.996), Kerala (0.898), and Uttarakhand (0.717) showed higher percentage for FHHs (NSSO Report 2012).

The composite social assets index demonstrated that Goa (0.805) had the highest social capital. More than one-third of the total states are still struggling to be equal to national average (0.284). Nine states were found in very low and low category with lowest being Bihar (0.155) (Fig. 1.4). Western and southeastern states belonged to medium and high levels of social assets, respectively. Interestingly, Haryana fell in very low category, largely on account of its gender disparity in terms of poor performance for female workforce participation rate and relatively very low female-headed households in the state (Table 1.3).

(d) **Natural Capital**: It refers to the natural resource stocks useful for livelihoods. India is endowed with great natural resources such as forest wealth, fertile agricultural land, water resources, etc. helping in the economic development of the country (GOI 2010-11). To measure this capital, four indicators were assessed. These included (i) percentage of forest area, (ii) soil testing laboratories, (iii) groundwater, and (iv) cropping intensity (Table 1.1).

Forest cover plays an important role in the socio-economic development of a country like India (ISFR 2011). However, it has recorded a sharp reduction of 2.27% in 2017 from 23.81% in 2011 (Forest State of India 2017). In India, only 21.54% of the total geographical area is covered under forests. Results depicted that Mizoram with 86.27% topped the list, followed by Manipur (77.67%), Goa (60.21%), and Kerala (52.38%). However, inter-state variations are considerable as Haryana reported the lowest forest cover with only 3.59%. Other states in the same line were Punjab (3.65%), Rajasthan (4.84%), and Uttar Pradesh (6.09%) (Forest State of India 2017).

Another indicator for natural capital was soil testing laboratories, which are the base for management decisions about fertilizer requirements. There were 15 states with their scores greater than the national average (0.676). Some of them were Sikkim (0.974), Himachal Pradesh (0.897), and Mizoram (0.832). Contrary to this, Jharkhand

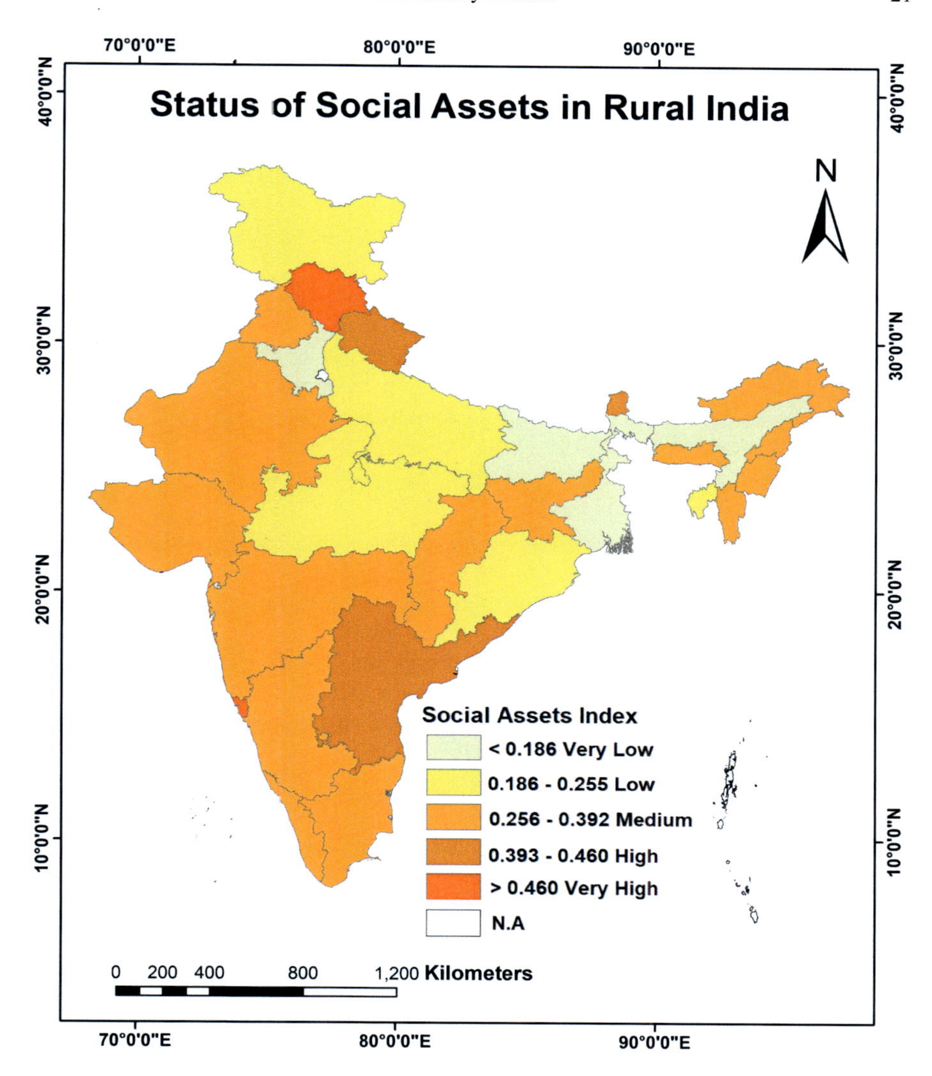

Fig. 1.4 Levels of social assets

(0.217), Nagaland (0.265), and West Bengal (0.470) were among the states lagging behind the national average.

Groundwater as a natural resource is very precious for all the states. The states with lower groundwater are adhered to the northeastern region of India with Arunachal Pradesh at the lowest (0.010), while Assam (0.081) at the highest among these seven northeastern states (Central Water Commission 2011; Jha and Sinha 2009). Although this region offers very little scope for groundwater storage owing to steep slopes and high runoff, nevertheless, it acts as the major source of recharge for the vast Indo-Gangetic and Brahmaputra alluvial plains (Sinha 2011).

The fourth indicator is cropping intensity. It refers to the ratio of gross cropped area to net cropped area. Odisha (0.166), Jharkhand (0.230), and Chhattisgarh (0.239) revealed poor cropping intensity (Agricultural Census 2011). However, Punjab (0.987), Haryana (0.939), and Uttar Pradesh (0.624) showed credible performance. National average for cropping intensity was recorded (0.464) and 15 states' scores were below the national average.

The composite natural asset index portrayed a sad picture about the status of natural capital in rural India. The situation becomes too worrying when the score of more than 15 states fell below the national average (0.438). States of central, south-eastern, and northeastern region except Jharkhand, Manipur, and Mizoram belonged to medium category (Table 1.3). While northern hilly states like Jammu and Kashmir (0.553), Himachal Pradesh (0.725), Uttarakhand (0.679), and Sikkim (0.943) were under high and very high categories (Fig. 1.5).

(e) **Financial Capital**: It refers to the capital base which is essential for the pursuit of any livelihood strategy. It shows economic status of the people and often employed to finance the ownership of tangible assets (Gupta, 2016; Sen 2003). For measuring the status of financial capital in rural India, five indicators were evaluated. These were (i) monthly per capita expenditure, (ii) population above the poverty line, (iii) main workers, (iv) cultivators, and (v) other workers (Table 1.1).

Monthly per capita expenditure is closely linked to individual's per capita monthly income. Average rural Monthly Per Capita Expenditure (MPCE) (MMRP)[4] was lowest in Odisha and Jharkhand (around Rs. 1000). In Bihar, Madhya Pradesh, and Uttar Pradesh, rural MPCE was about Rs. 1125–1160. All these states had their monthly expenditure capacity less than the national average of Rs. 1430 (NSSO 2011). The only three major states with MPCE above Rs. 2000 were Kerala (Rs. 2669), Punjab (Rs. 2345), and Haryana (about Rs. 2176) (NSSO Report 68th Round 2011–12).

According to the Tendulkar Methodology,[5] 79.1% population is above poverty line in India, comprising 74.3% of rural population (World Bank Report 2016; Planning Commission 2012). There were five states—Goa, Punjab, Himachal Pradesh, Kerala, and Sikkim—having more than 90% of their population above poverty line. Contrary to it, Chhattisgarh (55.39), Jharkhand (59.16), Arunachal Pradesh (61.07), and Madhya Pradesh (64.26%) recorded low share of population above poverty line. However, there are ten states which had their scores less than the national average (74.3%).

In rural India, only 45% of the total working population has worked for 180 and more days (main workers) (Census of India 2011). Maharashtra (86.49%) had highest

[4] Like the 66th round survey, the 68th round survey employed three different methods of measurement of Monthly Per Capita Expenditure (MPCE) at the household level—the URP (Uniform Reference Period), MRP (Mixed Reference Period), and MMRP (Modified Mixed Reference Period) methods. This paper has used MMRP method for analysis.

[5] Tendulkar Methodology for persons above poverty line (APL) refers to that methodology which identifies persons whose level of income is sufficient to meet their minimum living conditions.

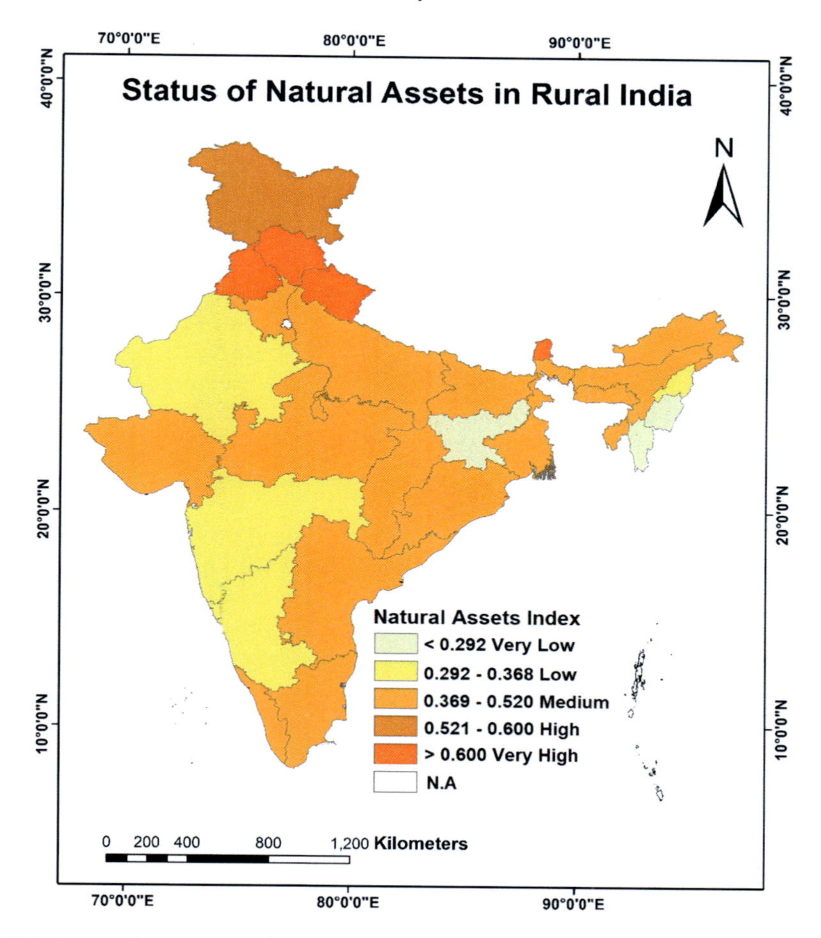

Fig. 1.5 Status of natural assets in rural India

share of main workers, followed by Mizoram (86.31) and Punjab (82.65%), while Jharkhand (45.34%), Jammu and Kashmir (53.64), and Odisha (57.10) recorded lower share of main workers.

The fourth indicator was cultivators. The decade 2001–11 has seen a decline in the number of cultivators. In India, it has declined by 6.77% from 2001 to 2011 (Census of India 2011). Mizoram held the first position with a score of 0.988. Other significant states were Himachal Pradesh (0.793) and Rajasthan (0.681). While there are nine states which had their scores less than the national average (0.355) (Census of India 2011). The lowest cultivators were reported from West Bengal (0.169) followed by Odisha (0.196) and Bihar (0.261).

Another indicator was other workers. They had increased from 37.59% in 2001 to 41.60% in 2011 (Census of India 2011). However, other workers also recorded wide inter-state gaps (Gupta et al 2008). The states which had higher percentages of other workers are Goa (75.33%), Kerala (71.29%), and Punjab (42.55%). However, there are 11 states with their scores lagging behind the national average (24.29%). Some of the poor-performing states were Chhattisgarh (11.38%), Madhya Pradesh (11.95%), and Bihar (16.99%) (Census of India 2011).

The composite index of financial asset revealed that the national average for this asset is 0.385, and surprisingly 20 states of India recorded their scores lesser than it. Central and eastern states of India such as Madhya Pradesh (0.262), Odisha (0.199), West Bengal (0.381), and Jharkhand (0.185) fell in very low and low category on account of their poor scores for other workers, cultivators, and persons above poverty line (Table 1.3). While all the southern states except Kerala (0.733) and Goa (0.728) belonged to medium category. Highest financial asset was recorded from Kerala (0.733) followed by Goa (0.728) (Fig. 1.6).

All these five assets explained above can be better understood in a glimpse with the help of a radar diagram, depicting composite scores of each asset in the pentagon (Fig. 1.7).

(B) **Economic Vulnerability**: It describes the external environment that the poor people live in, including critical trends, shocks, and seasonality. States reporting higher scores for economic vulnerability means poorly sustainable livelihoods (Table 1.3). To measure the status of this major component, four indicators were selected. These were (i) unemployment rate, (ii) population below poverty line, (iii) agricultural laborers, and (iv) marginal workers.

Young people are the great human resource, key agents for social. However, growing large number of unemployment rate is one of the serious problems faced by developed and developing countries. A report on Employment and Unemployment survey 2011–12 revealed that the lowest unemployment rate is in Gujarat with 0.010 followed by Maharashtra (0.017) and Karnataka (0.021). Contrary to it, highest unemployment rate was reported from Nagaland (0.978) and Tripura (0.488) (Ministry of Labour and Employment 2011).

The second indicator was population below the poverty line.[6] According to the Planning Commission (2012), 25.7% of the total population of rural India live below poverty line (BPL). Goa (6.81%), Punjab (7.66%), and Kerala (9.14%) belonged to the category of those states, where population living BPL was less than 10%. However, some states like Jharkhand (40.84%), Chhattisgarh (44.61%), and Madhya Pradesh (35.74%) had a large share of population living in BPL.

Agricultural workers constitute the largest worker class in Indian rural structure (Sajjad and Nasreen 2016). The rise of agricultural laborers is closely associated with the increment in economic vulnerability. Census of India (2011) revealed that the number of agricultural laborers had increased by 35.17% from 2001 to 2011. For the first time since independence, agricultural laborers have outnumbered cultivators. Bihar (56.86%) had the highest percentage of agricultural laborers, while Himachal

[6] It is an economic benchmark used by the Indian government to indicate economically disadvantaged people who need government's assistance and aid.

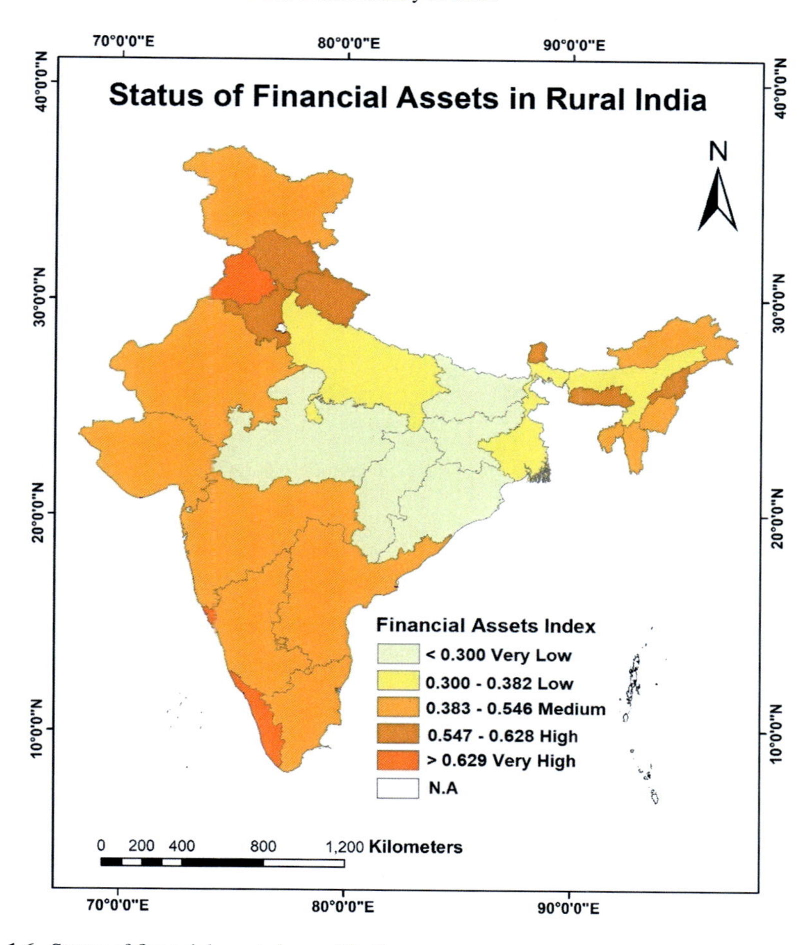

Fig. 1.6 Status of financial assets in rural India

Pradesh (5.18%) occupied the topmost rank among the states having lower share of agricultural laborers.

The fourth indicator was marginal workers. The Census of India classified workers into two groups, namely, main workers and marginal workers.[7] The analysis of census data showed that during the last two decades (1991–2011) the rate of growth in marginal workers was higher than that of main workers (Ramchandran and Swaminathan 2003). Maharashtra (13.51%) had the lowest share of marginal workers followed by Andhra Pradesh (16.90%) and Punjab (17.35%). However, Jharkhand (54.66%), Jammu and Kashmir (46.36%), and Odisha (42.90%) reported marginal workers greater than the national average (29.50%) (Census of India 2011).

[7] Marginal workers are those workers who get work for less than 183 days in a year.

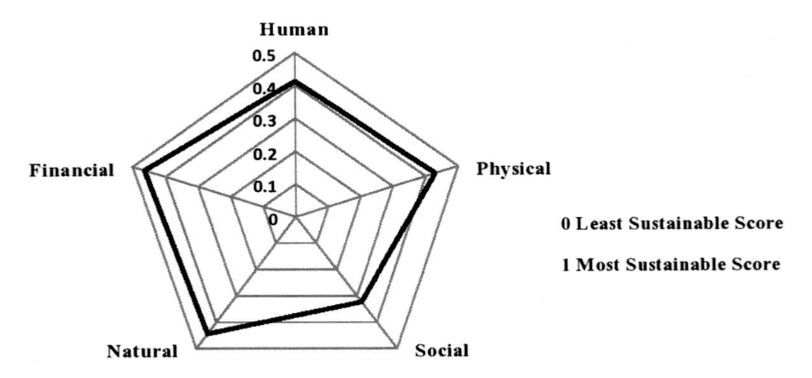

Fig. 1.7 Composite score of each asset in the pentagon. *Source* Data from different government agencies

The composite economic vulnerability index revealed that the national average of this component is 0.240 and there were seven states having their scores higher than this. Among them, states of central and eastern region like Odisha (0.313), West Bengal (0.247), Madhya Pradesh (0.290), Uttar Pradesh (0.264), Jharkhand (0.346), and Bihar (0.337) were reported. It is largely because of higher unemployment rate and increasing number of agricultural laborers recorded in these states (Fig. 1.8). However, entire southern India except Kerala and Goa fell into medium range, while Goa (0.113) followed by Sikkim (0.125) and Kerala (0.145) made their place in the very low economic vulnerability class (Table 1.4).

(C) **Sensitivity**: Sensitivity is the degree to which a system is affected either adversely or beneficially. Higher scores of sensitivity index ensure greater livelihoods sustainability of households. It comprised of three minor components. These were (i) housing condition, (ii) sanitation facility, and (iii) household assets (Table 1.4).

(a) **Housing Condition**: Everyone has the right to a standard of living adequate for the health and well-being of himself and his family. Nearly 53% rural population was reported to live in "good condition houses" (Census of India 2011). However, the inequality among states had widened as Odisha with 25.42% was ranked at the lowest ladder while Goa had 75% rural houses in the category of good condition houses. In India, about 56% of the total rural population had access to more than two dwelling rooms and electricity (Census of India 2011).

(b) **Sanitation Facility**: Sanitation encompasses formidable part of ensuring human dignity. It is not only an absence of garbage and waste materials strewn around but also access to toilet facility, safe drinking water, and connectivity to a drainage system (GOI 2016; Kumar and Das 2014). To measure this, five indictors were chosen. These are toilet facility, waste water outlets connection, tap water from treated source, water within premises, and safe drinking water.

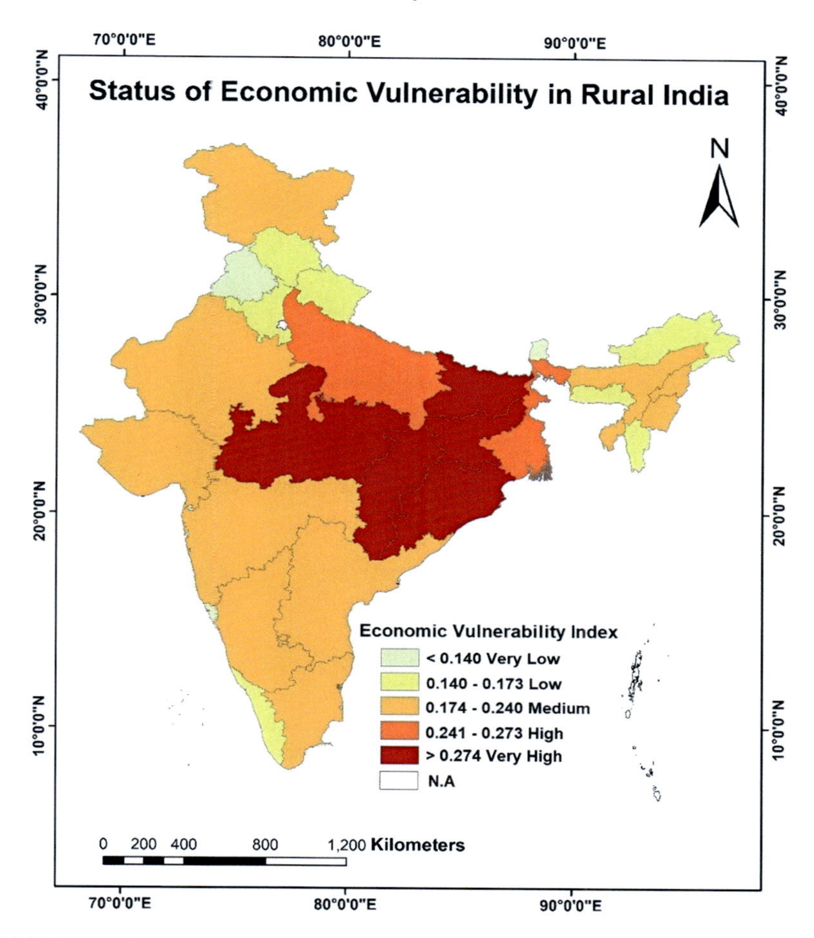

Fig. 1.8 Status of economic vulnerability in rural India

Only 31% rural households were having any toilet facility in their households (Census of India 2011). Nine states' percentage was even below the national average. Sixteen states were lagging behind the average of rural India (82.7%) for safe drinking water. Mostly northeastern states were found to be greatly deprived of this resource like Meghalaya (35.1%) followed by Manipur (37.5%) and Mizoram (43.4%). However, interestingly, poor-performing states such as Bihar (93.9%) and Uttar Pradesh (94.3%) had achieved relatively better position in this sphere.

Forty-two percent of Indian households had well-connected wastewater outlet, while only 38% of rural households had source of water within the premises.[8]

[8] Water Premises: Premises has been defined as a building along with the land and/or common places attached to it. It is used in context of availability of drinking water source, depending upon the distance at which it is available. (i) Within the Premises: where the household live. (ii) Near the

Table 1.4 Major component indices and sustainable livelihood index in states of India

States	Assets index	Economic vulnerability index	Sensitivity index	Sustainable livelihood index
Andhra Pradesh	0.437	0.212	0.552	4.662
Arunachal Pradesh	0.374	0.169	0.389	4.515
Assam	0.318	0.215	0.327	3.007
Bihar	0.222	0.337	0.312	1.586
Chhattisgarh	0.342	0.328	0.344	2.091
Goa	0.630	0.113	0.861	13.230
Gujarat	0.397	0.218	0.506	4.132
Haryana	0.407	0.160	0.724	7.071
Himachal Pradesh	0.646	0.150	0.816	9.737
Jammu and Kashmir	0.357	0.195	0.495	4.381
Jharkhand	0.230	0.346	0.284	1.487
Karnataka	0.417	0.201	0.516	4.633
Kerala	0.611	0.145	0.754	9.430
Madhya Pradesh	0.323	0.290	0.318	2.216
Maharashtra	0.435	0.200	0.524	4.791
Manipur	0.376	0.194	0.436	4.179
Meghalaya	0.398	0.144	0.291	4.778
Mizoram	0.449	0.148	0.399	5.723
Nagaland	0.438	0.193	0.378	4.228
Odisha	0.310	0.313	0.263	1.828
Punjab	0.491	0.129	0.794	9.982
Rajasthan	0.365	0.175	0.420	4.481
Sikkim	0.602	0.125	0.571	9.425
Tamil Nadu	0.499	0.205	0.597	5.338
Tripura	0.417	0.224	0.382	3.567
Uttar Pradesh	0.263	0.264	0.490	2.857
Uttarakhand	0.527	0.143	0.678	8.455
West Bengal	0.343	0.247	0.315	2.665
India	0.351	0.240	0.470	3.392

Source Computed from secondary data

However, 28% of households got access to tap water from treated source (Census of India 2011).

(c) **Household Assets**: Those households which have lesser access to assets are more vulnerable (Thakur and Mehta 2000). For measuring it, two indicators are selected. These are banking facilities and access to household assets. Fifty-eight percent of the rural households were availing banking facilities in India (Census of India 2011). In other household assets like access to bike, bicycle, mobile phones and television, Punjab (0.932) had the highest score chased by Goa (0.872), while Meghalaya (0.157) had the lowest score followed by Nagaland (0.185) and Arunachal Pradesh (0.251) (Census of India 2011).

The composite sensitivity index, calculated using 10 indicators (Table 1.1) reflecting housing conditions, sanitation facility, and household assets, revealed that the national average for this component is 0.470, where score of 14 states of India lied below the national average. Central, eastern, and northeastern states like Madhya Pradesh (0.318), Odisha (0.263), West Bengal (0.315), Chhattisgarh (0.344), Meghalaya (0.291), and Assam (0.327) belonged to low and very low categories (Table 1.4). However, entire southern and southwestern regions were under high and very high categories due to their better scores for housing conditions, sanitation, and household assets (Fig. 1.9). The analysis reveals that these three major components have varying scores, influencing livelihood sustainability of rural areas. In order to comprehend differences in their scores, a triangle diagram is utilized (Fig. 1.10).

1.6 Livelihood Sustainability Ladder

This ladder has been computed using 44 indicators from various domains of major and minor components (Table 1.1). Here, modified form of IPCC vulnerability index is incorporated to get Sustainable Livelihood Index (SLI) for the states in India (Table 1.4). This index is categorized into four sustainability ladders, namely, surviving, coping, adapting, and accumulating representing poor livelihood status to better livelihood status, respectively (Table 1.2). The analysis shows that the national score for livelihood sustainability for rural India is 3.392 and there are nine states with their scores less than this national average. Majority of these states belong to central and eastern regions, e.g., Madhya Pradesh (2.216), Uttar Pradesh (2.857), Bihar (1.586), Jharkhand (1.487), Odisha (1.826), Chhattisgarh (2.091), and West Bengal (2.665). They are grouped into "surviving" category (Fig. 1.11). This category refers to poor social, human, and financial assets as these states have lower assets base particularly for female literacy rate, work participation rate, kisan credit cards, and monthly consumer expenditure (Table 1.3). Secondly, they are economically vulnerable to varying stresses, shocks, trends, and seasonality, hindering their way to livelihood

---

Premises: If the source is located within a range of within a distance of 500 m in the case of rural areas. (iii) Away from the Premises: If the drinking water source is located beyond 500 m in rural areas.

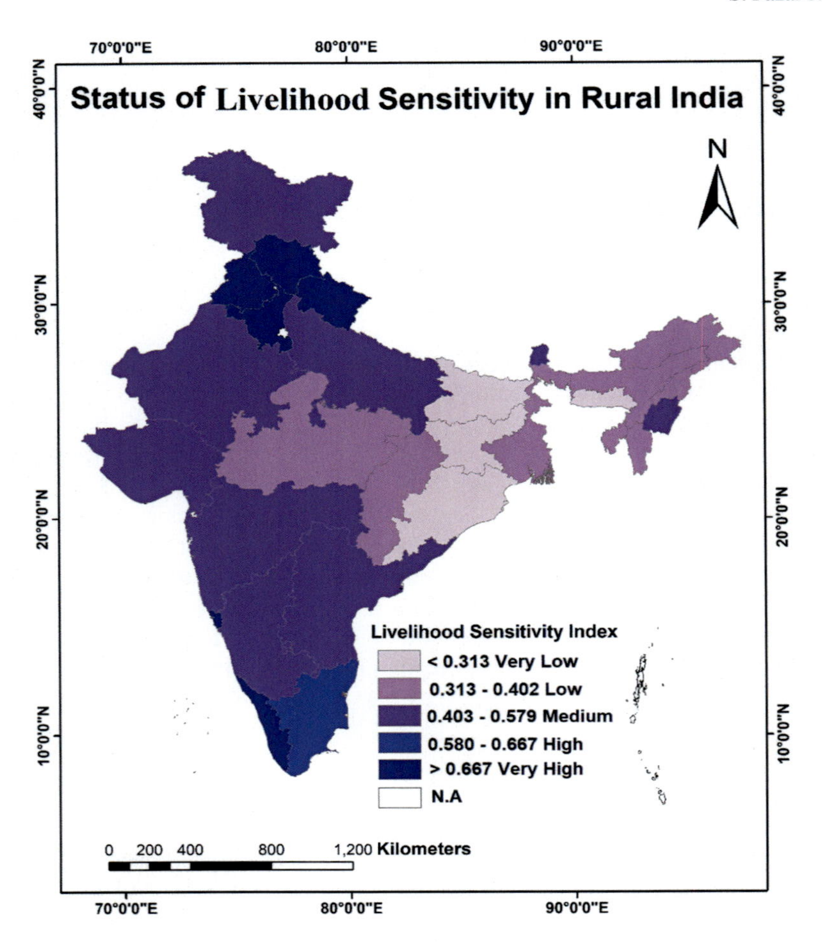

Fig. 1.9 Status of livelihood sustainability in rural India

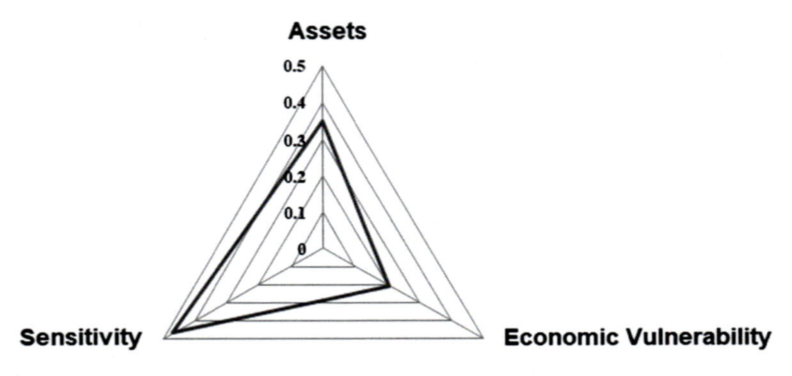

Fig. 1.10 Triangle depicting scores of assets, economic vulnerability and sensitivity. *Source* Secondary data

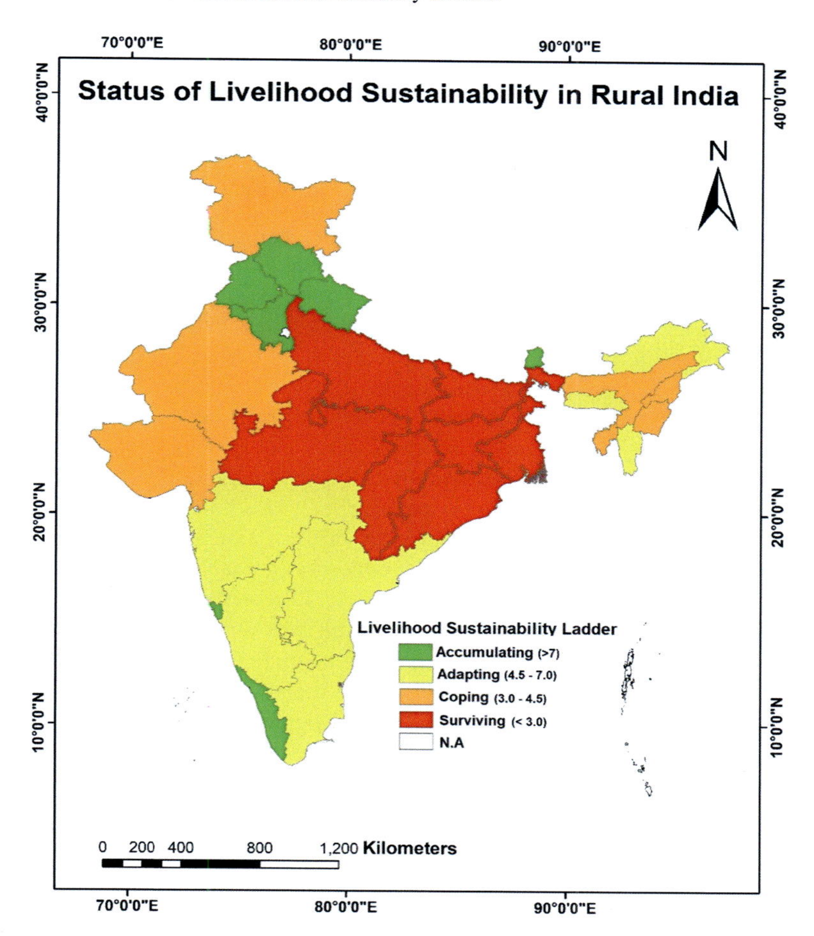

Fig. 1.11 Status of livelihood sustainability in rural India

sustainability, e.g., except West Bengal, all these six states reported more than 35% of their rural population living below the poverty line (Table 1.3).

Second livelihood ladder is "coping". It means that these states are also poor in assets and economically vulnerable. Majority of these states have lower scores for physical and social assets, particularly inefficient access to education, health facilities, rural banks, and gender disparity are the core concerns for these states (Table 1.3). However, this ladder is different from the "surviving" on account of its lesser severity for economic vulnerability, mainly for unemployment rate and population below poverty line. Their scores are above the national average (3.392). They are largely located in western and some northeastern states of India, Rajasthan (4.481), Jammu and Kashmir (4.381), Tripura (3.567), etc. (Fig. 1.11). Lesser scores for assets and increasing agricultural laborers are mainly responsible for placing these states into this category. Proper attention is required for identified poor assets among states to improve their position and thus they can also move upward in this ladder.

The third ladder of this rural livelihood sustainability is termed as "adapting", which is characterized by better scores for assets, especially for human, physical, and financial (Table 1.3). It has seven states mostly in southern region and three states from northeast India, namely, Maharashtra (4.791), Karnataka (4.633), Tamil Nadu (5.338), Mizoram (5.723), and Meghalaya (4.778) (Table 1.4). Though these states were reported to have easy and efficient access to various assets and household amenities, yet, they are placed in adapting category because these states are observing constant increase in the percentage of agricultural laborers and marginal workers since last few decades, disturbing their sustainability scores. Thus, it neutralizes the influence of higher assets and sensitivity, pushing these states out of accumulating ladder.

The fourth and most sustainable ladder is "accumulating". It means that "they have enough assets base to cope any stress and shocks". This category has seven states and most of them have better scores for most of the major and minor components (Table 1.1). Goa (13.230) has the highest livelihood sustainability, while Kerala has a score of (9.430). Apart from these two states, rest of them is located in northern hilly states except Jammu and Kashmir. These are Himachal Pradesh (9.737), Uttarakhand (8.455), Sikkim (9.425), and northwestern plains of Punjab (9.982) and Haryana (7.071) (Table 1.4). All these states could make their place in this category, owing to their higher scores for human, physical, and financial assets except Haryana and Punjab, which reported lower scores for human assets. Moreover, their scores for economic vulnerability is also low, thus making them more sustainable (Fig. 1.11).

1.7 Conclusion

India, surpassing different transformation stages, is still predominantly a rural country. Since independence, it has undergone improvements in different sectors such as health, elementary and higher education, transport and communication, public distribution system, per capita income, sanitation, etc. Despite all these progress, there are large-scale inter-state disparities, posing a serious challenge before the country to achieve livelihood sustainability in rural areas.

The idea of assets is fundamental to the sustainable livelihoods. They are broadly classified into five categories—human, physical, social, natural, and financial. Access to these assets enables poor people to undertake a range of livelihood strategies, activities, and choices, which finally determine their livelihood outcomes. The findings of this paper reveal that Goa, Kerala, Himachal Pradesh, and Sikkim are better performing states with their scores above the national average for most of the indicators of selected five capitals, particularly for human, physical, and financial capital. However, score for social and natural capital is a matter of concern even among these states. While Bihar, Jharkhand, Uttar Pradesh, and Odisha are placed at the bottom on account of their low scores for composite assets index (Table 1.3). Rest of the states fall in adapting and coping stage of sustainable livelihood ladder.

The second major component utilized in this study is economic vulnerability index. It is critical because it has a direct impact on the abilities of households to earn a living. It reflects negative association with livelihoods sustainability, i.e., higher the scores for economic vulnerability, lesser would be the sustainability of livelihoods and vice versa. Goa, Sikkim, Punjab, Uttarakhand, and Kerala experience very low economic vulnerability, which are much below the national score (Table 1.4). Here, it is worth mentioning that Kerala, which mostly outperforms for all indicators, slips down to fifth position. The reason for this slump lies in its higher unemployment rate, more than 9%, which is three times greater than the national average (Census 2011). Whereas Jharkhand, Bihar, Chhattisgarh, and Odisha reported their values above the national score, making them economically vulnerable. Another significant finding depicts that the northeastern states are better placed in terms of economic vulnerability. Surprisingly, Gujarat and Maharashtra are placed on twentieth and fifteenth positions, respectively. Although they recorded better scores for human, financial, and physical assets but because of high percentage of agricultural laborers, they are lying at lower ladders of livelihood sustainability.

The third major component is sensitivity index. A household having greater access to assets would be less vulnerable to varying shocks and stresses, meaning more secured livelihoods. The results for this index show that Odisha experience the lowest score for sensitivity followed by Jharkhand, Meghalaya, and Bihar with their scores for all its indicators much below the national average. However, the condition is poor for sanitation among these states (Table 1.4). On the contrary, Goa, Himachal Pradesh, Punjab, Kerala, and Haryana report higher scores for all the indicators of sensitivity, i.e., housing conditions, sanitation, and household assets and amenities. Twenty-two states lie between these two extremes and fourteen states record their sensitivity score greater than the national score (0.470).

Analyzing these three indices, livelihood sustainability ladder was developed to categorize states of India depending on their levels of livelihood sustainability. The findings reveal that Goa, Punjab, Himachal Pradesh, Kerala, Sikkim, Uttarakhand, and Haryana belong to the highest ladder of livelihood sustainability, named "accumulating", with their scores greater than the national average. These are the states which have better scores for both assets and sensitivity, while economic vulnerability is comparatively low. On the other hand, Jharkhand, Bihar, Odisha, Chhattisgarh, Madhya Pradesh, West Bengal, and Uttar Pradesh lie at the bottom of this ladder, named as "surviving", with their scores much below the national average (Table 1.4). They are poorly sustainable states owing to higher economic vulnerability and lower scores for human and financial assets and also reported poor household assets and amenities. However, Assam, Tripura, and Gujarat rest in coping category, whereas Karnataka, Maharashtra, and Tamil Nadu belong to adapting category (Fig. 1.11). Since the states like Gujarat, Karnataka, Maharashtra, and Tamil Nadu were reported to have easy and efficient access to various assets and household amenities, though they are placed in adapting and coping categories. The reason lies in the fact that economic vulnerability, i.e., percentage of agricultural laborers, marginal workers, unemployment rate, and people below poverty line are found to be high in rural areas of these states. Thus, it neutralizes the potent of higher assets and sensitivity, pulling these states out of accumulating ladder.

This study identifies poor-performing states under various domains of livelihoods and also in assessing varying needs for different states. Statistics presented and explained in this paper would help policy-makers and planners to design, rejuvenate, and strengthen the rural areas which require their assistance so that rural India may also contribute significantly to the nation's growth, escorting livelihood sustainability and well-being of rural areas.

References

Ashley C, Carney D (1999) Sustainable livelihoods: lessons from early experience. DFID, London, pp 4–16

Banu N, Fazal S (2017) A pragmatic assessment of livelihood status in the peri-urban interface: a case from developing India. Asian Geogr 9–15

Barah BC (2007) Strategies for agricultural development in the North-East India: challenges and emerging opportunities. Indian J Agric Econ 62:18–21

Bebbington A (1999) Capitals and capabilities: a framework for analyzing peasant viability, rural livelihoods and poverty. World Dev 27(12):2012–2044

Bhagat RB (2014) Rural and urban sanitation in India. Kurukshetra 11–13

Census of India (2011) Houses, household amenities and assets. Office of the Registrar General of India, Government of India, New Delhi

Chamber R, Conway G (1992) Sustainable rural livelihoods: practical concepts for the 21st century. Discussion Paper 296, Institute for Development Studies, University of Sussex, Brighton, United Kingdom

Dev SM, Venkatanarayana M (2011) Youth employment and unemployment in India. Working Paper, Indira Gandhi Institute of Development Research (IGIDR), Mumbai, pp 12–21

DFID (Department for International Development) (1999) Sustainable livelihoods guidance sheets. Department for International Development, London

Ellis F (2000) Rural livelihoods and diversity in developing countries. Oxford University Press, Oxford

Ellis F, Biggs S (2001) Evolving themes in rural development 1950s–2000s. Dev Policy Rev 19(4):437–448

FAO (Food and Agricultural Organization) (2012) The state of food insecurity in the world 2012. United Nations-FAO, IFAD and WFP, Rome. http://www.fao.org/news/story/en/item/161819/icode/. Accessed 5 Jun 2019

Fazal S (2000) Urban expansion and loss of agricultural land: a gis based study of Saharanpur city, India. Environ Urbanization 13:133–140

Gautam Y, Andersen P (2016) Rural livelihood diversification and household well-being: insights from Humla, Nepal. J Rural Stud 44:239–249

GOI (Government of India) (2010) Right to education. Department of School Education and Literacy, MHRD, GOI. http://mhrd.gov.in/rte. Accessed 6 Jun 2019

GOI (Government of India) (2011) Annual report of ministry of labour and employment. https://labour.gov.in/annual-reports. Accessed 15 Jun 2019

GOI (Government of India) (2010–11) Department of agriculture cooperation and family welfare. http://agricoop.nic.in/publication/agriculture-census. Accessed 10 Jun 2019

GOI (Government of India) (2013) National Food Security Act-2013. Department of Food & Public Distribution, Ministry of Consumer Affairs, Food & Public Distribution. Government of India. GOI. http://dfpd.nic.in/nfsa-act.htm. Accessed 6 Jun 2019
GOI (Government of India) (2016) Ministry of drinking water & sanitation. http://www.mdws.gov.in/
Gupta N (2016) Decline of cultivators and growth of agricultural labourers in India from 2001 to 2011. Int J Rural Manag 12:185–191
Gupta V, Singh B, Ranjan R (2008) An economic evaluation of Kisan credit card in Bhabua district of Bihar, India. Int J Curr Microbiol Appl Sci 7(02):5–7
Haan L (2012) The livelihood approach: a critical exploration. Erdkunde 66(4):345
Hahn MB, Riederer AM, Foster SO (2009) The livelihood vulnerability index: a pragmatic approach to assessing risks from climate variability and change—a case study in Mozambique. Glob Environ Change 19(1):74–88
IFPRI (International Food Policy Research Institute) http://www.ifpri.org/. Accessed 6 Jun 2019
IPCC (Intergovernmental Panel on Climate Change) (2001) Climate Change 2001: impacts, adaptations and vulnerability. In: Contribution of working groups II to the third assessment report. Cambridge University Press
ISFR (Indian State of Forest Report) (2011) http://fsi.nic.in/forest-report-2011. Accessed 12 Jun 2019
James KS (2011) India's demographic change: opportunities and challenges. Science 333:576–577
James KS, Goli S (2017) Demographic changes in India: is the country prepared for the challenge? Brown J World Aff 23:168–172
Jha BM, Sinha SK (2009) Towards better management of ground water resources in India. Q J Central Ground Water Board 24(1–4) (Ministry of Water Resources, Government of India)
Kumar A, Das KC (2014) Drinking water and sanitation facilities in India and its linkages with diarrhoea among children under five: evidence from recent data. Int J Humanit Soc Sci 3(4):52–55
Moran M, Wright A, Renehan P, Szava A, Beard N, Rich E (2007) The transformation of assets for sustainable livelihoods in a remote aboriginal settlement. Desert Knowledge CRC, Report 28, Australia
NSSO (National Sample Survey Office) (2014) Household consumption of various goods and services in India, 2011–12. Report No. 558. NSS 68th Round. National Sample Survey Office. Ministry of Statistics and Programme Implementation. Government of India
OXFAM (2006) The sustainable livelihoods approach. Oxfam International Report, Wales
Patidar H (2018) Livelihood security in rural India: reflections from some selected indicators. Forum Dev Stud 148–159
Patnaik S, Prasad SC (2014) Revisiting sustainable livelihoods: insights from implementation studies in rural India. Vision 18(4):353–360
Petersen EK, Pedersen ML (2010) The sustainable livelihoods approach from a psychological perspective: approaches to development. University of Aarhus, Denmark, Institute of Biology, pp 6–11
Planning Commission (2014) Report of the expert group to review the methodology for measurement of poverty. Government of India, New Delhi
Ramchandran VK, Swaminathan M (2003) Agrarian studies: essays on Agrarian relations in less-developed countries
Sajjad H, Nasreen I (2016) Assessing farm-level agricultural sustainability using site-specific indicators and sustainable livelihood security index: evidence from Vaishali district, India. Community Dev 47(5):602–619
Scoones I (1998) Sustainable rural livelihoods: a framework for analysis. IDS Working Paper, 72, Brighton, Sussex, pp 5–8
Scoones I (2009) Livelihoods perspectives and rural development. J Peasant Stud 172–180
Sen A (2003) Development as capability expansion. Reading in human development. Oxford University Press, New Delhi and New York

Singh PK, Hiremath BN (2010) Sustainable livelihood security index in a developing country: a tool for developing planning. Ecol Indic 10(2):440–448

Solesbury W (2003) Sustainable livelihoods: a case study of the evolution of DFID policy. Working Paper 217, Overseas Development Institute, pp 2–8

Talreja C (2014) India's demographic change: realities and opportunities. Indian J Labour Econ 57(1):141–144

Thakur F, Mehta S (2000) Analysis of education sector-study of Kerala and Jammu and Kashmir. J Humanit Soc Sci 23:46–48

UN (United Nations) (2018) Global issues. https://www.un.org/en/sections/issues-depth/global-issues-overview/. Accessed 7 Jun 2019

UNDP (United Nations Development Programme) (2015) Human development reports 2015: work for human development. http://hdr.undp.org/en/

United Nations (2000) United Nations Millennium Declaration, New York, Millennium Summit of the United Nations, 6–8 September 2000

WHO and UNICEF (2013) Progress on sanitation and drinking water: 2013 update. WHO/UNICEF Joint Monitoring Programme for Water Supply and Sanitation

World Bank Report (2016) India's poverty profile. https://www.worldbank.org/en/news/infographic/2016/05/27/india-s-poverty-profile. Accessed 9 Jun 2019

World Commission on Environment and Development (1987) Report of the world commission on environment and development: our common future. Bruntland Report, United Nations, pp 330–332. http://www.un-documents.net/our-common-future.pdf

World Economic Forum (2018) Global gender gap report. http://reports.weforum.org/global-gender-gap-report-2018/. Accessed 8 Jun 2019

Chapter 2
The Exodus in Times of Pandemic: Mobility, Migration and Livelihood of Informal Migrant Workers During COVID-19 Crisis in India

Twisha Singh and Anuradha Banerjee

Abstract This present study is an assessment of the impact of COVID-19 on mobility and livelihood of informal migrant workers in India. The paper is based on a theoretical background built upon the arguments of development economists and scholars of migration. Case studies based on media reporting and newspaper clippings, presentation by scholars have been used to highlight the context of the discussion. The analysis also rests on data collected from Government documents, NSSO and the Census of India, 2011. The focus is on the lower end of the diverse labour market in India, that on one hand constitutes insecure tail end jobs, unequal wages and poor access to basic amenities and infrastructure, deprivation of social security on one hand and on the other constitutes a major part of Government's nativist project, as it is an essential intra-national economic tool for development, however categorized by unevenness at its core. The paper places an exploratory effort to identify the institutionalized inequalities and mounting vulnerabilities during the 'exodus' of the informal migrant workers, amidst the nation-wide lockdown. The research finds that there is a necessity to rethink mobility and migration that has become apparent due to the COVID-19 crisis, and the fallout out of which has been extremely detrimental to the poorest section of the Indian workforce, i.e. the informal migrant workers, of which a vast section constitute the seasonal labourers. This also raises the issue of redistributive conflict and locating the problematic in the allocation of resources on one hand and on the efficacy of the Government schemes in reaching the defenseless 'poorest of the poor' on the other. The question here is also indicative of the inherent and persistent problems of data and reliable estimates involving this vast segment of the informal labour, required for enforcement of social safety nets.

Keywords Mobility · Migration · Livelihood · Inequalities · Exodus

---

T. Singh (✉)
Department of History and Classical Studies, McGill University, Montreal, Canada

A. Banerjee
Centre for the Study of Regional Development, School of Social Sciences, Jawaharlal Nehru University, New Delhi, India

N. C. Jana et al. (eds.), *Livelihood Enhancement Through Agriculture, Tourism and Health*, Advances in Geographical and Environmental Sciences,
https://doi.org/10.1007/978-981-16-7310-8_2

2.1 Introduction

The concept of *participatory socialism*[1] expounded by Prof. Thomas Piketty in his latest book, 'Capital and Ideology', highlights how a shift of balance of power to the workers would enable them to participate more in the governance of their companies and decide whether or not they have a share in capital. According to Piketty, this has been undertaken in a number of countries successfully including Sweden and Germany.[2] However, we cease to arrive at a solution to the problem of inequality in India.[3] Though participatory socialism remains an emancipatory idea with hints of utopian egalitarianism at its core, it becomes a difficult proposition to apply in case of India with exponentially higher rate of inequality as compared to the European nations. It is without a doubt that a global system of commodity production and exchange constitutes capitalism; with labour relations, production and organization at its core. Neil Smith argues that capital moves to specific places where economic advantages can be meted out.[4] Hence, spatial organization becomes crucial to understand conditions of labour and inequality. Geography of capitalism is characterized by spatial differentiation of rents, wages, production costs and so forth. These form important elements of differentiated systems of financial circulation inter-twined with social production further invariably feeding into geography of capitalism.[5]

> Uneven geographical development, which establishes discrete places differentiated from each other and at the same time pressures these places, across borders, into a single mould. Uneven development represents a forced yet contested, momentarily fixed, yet always-fluid resolution to this central contradiction of capitalism. The leveling tendency of capitalism continually gnaws at the radical differentiation of the conditions of exploitation of labour, and yet the corrosive differentiation of labour also eternally frustrates this 'annihilation of space by time'. The question of scale becomes absolutely vital here, because without a sense of the making of scale, it is impossible to grasp the expansion from Marx's largely temporal logic to the geographical logic inherent in uneven development.[6]

India has transitioned from a lived experience of colonial space that encompassed several aspects of production and trade but has been caught in an enduring flux of universalistic political economy and nativist project according to historian Manu

---

[1] Piketty and Goldhammer (2020).

[2] Ibid.

[3] In the work of Karl Marx, the question of social inequality and redistribution has been posed primarily in terms of opposition between capital and labour, profits and wages, employers and employees. Inequality is defined as contrast between those who own capital, that is, the means of production, and those who do not and must therefore make do with what they can earn from their labour. The fundamental source of inequality is thus said to be the unequal ownership of capital. As mentioned in, Piketty, Thomas, and Arthur Goldhammer 2020. *Capital and Ideology.*

[4] Smith (1984).

[5] Neil Smith argues that differentiation of places is more a question of inherent logic imbued in and inherent to capitalist modes of production, expounding Marx's temporal theory of capitalist crisis. He further quotes Jeffery Sachs an IMF consultant who attributes underdevelopment to ' a case of bad latitude'.

[6] Smith (1984).

Goswami.[7] Capital and Labour relations being at the core of this *uneven* intra-national as well as international economic development. Intra-national forms of spatial organization and production cannot be read without its international counterpart. However, this paper deals with the spatial divisions of labour that has altered the intra-regional forms of production and in turn generates a complex geography of production. Several inter-twined and sub-layered parameters complicate the Indian case with the question of inequality taking on precedence. Furthermore, informalization of labour markets and migration has led to a development of an all new Spatial Geography characterized by fragmentation of labour markets. This has led to an array of debates and discussions on inequality in India, especially inequality in the sector of labour market that has consistently increased for the last three decades.[8]

This uneven economic development can be further read in similitude with migration of labourers. Migration has had a significant impact on regional/intra-national development and its continually changing pattern. The unevenness of this development can be analysed by studying migratory patterns of labourers. Migration of labourers and its existing patterns has been further analysed in the second section of this paper. In India, migrant workers form an important element of this higher rate of inequality, as there is a considerable wage gap between informal and formal labourers in the organized sectors. This gap is often characterized according to gender, caste and class on a primary level and education and experience on a secondary level. As a result, poor women and low-caste workers form the lowest rung of this sector. The inequality arising out of social discrimination has a well-entrenched historical legacy, however it has assumed new forms conditioned by cultural reproduction of social discrimination as the workers have become more mobile and informalized.

The situation of unequal wages in the labour sector is not only multi-layered but also very complex. According to the Indian Wage Report 2018,[9] out of the four wage distribution indicators used, wage inequality increased for workers for three.[10] Ravi Srivastava denotes that the rising inequality of regular workers could be attributed to the rising wages of skilled workers (Rodgers and Soundarajan 2016; Srivastava and Manchanda 2015).[11] Amidst this existing and perpetuating gap in the wages it becomes even difficult to analyse the socio-economic status of seasonal migrant workers. Their precarious condition forms the crux of this paper. Severe data limitations deter the analysis of this segment of workers as highlighted by Srivastava and many other economists. Also, existing studies do not fully capture wage inequalities and conditions at work of this lowest rung of labourers in the urban, organized sector. The problem is worsened by the constant restructuring of labour

[7] Gyan Prakash, Manu Goswami. *Producing India: From Colonial Economy to National Space.* (Chicago Studies in Practices of Meaning.) Chicago: University of Chicago Press. 2004. Pp. xi, 401. Cloth $50.00, paper $20.00, The American Historical Review, Volume 110, Issue 2, April 2005, Page 457.

[8] Srivastava (2019).

[9] Ibid.

[10] Srivastava (2019).

[11] Ibid.

markets that is often characterized by the cultural reproduction of discrimination that upholds social hierarchies. Moreover, socially segregated classes like the Scheduled Caste and Scheduled Tribes, placed lowest in the social ladder are subjected to vulnerable working conditions and precarious wages. The working conditions and wages are even worse for women who are often subjected to double subordination, if from the lowest sections of society. Srivastava highlights how the contemporary labour markets in India are based on the blueprints of the traditional labour markets.[12]

The ever-increasing fragmentations of labour markets are deepened by the precarious nature of recruitment processes that are often informal, in some cases verbal. Hence, the data regarding the same is spurious to a large extent. Although the links between migration, poverty and underdevelopment are deeply context specific, movement of people for economic opportunities among others are a lot more commonplace than usually discussed. Arjan De Haan highlights how international literature on migration ignores the fact that most migratory activities remain concentrated in the Global South and within national borders.[13] There are varying propensities of migration that are affected by a number of reasons like worker's capabilities, opportunities and differentiated access. De Haan points out how politically there is acceptance to some degree regarding the mobility of people according to the demand for labour, however, there are significant strong voices to reduce the number of immigrants. He quotes the example of China, where migrants are the main catalysts in economic transformation at the same time the *hukao system* keeps migrants excluded from many public services in urban areas.[14] In the Indian case, considerable proportion of the labour force in urban areas is comprised of informal migrant workers. Since the reasons and patterns of migration are diverse across India, it is also varied and uneven. Hence, there is no monolithic strategy or a unitary social security plan that can be applied to this vast proportion of migrant workers uniformly in the face of any crisis.

On the eve of 24th March 2020, when the Indian Government announced a nation-wide lockdown as a preventive measure for the COVID-19 pandemic that currently affects the world at large; the fates of millions were held in uncertainty in India. This lockdown limited the mobility of the entire populace i.e. 1.3 billion people. This was post a voluntary fourteen-hour *jantacurfew* that ended on 22nd March 2020. This nation-wide lockdown continued till 31st of May, after which the Indian Government adopted a staggered mechanism for locking down regions according to the cases and spread of the virus. The lockdown comprised of an absolute restriction of movement and closure of primary, secondary and tertiary services.

[12] Ibid.

[13] De Haan (2011).

[14] De Haan also mentions the case of Europe that has shown opposite reversal, where the labour migrants of 1950's and 60's have come to be defines as the main societal problem of the 2000s: "the economic question of facilitating mobility is subordinated by nation states to the political issue of migrants as new citizens or as invaders". As mentioned in, De Haan, Arjan; '*Inclusive growth? Labour Migration and poverty in India*', Indian journal of Labour Economics, 2011.

2.2 Objectives of the Study

The aim of this paper is to focus on the lower end of the diverse labour market in India, that on one hand constitutes insecure tail end jobs, unequal wages and poor access to basic amenities and infrastructure, deprivation of social security on one hand and on the other constitutes a major part of Government's nativist project as it is an essential intra-national economic tool for development, however categorized by unevenness at its core. In the context of this backdrop, the effort has also been to rethink mobility and migration that has become apparent due to Coronavirus pandemic, and the fallout out of which has been extremely detrimental to the poorest section of the Indian workforce, i.e. the informal migrant workers, of which a vast section constitutes the seasonal labourers. The paper places an exploratory effort to identify the institutionalized inequalities and mounting vulnerabilities during the 'exodus' of the informal migrant workers, amidst the nation-wide lockdown. The aim here is to capture the multiple layers of disadvantages and discrimination faced by this segment of the workforce. This also raises the issue of redistributive conflict and locating the problematic in the allocation of resources on one hand and on the efficacy of the Government programmes in reaching the defenseless 'poorest of the poor' on the other. The question here is also indicative of the inherent and persistent problems of data and reliable estimates involving this vast segment of the informal labour, as data form the basis of initiation of ameliorative measures or enforcement of social safety nets.

2.3 Methods and Materials

Effort has been placed to build the problematique involving the informal migrant workers during the COVID-19 crisis in India, by drawing upon a theoretical background based on the arguments of development economists and scholars of migration. Case studies based on media reporting and newspaper clippings, presentation by scholars have been used to highlight the context of the discussion. The analysis also rests on data collected from Government documents, NSSO and the Census of India, 2011.

2.4 Discussion

2.4.1 Contextualizing COVID-19 Crisis in India: Rethinking Mobility and Migration

Migration has always defined and marked the human history of Indian subcontinent leading to economic advantages and producing cultural diversity. Patterns of labour migration have more intra-national attributes rather than international, while analysing India's overall economy and labour force. This is further characterized by neo-liberal policies followed by successive Governments in India, in the face of serious bottlenecks in Indian agriculture, unemployment, lack of sufficient employment generation, disparities in income, rapid growth of the informal economy and resultant migration particularly from the rural areas of less developed states to urban areas of the developed states. Migration has therefore been profoundly influenced by the pattern of regional development, the demand and supply of labour and by the existing social structures. The propensity of urbanization in Asia has also drastically altered the migration patterns, especially in India.[15] Due to factors like unaffordability in urban areas, only 3.3% of the urban households belonged to the migrant category according to the 64th round of 2007–08 NSS survey.[16]

Migration though has been in existence historically, but in the context of liberalization and globalization has assumed special significance. Since the inception of the development of migration literature several scholars, viz., Sjaastad, 1962; Lee, 1966; Harris and Todaro, 1970; Zelinsky 1971, Piore 1979, Sassen, 1988, De Haan, 2008, etc. have tried to link migration with several factors, of which economic reasons stand out as the predominant factor, others being demographic and social factors; Government policies, infrastructure, and prevailing technologies. Mabogunje (1970); Zelinsky (1971); Brown et al. (1985) strongly argue that factors deciding migration change according to a region's particular development level. Rhoda (1983) has shown in his study that factors such as increased levels of education, aspiration, awareness of urban opportunities and in general, level of modernization also stimulates migration. Evidences gathered from many countries in Asia suggest that often rural sectors characterized by low income and low farm yield foster higher out-migration (Connell et al. 1976; Simmons et al. 1977). In contrast to this, research based on African and Latin American countries reveal high rates of out-migration from rural areas are marked by comparatively high levels of income or education (Adams 1969; Caldwell 1969; Conning 1972; Riddell 1970; Byerlee 1974).

In India, several scholars have tried to link migration with development like Sovani (1964); Sundari (2005); Kumar and Sidhu (2005); Bhagat (2012) among others; mostly supporting the rural to urban stream of migration to be the most prominent one. Given the existing scenario of migrant workers and the past migration patterns, it becomes evident how complex and volatile the situation of migrants are, as there are several inter-twined factors at play beyond the push and pull. These reasons are shaped by economic opportunities, conditions at the origin, at the destination, patterns of recruitment, migration networks, family structure and so forth to name a few. In this context it becomes a colossal task to imagine a situation of pandemic in India, leave alone the idea of a pre-conceived and well-thought mitigation plan. The

[15] As mentioned in Kundu (2009), *Migration Employment Status and Poverty: An Analysis across Urban Centres*; Kundu analyses the urban growth rates, urban rural growth differences, and percentages of rural migrants in urban areas.

[16] NSSO (2010: 16–17). In both years, 1 per cent of rural households were classified as migrant households.

novel Coronavirus brought this dreaded imagination to life for millions of workers in India. It ushered in a different trail of mobility and migration that has not been experienced earlier. This is a case of heightened mobility marking forced reverse migration from developed districts to less developed ones and from urban to rural. The migrants in this case are the informal workers who form the backbone of the country's productive bases.

Coronavirus or popularly termed as COVID-19 has not been previously identified in humans. The first case was reported from Wuhan, a city in the Hubei province of China. Reports of the first COVID-19 cases started in December 2019, though researchers have pointed out that its existence may be even earlier. However, it was only on March 2020 that the World Health Organization (WHO) declared it to be a pandemic and that and the infection has the probability to spread rapidly to all regions of the world through person-to-person contact. Following this, many countries and organizations have responded to the impact of COVID-19 on migrants and the ways to provide support to migrants. According to International Labour Organization (2020) migrant workers and workers in informal sector have witnessed the worst possible situation being at the lowest rung of the labour sector. COVID-19 has transformed the world in a matter of days, this vast gap between the temporal and spatial spread of COVID-19 has resulted in a dire situation for millions. The regression that has been jolted because of this pandemic would take months if not years to be undone. For this, it is essential to read Informal Migrant Labour crisis and the COVID -19 crisis in India in simultaneity to understand the core of the issue being discussed in this paper.

2.4.2 Informal Migrant Labour in India and the COVID-19 Crisis: The Backdrop

A vast array of informal or unorganized labour characterizes the Indian labour market in contemporary times. According to the Ministry of Labour and Employment, Government of India, the Indian economy is predominated by the Informal Sector accounting for about half of the GDP. Again the employment scenario is dominated by informalization where about 90% of the workforce is engaged in informal work. In recent years, there is an increasing tendency of the informal sector getting progressively interlinked with the formal sector. The informal labour market is mostly characterized by migrant workers who lack in education, awareness and skills and are hence subjected to 'low paid-low end jobs'; often working under flexible production systems, under precarious conditions. Not only that, they are hardly protected by social safety nets or labour laws/regulations. Furthermore, this informalization of labour is not only characterized by economic opportunities based on the model of individual choices of workers postulated by the neo-classical and dualist models but there exists a forced nature of migration pattern. For example, conditions of bondage in migration processes amongst *Adivasis* in western India as described by David

Mousse (2002). This form of migration is usually through middlemen or labour contractors who could be moneylenders at the point of origin.[17]

In the wake of these complexities in migration patterns, India had about 500 reported corona positive cases, when on 24th March 2020, the Central Government under the aegis of Prime Minister Narendra Modi announced a complete nation-wide lockdown for 21 days, hardly with a four-hour notice. This came after a 14 h '*Janta Curfew*' or voluntary public curfew. The informal migrant workers remained at their workplace and the general consensus was that work would again start after the lockdown. However, this did not happen, and the lockdown went on being extended recurrently as many State Governments like that of Odisha, Punjab and later Maharashtra, Karnataka, Telengana and West Bengal recommended extensions of the state lockdowns, with the hope of breaking the chain of the spread of the virus. Hence the Prime Minister in consultation with the State Ministers extended the lockdown initially to 3rd May with a conditional relaxation from 20th April in regions with less spread of the disease. The lockdown was again extended further by two weeks to 17th May and all districts were mapped according to three zones marking the intensity of the spread of the pandemic, into red, orange and green zones. To add on the miseries of the migrant labourers things did not stop at that; as under the initiatives of the National Disaster Management Authority the Central Government declared a further extension of the lockdown till 31st May. There was yet again another announcement for a further set of lockdown, till 30th June, but this time it was meant for only the containment zones. Mumbai and Delhi, the topmost central places that act as destinations of several informal migrants hailing from the backward districts and states were the ones to be worst affected by the pandemic as well as the series of lockdowns. 8th June marked the period of a subtle unlock 1 in non-containment areas and the period of unlock 2 extended from 1 to 31st July.

According to the prediction of the International Labour Organization (ILO); about 400 million migrant labourers would come under extreme economic poverty arising out of the pandemic followed by the lockdown. To reflect on the consequence of lockdown for migrant workers it is important to analyse what it means for the workers impacted by this. The lockdown meant two things with immediate effect, firstly loss of income leading to loss of shelter and food. Secondly, it meant a complete halt of movement[18] leading to complete inaccessibility to even alternative resources that the workers could have banked upon. In the absence of these two factors it is impossible to assume that the entire purpose of lockdown that is *isolation* and to stop the spread of virus would be met without any difficulties. Self-isolation requires pre-thought strategies and workable plans from the micro level to a macro level, from individual perspective to a state-level perspective.

As was evident, the unending saga of the lockdown had extreme consequences on the informal migrant labourers particularly in the biggest cities like Mumbai and

[17] De Haan (2011).

[18] Transportation was suspended; Shramik Special trains are being operated primarily on the request of the states, which want to send migrant workers stranded due to the coronavirus lockdown to their native places, according to NDTV.com as per, 29th June 2020.

Delhi and on a lesser scale in other metropolises and urban centres of Maharashtra, Gujarat, Punjab and other States. The places of origin of these migrants were largely the States of Uttar Pradesh and Bihar, followed by the States of Rajasthan and Madhya Pradesh. According to the Census of India 2011, Uttar Pradesh is the largest out-migrating state followed by Bihar; whereas Maharashtra followed by Delhi is the largest in-migrating states. As their work came to a complete halt so also their meagre income. Many of the informal workers are daily wage earners, casual labourers and also self-employed persons, who were pushed into unending misery. A vast segment of the informal workers are engaged in construction industry, manufacturing industry, including small-scale manufacturing, trade, transport and hospitality sectors. They also include road side vendors, rickshaw pullers, household support works/domestic maids, security guards, handymen and daily wage labour attached to manufacturing supply chains, restaurants, shops and the sanitation sector. In rural areas the problem was the same for the agricultural sector of the developed states like Punjab and Haryana, where agricultural operations are done by daily wage labour from Bihar and Uttar Pradesh.

How do they work: Category of Employment

Type of employment	Rural	Urban	India
Self employed			
Own account worker/ employer	41	32.6	38.6
Helper in house-hold enterprises	16.7	5.7	13.6
All self employed	57.8	38.3	52.2
Regular wage/ salaried	13.1	47.1	22.8
Causal labour	29.1	14.7	24.9

Source: PLFS 2017-18 (NSSO)

Percentage of regular wage/ salaried employees with no written contract

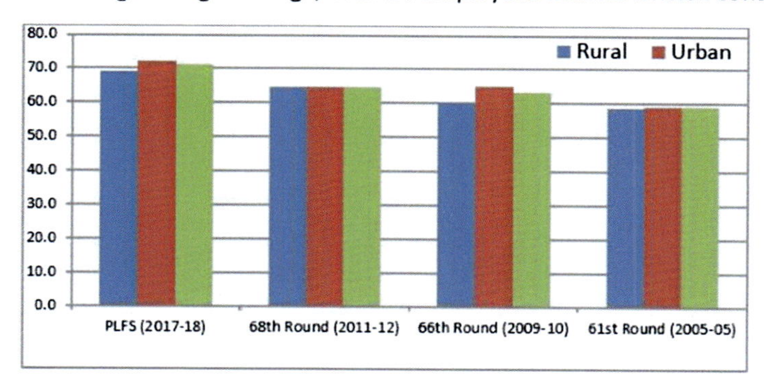

SourceReshmi Bhaskaran "India's migrant labour exodus and the missing trade unions" Policy Circle, June 12, 2020

Job loss left them indefensible, incapable of buying food or paying accommodation rent, recurrent eviction reinforced by shut down of public transport system during lockdown, resulted in aggravating their existing vulnerabilities and insecurities. These paved the roadmap of a mass exodus of migrant labourers inevitably defying the basic purpose of lockdown, but out of sheer helplessness as it had become dire. Before the virus could spread, it was hunger that took its toll and propelled this mass exodus. Migrants mostly in the informal sector left for their native villages, on bicycles or often by foot, thereby resulting in tragic accidents and deaths.

2.4.3 *COVID-19—Extenuating Institutionalized Inequalities and Exacerbating Existing Vulnerabilities Among Informal Migrant Workers: The* Exodus

As argued by economist Arjan De Haan, the presence of migrant workers is increasing among the lower bracket of the labour workforce because of poor asset base and poor human capital endowments.[19] He further expounds,

> Although this could be a result of discrimination, this may not be a product of deliberate employer discriminatory strategies. However, there is overwhelming evidence (a) that within the kind of segmented labour markets described, low-status workers are a preponderant share in work that is regarded as low status, dirty and dangerous; (b) that after accounting for other attributes, there are persistent (unexplained) differences in wages between male and female workers, and low-status castes and high-status castes; (c) that the evidence of discrimination is less in public sector establishments which follow an affirmative action policy based on quotas, both in terms of persons employed and in terms of unexplained wage gaps.[20]

The pandemic has reconfigured and reinforced these vulnerabilities and inequalities among the informal migrant workers and that too at a tremendous speed that came almost as a shock to this segment of the Indian population. The salient characteristics of the shock came in the form of loss of livelihood, harassment by landlords, hunger, health hazards in the wake of COVID-19 arising out of overcrowding; and lack of social distancing arising out of poverty and insanitary conditions; and restrictions imposed immobility leading to loss of lives. The above-mentioned factors led to a mass exodus of thousands of migrant workers, who out of desperation in the absence of public transport system started their journey back home to their native villages bare foot, or on cycles often on empty stomach. In the following section, we are going to highlight a few primary newspaper sources to emphasize on the current scenario and the plight of migrants.

> 'Huge gatherings of such workers and group mobility defied all precautionary measures of social distancing, as they feared death due to starvation more than the Coronavirus. Hence

[19] De Haan (2011).

[20] Ibid., pp. 165.

they gathered in large congregations at train termini and bus stops', in a desperate attempt to go back to their native places.

—Dutt, Barkha (15 May 2020).[21]

Out of desperation many of the migrant workers flouted the lockdown regulations, only to be subjected to police brutality in the form of 'lathi-charge' or arrested while crossing territories, forested tracts and even rivers.

—Babu, Venkatesha; Saini, Sachin; Swaroop, Vijay (8 May 2020).[22]

Deaths have now become a common norm associated with the flight of the migrant workers. This was either due to exhaustion instilled by hunger, malnutrition and tropical heat or by road accidents.

'She was just 12, the only child of her parents. Extreme poverty had pushed the minor to work as a daily wage earner in the Indian State of Telengana. On April 18 young Jamalo Madkam died, after she had lost her job amidst India's COVID-19 lockdown. Realizing that the extensions of the lockdown meant no job, more days without food, she and a group of 12 other migrant workers decided to return home on foot, to the Indian State of Chattisgarh from Telangana where they worked. For Madkam, the 100km journey ended in death due to electrolyte imbalance and exhaustion, barely 11 km from home'.

—Elsa, Evangeline (21 April 2020).[23]

More than any other incident in over 40 days of a lockdown that is reputedly the world's most stringent, it was the mowing down of 16 migrants sleeping on a rail track in Aurangabad, Maharashtra, on Friday that encapsulated the human tragedy unleashed by the COVID-19 pandemic. The migrants, who had walked 36km on the first stretch of what would have been their journey home to Madhya Pradesh, had fallen asleep on the rail track, tired. A goods train ploughed through them at day break. The COVID-19 pandemic has hit the poor the hardest in India, with locked factories and other workplaces triggering the biggest internal migration since Partition.

—Nandi, Shreya; Bhaskar, Utpal (8 May 2020). "Migrants' deaths on the tracks a wake-up call for India". Live Mint. 9 May 2020[24]

Twenty-four migrant workers had died and 36 injured on Saturday early morning when a trailer truck carrying sacks of lime along with 43 people rammed into the back of a stationary truck with many other migrant workers sitting in it near a roadside eatery on the national highway in Auraiya. In a statement issued here, the Auraiya police said the accident took place near Shivji Dhaba in Tikauli village involving a trailer truck and a mini-truck with both the vehicles falling into a ditch by the roadside. In the accident, 26 migrant labourers died, while 34 others sustained injuries," the statement said while updating the casualty figures.

—'Auraiya road accident: Two more migrant workers die, toll rises to 26'.[25]

A 15-year-old girl carried her ailing father on a bicycle for 1,200 kilometres (750 mi) from Bihar to Gurugram over the course of a week. She was later approached to try out for the

---

[21] "There is a humanitarian crisis in India. Lift the lockdown, now". Hindustan Times. Retrieved 22 May 2020.

[22] "Across the country, migrants still forced to walk thousands of miles". Hindustan Times. Retrieved 9 May 2020.

[23] Elsa, Evangeline (21 April 2020) home, Gulf News. Retrieved 9 May2020.

[24] Nandi, Shreya; Bhaskar, Utpal (8 May 2020). "Migrants' deaths on the tracks a wake-up call for India". Live Mint. 9 May 2020.

[25] Hindustan Times. 17 May 2020. Retrieved 19 May 2020.

National Cycling Academy by the Cycling Federation of India,and received praise from Ivanka Trump.

—Gettleman, Jeffrey; Raj Suhasini, (22 May 2020)[26]

There had been attempts by certain states to bring back the migrants and Uttar Pradesh has been one of these states to arrange for buses. Huge congregations of informal migrants at bus stops had become the order of the day in cities like Delhi; (Anand Vihar bus stop). However, because of the extension of the lockdown, permission of plying buses by the State Governments started only by the end of April. Social distancing could neither be maintained either at the bus stops or in the buses.

The aftereffect of the recurrent lockdowns, loss of jobs and irresponsible attitude of the society resulting in innumerable suffering, was so strongly felt by the migrant labourers that even after the inauguration of special trains or '*Shramik Special*' trains by the Government in late May, the workers were desperate to travel back to their native places and take up any available petty jobs in the villages or under MGNREGA. Many travelled in the cargo compartments of containers and trucks in large groups instead of availing the Government arranged trains. This was also because their number was very large and the modalities of availing the service provided by the Government were cumbersome for them. Considerable problems were also associated with the Shramik Special trains launched by the Government for travel by migrant labourers to their villages. The complications in this case were the fares charged, sudden cancellation of trains like that by the Karnataka Government, extensive delays, sometimes as high as 58 h; insanitary conditions in the trains, lack of food and water, and other such miseries. "*In the cities they treat us like stray dogs. Why would they treat us any better now?*"[27] Migrant workers already exhausted by innumerable problems in the places of work and persistent problems, or those travelling with a disease burden that have already invaded them, often died while travelling by these trains. The Railway Protection Force had reported almost 80 such deaths between 9 and 27 May 2020.[28]

2.4.4 Institutionalized Inequalities and Harassment

Srivastava and other economists have indicated that migrant labour market is highly differentiated. Labour market comprises of poorer migrants in different ways. At the lower end with seasonal labourers, migrants enter labour circuits on contractual basis. They are also fragmented along the lines of gender, ethnicity and age. Migrant labourers often work for long and odd hours and are not paid on time. All these conditions lead to their precarity that has been worsened by this lockdown due to

---

[26] Nation. The New York Times. ISSN 0362-4331. Retrieved 29 May 2020.

[27] A migrant worker describing the treatment he received on a "Shramik Special" train Bhowmick, Nilanjana (27 May 2020). "They treat us like stray dogs': Migrant workers flee India's cities". National Geographic. Khandelwal, Saumya. Retrieved 27 May 2020.

[28] Dutta (2020).

COVID-19. There are innumerable incidences of harassment faced by the migrant workers, both during the exodus and after. Migrant workers form the backbone of the economy in big cities. During lockdown the same workers had to bear the brunt from the society, be it their employers, landlords, neighbours or the police. They were the ones blamed to be responsible for the disease spread and were even assaulted by the neighbours or had to face police brutality if they ventured out to buy food or milk for their little ones. Coronavirus pandemic and the process of lockdown in India have suddenly made visible—poverty, vulnerability, malnutrition and existing inequalities persisting in our society, even more apparent than ever before.

Alpa Shah and Jens Lerche (2020) report the superimposition of the COVID-19 crisis on the already embedded inequalities characterizing migrant labour particularly seasonal migrants belonging to the Scheduled Caste or Dalit communities, who happen to be the poorest of the poor.

> First, business and industry is dependent on migrant labour that is paid less, works longer and harder, and is more flexible than local labour. Though, in many parts of the world, such a precarious migrant workforce travels across national borders; in India, it is a huge internal migrant force traversing state borders for informal contract work in more developed parts of the country where they are treated as second-class citizens. Usually unable to speak the lingua franca of where they migrate to, rarely represented by any union or social movement, they are easily harassed by employers, Government institutions and by other workers. This vulnerability makes them more easily controlled, cheap and dispensable. Second, in India, many of these migrants — about 100 million in total — work seasonally and circulate between their rural homes and faraway work sites for a part of the year. Third, workers treated the worst, often come from regions of India like Jharkhand, Odisha or Chhattisgarh which have long suffered a form of internally oppressive structures as their indigenous wealth — minerals, forests, other natural resources — has been extracted by outsiders, leaving little but high levels of poverty for the locals. Fourth, India's historically disadvantaged minorities do the hardest work in the worst living conditions. Dalits and Adivasis are overly represented as seasonal labour migrants; they make up more than 40 per cent of the seasonal migrant workforce even though they are only 25 per cent of the population. All these migrant workers are subject to the stigmas of caste and region, treated as jungli (filthy or savage), including by the local workforce. Fifth, these seasonal migrant workers are, in turn, supported by a further invisible economy — the household. Seasonal migrants can only be workers because of all the work undertaken across generations at home; including care provided by the spouse, children, siblings and elderly parents. Seasonal migrant workers are, therefore, not just exploited, they are super-exploited.
>
> —Shah, Alpa; Lerche, Jens (13 July 2020).[29]

Arpa Shah and Lerche opine how invisible the contribution of migrants is towards economic growth and development. Moreover, these 'economies of care' are not just invisibilized but downgraded and pushed to the periphery of the socio-cultural matrix of urban societies. These migrant workers not only form a considerable percentage of the workforce but also live in grave precarity on a day-to-day basis without adequate and systemic representation in the occupation sector and society.

Therefore, COVID-19 had not just exacerbated these pre-existing concerns but worsened these at the same time. Once at destination i.e. their native villages, the

[29] "The five truths about the migrant workers' crisis | Opinion". Hindustan Times. Retrieved 15 July 2020.

migrant workers had to face another set of ordeal. The migrant workers who are already treated as 'dangerous' 'outcastes' and 'filthy' were also seen as a subsequent threat that they will bring the infection with them from the cities. Hence, they were treated with extreme apathy, discrimination and were harassed and discriminated in many instances. Sometimes they were hosed down by disinfectants or treated with utter negligence in the State designed quarantine centres. The plight of many of these quarantine centres provided by the State, needless to say, is extremely deplorable. There have been regular complaints about the same and in certain cases. With no outlet and no social security the migrants sometimes have been forced to run away and escape to their ill fate. The stories of their sufferings are therefore never ending.

2.4.5 Migrant Workers, COVID-19 Crisis and Mental Health

To sum up, the pre-existing conditions of the migrant workers in India, as mentioned in the previous sections, while there is deep-seated horizontal/vertical segmentation in the capitalist sector of the economy, it is deepened by formal and informal segregation. This is further characterized by social discrimination, which is closely interlinked with this fragmentation resulting in a great deal of precariousness for the migrant workforce. In the face of COVID-19 pandemic, it is essential to discuss mental health amidst such deep-rooted complexities and problems the migrants face on an everyday basis. The COVID-19 pandemic has instigated challenges in mental health of migrant workers, in addition to their fragile physical health in many cases that is already in place due to their poverty, hazardous working conditions and insanitary living environment. The series of adversities faced by them explained in the previous sections have led to a series of negative cognitive responses and emotions, the fallout of which is extremely worrisome. The daily wage labourers have been the worst hit in this respect. Congested dwellings like that of slums in mega cities like Delhi and Mumbai had already set the ground for mental stress, just to be aggravated by the disillusionment imposed by the pandemic, lockdown and ban on travel, loss of jobs and above all extreme poverty. Overall disillusionment has led to depression, anxiety, and panic attacks. This has also led to hostilities and irrational behaviour including suicides.

According to Ranjana Choudhari (2020);

> Internal migrant worker is a vulnerable community for the development of severe, acute and chronic, adverse mental health consequences due to COVID-19 pandemic, through various multi-dimensional factors, many acting concurrently to cause physical, mental, and socio-economical adversities. Besides, the restrictive measures adopted during lockdown and containment COVID-19 policy, associated down gradation of the legislations and laws of occupational safety and health in India, has the potential to aggravate and precipitate the adverse effects on the psyche of internal migrant workers. Considering the detrimental occupational angle, which enhances the vulnerability and mental health; community should prepare themselves for handling the challenge of an upsurge in the psychological illnesses among this occupational community. Mental health is a critical aspect that needs to be addressed, making it imperative to initiate steps against the psychological ill effects due to

pandemic through generating awareness and psychological preparedness among the internal migrants.

—Ranjana Choudhari, Indian Council of Medical Research (ICMR).[30]

According to the Ministry of Health and Family Welfare, Government of India, migrant workers are instilled with a new uncertain situation that has led to considerable increase in the level of anxiety and fears and are in a desperate need of psychosocial support. In a country like India, the pre-existing awareness of mental health conditions is deeply imbued on social discriminatory practices and taboos. The socio-cultural matrix is currently transitioning in order to adopt and adapt to the newly emerging knowledge in the field of mental health. However, this is only limited to urban pockets of the country. Informal Migrant workers form a considerable section of these urban pockets further stratified by caste and class discrimination and have socio-cultural ties from their places of origin, which are in fact the rural pockets of the country. There is considerable possibility that they are unaware of these mental health issues despite facing them on a day-to-day basis. It has been greatly debated and argued that there must be a wide-ranging perception of mental health issues as an urban health condition. However, the increase in the cases of suicides of these migrant workers point towards a totally different trajectory. Anxiety and stress imbued due to Coronavirus is an all-pervasive phenomenon, migrant workers being at the lowest rung of the occupational social order face the brunt on a colossal scale. As Ranjana Chowdhuri argues migrant workers are pre-disposed to experience adverse psychological conditions that exist at an interface of various factors such as poverty, malnutrition, cultural bereavement, loss of religious practices and social protection systems, absence of family support during crisis, limitations to follow rules and regulations, social exclusion of personal safety, language difficulties, changes in identity, substance abuse and poor access to health care systems; all this while constantly dwelling in a grave financial concern and poor living conditions.[31] These factors are multi-dimensional and often exist as an interface to the other. The mental health community should increase their preparedness in handling such a crisis. Nodal Mental Health Institutes should make it imperative to initiate steps in order to fight the psychological ill effects caused due to and as a result of the pandemic situation in the country keeping these multi-dimensional factors in mind. However, in order to address redistribution of aid and allocation of resources a lot of vertical and horizontal factors have to be analysed and problematized first. The idea of redistributive conflict is essential to discern which we will be addressing the subsequent section of our paper.

[30] Chowdhuri (2020).

[31] Ibid.

2.4.6 *Idea of Redistributive Conflict: Locating the Problematic in Allocation of Resources and Government Response to Informal Labour and Massive Reverse Migration*

Redistributive conflict arises where redistributions are not alike. It is not a matter of *who* as it is too simplistic to read that way, it is a matter of *how*. It is important to consider the effects of proposed redistribution on the economic system as a whole, especially towards the sector redistribution are aimed for. We have previously analysed problems like social discrimination, informalization and fragmentation of the labour sector that has significantly contributed to the inequality discussed at the very beginning of this paper. COVID-19 led to worsening of all of these issues to a greater extent but also has initiated a process of *regression* that might take considerable time to overcome. Hence, the Government response needs to be analysed against a redistributive conflict, as simplistic schemes of allocation cannot deal with this multi-layered and fragmented problems relating to the migrant labour sector. There was a series of measures undertaken by the Central Government to help the poor. These include the announcement of the Finance Minister Nirmala Sitaraman, the sanction of an amount of Rupees 1.7 lakh crore, (US$24 billion)[32] for the poor through cash transfers and to revamp food security. Funds were also provided to the NDRF (National Disaster Response Fund) to tackle food and shelter security to the poor. On 13th May 2020, Rupees 1000 Crores were granted to the migrant labourers from the PM CARES Fund.[33]

Another announcement on 14th May by the Finance Minister allocated Rupees 35 billion (US$490 million)[34] for free supply of food-grains to 80 million migrant workers. A rural public works scheme, i.e. the Garib Kalyan Rojgar Abhiyan was launched by the Government to help the COVID-19 affected migrant workers in the country, in 116 districts in 6 states with a budgetary allocation of Rupees 50,000 crore (US$7.0 billion).[35]

Initially, the Indian Government ensured to retain migrant informal labour in their states of destination during the lockdown that was initiated on March 27th 2020. Providing food and shelter was relegated to the State Governments that were permitted to utilize the National Disaster Response Fund for this purpose. Moreover, the Government issued orders directing the employers not to deduct the wages and to the landlords to house the migrant labourers without charging rent. There was a withdrawal of the payment of the regular wages during the lockdown issued on 17th May 2020. In order to address the exodus, the Central Government directed all State

---

[32] Beniwal and Srivastava (2020).

[33] Sharma and Prabhu (2020).

[34] "*India to Provide Free Food Grains to Millions of Migrant Workers*". The New York Times. Reuters. 14 May 2020. ISSN 0362-4331. Retrieved 16 May 2020.

[35] "*PM Modi launches Rs 50,000-crore Garib Kalyan Rojgar Abhiyaan to generate jobs*". Hindustan Times. 20 June 2020. Retrieved 20 June 2020.

Governments facing reverse migration to provide food and shelter to the stranded workers by setting up relief camps. Some NGOs and religious institutions came forward in providing food to the workers, in addition to the provision of essential medicines, sanitizers and masks. *'By 5 April, 75 lakh people were being provided food across the country in food camps run by the Government and NGOs'*.[36]

However, all these efforts could not retain the migrant labourers as there are a multitude of reasons as discussed above, needed to be addressed instead of perfunctory schemes that were provided like provision of food, shelter and setting up of relief camps. Not to forget that all these are mandatory requirements in the face of a crisis like COVID-19, especially when all movement has been curtailed. However, the larger question of variability and the inter-twined risk factors as discussed above is the kernel of the exodus that needs to be dealt with. After all, migrant workers were still dwelling in darkness with the economic and social uncertainty that was looming large. However, the Government, at a later date arranged buses initially (in April) followed by Shramik Trains (1st May onwards). The lack of proper analysis of the informal migrant's situation was evident again as innumerable problems were associated from the inception of the Shramik Trains scheme. For example, problems arose in the form of registration of workers to avail these, unaffordable fares in certain cases, problems of service delivery, quality of service being provided, lack of information, social distancing being flouted and deaths under inhospitable circumstances. Moreover, at the initial level, the train service was not free, due to opposition, the Central Government offered an 85% subsidy and State Government offered15% subsidy to the train fare. Even then, in many cases these directives were flouted and not imposed rigorously. *'4,277 Shramik Special trains had transported about 60 lakh people, as of 12 June'*.[37]

Due to the economic derailment of the informal migrant workers, the Government of India decided to provide relief to the migrants by providing additional ration. Since the migrants were stranded due to lockdown, the entire question of accessibility became a colossal concern as the question arises of *how* will they go to the ration shops? Is the idea of *'redistribution'* enough to come up with a solution? *How* to deal with the underlying and inter-twined factors? As a 'solution' the Government of India in mid April implemented 'One Nation, One Ration Card'. According to the Food Corporation of India, India had enough rations for the poor that would last for a year and a half. This brings us back the main question of adequacy of the Government schemes that in turn brings forth the idea of redistributive conflict. This conflict majorly arises out of three reasons for the informal migrants, (A) Internal fragmentation of the informal labour market (as discussed in the first section), (B)

[36] Mathur, Atul (28 March 2020). "*Delhi lockdown: Over 500 hunger relief centres set up for 4 lakh people*". The Times of India. Retrieved 3 April 2020.

"*As of 12 April, 37,978 relief camps and 26,225 food camps had been set up" Over 75 lakh being fed at food camps: MHA*". The Tribune. 6 April 2020. Retrieved 12 April 2020.

[37] Nandi (2020).

Economic and social segregation of migrant workers,[38] and (C) Disparity in allocation of schemes and resources. The Public Distribution System in India has been in debate due to these three reasons that have heightened the redistributive conflict for the migrant workers to another level during COVID-19 crisis. This is due to the pre-existing institutionalized bottlenecks. As a result, the functioning of the Public Distribution System (PDS) proved to be ineffective, leaving the poorest of the poor marginalized, even after the implementation of 'One Nation, One Ration Card'. There were several problems associated with the PDS. Firstly, the migrant workers were not aware of the fact that they can avail food-grains from any place. Secondly, the biometric authentication required for availing this was discontinued due to the spread of the Coronavirus. Additionally, many of the workers did not have the Aadhar Card (Unique Identification Authority of India) and in many States the workers did not receive the stipulated amount of ration, or if they did, they received inferior grains not fit for consumption, highlighting the malpractices existing in the society. Hence these underlying concerns form a larger question of the provision of security to informal migrants that needs to be discerned before providing aid and security that is *adequate* enough for survival. If not dealt within time, this will result in widespread starvation and exhaustion that is becoming a harsh reality for the considerable section of the informal migrants as we have already discerned.

The Scheduled Caste or Dalit informal migrant workers, who are at the lowest rung of the informal labour market as well as the social ladder, bore the paramount brunt of the entire chain of adversities and human sufferings. The National Herald on 21st May, 2020 reports that 'in order to draw attention to their plight of starvation or "*Bhookh*", the villagers of Kherwa, Banda District have placed a banner to launch their "hunger strike", being tired of waiting for ration cards and also that of false assurances'.[39] The newspaper article highlights,

> It has been a long wait for the villagers of Kherwa in Banda. For the last two months they have been hearing that they were eligible for subsidized food-grains. But they needed to complete formalities, fill up forms and wait for provisional ration cards. Some villagers do seem to have received them. Some, for reasons that are still not clear, have not. One of the banners read, *'Huzoor, Bhookh se mar rahe hain",* this has been put up on the main village road. Women, said activists from Banda, were holding fort, banging thalis (plates) demanding food-grains and drawing attention to their plight. In the Kherwa village, 42 Dalit families belonging to the community of *Kucchbandiyas* have been denied ration cards. Their livelihood was dependent on *'Pheri'* – selling small household tools and delivering services while roaming from one village to another. They used to earn on an average around Rupees 200 per family per day. Since, the lockdown began on the 25th of March, they got stuck at home, were not allowed to move and have not earned a single paisa. The entire village is worried about its survival. Despite being eligible for all of the Government schemes meant

[38] This social segregation also operates at the intersection of Gender. Therefore, it is not only the caste and class lines that needs to be analyzed and discerned before implementing redistributive government schemes but it is also the gender aspect that operates at the apex of social and economic discrimination that needs to be put into account. The Covid-19 crisis has worsened the double subordination faced by the women migrant laborers, as there have been rampant cases of domestic violence and assaults on the top of the sexual and social discrimination they face on a pre-exiting and everyday basis.

[39] National Herald on 21st May, 2020.

for the poor, they are not getting the food. *Raja Bhaiya* of *Vidya Dham Samiti*, an NGO, state that these poor families are severely hit by the lockdown as there is no food for even one proper meal a day. The NGOs working in the area are helping them to survive. Families who have ration cards are also struggling, he informed. The situation is similar in other villages of Banda. In Khamora village, villagers staged a protest demonstration, while in Khera village, people have written Bhookh on the walls of the Panchayat Bhawan. Incidentally all these villages are dominated by Dalits.[40]

A lot of malpractices are found to operate in the ration shops and PDS outlets in rural areas. These outlets do not adhere to the standardized rules and regulations and many shops siphon rations and sell it in the open markets by manipulating their records.

"While a family with eight members are eligible for 5 kg of food-grains per member for the entire month, many such families receive only 10 kg in all. Such a gap in the system and the strict rules of lockdown are making it difficult for people to survive,' says the activist. The *Kucchbandiya's* also fear that post-lockdown even if the spread of the virus slows down, life is not going to be the same as before. The villagers have written an application to District Magistrate (DM) about the state of hunger in their villages. In the application names of those who do not have ration card was also mentioned. It was demanded that the ration card of all the villagers be made and groceries be delivered immediately. A complaint about the issue was also made on the Portal of Chief Minister of Uttar Pradesh. The villagers wrote in the application, - if any one of us dies because of hunger the DM will be responsible for the death'.[41]

They are still waiting for response from the Government. At this point, after talking about the redistributive conflict of allocation of resources, it becomes essential to analyse the efficacy of such schemes by the National and State Government in simultaneity. Questions of analysis arise regarding working in conjunction to both the Governments. Is there a disparity? Are there any efforts of uniformalization of data on migrants during this crisis? What are the programmatic interventions of the National and the State Governments based on? Is there any nodal institute at work? Or is the date scattered and spurious?

2.4.7 *Efficacy of Programmatic Interventions by National and State Governments During the Pandemic and Their Impact on Migrants*

The exodus of informal migrant labourers in the unorganized sector has unfolded the saga of human suffering pointing out to the efficacy and bottlenecks of the programmatic interventions by the Central and the State Governments. The Hon'ble Supreme

[40] https://www.nationalheraldindia.com/india/starving-villagers-put-up-banner-launch-hunger-strike-to-draw-attention-to-theirplight?fbclid=IwAR2Q2asxwEbZCfMMqX8Y7007RZWod2maTTLmtiLOVn5YmK8N8yHj6vVdEns.

[41] https://www.nationalheraldindia.com/india/starving-villagers-put-up-banner-launch-hunger-strike-to-draw-attention-to-theirplight?fbclid=IwAR2Q2asxwEbZCfMMqX8Y7007RZWod2maTTLmtiLOVn5YmK8N8yHj6vVdEns.

Court itself on 26th May 2020 stated that the problems of the migrants had still not been solved and that there had been 'inadequacies and certain lapses' on the part of the Governments. This clearly indicated the disparity between Centre and the State Governments in implementing programmes and schemes that it inevitably altering the efficiency of the schemes already at work.

Another problem facing the migrant informal workers is the lack of organization. Since they are not protected by any Trade Unions and collective bargaining is conspicuously absent, they are at the losing end at times of adversities.

> They did not have able leaders and trade unions to engage in collective bargaining and conduct negotiations with their employers and the state. Years of neoliberal policies have weakened India's labour/trade unions through a process of informalization of the work place and workers. The labour laws were relaxed in favour of employers, and the image of trade unions took a beating. The exodus during COVID-19 lockdown exposed the vulnerability of workers in India in the absence of trade unions. They were pushed to the corner after a gradual, but systematic decimation of trade unions. The workers in the unorganized sector failed to collectively bargain with the Government for transportation facilities. Without the power to assert their human and labour rights, they were left at the mercy of the State and charity of others for food, water, stay and transport. Simply, the system converted informal workers into beggars and forced them to walk to their homes hundreds of kilometers away.[42]

2.4.8 Need for Data and Proper Documentation of Migration Histories Imposed by COVID-19

Another major obstacle in programme implementation and effective targeting of the poor migrant worker is the *paucity of data* both regarding informal work and that regarding their migration and in particular seasonal migration. Data on seasonal migrants are not available in the Census of India. Though the NSSO provides data on seasonal migrant but all segments of informal work are not captured. The representation of these groups of workers is mostly through estimates.

Today, many states that form the origin of migrant labour, are struggling to deal with the large-scale reverse migration that is the most visible outcome of the COVID-19 crisis. A major deficiency that has become evident during the pandemic is the paucity of data on migrant labourers. States like Jharkhand, Uttar Pradesh and Madhya Pradesh have worked out certain estimates to deal with the migrants in an effective manner. However, this is just at the tip of the iceberg, since a lot has to be done to create a proper database to this effect. Hence for the successful implementation of any scheme, estimates are not just sufficient. Both the Central and the State Government should create database for drawing plans to ameliorate the condition of informal migrants both at the origin/source as well as at the destination of the migration trails.

---

[42] Bhaskaran, Reshmi: *India's migrant labour exodus and the missing trade unions,* 12th June 2020, Policy Circle.

2.4.9 *Need of the Hour: Effective Labour Migration Governance*

It is clear from the previous discussion that the Coronavirus pandemic has brought to the fore the forgotten victims of the times, i.e. the innumerable informal workers of the unorganized sectors, particularly the migrants, who are already disadvantaged due prevailing inequities, their petty incomes, lack of safety, security, nutrition, basic amenities including housing and sanitation; but form the productive bases of the Indian economy. This segment of workers should form a primary concern of both the National and the State Governments. Both the sending and the receiving states should be involved in providing safe mobility and proper allocation of resources to these disadvantaged groups. The Government should therefore come up with both short and long term strategies to intervene the problematic and for protecting the life along with the livelihood of these workers, living at the margins. Effective targeting should be for the most disadvantaged groups be it the Scheduled Castes/Dalits, Scheduled Tribes/Adhivasis and other marginalized segments. Post COVID period would also prove to be very crucial for the provisioning financial and technical support in terms of providing work, nutrition, safety and in all round facilitation of workers to return to work either in their native places or else back to their work places in the cities. The lockdown, loss of jobs, lack of facilitation in mobility and other adversities has demoralized thousands of migrant workers across the country. It is time to instill mechanisms that can revitalize and mainstream them with other workers in the economy.

Mobility and migration are interlinked aspects of livelihood. In India, due to the stagnation and distress in the agrarian sector, human poverty often compels the population to move to the big cities in search of jobs. These migrants often lack in education, training and skills and hence get absorbed in any odd job in the informal sector that is often hazardous. They are therefore already exposed to risks and vulnerabilities. The COVID-19 pandemic has rejuvenated their already vulnerable positions and they have thus become the worst hit victims lacking protection and bargaining power. The resulting exodus has therefore been an eye-opener for not only the Government but also the civil society at large. The need of the hour demands the formulation of a holistic policy by both the Centre and the State Governments. Effective labour governance should therefore address the needs of this diverse group of migrant informal workers, including men and women in productive ages.

> Labour migration can yield many positive benefits for all, when it is well-governed. This is not illusion, this is not utopia, and this is entirely possible. But at the same time, policies that are not firmly grounded in respect of human rights, including labour standards, present high risks and costs for migrant workers, for businesses and for the countries concerned. They also run the risk of reducing the status of migrant labour to that of a commodity.[43]

[43] Guy Ryder, ILO Director-General, April 2020.

Such a governance should not only aim to increase education, awareness, skills and competency among workers; but simultaneously instill mechanisms of protection of their rights based on equity and equality. There is also an impending need to revamp the existing social security schemes and eliminate structural inequalities and iniquities. These include the Public Distribution System (PDS), Mahatma Gandhi National Rural Employment Guarantee Act (MGNREGA), Mid-Day Meals at *Anganwadis*; Old Age Pensions, etc.

> The urgent need for effective social security measures makes it all the more important to avoid a loss of nerve. The way things are going today, it will soon be very difficult for some State Governments to run the Public Distribution System or take good care of drinking water. That would push even more people to the wall, worsening not only the economic crisis but possibly the health crisis as well. This is not the time to let India's frail safety net unravel.[44]

In this respect several economists and scholars in India working on migrant informal worker crisis during COVID-19 have indicated a paradigm shift in our development model, by making it more robust and holistic by tying up the inter-twined needs of both the rural and the urban areas.

> From workers walking for days to reach home to the long queues for a single meal, the COVID-19 crisis has reiterated the perilous situation of informal workers. Neither their rights as labour nor their rights to state welfare are adequately addressed by the existing approach. Only a radically altered development model, which addresses the conditions that foment informalization, can ameliorate these conditions. These would include significant investment in agriculture, ensuring stable livelihoods in the villages to prevent the hunt for precarious jobs by the rural masses; formulating new state policies that address the increased dependence on metropolises; increasing state capacity to implement existing laws covering the informal sector.
>
> But all these measures would require the principle of equity as a guiding force of state policy. Unfortunately, the opposite has been the case. Several states have issued ordinances to weaken existing labour laws, thereby deepening the precarious existence of informal labourers. The task at hand, therefore, is a political one. As this crisis prolongs and the state turns a blind eye to the suffering of workers, forces aligned with labour have no option but to take up this task.[45]

2.5 Conclusion

The banal pattern of migration as depicted in the present paper could also be analysed as 'mobility by default' as argued by Sunanda Sen.[46] The reasons as discussed above are myriad, complex and often inter-twined. COVID-19 has put the highlighted migrant workers' fragile working conditions. First, the sudden and consecutive lockdowns that left many workers jobless, subjecting them to the arduous task of walking

[44] Jean Drèze, IGC in conversation with Jean Drèze: *The impact of COVID-19 on migrant workers in India*, 15th May, 2020.

[45] Jenny Sulfath and Balu Sunilraj, senior researchers at the Centre for Equity Studies, New Delhi "Covid-19 Crisis Exposes India's Neglect of Informal Workers". News Click 12 May 2020.

[46] Sen (2020).

hundreds of kilometres to their villages of origin. Now, as a result of the novel coronavirus, states are relaxing the labour laws in an effort to boost and foster their economies. Resultantly, migrant labourers are forced to even longer hours.[47] The existing acts[48] and social safety nets have proved to be of little workability in dealing with the colossal crisis triggered by the pandemic where migrants have been literally pushed out of urban centres.

The current situation leaves us with a few important questions like; how will the economic outcomes of the health crisis re-structure the informal labour market in India and how will it affect the future labour migration policies? How will the increase in unemployment impact human mobility in India in years to come? How will the lack of seasonal migrant workers impact certain sectors such as construction, manufacturing, trade, transport, agriculture and so forth? In a new era of an expanding COVID-19 crisis and its resultant travel restrictions and required medical testing of migrants—how effectively can the situation be handled given the fact that there is fear of catching the virus at every step on one hand and the fear of zero income and starvation on the other? All these questions remain to be answered as the crisis worsens under the impact of the several waves of the pandemic and the Indian economy is struggling to handle the backlash.

References

Beniwal V, Srivastava S (2020) India unveils $22.6 billion stimulus plan to ease virus pain. Blooomberg Quint

Chowdhuri R (2020) COVID-19 pandemic: mental health challenges of internal migrant workers of India. Elsevier Public Health Emergency Collection, 19 Jun 2020 Date Written: April 23, 2020

Dutta A (2020) Railway Protection Force reports 80 deaths on Shramik trains. Hindustan Times. Retrieved 1 June 2020

De Haan A (2011) Inclusive growth? Labour migration and poverty in India, vol 513

Nandi S (2020) Indian Railways receive request for 63 Shramik special trains. Livemint. Retrieved 14 June 2020

Piketty T, Goldhammer A (2020) Capital and ideology

Sell RR, DeJongGF (1978) Toward a motivational theory of migration decision making. J Population 1(4):313–35. http://www.jstor.org/stable/27507584

[47] Exemplified by Case Studies through media reportings.

[48] "There is the Contract Labour Regulation and Abolition Act 1970, which conferred casual labor with a legal status by providing a mechanism for registration of contractors engaging 20 or more workers. Failing registration, the employer was directly responsible for the employment provided. One can also mention the Inter-state Migrant Workmen Act 1979, the National Disaster Management Act 2005 and the Street Vendors Act, 2014 to regulate street vendors in public areas and protect their rights. More recently, the Code on Occupation, Health, Safety and Working Conditions, sought to regulate health and safety conditions of workers in establishments with 10 or more workers, and replace the 13 prevailing labour laws. The Code was referred to a Standing Committee of the Parliament in July 2019 that responded positively on a date as recent as February 11, 2020". As mentioned in, Sen, Sunanda; Rethinking Migration and the Informal Indian Economy In the Time of a Pandemic; The Wire, 1st June 2020.

Sen S (2020) Rethinking migration and the informal Indian economy in the time of a pandemic. The Wire, 1st June 2020

Sharma A, Prabhu S (2020) Rs 3,100 crore from PM CARES fund allocated for ventilators, migrants. NDTV. Retrieved 15 May 2020

Smith N (1984) Uneven development: nature, capital, and the production of space. Blackwell, New York, NY

Srivastava R (2019) Emerging dynamics of labour market inequality in India: migration, informality, segmentation and social discrimination. Indian J Labour Econ 62:147–171. https://doi.org/10.1007/s41027-019-00178-5. August 2019

Chapter 3
Challenges in Livelihood of Residents in Kilinochchi District, Sri Lanka Due to Water Scarcity

Kirishanthan Punniyarajah

Abstract Water scarcity is the main constraint and it has become a serious threat in meeting domestic purposes as well as the livelihood as a whole of the resettled people in Kilinochchi district, Sri Lanka. These current scenarios, therefore, call for an investigation on challenges in livelihood due to water scarcity and its impacts on sustainable development goals. This research is carried out in Karachchi Divisional Secretariat Division that belongs to Kilinochchi district. The study manipulates a combined approach of both qualitative and quantitative methods. It relies mainly on primary data that were collected through household survey, structured interviews and observation, while secondary data from certain sources were also used. The questionnaire survey was conducted with randomly selected two hundred and fourteen households from the purposively selected five *Grama Niladhari* Divisions. Qualitative data and quantitative data of the study were analysed using content analysis, and descriptive statistics respectively. Frequency distribution tables, pie charts and graphs were used to display the findings of the data analysis. Findings of the study include various livelihood challenges: monthly income, cost of living, agriculture, home gardens, livestock, inland fishing, unemployment, and small business which are negatively impacted by water scarcity in the study area. Further, the study indicates that the sustainable development goals no: 1-No Poverty, 3-Good Health and Wellbeing, 5-Gender Equality, 8-Decent Work and Economic Growth and 9-Industry, Innovation, and Infrastructure are highly constrained to achieve due to the negative impacts on livelihoods caused by water scarcity. Moreover, SDG 6-clean water and sanitation and, SDG 2-Zero Hunger are extremely threatened in the study area due to water scarcity. Therefore, relevant stakeholders should take urgent actions to ensure the availability and accessibility of water to address the challenges faced by the households in Karachchi, Kilinochchi district, in terms of livelihoods and sustainable development.

Keywords Agriculture · Household · Livelihood · Kilinochchi · Sri Lanka · Water scarcity

---

K. Punniyarajah (✉)
Department of Geography, University of Colombo, Colombo, Sri Lanka
e-mail: krishanthan@geo.cmb.ac.lk

N. C. Jana et al. (eds.), *Livelihood Enhancement Through Agriculture, Tourism and Health*, Advances in Geographical and Environmental Sciences,
https://doi.org/10.1007/978-981-16-7310-8_3

3.1 Introduction

Water is life and it is difficult to live without it. Water is required for farming, food protection, feeding livestock, maintaining organic life, industrial development, household needs, recreational activities, as well as biodiversity, and environmental conservation. However, due to the rising population and unhealthy lifestyles, water has become a vulnerable source due to reckless misuse and increased demand. Water scarcity is characterised as the lack of a sufficient amount of available water at the necessary time and location for human and environmental consumption. The Falkenmark indicator, which relates the more or less fixed amount of renewable freshwater resources in the world to the population using a per capita estimate of water needed to satisfy domestic, agricultural, and also industrial needs, is the most common quoted indicator or measure of water scarcity. According to this method, a country with less than 1700 m^3 of water per person per year, would face water stress, which will become acute at less than 1000 m^3 per person per year.

Water shortage is a well-known concept that has impacted every continent. About 1.2 billion people, or approximately one-fifth of the global population, reside in areas of physical water scarcity, and 500 million people are prone to this situation. In addition, another 1.6 billion people, or nearly a quarter of the global population, are affected by economic water scarcity (due to a lack of infrastructure to extract water from rivers and aquifers) (UN-Water 2012). Water shortage is expected to be a major challenge for most of the region as a result of increased water demand and poor management, according to the IPCC's fifth assessment report (AR5 - 2001). It also emphasises that water resources are vital in Asia due to large population and that demand varies by region and season. In addition, the report highlights that increasing population and rising demands resulting from higher living standards could intensify water shortage in many parts of Asia, impacting many people in the future. Further, it pinpoints the need for integrated water management strategies, such as implementing water-saving technology and rising water productivity and water, to help respond to climate change. Many developing countries, including Sri Lanka, will face a water shortage crisis in the immediate future if appropriate measures are not taken.

Some researches indicate that Sri Lanka has observed water scarcity in various forms: physical scarcity, economic scarcity, and institutional and political scarcity. According to Sri Lanka National Water Development Report (2006), 25% of people in Sri Lanka are deprived of safe drinking water. Especially the dry zone of the island is affected seriously. Temporal and spatial water scarcity has been a serious issue in the dry zone areas from the past and the wet zone does not experience much water scarcity quantitatively, when compared with the dry zone of Sri Lanka. Kilinochchi falls under the dry zone of Sri Lanka, where the availability and quality of the water sources are limited. Water scarcity, referred to in this study, is crucial for basic household needs such as washing, cooking, drinking, sanitation, and livelihood activities. This situation depicts an inadequate availability of water in Karachchi DS division. As a result of water scarcity, the people of Karachchi DS division are experiencing severe socio-economic and environmental problems, which may lead to negative impacts on the future development of the district. It prompted the researcher to study the water scarcity and its challenges in livelihood in Karachchi DS division, Kilinochchi.

3.2 Literature Review

Less than 1% of the world's freshwater (0.007% of all water on the planet) is available for direct human use (Gleick 2000). Despite, the fact that freshwater is a renewable resource, the world's supply of safe, freshwater is slowly diminishing. In many areas of the world, water demand already exceeds supply, and as the global population grows, the demand of water also rises (Chartres and Varma 2010). According to UN-Water (2006), water scarcity is defined as the lack of sufficient available water resources to meet the demand of water usage within a region. When annual water supplies per person fall below 1700 m^3, an area is said to be experiencing "water stress." A country or region faces "water scarcity" when its annual water supply is less than 1000 m^3 per person per year, and "absolute scarcity" may occur when it is less than 500 m^3 (UN-Water 2012).

Two distinct forms of water scarcity could be identified namely physical scarcity and economic scarcity. Economic water scarcity is described as a situation in which there is a lack of infrastructure or technology to extract water from rivers, aquifers or other water sources, or a lack of human capacity to meet the water demand. Physical water scarcity is characterized as a lack of natural water resources to satisfy both human and environmental flow requirements. Extreme environmental degradation, dwindling groundwater and water allocations that favour some groups over others are all signs of physical water scarcity (FAO 2012). The problem of water scarcity is a growing one. Currently, water scarcity affects more than 40% of people from all continents around the world, a proportion set to reach two-thirds by 2050 (FAO 2015). By 2025, 1.8 billion people will live in countries or areas where water is in absolute scarcity. In addition, by 2030, water scarcity in some arid and semi-arid regions would displace between 24 and 700 million people. Sub-Saharan Africa has the most countries with water scarcity of any region in the world. Similarly, water scarcity affects most countries in the Near East and North Africa, as well as Mexico, Pakistan, South Africa, and large parts of China and India (UN-Water 2006).

According to Seckler et al. (1998) water scarcity is the single greatest threat to food security, human health as well as natural ecosystems. Further, it demands an increased focus by policymakers and professionals on the continuous depletion of groundwater, because of the major threat it will have on food security. Over the next few decades, expected population growth could drive another two dozen countries to the verge of water scarcity, adding to the fact that anticipated changes in global climate, as well as increased water depletion, would limit water distribution and exacerbate the challenges ahead (Lefort 1996).

In Sri Lankan context, water is likely to be one of the most critical resource issues for development in the twenty-first century. In recent years, the water demand for food production, human needs (drinking, sanitation, commercial and industrial requirements), and environmental values has been growing in many parts of the country. Competition for limited water supplies and conflicts in sharing water resources among the different water users has emerged and is growing (Amarasinghe et al. 1999). De Silva (2004) reports that Sri Lanka receives 108,000 MCM of water per year from rainfall. As a result of evaporation and evapotranspiration, about 64,000 MCM of water escapes to the atmosphere. Therefore, Sri Lanka's gross annual run-off water is 44,000 MCM, with 40,000 MCM flowing into 103 rivers and 4000 MCM being used to recharge the groundwater. Further, the estimated per capita water availability (as of 2006) is 2200 m^3, and by 2025, with a stabilized population of 23 million, the per capita water availability would be 1900 m^3. As a result, even in 2025, Sri Lanka's annual per capita water supply will surpass the UN's per capita water adequacy limit of 1600 m^3 per year.

According to Nakagawa et al. (1995), the World Bank estimates that, of the 85% of Sri Lanka's population live in rural areas, only 10% have access to treated tap water and every fifth person relies on unprotected water sources for drinking and ~30% of those living in rural areas and small towns do not have access to sanitary latrines. Temporal and spatial water scarcity has been a key issue in the dry zone areas from ancient times. Recent studies show that the dry zone accounts for more than 90% of current water withdrawals (mainly due to the higher share of irrigation demand), whereas only 44% of the population lived there in 1991. Demand projections for 2025 show that the dry zone will continue to absorb over 90% of total water withdrawals (IWMI 2007). Further, this study has emphasised that the increasing demand for water is caused by various sectors such as the domestic sector, hydropower, and industry along with population growth. This leads to conflict within and between water users of sectors. In addition, water quality deterioration due to urban pollution, hazards like the tsunami, and agrochemical use are also reported from many areas of the country leading to a scarcity in usable water. Studies also reveal institutional problems and weaknesses that lead to inefficient use of water in many irrigation schemes.

Jayatillake et al. (2005) the average annual rainfall received in the districts from 1931 to 1990 indicates that belonging to the upper and middle watersheds has fallen significantly, causing major impacts on downstream water users in the dry zone regions. In addition, Chandrapala (1996) revealed that during the period 1961–1990 average annual rainfall in Sri Lanka decreased by about 144 mm, or about 7%, as compared to the period 1931–1960. Since the 1930s, Sri Lanka has endured droughts once every ten years or so; however, droughts have tripled since 2000, suggesting a rise in the frequency of droughts (Ariyabandu and Madhavi 2005). The rainfall expected from the North-East monsoon is expected to be reduced by about 26–34% in the future, particularly in the dry zone districts of Anuradhapura, Trincomalee, Batticaloa, Jaffna, Mannar and Vavuniya, with the change of future rainfall pattern as wet areas become wetter and dry areas become drier.

The Tsunami disaster in 2004 has damaged about 40,000 dug wells and the long term sustainability of these wells in the coastal sand aquifers has been questioned (Illangasekare et al. 2006). Furthermore, Freshwater resources in Sri Lanka are polluted by the extensive use of fertilizers, herbicides, and insecticides in agriculture, grease and toxic chemicals from urban run-off; sediments from mines and construction sites, and dumping of solid waste in open spaces. Besides, the above-mentioned causes, the population growth and economic development of Sri Lanka have registered an upward trend and the per capita income is also on the increase following the strides made towards development. Therefore, more people have reached a higher standard of living which in turn has made an upward swing in the per capita water demand. If the current economic growth continues, there would be major changes in the lifestyles of the people, necessitating higher water requirements (Amarasinghe et al. 1999). Furthermore, areas that are stricken with poverty are also areas that have shown a high demand for water (Falkenmark et al. 1989).

There are many studies, which have focused on the impacts of water scarcity on livelihoods. Water scarcity may have adverse effects on economies by diminishing production in both agriculture and the industrial sectors. Ultimately unemployment would eventually lead to poverty, which would harm society. Water scarcity can also result in fewer crops, which would have a detrimental effect on communities. The negative impacts of water shortage may lead to poverty. Plants, animals and human beings become vulnerable and face difficulties in survival by the shortage of water. Also, in terms of the above source shortage of water directly affects livelihood in many ways. Moreover, United Nations (2018) reported that about 1.4 billion livelihoods globally are directly dependent on water, including jobs in the food and beverage industry, the energy industry, and the water industry. Millions of smallholder farmers in developing countries rely on water for irrigation and livestock farming for their livelihoods.

Rassul (2011) highlighted that water scarcity has negative impact on the livelihoods of people, especially in agricultural societies. Further, the decline of the availability of water will be a significant cause of poverty or be a root of further growth in poverty levels. People, who experience a decline in their ability to feed themselves and their families will naturally experience fear and desperation and become more vulnerable to different types of exploitation. Freshwater, in sufficient quantity and quality, is essential for all aspects of life and sustainable development, and also water resources are embedded in all forms of development (e.g. food security, health promotion, and poverty reduction), in sustaining economic growth in agriculture, industry and energy generation, and in maintaining healthy ecosystems (United Nations 2018).

Water scarcity affects the health well-being of communities and can potentially be a pandemic if not addressed by the government and health officials. Due to the lack of water supply for sanitary usages, significant aspects of human existence such as hygiene become affected. This also consequently leads to the spread of diseases such as malaria or cholera and foodborne illnesses. Long term physical damages to the body can occur as a result of water scarcity, which affects both adults and children who have to bear heavy loads of water while travelling long distances to and from their homes (Pruss et al. 2002). Moreover, cost of buying water; poor communities without access to water supplies (piped water, boreholes, and shallow wells), particularly in urban areas, often have no option but to spend money they can hardly afford to buy water from expensive water vendors who can get their water from unaccountable sources (UNEP 2012).

3.3 Statement of the Problem

Water scarcity has been one of the most pressing issues in the many areas of the Northern Province, Sri Lanka, where people face shortages even for drinking and domestic purposes. The drinking water coverage of Kilinochchi district (2%) is very little (District Secretariat Killinochchi 2018) and Amarasinghe et al. (1999) rightly pointed out that in 1991 Kilinochchi faced moderate water scarcity and in 2025 the prediction is that it would lead to severe absolute water scarcity. Thus, limited water resources are a threat to the survival of the people. Kilinochchi district was ravaged by civil war between minority Tamils and majority Sinhalese for nearly three decades. Life and livelihoods of the people of the study area were among the victims of this war. Natural resources including water bodies were no exception. Many rehabilitation and development activities have been implemented in the post-war Kilinochchi district. These development activities and projects cause an acute rise in water demand. The current situation clearly reveals that water scarcity, due to increased demand versus decreased supply, is gradually becoming a burning issue in domestic, agricultural, industrial, and development sectors especially in the livelihoods of the population. Karachchi DS division is not an exception, as many households do not receive adequate units of water to suffice their demand in various sectors. The problem is escalating at an alarming rate as the number of houses prone to challenges of water scarcity increases gradually. These current scenarios, therefore, call for a study on investigating in particular, the primary challenges in livelihood due to water scarcity and its impacts on sustainable development goals in Karachchi DS division. So that solutions can be prioritised as per the requirements of residents ranging from the highest to the lowest. Hence, the primary aim of this research was to investigate the household level water scarcity and challenges in livelihood due to the water scarcity in Karachchi, Kilinochchi district, Sri Lanka.

3.4 Significance of the Study

The problem of water scarcity is not just a local, provincial, or national phenomenon; it's a global issue that affects individuals, industries as well as economies at large extent. Hence, this study is significant since it seeks to address the problems of water scarcity in Karachchi DS Division, where there is a lack of information and study on water scarcity and its impacts on livelihood. The findings from this study would be very crucial as they provide input in the future planning by the government. Moreover, the findings could be useful to the local authorities and water supply and irrigation departments in making informed decisions related to water supply in the near future. Further, the study may also help to contribute to the already existing body of knowledge as well as to inform policymakers in government. This would further enable them to enact laws and regulations on water management to address the negative impacts of water scarcity. Finally, this research will definitely assist and guide in ensuring and achieving the sustainable development goals in both war affected and resettled areas like Karachchi, Sri Lanka.

3.5 Study Area

3.5.1 Geographical Location

The study area is located in Kilinochchi district, the Northern part of Sri Lanka. The latitude and longitude of Kilinochchi district are 9.3807°N and 80.8770°E, respectively. Kilinochchi district is divided into four divisional secretary divisions namely; Karachchi, Kandavalai, Poonakari, and Pachchilaipalli. Among them, Karachchi Divisional Secretariat Division was selected for this study. The total area is 436.93 km^2 and it consists of 32% of the total land of the district (District Secretariat Killinochchi 2018). Figure 3.1 illustrates the administrative boundaries of the district and the study area.

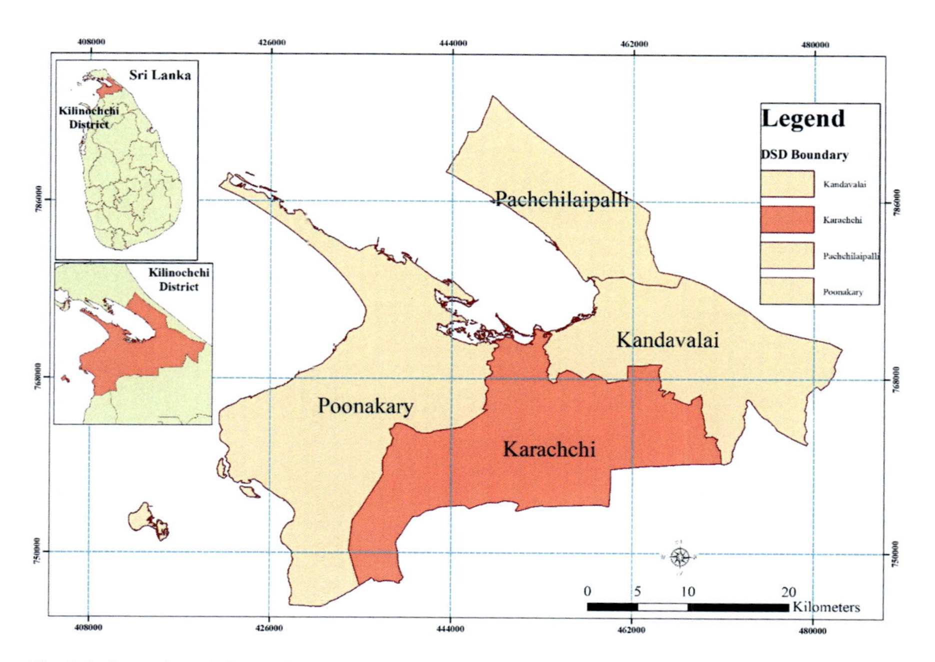

Fig. 3.1 Location of the study area. *Source* Survey Department, Sri Lanka, 2018

3.5.2 Geographical Characteristics of the Study Area

Geographically most of the areas of the district lie on the main land and the climatic conditions can be mainly categorised into the dry humid and tropical. The mean annual rainfall is 1520.57 mm. Out of the precipitation, 75% of the rainfall is received during the period from September to December by North-East monsoon periodical wind. The remaining period other than aforesaid of the year is dry and warm. Hence, every year from June to August is considered as the dry season due to the warm temperature. The monthly average temperature range is 25–30 °C. Kilinochchi district mainly depends on groundwater. Pallavarayan Aru, Mandakal, Akkarayan, Kalakaluppu Aru, Kanagarayan Aru are the major river basins of the study area. Most of them are became dry during the dry season. In addition, there are many tanks, namely, Iranaimadu Kulam, Vannerikulam Kulam, Puthumurippu Kulam, Kanagambikai Kulam and Akakarayan Kulam which are utilized for irrigation purposes. Irranaimadu tank is the major tank in the district and irrigates approximately 8900.81 ha area.

3.5.3 Demographic and Socioeconomic Characteristics

The study area is one of the brutally affected areas in Kilinochchi district, Sri Lanka due to the nearly three decades of civil war. The total population of Karachchi was 76,325 in 2015. Among them, 39,503 were females (51.76%) and 36,822 (48.24%) were males. In 2015, the total number of families in Karachchi was 23,381. The population of the district is exclusively Sri Lankan Tamil (District Secretariat Kilinochchi 2018). Further, considering the livelihood of the study area, agriculture is the primary source of income. The majority of the people are engaged solely in paddy cultivation while others do paddy cultivation together with subsidiary food crop cultivation. In addition, many are rearing cattle. Next to agriculture, fishing takes second place and helps the people for their livelihood. The living standard of the resettled people is gradually facing a rise since 2009 when the resettlement programme was commenced unitedly with the assistance of the ministry of resettlement, ministry of economic development, UN agencies, and non-governmental organizations. The livelihood of the people, who were affected by the conflict, is being rebuilt.

3.6 Methodology

To investigate the research problem, the study's methodology incorporated both qualitative and quantitative methods. The primary data were collected through household surveys and structured interviews in order to allow for more probing by the researcher and the respondents to freely express their opinions.

3.6.1 Sample and Sampling Procedure

After having a few lengthy discussions with engineers and directors of the National Water Supply and Drainage Board and Department of Irrigation of the district the following *Grama Niladhari* (GN) divisions (it is the smallest administrative unit in Sri Lanka that typically comprises between 100 and 500 households living within one to three villages), namely, Anaivilunthankulam, Konavil, Ponnagar, Ambalnagar, and Uruthirapuram North were selected from Karachchi Divisional Secretariat for the study using a purposive sampling method. The selected GN divisions are identified as more vulnerable areas to water scarcity by the National Water Supply and Drainage Board. Figure 3.2 shows the selected GN divisions for the study.

The total households of the selected five GN divisions were 3563. Two hundred and fourteen (214) households were randomly sampled from the five selected GN divisions of the study area for the household questionnaire survey. Table 3.1 provides the details about sampled DN divisions and sample size.

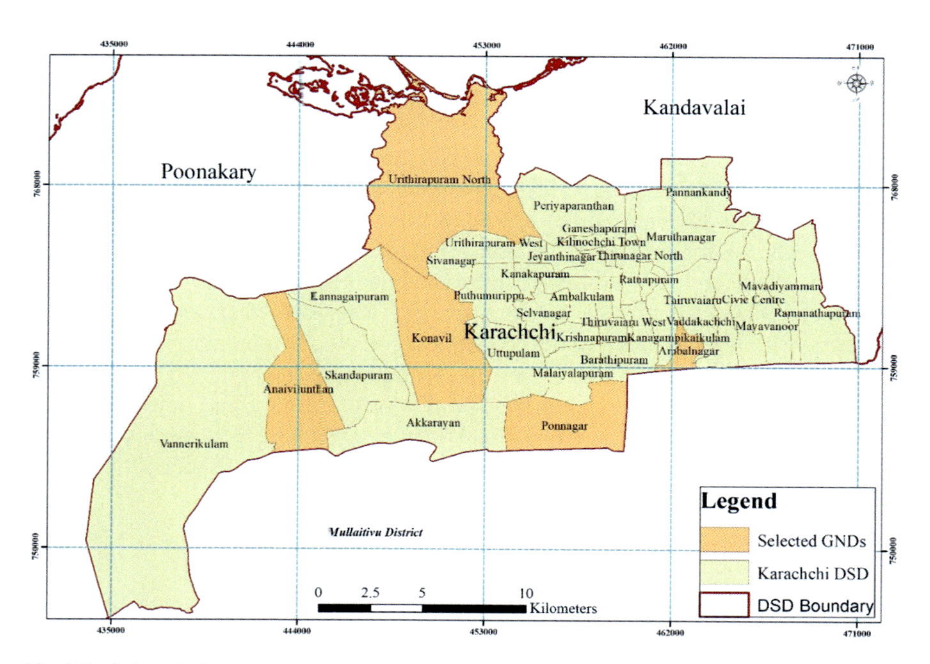

Fig. 3.2 Selected *Grama Niladhari* (GN) divisions for the study. *Source* Survey Department, Sri Lanka, 2018

Table 3.1 Sampled Grama Niladhari divisions (GN divisions) and sample size

Name of the GN division	Total households	Sample size
Anaivilunthankulam	401	24
Konavil	1070	64
Ponnagar	566	34
Ambalnagar	1015	61
Uruthirapuram North	511	31
Total	3563	214

3.6.2 Data Sources

Both qualitative and quantitative approaches were used for data collection, it mainly included household questionnaire survey, structured interviews, and observation. The household questionnaire survey was used to collect quantitative data and was administered by the researcher randomly so as to ensure correct responses. The questionnaire had both open-ended and closed-ended questions. The data collected covers the full range of issues in line with the study objectives. Approximately 20–30 min were spent in each household to complete the questionnaire survey.

In addition, structured interviews were also conducted. The researcher scheduled interviews to gather further information from the participants. Interviews were conducted with twenty relevant officers, namely, Engineers (water supply and irrigation), Directors of Planning, Project Development officers, Public Health Inspectors (PHI), *Grama Niladhari,* (Village officers), Village leaders, and Drivers of the water supply board. Moreover, the observation method was mainly used to understand the issues, which are related to water scarcity and its impacts on livelihood at the household level. The observations were recorded and photographed on activities revolving around the water sites taken during the field survey. Furthermore, the study used government documents, professional journals, and books to collect relevant secondary data. The key sources of the secondary data were as follows: annual reports of the Secretariat Kilinochchi, and statistical abstracts published by the planning divisions of the Divisional Secretariat Divisions.

3.6.3 Data Analysis

The data analysis was carried out in accordance with the research objectives. Following data collection, the researcher identified any incomplete or inaccurate responses, and those were corrected to improve the accuracy of the responses. The responses from the questionnaires and interviews were coded into Microsoft Excel 2019 (v16.0) and Statistical Package for Social Sciences (SPSS)–(v 18.2) for analysis.

Content analysis was used to examine qualitative data, which was based on an examination of the meanings and implications derived from respondents' information and documented data. It was summarized into meaningful statements, which were used to supplement the quantitative data to enrich the interpretation of the findings. In addition, descriptive statistics were used to evaluate quantitative data from questionnaire surveys, in which data was subjected to frequencies and percentages because frequencies are easy to interpret, understand, and compare. Frequency distribution tables, pie charts and graphs were used to display the findings of the data analysis.

3.7 Findings

The first part of the findings focuses on the characteristics of water scarcity at the household level in Karachchi, in order to obtain the background knowledge and information to investigate the challenges in livelihood due to water scarcity in Karachchi area. The study revealed that dug wells from the communal sources, wells at neighbours' houses and tube wells from communal sources are identified as the primary sources of water for household purposes in the study area. According to the study, nearly 74% of the households obtain water from dug wells, while 26% of them collect from tube wells for domestic purposes. Further, in reference to the ownership of the water sources in the study area, about 72% and 20% of the households depend on communal water sources and private sources respectively, whilst the rest 8% of the households depend on their neighbours' water sources (Fig. 3.3).

This study identified that nearly 44, 32, and 24% of the surveyed households rely on tube wells, dug wells, and the National Water Supply and Drainage Board (NWSDB), respectively, for portable water as illustrated in Fig. 3.4. Although, the

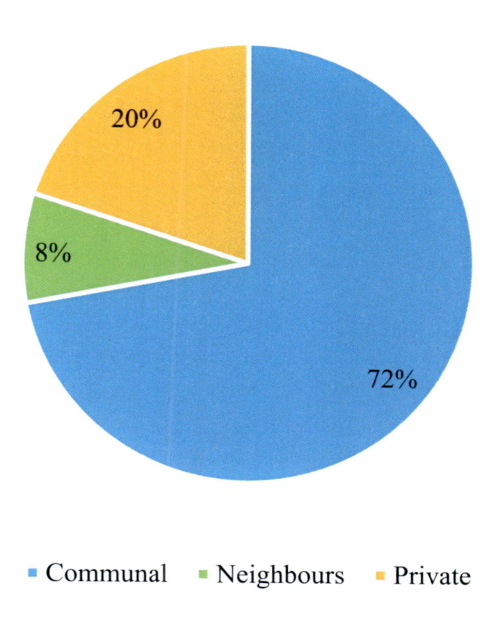

Fig. 3.3 Ownership of water sources for domestic purposes. *Source* Field study, 2018

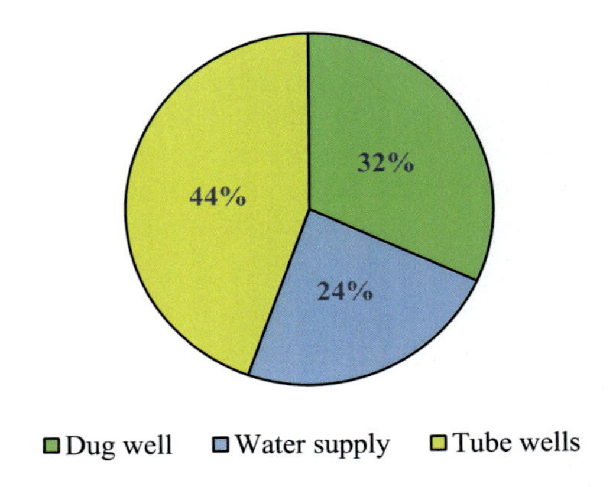

Fig. 3.4 Source of drinking water. *Source* Field study, 2018

proportion of households utilizing tube and dug wells are higher (76%) compared to that of those who obtain water supply from the local government bodies. Residents of the study area reported that the quality of the drinking water from their sources is polluted that it cannot be consumed for drinking purposes when compared to drinking water provided by the water board (Fig. 3.5).

The National Water Supply and Drainage Board, Kilinochchi provides drinking water to households daily in the few areas of the study. Generally, they fix a 1000 L plastic water tank at public places like temples, road junctions, community centres, or near the houses of villagers or shops. The total water of a tank has to be shared among nearly 20 households. Thus the units of water allocated to each household are very limited and allow only 50 L per household per day. In order to meet the household demand for water, the water board has to supply daily. However, only very few villages, namely, Ariviyalnagar, Anaiviluthan, Ponnagar Centre, and Sakthipuram receive water supply on daily basis. On the other hand, Santhapuram, Magalirthiddam, and Unionkulam receive water supply only two or three times per week. However, many areas are mostly depend on the nearest common or neighbour's water sources for their daily household needs.

3.7.1 *Quantity of Water Collection*

The quantity of water per household varies among the households. Nearly 37%, 20%, 14%, 10%, 7% of the households fetch approximately 300–400 L, 400–500 L, 200–300 L, 100–200 L and less than 100 L water per day respectively (Fig. 3.6). The rest of the households (12%) collect more than 500 L per day. There is a direct proportion between the quantity of water and the size of the family, because large families require a huge amount of water, while families with a few members require less amount. Furthermore, people mentioned that they need more water than a usual day if visitors or relations need to be accommodated.

Fig. 3.5 **a**, **b** are communal wells, **c**, **d** are communal tube wells, **e**, **f** are unprotected and un-safe water sources and **g**, **h** are dried wells in the study area. *Source* Field survey, 2018

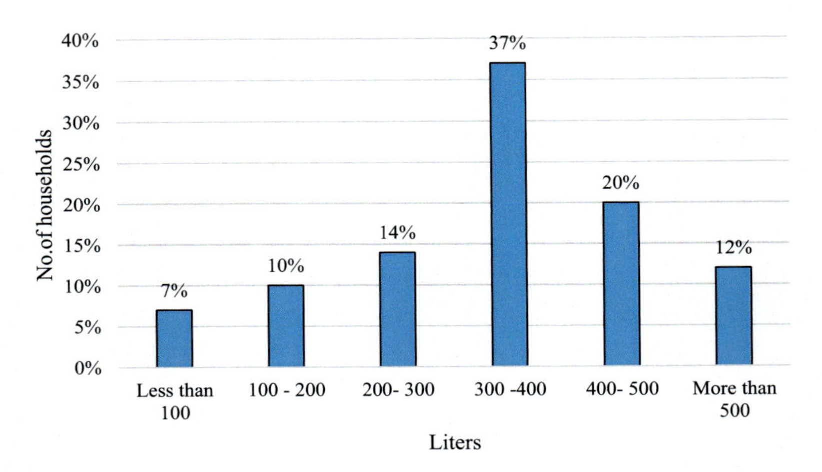

Fig. 3.6 Amount of water fetching-per day in litres. *Source* field survey, 2018

The distance from the water source is an equally significant aspect when considering water accessibility. This study found that the distance travelled to access water sources varied among the households. The majority (41%) of the water sources are located less than 50 m away from households, while 17%, 15%, 11%, and 10% of the households travel about 50–100 m, 100–200 m, 200–300 m, and 300–400 m, respectively, to collect water (Fig. 3.7). The rest of the households (6%) travel greater than 400 m daily to fetch water. Thus, the households, which are situated far from the water collection points need to spend more time to collect water and it holds a negative impact on their productive time, that could be utilized for other activities such as household chores, home gardening, and small entrepreneurial activities. In

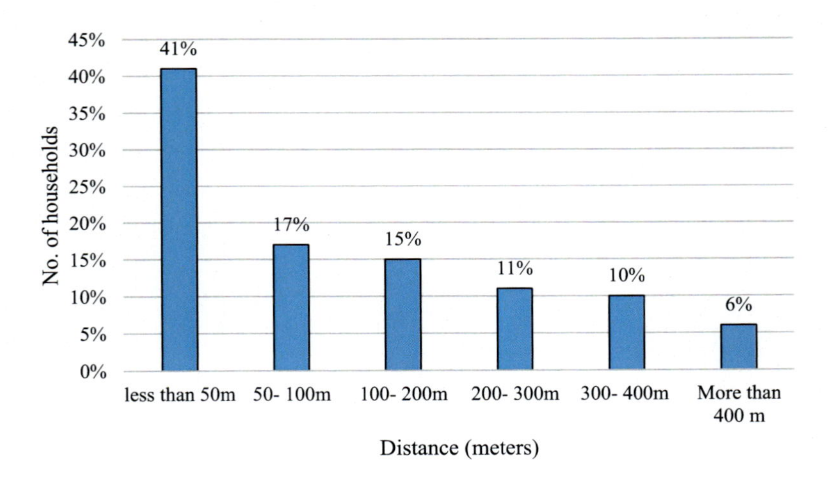

Fig. 3.7 Distance of water source. *Source* Field study, 2018

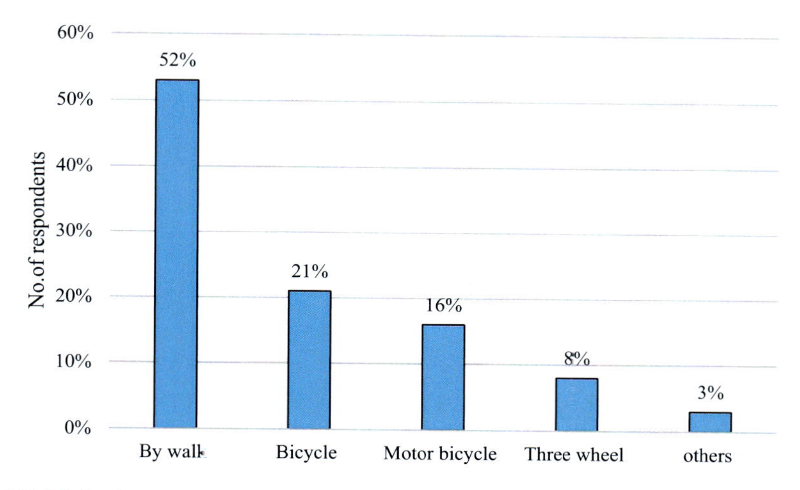

Fig. 3.8 Mode of transportation for water collection. *Source* Field survey, 2018

addition, people stated that they have to travel more than 1 km to the nearest water points to fetch water during the dry season due to the absence of a proper drinking water supply.

3.7.2 Modes of Transportation for Water Collection

This study revealed that people of this area, use various modes of transportation for water collection. Figure 3.8 clearly illustrates it. In the study area, more than half of households (52%) carry water from source to home by walk, while 21%, 16%, and 8% of households carry using bicycles, motor bikes and three wheels, respectively. Only 3% of households use other types of transportation such small water bowsers and tractors.

3.7.3 Relationship Between Water Collection and Gender

This study indicates that women travel longer distances for water, since they are involved in most of the house chores daily. Figure 3.9 shows the gender inequality in terms of water collection in the study area. According to the study, women are the primary water collectors in many households in the study area, where nearly 76% of women are responsible for the collection of water and among them 14% were girls. Thus, it may be lead to a decrease in women participation in economic activities in the study area. Only 24% of men are engaged in water collection, among them 17% were boys under 18 years.

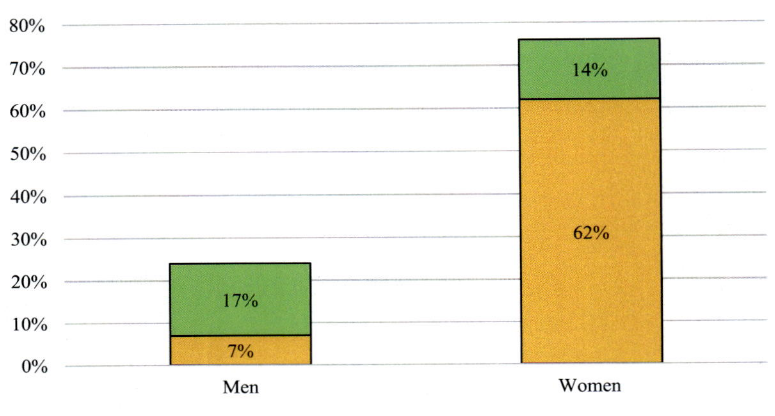

Fig. 3.9 Relationship of water collection and gender. *Source* Field study, 2018

3.7.4 Frequency of Water Collection

Every household needs a sufficient amount of water to fulfill day-to-day activities without fail. If a household has more members in their family, they need more water than a small family. Therefore, the water collector of a household has to travel several times to collect water within a day. This study depicted that the water collector of a household has to travel an average of three times per day to collect water for household purposes. Further, the majority 30% of the households travel four times per day, while among the rest of (70%) the households nearly 25%, 17% 14% and 8% travel three, two, five and one times per day respectively. However, the rest of the households (6%) fetch water more than five times per day. Figure 3.10 illustrates

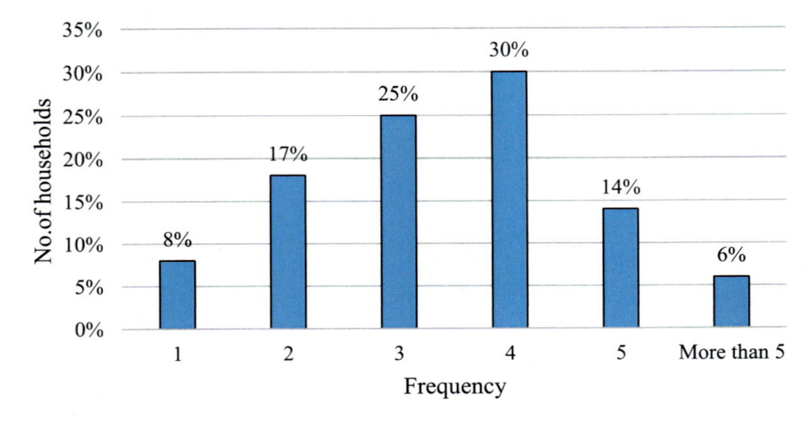

Fig. 3.10 Frequency of water fetching-per day. *Source* Field study, 2018

the frequency of water fetching per day by households.

3.7.5 *Spending Time for Water Collection*

The study confirmed that people of this study area spend a considerable amount of time to fetch water. About 34%, 24%, 19%, 13%, and 10% of the households spend an average of two hours, one hour, three hours, less than one hour, and more than 3 h per day for fetching water respectively. This translates to a good amount of time spent to access (distance covered and time spent in the queues) and fetches water which impacts negatively on livelihood activities in Karachchi DS Division (Fig. 3.11).

Thus, this study revealed that uninterrupted availability of water throughout the year is a major challenge in Karachchi area especially in Ponnagar, Konavil, and Ambalnagr GN divisions that are facing drinking water scarcity almost throughout a year due to unsafe water for drinking. Moreover, in the study area, nearly 68% of the households face seasonal water scarcity, especially during the dry season (April to August), while 32% of the households face water scarcity throughout the year, because of the less availability of water. Further, considering the interviews of public and officers water scarcity has been identified as a seasonal problem in many parts of the study area.

3.8 Challenges in Livelihood Due Water Scarcity

The second part of the analysis focuses on challenges in livelihood at the household level due to water scarcity in Karachchi. The study found that entire households primarily use water for their household purposes, including drinking, cooking,

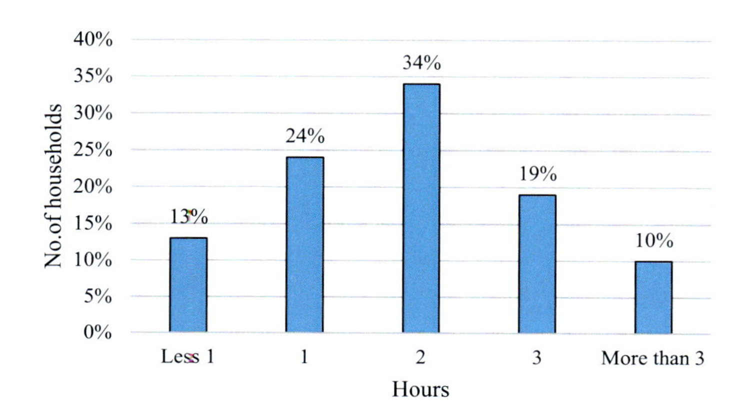

Fig. 3.11 Time spend for water collection per day. *Source* Field survey, 2018

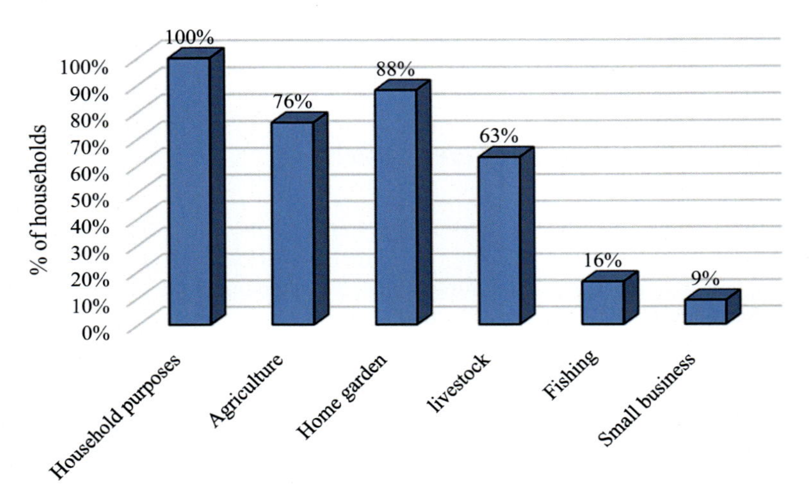

Fig. 3.12 Water usage patterns of the study area. *Source* Field survey, 2018

washing, sanitation, clearing, and other daily needs. Next to household purposes, 88%, 76%, and 63% of the households utilize water for home gardening, agricultural, and livestock activities, respectively. The minority of the surveyed households use water for inland fishing (16%) and small business (9%) (Fig. 3.12).

According to the Integrated Water Resources Management (IWRM) "Water has an economic value in all its competing uses and should be recognized as an economic good". Recognizing water as an economic good is a crucial decision-making tool for distributing water among various economic sectors and users within sectors. It's especially critical when the water supply can't be increased. Thus, inadequate water can be a cause for economic impacts on a society. This study also has identified the key livelihood challenges, which are either directly or indirectly linked to water scarcity. Figure 3.13 illustrates the key challenges on livelihood at the household level due to water scarcity in Karachchi area.

With reference to Fig. 3.13 various livelihood challenges: negative impacts on monthly income, increase in the cost of living, negative effect on agriculture, home gardening, and livestock, reduction in fishing activities, unemployment, and negative impact on small business are identified as key issues on account of water scarcity in the study area.

3.8.1 Challenges in Income

In line with the study, water scarcity has negatively affected 82% of the surveyed households' monthly income, while the rest 18% of households remains unaffected. People asked to explain how water scarcity has affected their monthly income. As

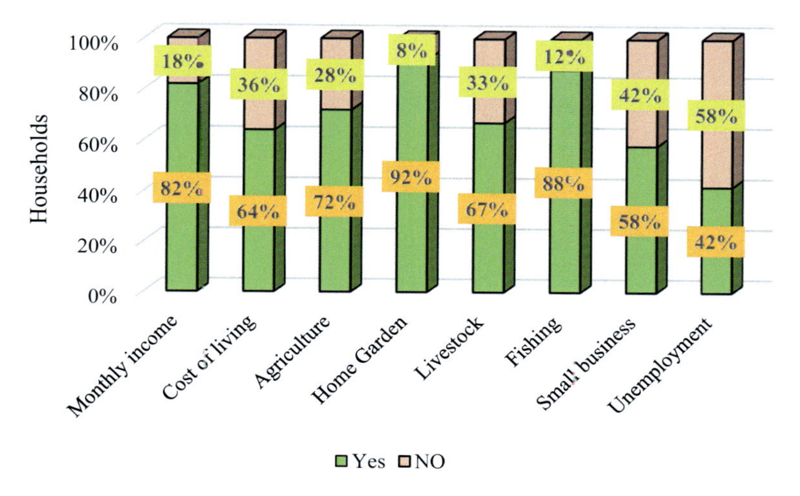

Fig. 3.13 Livelihood challenges due to water scarcity. *Source* Field survey, 2018

most people pointed out the direct adverse effect of water scarcity is on agricultural activities, which in turn causes a decline in income because agriculture is the primary occupation of the majority of the people in the study area. Unfortunately, agricultural practices could not be carried out currently due to water shortage. In addition, a significant decrease in monthly income may lead to poverty and unemployment, which hold either direct or indirect connections with water scarcity in the study area. Moreover, people mentioned that job opportunities for male residents run short during the period of severe water shortage. Further, very few households put forth the extra time spent on fetching water from distant water points as the reason for inability of their male members to attend work. These cases were rare though.

3.8.2 Challenges in Agricultural Activities

Insufficient water is always one of the primary threats to agriculture, especially in dry climatic areas. This study also found that water scarcity has catastrophic effects on agricultural activities in the study area. Agriculture related activities need a huge amount of water for the growth of crops and good harvesting. Appropriately 72% of households in the study area, experienced unfavourable consequences in agricultural activities as a result of inadequate water for farming. Similarly, participants underlined the fact that an adequate amount of water will not be available to irrigate the field, especially during the dry season of the year, as a result of the decline in water level of most of the irrigational tanks, agro wells, and tube wells and those became dry or empty during the dry season. In addition, the high cost of petrol to generate water pumps and/or increase in electricity bill payments caused by using electric motor pumps to pump water from the deep depth of the water sources (agro-wells) for the agricultural activities add to the struggle.

Furthermore, in recent years, nearly 32% of the households have decided to discontinue agricultural activities, because of the water shortage in the study area. Similarly, farmers also highlighted that lack of water availability and supply have forced them to reduce their total farming land area and also restricted cultivation with few crop varieties which in turn has declined their yields. Moreover, droughts for several months also severely affect paddy fields and other crops in the study area, as well as unexpected floods, also destroy their crops. Therefore, farmers who depend on irrigation water or rain fed agricultural activities are more vulnerable because of water scarcity in the study area.

3.8.3 Challenges in Home Gardening

Home gardening is one of the key elements in any rural household arrangement, and it can be considered as a key means of household food security as well as an income generated source at the household level because people can sell their food products to neighbours as well as local markets. In the study area, soon aftermath of the resettlement many households started home gardening with the aid of NGOs and family members, especially women, who actively participated. However, later they gradually started to give up the home garden practices as a result of water scarcity, especially during the dry months of the year. This study indicated that nearly 92% of the household's home gardening has faced problems to the level of giving up the idea due to water shortage because water availability plays a vital role to maintain a home garden at the household yards. During the study period most of the home gardens were abandoned without sufficient water. Undoubtedly the loss of home garden practices leads to food insecurity at the household level.

3.8.4 Challenges in Food Security

The abandonment of home gardens and agricultural practices due to water scarcity have direct correction with the decline of food production and food security at the household to the local level. About 88% of the surveyed households mentioned that water scarcity has affected their household food production and also food availability. Among them most of the households were depending on self-sufficient agriculture and home gardens for their food. In common the following food crops: vegetables (chilies ginger, pumpkin, tomato, peas, brinjal, beetroot, bottle gourd, bitter gourd, snake gourd, ladies' fingers, and drumsticks) grams (cowpea, green gram, black gram, groundnut, and maize), fruits (banana, lime, mango, guava, papaw, pomegranate, and jack) and other crops such as coconut are harvested in home gardens in the study area. Moreover, they added that if and when sufficient water is available more vegetables, fruits, grams, and herbs could be cultivated at their home yard to fulfil their household food needs as well as to make revenue by marketing the rest. However, due to water shortage households are unable to practice home gardens continuously. It has decreased the income of a few households, where women earned a little money from their home gardens. In turn, it has forced people to spend extra money to purchase food, especially vegetables and fruits from outside or markets.

3.8.5 Challenges in Livestock

The next challenge due to water scarcity is livestock, this study found that nearly 67% of the households are unable to engage in livestock activities, especially in cattle, goats and pig farming as a result of water scarcity in Karachchi area, because during the period of water scarcity, the water bodies get dried and vegetation covers are decreased resulting in a shortage of food and water for cattle. Thus, people struggle to provide pasture and water to animals. As a result, most of the households have left the livestock activities recently. They have even given up growing hens and chickens because the dry season is unfavourable for their survival and they die due to water shortage and heat. Nearly, 48% of the households no longer engage in livestock activities, because of water shortage and reduction of grazing areas negating turn negatively affects the household's income, food production (milk, meat, egg) as well as increases the cost of living.

3.8.6 Challenges in Inland Fishing

In the study area, a minority 16% of the people engage in inland fishing activities as the primary means of their livelihood. The inland fishing commonly takes place in irrigation tanks and intermittent rivers, which are situated in the study area. However, among the fishing households, nearly 88% expressed water scarcity as a hindrance to fishing activities during the dry months, as these tanks and intermittent rivers get dried and water is no longer available for fish species and consequently for fishing activities. As a result either fishing is reduced or most of the time stopped. Finally, it affects the household income of fishermen' families and subsequently the food supply in the study area as a whole.

3.8.7 Challenges in Small Business Activities

Accordingly, the study found that a very few surveyed households (9%) are currently involved in small business activities such as construction work, bricks production, grocery shops, vehicle wash shops and saloons all of which require water at varying levels. Among the households which engage in such businesses, nearly 58% face difficulties in carrying out their business due to water scarcity. People, who involve in bricks production, running saloons, and restaurants (small food shops) commented that inadequacy of water causes various struggles to successfully run their business and that they either had to discontinue or entirely close their business activities if water is not sufficient especially during the dry season. Similarly, bricks producers and construction workers (masons) added that they require more water than other businesses. In addition, restaurants that are mostly owned by women enterprises, saloons, and vehicle washing centers also require water in order to function and provide their services well. Nevertheless, due to the current water scarcity, they are unable to proceed, and in some cases, their businesses are forced to close during the dry season when water is became scarce. As a result, people need to find alternative jobs to run their families. During the interviews with restaurant and saloon owners, it was found that the shops had to be closed during the absence of water supply. Thus, water scarcity has resulted in the reduction of income of the households involved in small business in Karachchi area.

In contrast, water scarcity also has created a few new business opportunities; tendering water sources, water supply, water purification and hiring vehicles for water supply or delivery. Nowadays these kinds of businesses are emerging in Karachchi areas. Currently, only 4% (8 households) are doing these types of business in the study area. Few people, who owned more than enough water in their private wells have tendered them for water supply, and people who own vehicles have started to lease them for water delivery services to generate income. As well as, few shops purify drinking water and sell it for money and some of them do water delivery to households. These days few branded water supply companies are also targeting the drinking water-prone areas to market their drinking water. Moreover, a few neighbour houses also have started to charge for drinking water. However, businesses that deal with water for a profit or money have negative effects on poor people, who are unable to pay for water, and also it might increase their cost of living.

3.8.8 Challenges in Employment

There is a link between employment opportunities and the availability of water, as water can open up many occupations such as farming, home garden, and small business. The study revealed that water scarcity has detrimental impacts on employability, especially among the young in the study area. According to this research, water scarcity has reduced employment opportunities of nearly 42% of the surveyed households, as they are unable to participate in self-employed activities such as farming, small business, and construction work. Therefore, people used to switch from their livelihood activities and move to the nearest areas (neighbouring villages to find daily jobs) or migrate to other areas (Mullaitivu, Vavuniya, and Mannar for seasonal jobs) or major cities (Colombo, Gampaha, and Jaffna) to find new jobs. Meanwhile, some have already migrated to the Middle East counties for work. The migration of the young labour force could create stagnation in the development of the households as well as local development in Karachchi area.

3.8.9 Increasing Cost of Living

Further, the study found that, nearly 64% of households experience a slight increase in the cost of living as a direct and/or indirect effect of water scarcity. It is crystal clear that water scarcity has negative impacts on people's livelihoods and also sources of income in the study area. The following includes the ways in which water scarcity raises the cost of living of the households in Karachchi area. Currently, 37% of total households spend money on drinking water. In addition, the total household (100%) spends a considerable amount of money on vegetables and fruits purchased from outside markets. These two, coupled together are added as extra expenditure into the monthly expenditure of the households in the study area. Moreover, few households have already invested capital for constructing well or tube well, or to increase the depth of wells by drilling, while some residents have built rainwater harvesting tanks with the support of NGOs all of which require monetary support to a certain extent. The implication thus is that water scarcity has either directly or indirectly increased the cost of living of the poor people in Karachchi area. Unless these challenges are addressed in due course the situation may worsen in the near future.

Finally, the surveyed households were inquired about the different types of livelihood activities that could be carried out by them if the water is sufficient in terms of availability and accessibility. Among the surveyed households 33% are willing to engage in home gardening and the main reasons for selecting it are to improve and ensure household food security and to reduce the cost spend on food. Secondly, 29% of the households would like to involve in agricultural activities to generate more income as well as to produce food for household needs. Further, among the surveyed households 24, 11, and 3% expressed a desire to engage in livestock, small business, and inland fishing activities, respectively, as a means of their primary income (Fig. 3.14).

3.9 Challenges in Sustainable Development Goals Due to Water Scarcity

Finally, this study attempted to identify the perceptions on how the identified livelihood challenges in the study area would affect achieving sustainable development goals (SDGs). Table 3.2 illustrates how water scarcity would influence achieving sustainable development goals in the resettled areas in Karachchi Divisional Secretariat Division, Kilinochchi, Sri Lanka.

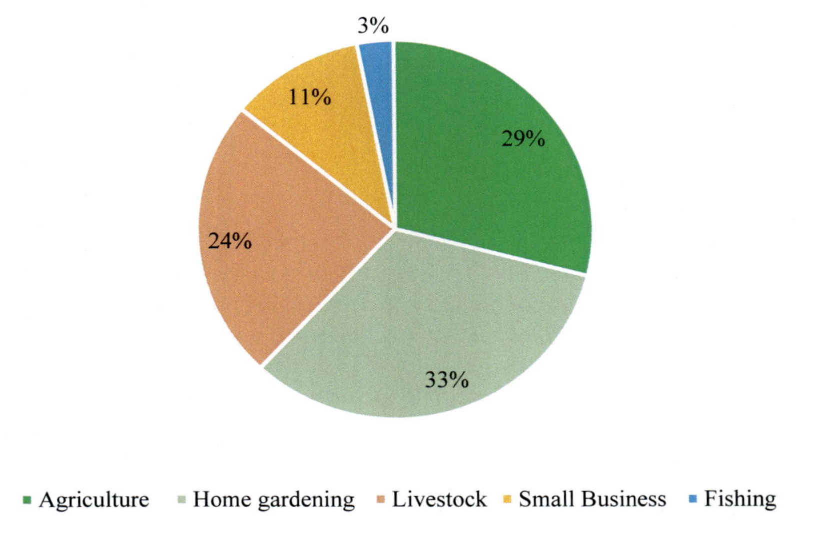

Fig. 3.14 Livelihood activities that would be carried out by the households. *Source* Field survey, 2018

The results depicted that sustainable development goals: 1, 2, 3, 5, 6, 8, 11, 13, 14, and 15 are more vulnerable (directly affected) in the study area due to water scarcity. Among them the goals: 1, 2, 3, 5, 8, and 9 are closely associated with the negative consequences on livelihood. Notably goal no. 6: Clean Water and Sanitation is the most threatened goal because of water scarcity in Karachchi. In addition, goals 2, 4, 9 and 10 may be challenged in terms of achievement in the study area due to the negative effects of water scarcity on livelihood means. Further, 64% of people reported that water scarcity has an indirect impact on zero hunger, which is closely linked to the negative impacts of water scarcity on agriculture, home gardening, and livestock activities, which are the means to ensure household food security at both the household and local level.

3.10 Conclusion

In conclusion, this study evidently highlighted that challenges in livelihood due to water scarcity are at an alarming level in Karachchi DS Division, Kilinochchi. The majority of the people in study area are experiencing severe effects of water scarcity in their daily life, especially in livelihood means. The primary activities such as agriculture, livestock, and inland fishing are the main sources of livelihood in Karachchi DS division. Agriculture is the key income source of many households in the study area. Totally 72% of the surveyed household's agricultural activities have been negatively affected, while 32% of the households have already left agricultural

Table 3.2 Impacts of water scarcity in achieving SDGs

SD goals	Opinions of participants			
	Directly affect (%)	Indirectly affect (%)	Does not affect (%)	No idea (%)
Goal 1: No poverty	58	42	–	–
Goal2: Zero hunger	23	64	13	–
Goal 3: Good health and well-being	83	17	–	–
Goal 4: Quality education	11	89	–	–
Goal 5: Gender equality	79	18	3	–
Goal 6: Clean water and sanitation	100	–	–	–
Goal 7: Affordable and clean energy	23	17	31	29
Goal 8: Decent work and economic growth	72	18	6	4
Goal 9: Industry, innovation and infrastructure	19	46	23	12
Goal 10: Reduced inequality	37	43	13	7
Goal 11: Sustainable cities and communities	42	21	22	15
Goal 12: Responsible consumption and production	32	26	22	20
Goal 13: Climate action	52	30	11	7
Goal14: Life below water	96	4	–	–
Goal 15: Life on land	95	5	–	–
Goal 16: Peace and justice strong institutions	38	24	18	20

(continued)

Table 3.2 (continued)

SD goals	Opinions of participants			
	Directly affect (%)	Indirectly affect (%)	Does not affect (%)	No idea (%)
Goal 17: Partnerships to achieve the goal	37	18	35	10

Source Field survey, 2018

activities due to water scarcity. Further, water scarcity has also negatively impacted on income, household food security and employability in the study area. These negative impacts have challenged on reduction of poverty and hunger in the rural settings, especially in the war affected areas like Kilinochchi. Besides poverty, unemployment, and the inability to access water resources would result in violence or conflict among community members or between intra communities while leading to social crimes as well.

Furthermore, both male and female labour migration for livelihood can break up family ties and negatively affect education and the future of their children. In addition, it also literally affects local development. In reference to the IWIM study, in 2025 Kilinochchi will face absolute water scarcity. Climate change and rapid development activities would increase these scenarios in the near future. Thus, livelihood challenges due to water scarcity have become an acute danger in some parts of Karachchi area and it could be one of the key threats to achieving sustainable development in Kilinochchi district in the foreseeable future. Further, the study indicates that SDGs: 1, 3, 5, 8, and 9 are more prone to be unachievable due to the negative impacts on livelihood as a result of water scarcity. Moreover, SDG 6: Clean Water and Sanitation is the most threatened goal as it is the key direct consequence of water scarcity in Karachchi. Therefore, relevant stakeholders should take immediate steps to ensure the availability and accessibility of adequate water in the resettled areas of Sri Lanka to resolve the livelihood challenges as well as achieve sustainable development goals.

3.11 Recommendations

It is crucial to recognize the importance of enhancing water use efficiency in addressing the current and potential challenges that water scarcity poses to rural livelihoods. According to this survey, the researcher proposes the following recommendations for relevant stakeholders to mitigate the livelihood challenges caused by water scarcity.

3.11.1 Government Institutions

- Local authorities such as Pradeshiya Sabha, and the water board should do a needs analysis to identify the vulnerable areas in terms of water supply and sanitation.
- A better suggestion for government to reduce drinking water issues would be to take proper measures to provide safe water through the water supply or pipeline water.
- Members of the community should generate innovative ideas to supplement the current water sources and make a communal contribution to support the same, e.g. drilling more wells and tube wells and implementing any water conservation techniques that may be beneficial.
- The majority of the participants mentioned that the rural areas which are far from the town do not have adequate access to drinking water. Besides, there are disparities in the units of water accessed by various areas and sectors. Some access a few units while some receive hardly any. Therefore, the local authority should assure there is equal distribution despite a water shortage.
- The public should be educated on water scarcity and awareness should be created about the imminent threat of water scarcity. Based on the above findings, the recommendation is that the local authority in Karachchi should be in consultation with the community.

3.11.2 Non-government Organizations

- NGOs could research water scarcity and its prime concerns from a multidisciplinary perspective. It will provide clear insights into the problems of water scarcity.
- NGOs could contribute and support financially to the government bodies to initiate and implement community water supply programmes. Innovative techniques based on strong science and technology will be needed to eliminate the pollution of surface and groundwater resources, improve water quality and step up the recycling and re-use of water.
- Provision of communal tube wells and financial helps to build wells in the household for the poor community could be another means by which NGOs can contribute to minimizing hazardous effects of water scarcity. In addition, water tanks and sanitation facilities could be provided to public service centres such as schools, hospitals and public toilets. Organizing awareness programmes about water scarcity and its adverse effects could also be carried out by NGOs.

3.11.3 Households

- The households should embrace maximum use of roof water harvesting in most buildings so as to collect as much water as possible during the rainy seasons. In some areas rainwater harvesting could be carried out, however only a very few are implementing it properly, nor does the government take any effort to initiate the process.
- Managing wastewater is essential for several reasons. First, wastewater is often discharged in places where it cannot be re-used, or directly to the sea, thus losing an opportunity for beneficial use. Second, wastewater is often rich in plant nutrients and these and the residual water can both be put into beneficial use through irrigation. Therefore households can practice wastewater re-use with the support and help of the government and NGOs.
- It is significant to raise awareness among the residents about the effects of polluted water. Water filtration at the household level is a relatively simple and cheap option to reduce the illnesses resulting from the consumption of un-safe water.

References

Amarasinghe UA, Mutuwatta L, Sakthivadivel R (1999) Water scarcity variations within a country: a case study of Sri Lanka. International Irrigation Management Institute, Colombo

Ariyabandu, Madhavi (2005) Hazard risk and water resources management. In: Wijesekara NTS et al (eds) Proceedings of the preparatory workshop on Sri Lanka National Water Development Report, World Water Assessment Programme, Paris, France

Chandrapala L (1996) Calculation of arial precipitation of Sri Lanka on district basis using Voronoi Tessalation (Thiessen Polygen) method. In: Proceedings of the national symposium on climate change, Colombo, Sri Lanka

Chartres C, Varma S (2010) Out of water. From abundance to scarcity and how to solve the world's water problems. FT Press, USA

De Silva KSR (2004) Water sector reforms in Sri Lanka. Seminar on water supply and sanitation challenges and opportunities. CECB, Colombo, Sri Lanka

District Secretariat Kilinochchi (2018) Statistical hand book. District Secretariat Kilinochchi, Sri Lanka

Falkenmark M, Lundqvist J, Widstrand C (1989) Macro-scale water scarcity requires micro-scale approaches: aspects of vulnerability in semi-arid development. Nat Resour Forum 13(4):258–267

FAO (2012) Coping with water scarcity: an action framework for agriculture and food security. Food and Agriculture Organization of the United Nations, Rome. Retrieved from http://www.fao.org/docrep/016/i3015e/i3015e.pdf

FAO (2015) The state of food and agriculture social protection and agriculture: breaking the cycle of rural poverty. Food and Agriculture Organization of the United Nations, Rome. Retrieved from http://www.fao.org/3/a-i4910e.pdf

Gleick P (2000) The world's water, the biennial report on water resources. Island Press, Washington DC

Illangasekare et al (2006) Impacts of the 2004 tsunami on groundwater resources in Sri Lanka. Water Resour Res 42:W05201

Intergovernmental Panel on Climate Change (2001) Climate change: impacts, adaptation and vulnerability. Intergovernmental Panel on Climate Change. Retrieved from https://www.ipcc.ch/site/assets/uploads/2018/03/WGII_TAR_full_report-2.pdf
International Water Management Institute (2007) Water matters: news of IWMI research in Sri Lanka issues: 2. International Water Management Institute, Sri Lanka. Retrieved from https://www.iwmi.cgiar.org/News_Room/Newsletters/Water_Matters/PDFs/Water%20Matters_Issue2-Final.pdf
Jayatillake HM, Chandrapala L, Basnayake BRSB, Dharmaratne GHP (2005) Water resources and climate change. In: Wijesekera NTS, Imbulana KAUS, Neupane B (eds) Proceedings of workshop on Sri Lanka, national water development report. World Water Assessment Programme (WWAP), Paris
Lefort R (1996) Down to the last drop. UNESCO sources, vol 84. Retrieved from https://unesdoc.unesco.org/ark:/48223/pf0000104613
Nakagawa K, Edagawa H, Nandakumar V, Aoki M (1995) Long-term hydrometeorological data in Sri Lanka. In: Data book of hydrological cycles in humid tropical ecosystems, Part I, University of Tsukuba, Japan
Pruss et al (2002) Estimating the burden of disease from water, sanitation, and hygiene at a global level. Environ Health Perspectives 110:537–542. Retrieved from https://www.ncbi.nlm.nih.gov/pmc/articles/PMC1240845/pdf/ehp0110-000537.pdf
Rassul K (2011) Water scarcity, livelihood & conflict, cooperation for peace unity
Seckler D, Upali A, Amarasinghe MD, De Silva R, Baker R (1998) World water demand and supply 1990 to 2025: scenarios and issues. Research Report 19. International Irrigation Management Institute, Colombo
United Nations (2018) Sustainable development goal 6 synthesis report on water and sanitation. United Nations, New York
United Nations Environment Programme (2012) Fresh water for the future: a synopsis of UNEP activities in water. United Nations Environment Programme
UN-Water (2006) Coping with water scarcity: a strategic issue and priority for system-wide action. United Nations, New York
UN-Water (2012) Coping with water scarcity: a strategic issue and priority for system-wide action. United Nations, New York

Chapter 4
Tobacco Cropping Increases Sediment Delivery in a Subtropical Agricultural Catchment in Southern Brazil

Edivaldo Lopes Thomaz, Fátima Furmanowicz Brandalize, Valdemir Antoneli, and João Anésio Bednarz

Abstract This study shows how the agricultural calendar affects the concentrations of suspended sediments both spatially and temporally in an agricultural catchment. A nested monitoring design was deployed with a group of small headwaters within a second-order catchment. Suspended sediment concentration (SSC) was performed at each monitoring site with a set of rising-stage sampling collectors. Analysis of the land use dynamics showed a clear intra-annual sediment transfer into the aquatic system. The SSC in the catchment differed in summer under tobacco cropping (*Nicotiana tabacum* L) compared to that in winter under oat cropping (*Avena sativa* L). The accumulated sediment in summer was 67% higher than in winter. We found that the structural and functional hydro-geomorphic connectivity in an area with tobacco crops in the channel expansion zone caused significant hillslope–channel sediment transfer. In addition, the tobacco areas have plenty of connectors causing disruptions and enhancing the sediment delivery into the stream.

E. L. Thomaz (✉)
Department of Geography, Soil Erosion Laboratory Universidade Estadual do Centro-Oeste, UNICENTRO, Street Alameda Élio Antonio Dalla Vecchia, 838, Guarapuava, Paraná CEP 85040-080, Brazil
e-mail: thomaz@unicentro.br

F. F. Brandalize
Escola Pública Municipal, Escola Esperança C. Chuilki, Irati, Paraná, Brazil

V. Antoneli · J. A. Bednarz
Department of Geography, Universidade Estadual do Centro-Oeste, UNICENTRO, Campus de Irati, Irati, Paraná, Brazil

N. C. Jana et al. (eds.), *Livelihood Enhancement Through Agriculture, Tourism and Health*, Advances in Geographical and Environmental Sciences,
https://doi.org/10.1007/978-981-16-7310-8_4

Graphical Abstract

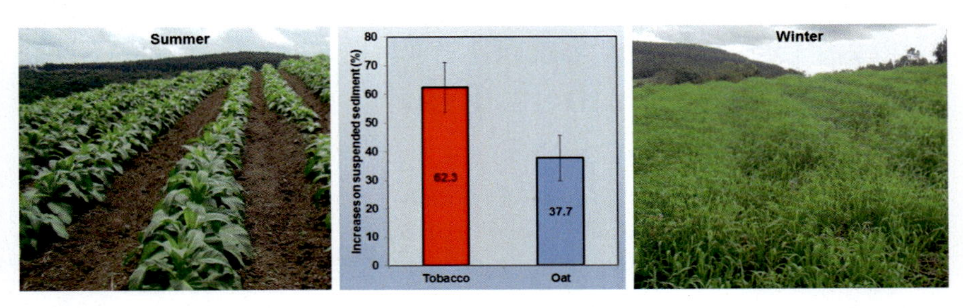

Seasonal sediment transfer

Keywords Conventional tillage · Critical source · Sediment transfer · Transport pathway · Riparian zone · Aquatic system

4.1 Introduction

Surface runoff and sediment transportation from hillslopes into the river are extremely important in order to understand the hydro-geomorphological catchment dynamics and landscape evolution (Collins and Walling 2004; Egozi and Lekach 2014). In fact, the concept of hydro-geomorphic (dis) connectivity has been highly valued as an approach to better understand the fluxes of water and sediments within the catchment (Bracken et al. 2015; Brierley et al. 2006).

Hydro-geomorphic connectivity is based on the structural (i.e., the slope and landscape form) and functional (i.e., the process) characteristics of the catchment. However, the flow of water and sediment within a catchment are affected by multivariate controlling factors such as rainfall regime, infiltration, soil type, slope, land use dynamics, vegetation cover, and soil management (Bracken et al. 2015; Fiener et al. 2011).

Despite advances in understanding the pathways and sediment delivery in a geomorphic system, some knowledge gaps exist. For example, the effects of soil use management and anthropogenic activities on hydro-geomorphic processes and sediment transfer, especially in the tropical regions, have not been thoroughly understood (Egozi and Lekach 2014; Fiener et al. 2011).

Agricultural catchments with diversified land uses produce field patches with seasonal crop growth. Hence, spatio-temporal landscape organization and interaction could be expected between land use and soil management patches (e.g., conventional and no-till system) affecting runoff, sediment sources, and their transfer through the catchment (Pradhanang and Briggs 2014). In addition, spatial organization of patches and natural and manmade hydro-geomorphic connectivity require further investigation (Fiener et al. 2011).

The research related to the hydrological and sedimentological processes in rural catchments are biased since greater number of studies is focused in the northern hemisphere than that in the south (Fiener et al. 2011; Burt and McDonnell 2015). Likewise, field studies on the hydro-geomorphological aspects are scarce, especially that are related to headwaters where water pathway and runoff is generated. These studies indicate that the impact of watershed intervention on hydrological, geomorphological, and ecological processes require urgent investigation (Burt and McDonnell 2015).

Sediment delivery is generally monitored only at the catchment outlet (i.e., one point) (Vente et al. 2007; Walling 1983) because monitoring several headwaters to evaluate the spatial and organizational effects in sediment exportation could be expensive and difficult to apply (Bracken et al. 2015; Didoné et al. 2015; Gumiere et al. 2011). Nevertheless, sediment transfer at the catchment scale is dependent on the landscape's spatial organization, and therefore, management practices are critical for sediment delivery (Collins and Walling 2004; Fiener et al. 2011; Gumiere et al. 2011).Measuring sediment delivery at the catchment outlet loses the critical information on internal spatio-temporal variability of the sediment sources, connector pathways, and sediment transfer within the catchment.

The spatio-temporal patterns in land use and soil management that affect surface runoff response on agricultural catchments are extremely important to soil–water conservation and hydrological modeling, especially the influence of spatial organization of patches (e.g., field crop diversity) and interaction with landscape morpho-structures (i.e., connectors) (Fiener et al. 2011).

In addition, understanding the connectivity and sediment transfer are particularly important in catchment under tobacco cropping since several studies have pointed out the detrimental effect to the ecosystem in tobacco areas cultivated in the conventional system (Antoneli et al. 2018; Olivet et al. 2014).

The aim of this study was to evaluate the spatial and temporal changes to suspended sediment concentrations (SSC) in a small catchment with diversified land use following the agricultural calendar. The specifics objectives of this study were (1) to assess the effect of seasonality (winter–summer) on sediment transfer and (2) to understand qualitatively the hillslope–channel coupling when the catchment is covered by the winter-summer crops. Further understanding of spatial and temporal sediment transfers in catchment systems is extremely important for the conservation of soil and aquatic systems, especially in tropical areas with diversified land use, especially with tobacco crop.

4.2 Material and Methods

4.2.1 Study Area

The Guamirim de Cima catchment (~1 km^2), is in the rural area of the Irati municipality in the southeastern state of Paraná—Brazil, located between 25°34′ and

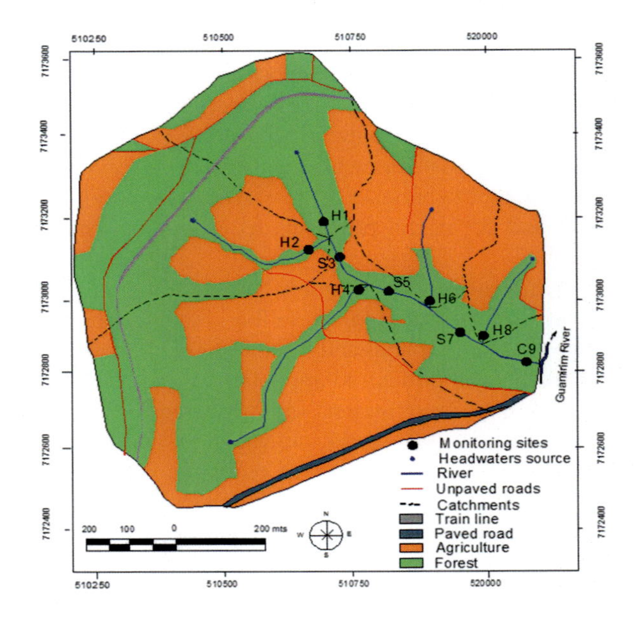

Fig. 4.1 Land use distributions and nested monitoring approach with headwaters and catchments. Explanation of the monitoring sites is provided in Table 4.1

25°33′S and 50°54′ and 50°53′W (Fig. 4.1). The study area is characterized by small farms (<30 ha) firstly, due to the fragmentation of land during colonization by the European immigrants and, secondly, due to adverse physical conditions such as poor and shallow soils, and steep terrains. These characteristics contribute to the choice of tobacco farming as one of the main agricultural activities in these small farms. Tobacco cultivation is mainly carried out using rudimentary equipment due to the terrain conditions such as slope, shallow soil, and subtropical weather with a cold winter (June–August) (Antoneli and Thomaz 2014; Bender et al. 2018).

The rocks are formed of siltstones and argillites originating from the Serra Alta formation during the Permian period. The soil texture consists of 20% sand, 30% silt, and 50% clay. Three soil types (IUSS Working Group 2006) exist in the catchment of the study site. Haplic and humiccambisols occupy around 70% of the catchment. This soil is shallow with the B-incipient horizon development to a depth of 50–130 cm and has moderate, convex hillslopes (slopes <20%). Most of this soil type is used to grow mechanized soybean. The second soil type is Regosols representing 25% of the total catchment area with a lower profile development (<50 cm in depth) displaying restrictions for some land use due to lower water holding capacity, exposure of rocks, and poor nutrient content. It predominantly has a steep terrain (slope >20%) with concave hillslopes and used for corn, beans, and tobacco cultivation. The third soil type is Haplic gleysols which represents 5% of the area and is distributed around the riparian zone.

The local climatic conditions influence the dynamics of the agricultural practices. In the spring and summer (September–February—hereafter, summer), farmers grow soybean, corn, black bean, and, particularly, tobacco (average temperature 23–24 °C).

Table 4.1 Measurement sites, land use characteristics and land physical characteristics

Measurement sites	Land use area (%)	Physical characteristics
H1	Forest (82%), corn (15%), and roads (3%)	Haplic Cambisols, concave hillslopes, slope average 21%, and riparian zone preserved
H2	Forest (46%), black bean (50%), and roads (4%)	Haplic Cambisols, convex hillslopes, slope average 23%, and riparian zone preserved
S3	Riparian zone (100%)	Gleysols, slope average <3% and riparian zone preserved
H4	Soybean (70%), forest (27%), and roads (3%)	HumicCambisols/Haplic Cambisols, convex hillslopes, slope average 13% and riparian zone preserved
S5	Riparian zone (100)	Gleysols, slope average <3% and riparian zone preserved
H6	Tobacco (90%), forest (9%), and roads (1%)	Regosols, concave hillslopes, slope average 25% and riparian zone is not preserved
S7	Riparian zone (100%)	Gleysols, slope average 4% and riparian zone preserved
H8	Tobacco (70%), forest (27%), and roads (3%)	Regosols, concave hillslopes, slope average 18% and riparian zone is not preserved
C9	Riparian zone (100%), catchment outlet	Gleysols, slope average <3% and riparian zone preserved

Monitoring sites: H = Headwater; S = Stream; C = Catchment

After the harvest of these crops in autumn and winter (March–August—hereafter, winter), farmers prepare the soil to sow oats (average temperature 13–14 °C), which helps in soil conservation during the fallow period. The annual precipitation ranges from 1600 to 1800 mm and annual temperature ranges from 17 to 18 °C (Nitsche et al. 2019).

4.2.2 Land Use Mapping and Monitoring Design

A global positioning system and the ArcGIS 9.2® software were used to produce the maps of land use and the agricultural calendar. The total area of the catchment is 99.8 hectares, out of which 48% is covered by forests (native and pine forests), 47% by agriculture, and 5% by roads (Table 4.1). The catchment consists of 4.9 km of rivers; 2.3 km out of which is formed by first-order rivers and 2.6 km by second-order rivers. The drainage density is 1.29 km km^2 (km of river per square km of area). The catchment drainage system is classified as "second-order" and is composed of five headwaters of first-order rivers (Fig. 4.1). This description is important to understand the nested monitoring scheme and procedures.

Supported by a land use map, we carried out a field survey in the catchment to identify monitoring sites for suspended sediment collection. Nine monitoring sites, forming a group of small headwaters within the whole catchment were organized in a nested monitoring design. The collectors were installed at five sites (H1, H2, H4, H6, and H8), at the end of the first-order rivers. They were also installed at four other sites in the main river (S3, S5, S7, and C9) to register the effect on the suspended sediment concentration of each headwater connection (discrete or cumulative), along the main catchment (Fig. 4.1).

Field-walking and visual appraisal of the sediment sources during a rainstorm were conducted during the year and potential connectors were also observed. In addition, the soil tillage processes along the year were followed constantly from a visual appraisal and photograph registration of each phase. The potential hillslope-suspended sediment sources were classified as "primary" due to being generated by sheetwash, rill, and gully erosion. These sources are close to river channels, i.e., hillslope–channel coupling (Collins and Walling 2004). Therefore, no hysteretic effects are expected, particularly, on the first-order stream.

4.2.3 Suspended Sediment Collection and Calculations

An indirect data collection procedure was applied (i.e., single-stage suspended sediment sampler), to document the catchment's suspended sediment sources and their temporal and spatial erosion patterns (Edwards et al. 1999; Gordon et al. 2004).

A single-stage suspended sediment sampler is generally used to automatically collect samples at the rising limb from a flashy stream (Gordon et al. 2004) like that of the study site, where all the headwaters are classified as a first-order stream. The drainage basins and the hydrophysical processes and morphology of most streams are quite similar (e.g., area of the drainage basin, length of overland flow, channel slope, hydraulic channel geometry, etc.) (Horton 1945). Rainfall at the study site does not have a seasonal trend over the year. Therefore, the catchment discharge is relatively uniform over the year. Finally, the stream response in small catchment is dominated by hillslope response as well as land use (Knighton 2014).

Despite the limitations of the single-stage suspended sediment sampler (e.g., samples cannot be collected on falling stages or on secondary rises), it fits well with the objectives and methodology of this study (Edwards et al. 1999). The focus here is to compare the SSCs in small streams originating from the hillslope with differing land use per unit of rainfall event. In terms of absolute sediment load, quantifying the total soil erosion and river system sediment transportation were out of the scope of this study.

At each of the monitoring sites (Fig. 4.1), the equipment adapted from the U-59 (SS-59) model (Edwards et al. 1999) termed as the "single-stage sampler" was installed. This device is useful for obtaining sediment data from streams where remoteness of site location and rapid changes to it (e.g., flashy stream) may make it impractical to use a conventional depth integrating sampler (Edwards et al. 1999). This equipment was chosen as per the appropriateness of the conditions of the study area.

The sampler collectors were installed 5 cm above the water level under normal flow conditions (without rain). The distance between the collector ends was of 5 cm. Samplers installed in the headwater of first-order rivers (i.e., H1, H2, H4, H6, and H8 sites) had six collector bottles because the riverbank height was around 70 cm, whereas the mean depth of the running water at those monitoring sites was shallow. The samplers installed in the main river (S3, S5, S7, and C9) were composed of ten collectors bottles because of the river hydraulic geometry.

Following rainfall events, the bottles were removed and carried out to the laboratory. The water evaporation method was used to analyze the sediment concentration stored in each bottle. The samples were evaporated at 105 °C for 24 h. After drying, the sediment was weighed on an analytical scale with 0.001 g of accuracy. The SSC was displayed in g L^{-1}, relative concentration (%) and total sediment mass (g) as per the analysis were performed. In addition, the rising-limb samplers were integrated across an entire event in each collector.

The rainfall was measured through three manual rain gauges and one automatic rain gauge (i.e., pluviograph) installed at different locations at the catchments. Finally, statistical analysis with the Mann–Whitney test ($p < 0.05\%$) was applied to compare the sediment concentration medians between the summer and winter phases.

4.3 Results

4.3.1 Agricultural Calendar and Land Use Dynamics

Table 4.2 shows the changes to the land use at the Guamirim de Cima catchment over the study period. During the summer, the agricultural land area is occupied by 46% of annual crops such as soybean, tobacco (Fig. 4.2a, b), black bean, and maize. In contrast, in the winter, all the annual crop area is covered by oat (Fig. 4.2c). Oat is sown to protect the soil against water erosion during the winter fallow and to recover soil organic matter. Other land uses (54%) were stable during the year. In addition, the most permanent land use is composed of native and secondary forests, especially following the riparian zone (Table 4.2).

The conventional tillage system was used for crop cultivation in which the soil is prepared using plows and hoes, which causes soil disturbance (Figs. 4.2a and 4.4). The most critical period of soil disturbance occurs during the initial crop-sowing phase because the soil becomes exposed until the crop grows and covers it with its canopy. This cycle repeats continuously, resulting in phases with protected soil and phases in which it becomes more vulnerable to water erosion.

Tobacco is the most important crop related to soil disturbance and exposure. It is sown over a row-ridged framework forming a gradient channel, which is susceptible to become a rill. Continuous mechanized tillage is common in tobacco cropping that is used to eliminate weeds, soil crust, and repair the rows of the ridges (Fig. 4.2a).

4.3.2 Rainfall Characteristics and Its Effect on Sampling Suspended Sediment

Precipitation during the study period was 2,023 mm, 27% above the historical rainfall average (Table 4.3). June was the wettest month (365 mm), with precipitation three times higher compared to the historical average (120 mm) (Nitsche et al. 2019). September and November were the driest months. In total, 81 rain events with different characteristics of depth, intensity, and duration were registered which landed on the crops at their different developmental stages. Precipitation resulting in hydrological responses in the headwater varied from 20 to 110 mm and their intensities were between 5.2 and 34.0 mm h^{-1} (Table 4.3).

Sixty-six rain events (81%) did not generate runoff and sediment response at the collectors in any headwater. Generally, precipitations lower than 24 mm in depth did not result in sample collection. Only 15 rain events with eight occurring in summer and seven in winter, produced discharge of water to reach at the bottom sampler (Table 4.3).

The sample collection in the headwaters was representative of an annual based agricultural calendar in the region and it was collected around 42% of the total precipitation registered during the monitoring period (Table 4.3). The summer phase accumulated 379 mm of rain, whereas 482 mm accumulated during the winter phase (i.e., 27% higher). In addition, average rain intensity was 17.6 and 13.8 mm h^{-1} in the summer and winter, respectively ($P > 0.05$) (Table 4.3).

Table 4.2 The land use in the Guamirim de Cima catchment along the year

Land use (summer)	Land use (winter)	Area (ha)	Area (%)
Unpaved road	Unpaved road	2.3	2.3
Railroad	Railroad	0.9	0.9
Highway	Highway	1.9	1.9
Tobacco (H6 and H8)	Oat (H6 and H8)	10.5	10.5
Corn (H2)	Oat (H2)	6.1	6.1
Black bean (H2)	Oat (H2)	3.8	3.8
Soybean (H4)	Oat (H4)	26.1	26.1
Shrubs	Shrubs	7.0	7.1
Native and secondary forest (H1)	Native and secondary forest (H1)	31.3	31.4
Eucalyptus forest	Eucalyptus forest	8.6	8.6
Pinus forest	Pinus forest	1.3	1.3
Total (ha)	**Total (ha)**	**99.8**	**100.0**

Note The land use dynamic could be observed in Fig. 4.1 (land use map), since only the agriculture land use change from summer to winter (~46%)

Fig. 4.2 Land use dynamic agricultural calendar for the year (summer and winter): **a** tobacco in the early stage of growing; notice the transversal furrows crossing the hillslope and the disturbed topsoil; **b** tobacco in the last stage of growing; however, the soil exposure between crop stripes persist; **c** oat winter crop covering the topsoil

Table 4.3 Number of rain event, rainfall physical characteristics, and number of samples collected

Months	Number of rain events	Total rainfall (mm)	Number of sampling collection	Sampling rank and precipitation volume (mm)	Rain intensity (mm h^{-1})	Proportion in relation to the total monthly precipitation (%)
Summer						
September	05	74	nr	nr	Nr	nr
October	09	222	01	1ª (84)	28.0	36
November	06	71	nr	nr	Nr	nr
December	10	203	01	2ª (29)	34.0	14
January	09	156	02	3ª (40)	8.0	25
				4ª (20)	25.0	13
February	13	275	01	5ª (50)	12.5	18
March	07	235	03	6ª (25)	10.0	11
				7ª (85)	8.0	36
				8ª (45)	15.2	19
Winter						
April	03	127	01	9ª (90)	22.5	71
May	06	119	01	10ª (55)	13.7	46
June	06	365	03	11ª (48)	22.0	13
				12ª (110)	10.0	39
				13ª (70)	5.8	19
July	03	120	01	14ª (70)	5.2	58
August	04	56	01	15ª (39)	17.9	69
Total	**81**	**203**	**15**	**860**		

Note nr = non-recorded

4.3.3 *Intra-annual Sediment Transfer: Summer and Winter*

The SSC during the summer increased gradually along the main stream from the headwaters (e.g., H1) to the S5 sampler. However, a sharp increase in the total SSC occurred in the catchment due to the H6 and H8 headwater connection, registered at S7 and C9 samplers, respectively (Fig. 4.3a). After the H6 connection, the SSC in the catchment increased by 1.6 times and doubled after the H8 connection to the main stream. Even though, with a dilution registered to C9, the headwaters with tobacco were responsible for the increase in SSC, especially due to H8. Headwater H1, covered mostly by forests (~80%), produced minimal SSC per rain event (0.87 g L^{-1}), followed by H4 cropped with soybean (4.9 g L^{-1}) and H2 with corn and black bean (9.3 g L^{-1}). Overall, H6 and H8 headwaters with tobacco produced 19–195 times more suspended sediment, respectively, than H1 headwater, covered mostly by forests (see Appendix 1).

In contrast, in the winter crop phase, when 46% of the catchment area was covered by oat, a lower total SSC was produced from H1 to C9 than that compared to the summer (see Appendix 2 for details), although both headwaters H6 and H8 still played a major role. In the winter, H6 and H8 headwaters with oat produced 10–64 times more suspended sediment, respectively, than H1 headwater, covered mostly by forests. However, the SSC in H8 decreased by more than five times from summer in comparison to the winter ($p = 0.10$, Mann–Whitney test) (Fig. 4.3a).

The accumulated suspended sediment concentration (ASSC) in both summer and winter increased gradually from H1 to S7 (Fig. 4.3b). Moreover, there was no difference in SSC on this catchment stretch. However, a clear distinction in SSC was observed when the H8 headwaters connection occurred. The SSC of the catchment differed in summer (63%) than that in winter (37%). In short, the ASSC in summer with tobacco was 67% higher than during the winter with oat ($p = 0.002$).

The relative SSC contribution of each headwater varied from upstream (H1) to the catchment outlet (C9) (Fig. 4.3c). The catchment stretches from H1 to S6 produced a lower relative contribution. In the summer, the sediment relative contribution ranged from 0.6% (H1) to 5.7% (H2) and the average contribution was around 3% on this catchment sector. H6 produced 9.6% of the relative SSC in the catchment. The major sediment contribution to the catchment originated from H8 headwater (Fig. 4.3c). The relative SSC contribution of this headwater ranges from 20 to 95.6% per rainfall event, with an average of 52.8% during the summer.

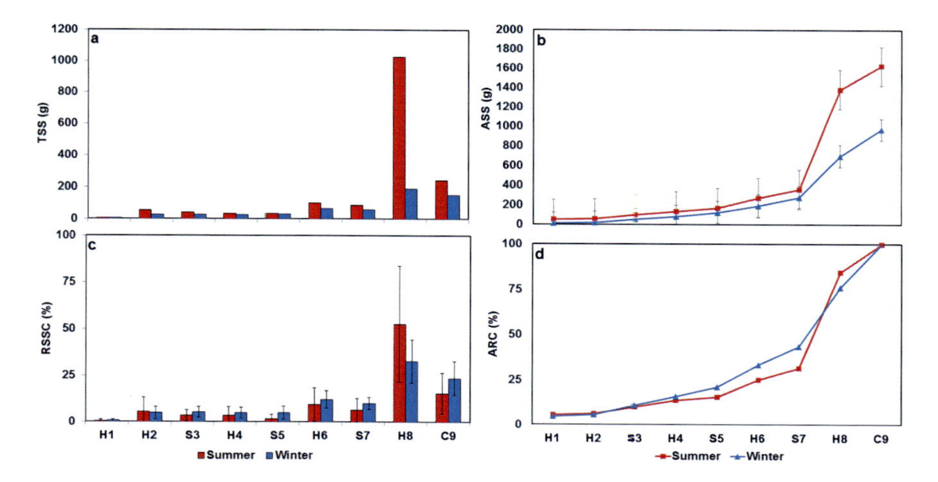

Fig. 4.3 Sediment concentration measured along the catchment system in the summer and the winter phase: **a** total suspended sediment (TSS); **b** the accumulated suspended sediment in the summer and winter (ASS); **c** the relative contribution of each headwater, stream, and catchment system (RSSC); **d** the accumulated relative sediment contribution (ARC) of each headwater, stream, and catchment system in the summer and winter

Similarly, the catchment stretches from H1 to S6 produced a lower relative contribution in the winter. In this catchment sector, the sediment relative contribution ranged from 0.8% (H1) to 5.0% (H4) with an average contribution of 4.2%, while H6 produced 12.3% of the relative SSC in the catchment. The major sediment contribution to the catchment originated from H8 headwater ranging from 19.7 to 52.6% per event, with an average of 52.8% during the winter.

The sectors from H1 to H4, which accounted for ~70% of the catchment area, were responsible for only ~15% of the ARC (Fig. 4.3d). Despite the small area, H6 was responsible for 19% of the ARC in the catchment; a proportion that was superior to that registered in the H1–H4 catchment sectors.

Moreover, H8 was in the lower course of the Guamirin de Cima catchment and represents no more than ~10% of the total catchment area. However, in both periods, i.e., summer and winter, this small headwater was responsible for 63% of the sediment generated in the catchment. The two headwaters (H6 and H8) mainly meant that the tobacco crop was responsible for more than 80% of the relative sediment contribution in the catchment.

The relative contribution of the SSC in the catchment displayed a similar pattern in both summer and winter (Fig. 4.3d). The percentage of the contribution of each headwater for the whole catchment was preserved in both periods. Therefore, changes occurred only to the absolute amount of sediment produced in summer, with tobacco, that was much higher than that registered in winter under oat crop.

4.4 Discussion

4.4.1 Agricultural Calendar and Sediment Transfer on the Tobacco Headwaters

In the present study, a spatial and temporal variation on the suspended sediment was observed in the summer and winter crop phases. In the summer, 46% of the catchment area was under different annual crops such as corn, black bean, soybean, and tobacco. In contrast, during the winter, all the agricultural areas were under oat cropping. This intra-annual crop succession in the catchment played a major role in sediment transfer from hillslope into stream.

In addition, permanent land use (54%) composed of native and secondary forests, especially in the riparian zone, was stable throughout the year. This sort of permanent land use is assumed here as a not important sediment source, since the canopy and a thick litter layer on the forest floor reduces rainsplash and increases soil–water infiltration. Practically, limited erosion is expected in the forest areas (Blanco-Canqui and Lal 2010; Morgan 2009). Headwaters with soybean, corn, and black bean were relatively lesser important sediment sources with respect to the monitoring in this study. Additionally, these crops did not create direct connectors through riparian zone, since the average percentage of ground cover of corn and soybean is much higher than that of tobacco (90% and 63%, respectively). The average percentage cover of tobacco ranges from 11 to 54% during the entire crop development cycle. Indeed, the tobacco crop displays the worst cover management factor compared to that of other crops. Therefore, the tobacco crop is prone to a great amount of soil loss compared to that of oat, wheat, corn, and soybean (Blanco-Canqui and Lal 2010; Morgan 2009; Bakker et al. 2008).

Tobacco is cultivated worldwide in three different ways: no-till, strip-till, and conventional tillage. Conventional tillage is the most disadvantageous with respect to soil erosion and degradation. Conventional tillage on tobacco crop necessitates several mechanized operations (e.g., subsoiling, moldboard plowing, chisel plowing, and multiple disking) before sowing and during crop development. Most multiple disking is conducted to control weeds, prevent soil crusting, and repair row ridge strips (Olivet et al. 2014; Seebold et al. 2013). Tillage, therefore, is one of the major causes of soil erosion as noticed here.

Conventional and no-till tobacco crop display a similar runoff; however, soil loss in conventional tillage is 20–90% higher compared to that in no-till (Wood and Worsham 1986). Therefore, conventional tillage produces much more sediments in comparison to the no-till tobacco crop system (Wood and Worsham 1986; Benham et al. 2007) and soil degradation (Olivet et al. 2014).

The studies reported above displayed soil loss in tobacco crop over gentle slopes <8%. Here, the soybean, corn, and black bean crops occupied gentle slopes, with declivity <20% in Cambisols. It is usual to find tobacco cultivation in areas with more dissected terrain with Leptosols and a slope ranging from 20 to 45%. Therefore, the highest soil erosion in the study area is well expected (Fig. 4.1).

Soil erosion experiment performed in the study area at plot scale (i.e., H6 headwater) displayed a great contrast with runoff and soil loss between oat and tobacco soil cover. During oat cropping, the coefficient of runoff was 15%, whereas the soil loss reached 14 t ha^{-1}. However, the coefficient of runoff was only 7% during tobacco phase, although the soil loss reached 74 t ha^{-1}. In short, tobacco generated lower runoff, but soil loss was five times greater in comparison with oat soil cover (Bednarz 2013). Similarly, in southern Brazil, the intensification of tobacco production increased the sediment yield by 25% in a small rural catchment (Merten and Minella 2006).

Here, the sediment transfer in the catchment differed in the summer when it was covered by tobacco crops and produced 63% of the total SSC; in comparison to winter, when only 37% was produced with soil covered by oat. Type of management crop (C factor from USLE) is extremely important to mitigate soil erosion and the C factor for winter crops (e.g., wheat, barley), including oat (0.18), is three times lower (Mari 2015) than that for the tobacco crop (0.49) (Bakker et al. 2008). In comparison, bare soil displays an annual C factor average of 1.0 (Morgan 2009). In summary, the closer the C factor to 1.0, the more erosion is expected.

In the study area, the most usual system is a succession of tobacco by a winter crop, particularly oat. Oat seeds are spread over a plow-harrowed seedbed that incorporates the tobacco crop residues. After crop development, oat gives soil cover protection against raindrop impact and overland flow. At the end of the crop cycle, oat is desiccated through herbicide and incorporated into the soil by chisel plowing and harrowing. Next, the tobacco is sowed, continuing the annual crop cycle. Consequently, there is a strong relationship between the agricultural calendar, hillslope–stream coupling, and sediment transfer in the catchment due to tobacco succession and the winter crop. However, the hydro-geomorphic connectivity is not straightforward, since 66 rainfall events (81%) did not generate sediment collection in the samplers. Generally, a precipitation of ~20 mm in depth and intensity $\geq$5 mm h^{-1} (i.e., the threshold) are necessary to generate sediment delivery into the stream.

4.4.2 *Tobacco Areas Have Plenty of Connectors Facilitating Sediment Transfer into the Stream*

Despite the importance of the agricultural calendar and tillage processes in providing plenty of sediment on-site, the structural and functional landscape of hydro-geomorphic connectivity is important to transfer the sediments off-site in the study catchment (Fig. 4.4).

Fig. 4.4 Overview of the structural and functional hydro-geomorphic connectivity in the catchment: **a** H4 with wide riparian zone and channel expansion zone preserved; **b** H6 and H8 displaying the channel expansion zone occupied by conventional tillage. The black arrows as fishbone indicating furrows convergence connected to a gully; **c** Details of H8 with gully collecting furrows and connecting hillslope into first-order stream; **d** Tobacco crop growing over a gully in H8

The type of linkage within the landscape, i.e., landform scale, should be considered in the catchment's context. The downslope sediment delivery along the catena, especially in zero and first-order systems, plays a major role in hillslope–channel coupling in small catchments. In addition, lateral–longitudinal sediment and flow linkages, slope angle, morphology, sediment supply, and the magnitude–frequency of hillslope process are some of the main controlling factors of landscape (dis)connectivity (Brierley et al. 2006).

Headwater zones, i.e., hillslope and channel are mostly coupled because tributaries and trunk streams are connected (Fig. 4.4). Therefore, a hydro-geomorphic context with efficient flow (i.e., overland and subsurface flow) and longitudinal sediment transfer is expected in headwater zones as the study catchment (Fryirs 2013). Likewise, the upslope expansion of water and sediment-contributing areas, particularly on hollow depressions, seems to be an important factor of runoff and sediment generation on the hillslope system (Cammeraat 2004).

Here, we observed that the tobacco crop occupies the channel expansion zone, i.e., the hollow concavities (Fig. 4b,c). The structural (hillslope form) and functional (rill processes) connectivities are active in the small catchment, especially on the tobacco headwaters. Moreover, tobacco crop ridge rows in gradient are distributed transversally through the hillslopes. The crop row channels are connected to the concavities (e.g., zero and first-order streams), which are the channel expansion zones (Fig. 4c). This structure results in an extreme hillslope–channel coupling; sometimes, the channel expansion zone is connected by rills and gullies (Fig. 4c,d).

Whereas, the other headwaters (e.g., H1–H4) did not display such hydro-geomorphic connectivity and the riparian zone acts as a buffer retaining the sediments (Fig. 4.4a). In contrast, a headwater with tobacco crop exhibits an intensification of the hydro-geomorphic connectivity (i.e., conventional tillage), causing a direct connection of sediment from hillslope into the stream (e.g., H6 and H8). Even during winter crop growth, these connectors continue to contribute to hillslope–channel coupling and sediment delivery.

Here, the conventional tillage system is expected to transfer a significant amount of nutrient with the sediments, especially phosphorus (Fig. 4.4d). Overall, the agricultural landscape generates sediment and nutrient exportation from the agricultural areas (Alemu et al. 2017). Conventional tobacco produces three times more nitrogen and four times more phosphorus than the no-till system (Benham et al. 2007). In addition, around 88% of the transfer of phosphorus occurs mainly in the particulate form due to a high affinity for this element and mineral colloids (Bender et al. 2018).

In the present study, a clear enhancement of sediment transfer from tobacco headwater was noticed; consequently, the water quality may have been affected seasonally in the catchment. Further studies are warranted to understand the effect of tobacco crop disturbance on riparian and aquatic ecosystems; because from the perspective of water quality in the agricultural landscape, the riparian zone and associated buffer strips are important for retaining sediments, nutrients, and contaminants from hillslopes and keeping them away from the aquatic system (Alemu et al. 2017; Dabney et al. 2006).

4.5 Conclusions

(1) The agricultural calendar affects sediment delivery in the catchment (i.e., soil management). During the summer, under tobacco cover, the suspended sediment concentration was around 63%, whereas, in winter with oat cover, it accounted for only 37% of the total suspended sediment. Therefore, the tobacco-oat succession results in a temporal hydro-geomorphic connectivity variation affecting sediment delivery in the catchment.
(2) Conventional tobacco systems with multiple disking operations along crop growth increase sediment availability. In addition, tobacco occupies areas with strong structural and functional hydro-geomorphic connectivity, i.e., the channel expansion zone on hollow concavities. Therefore, the characteristics of tobacco tillage processes (e.g., row ridge crop) and the linkage within the landscape compartment results, to a certain extent, in permanent connectors. The persistence of hillslope–channel coupling occurs even when the catchment is covered by the winter crop. However, the sediment delivery decreases because oat crop causes sediment decoupling due to its spatio-temporal soil cover protection.

Acknowledgements This study was supported by the Brazilian Research and Development Council (CNPq) through the productivity fellowship to the first author (Grant 301665/2017-6).

Appendix 1

See Appendix Table 4.1.

Appendix 2

See Appendix Table 4.2.

Table 4.1 Suspended sediment concentration according to monitoring sites in summer

Samplings headwaters	1ª (80 mm) g L^{-1}	2ª (29 mm) g L^{-1}	3ª (40 mm) g L^{-1}	4ª (20 mm) g L^{-1}	5ª (50 mm) g L^{-1}	6ª (25 mm) g L^{-1}	7ª (85 mm) g L^{-1}	8ª (45 mm) g L^{-1}	Total g L^{-1}
H1	0.29	2.90	nr	0.37	0.45	0.89	nr	0.36	**5.26**
H2	17.14	21.71	nr	0.38	1.18	10.38	nr	4.78	**55.57**
S3	16.79	8.80	1.05	nr	nr	3.49	7.79	3.00	**40.92**
H4	0.83	3.67	nr	1.16	2.20	3.57	20.39	2.61	**34.43**
S5	22.75	2.11	nr	nr	2.53	5.71	nr	Nr	**33.10**
H6	26.13	1.02	nr	7.12	14.02	14.96	31.93	7.06	**102.24**
S7	41.61	3.22	nr	nr	1.87	11.00	21.88	7.01	**86.59**
H8	390.04	31.58	37.53	445.84	54.13	25.28	29.31	13.23	**1026.94**
C9	132.72	27.05	2.33	11.08	Nr	20.59	35.08	11.92	**240.77**
Total g L^{-1}	**648.3**	**102.06**	**40.91**	**465.95**	**76.38**	**95.87**	**146.38**	**49.97**	**1625.82**

nr = non-recorded

Table 4.2 Suspended sediment concentration according to monitoring sites in Winter

Samplings headwaters	9ª (90 mm) g L^{-1}	10ª (55 mm) g L^{-1}	11ª (48 mm) g L^{-1}	12ª (110 mm) g L^{-1}	13ª (70 mm) g L^{-1}	14ª (70 mm) g L^{-1}	15ª (39 mm) g L^{-1}	Total
H1	1.65	0.37	nr	3.26	0.23	1.05	nr	**6.56**
H2	7.94	1.023	0.87	6.82	3.78	4.98	1.02	**26.43**
S3	12.57	1.32	0.88	8.32	3.68	4.36	1.20	**32.33**
H4	9.99	1.87	0.92	11.51	2.79	3.06	0.67	**30.81**
S5	20.39	2.47	nr	12.62	2.75	2.99	1.34	**42.56**
H6	16.32	3.89	2.32	23.51	3.84	10.84	5.89	**66.61**
S7	44.45	2.18	2.34	25.28	2.94	5.94	3.62	**86.75**
H8	307.45	6.74	4.72	51.51	6.25	27.64	15.43	**419.74**
C9	163.58	3.89	3.62	46	5.42	43.01	5.32	**270.84**
Total g L^{-1}	**584.34**	**23.75**	**15.67**	**188.83**	**31.68**	**103.87**	**34.49**	**982.63**

nr = non-recorded

References

Alemu T, Bahrndorff S, Alemayehu E, Ambelu A (2017) Agricultural sediment reduction using natural herbaceous buffer strips: a case study of the east African highland. Water Environ J 31(4):522–527

Antoneli V, Thomaz EL (2014) Soil loss in tobacco cultivation on different tillage system in the southeastern of Paraná (*in* Portuguese). Revista Brasileira De Geomorfologia 15(3):455–469

Antoneli V, Lenatorvicz HH, Bednarz JA, Pulido-Fernández M, Brevik EC, Cerdà A, Rodrigo-Comino J (2018) Rainfall and land management effects on erosion and soil properties in traditional Brazilian tobacco plantations. Hydrol Sci J 63(7):1008–1019

Bakker MM, Govers G, van Doorn A, Quetier F, Chouvardas D, Rounsevell M (2008) The response of soil erosion and sediment export to land-use change in four areas of Europe: the importance of landscape pattern. Geomorphology 98(3):213–226

Bednarz JA (2013) Hydrogeomorphological processes in a growing area of tobacco in the Irati municipality. Universidade Estadual do Centro Oeste, Guarapuava–Paraná, Brazil. Master Dissertation (unpublished) 92p

Bender MA, dos Santos DR, Tiecher T, Minella JPG, de Barros CAP, Ramon R (2018) Phosphorus dynamics during storm events in a subtropical rural catchment in southern Brazil. Agr Ecosyst Environ 261:93–102

Benham BL, Vaughan DH, Laird MK, Ross BB, Peek DR (2007) Surface water quality impacts of conservation tillage practices on burley tobacco production systems in southwest Virginia. Water Air Soil Pollut 179(1):159–166

Blanco-Canqui H, Lal R (2010) Principles of soil conservation and management. Springer Science & Business Media

Bracken LJ, Turnbull L, Wainwright J, Bogaart P (2015) Sediment connectivity: a framework for understanding sediment transfer at multiple scales. Earth Surf Proc Land 40(2):177–188

Brierley G, Fryirs K, Jain V (2006) Landscape connectivity: the geographic basis of geomorphic applications. Area 38(2):165–174

Burt T, McDonnell J (2015) Whither field hydrology? The need for discovery science and outrageous hydrological hypotheses. Water Resour Res 51(8):5919–5928

Cammeraat ELH (2004) Scale dependent thresholds in hydrological and erosion response of a semi-arid catchment in southeast Spain. Agr Ecosyst Environ 104(2):317–332

Collins AL, Walling DE (2004) Documenting catchment suspended sediment sources: problems, approaches and prospects. Prog Phys Geogr 28(2):159–196

Dabney SM, Moore MT, Locke MA (2006) Integrated management of in-field, edge-of-field, and after-field buffers. J Am Water Resour Assoc 42(1):15

De Vente J, Poesen J, Arabkhedri M, Verstraeten G (2007) The sediment delivery problem revisited. Prog Phys Geogr 31(2):155–178

Didoné EJ, Minella JPG, Merten GH (2015) Quantifying soil erosion and sediment yield in a catchment in southern Brazil and implications for land conservation. J Soils Sediments 15(11):2334–2346

Edwards TK, Glysson GD, Guy HP, Norman VW (1999) Field methods for measurement of fluvial sediment. US Geological Survey Denver, CO

Egozi R, Lekach J (2014) Stream catchment dynamics. Geomorphology 212:1–3

Fiener P, Auerswald K, Van Oost K (2011) Spatio-temporal patterns in land use and management affecting surface runoff response of agricultural catchments—a review. Earth Sci Rev 106(1–2):92–104

Fryirs K (2013) (Dis)Connectivity in catchment sediment cascades a fresh look at the sediment delivery problem. Earth Surf Proc Land 38(1):30–46

Gordon ND, Finlayson BL, McMahon TA, Finlayson BL, Gippel CJ, Nathan RJ (2004) Stream hydrology: an introduction for ecologists. John Wiley and Sons

Gumiere SJ, Le Bissonnais Y, Raclot D, Cheviron B (2011) Vegetated filter effects on sedimentological connectivity of agricultural catchments in erosion modelling: a review. Earth Surf Proc Land 36(1):3–19

Horton RE (1945) Erosional development of streams and their drainage basins; hydrophysical approach to quantitative morphology. Geol Soc Am Bull 56(3):275–370

IUSS Working Group W (2006) World reference base for soil resources. World Soil Resources Report, 103

Knighton D (2014) Fluvial forms and processes: a new perspective. Routledge

Mari F (2015) The evaluation of soil erosion C factor. Int J Appl Sci Technol 5(6):30–38

Merten GH, Minella JPG (2006) Impact on sediment yield due to the intensification of tobacco production in a catchment in Southern Brazil. Ciência Rural 36:669–672

Morgan RPC (2009) Soil erosion and conservation. John Wiley & Sons

Nitsche P, Caramori P, Ricce S, Pinto L (2019) Atlas Climático do Estado do Paraná. Londrina, PR: IAPAR

Olivet YE, Sanchez-Giron V, Hernanz JL (2014) Reduced tillage for tobacco (*Nicotiana tabacum* L.) production in East Cuba. Soil physical properties and crop yield. Span J Agric Res 12(3):611

Pradhanang SM, Briggs RD (2014) Effects of critical source area on sediment yield and streamflow. Water Environ J 28(2):222–232

Seebold KW, Pearce RC, Bailey WA, Bush LP, Green JD, Miller RD, Powers LA, Snell WM, Townsend LH, Purschwitz M (2013) 2013–2014 Kentucky & Tennessee tobacco production guide. Agric Nat Resour Publ. Paper 74

Walling DE (1983) The sediment delivery problem. J Hydrol 65(1–3):209–237

Wood SD, Worsham AD (1986) Reducing soil erosion in tobacco fields with no-tillage transplanting. J Soil Water Conserv 41(3):193–196

Chapter 5
Digital Elevation Model and Irrigation Management Planning in Bangladesh

M. Manzurul Hassan and Md. Ashraf Ali

Abstract Digital elevation model (DEM) is thought to be an important aspect of irrigation management. In order to ensure food security in a land-hungry country with a high population density, the Government of Bangladesh has decided to use every inch of her agricultural land for cropping. Therefore, DEM is essential for proper irrigation management in Bangladesh. The main objective of the study is to prepare DEM to explore the areas suitable for irrigation in Kishoreganj upazila/subdistrict (205 km^2 in area) under Nilphamari district with water supply from the surface water source. Printed topographic maps prepared between 1960 and 1964 with contour lines (with 1:15,840 for 1-foot contour maps) by Bangladesh Water Development Board (BWDB) were used for preparing the DEM. The collected topographic maps for the study site were digitized with ArcGIS (version 10x) format. Field investigations were carried out using a global positioning system (GPS, Model: Garmin eTrex 30) for relevant positional information, levelling devices (G2-32X) for exploring spot height within a selected location of interest (LOI), and Google Earth images for the identification of physical features. The study site is mainly a gentle slope with a reduced level (RL) between 30.0 and 48.0 m from the mean sea level (MSL). Soil property for agriculture and agricultural productivity is very suitable in the study site. The DEM map, area-elevation-discharge curve, and GPS data show different land features in the study area for irrigation. Our study shows that about 11,460 ha (55.9%) of land are available for irrigation subject to providing sufficient water during the dry season with developing new canals from the nearby perennial Teesta River. The irrigation suitable area covers a very gentle slope within the RL of 37.0–44.0 m from MSL. The prepared DEM shows different physiographic characteristics and suitable irrigation areas in the study site. Suitable land for irrigation covers a very gentle slope and there is a high opportunity to increase the suitable land area by developing new canals from the nearby perennial river. The prepared DEM will be helpful for future

M. M. Hassan (✉)
Department of Geography and Environment, Jahangirnagar University, Savar, Dhaka 1342, Bangladesh

Md. A. Ali
Design Engineer, Water and Environment, Mott MacDonald Ltd., Dhaka, Bangladesh

N. C. Jana et al. (eds.), *Livelihood Enhancement Through Agriculture, Tourism and Health*, Advances in Geographical and Environmental Sciences,
https://doi.org/10.1007/978-981-16-7310-8_5

irrigation and agricultural development planning. The delineated small, narrow, and terrain features in the study site can be utilized for potential irrigation.

Keywords DEM · Irrigation · Cropping intensity · GIS · GPS · Bangladesh

5.1 Background

Irrigation performs a key role to increase crop production ensuring food security (Araya et al. 2015; Githui et al. 2020; Grogan et al. 2015; Jin et al. 2016; Manaswi and Thawait 2014; Rahman et al. 2017; Vergine et al. 2017). Improving irrigation productivity with available water resources is vital for food security and sustainable development in a land-hungry country like Bangladesh. The available water from different sources, e.g., surface water, groundwater, and rainwater can be used for agriculture and irrigation. The potential for the use of available surface water for irrigation is essential for sustainable agricultural development and planning (Dawit et al. 2020; Gao et al. 2018; Kharrou et al. 2013) in Bangladesh, where groundwater is heavily contaminated with excessive levels of arsenic (Atkins et al. 2006; Ciminelli et al. 2017; Hassan and Atkins 2011; Mihajlov et al. 2020).

DEM is vital for cropland irrigation management. The term DEM is a generic description for digital imagery of elevation, topography, and sometimes digital imagery of bathymetry. Commonly, DEM is considered as a "model" with two aspects: (a) a pixel-based "modelled representation" of the earth's surface, where each pixel of a DEM represents an elevation value; and (b) model of three-dimensional (3D) topography (Croneborg et al. 2015; Val et al. 2015). The term DEM is also known as digital terrain model (DTM) where "terrain" refers to the bare-earth surface or ground, digital surface model (DSM) where "surface" typically refers to the top-most (radar reflective) surface for a given area that includes all exposed objects or surfaces in the scene. Triangulated irregular network (TIN) is defined as a digital elevation surface that is signified as a connected series of non-overlapping triangles, not as a grid by pixels, and canopy height model (CHM) refers to a height for determining vegetation structure or height that can be expressed as the difference between the top canopy surface (DSM) and the underlying ground topography (DEM or DTM) (Croneborg et al. 2015).

DEM is widely used in irrigation, hydrological studies, topography mapping and orthoimage generation, overland flow modelling, viewshed analysis, and other terrain surface-influenced phenomena (Anwar et al. 2016; Gillies and Smith 2015; Lohani et al. 2020; Wilson et al. 2000; Zhou et al. 2016). In addition, DEM has been used to identify and illustrate geomorphologic features in different environments such as landslides, drainage networks, surface morphological analysis, slope processes, and fluvial geomorphology (Boulton and Stokes 2018; Oh et al. 2018; Fenton et al. 2013; Hosseinzadeh 2011; Imran and Rehman 2019; Wu et al. 2019). DEM, in Bangladesh, can be utilized in the irrigation process including surface morphology, fluvial morphometry, water availability, climate change impacts, and so on. The resolution and accuracy of these data sources are of utmost importance in modelling land-driven processes. DEM leads to a new direction for mapping irrigated areas to better support water resources and agricultural development (Gabr et al. 2020; Bashir et al. 2020; Dehghani and Changani 2020; Gallardo et al. 2020).

DEM can be prepared with different systems, e.g., optical multi-bands classification method and the radar-based approach as well as GIS-based interpolation approach. A plethora of literature has shown the capability of optical multi-bands remote sensing for irrigation mapping (Gao et al. 2018; Akbari et al. 2006; Beltran and Belmonte 2001; Biggs et al. 2006; Gowing et al. 2020; Kamthonkiat et al. 2005). Gao et al. (2018) demonstrated that a single-date image can be used for distinguishing irrigation fields, but it is not reliable to calculate the temporal variation of cropping intensity. Therefore, the multi-temporal analysis provides a better potential to define irrigation with different cropping intensities. Gowing et al. (2020) assessed irrigation management planning with groundwater resources and satellite data in combination with modelling at the national or regional scale. Thenkabail et al. (2004) used moderate resolution imaging spectroradiometer (MODIS) time-series data to produce land use/land cover (LULC) map of irrigated areas for the Ganges and Indus river basins.

Numerous vegetation indices, such as the Normalized Difference Vegetation Index (NDVI), the Normalized Difference Wetness Index (NDWI), and the Green Vegetation Index (GVI), derived from multi-bands satellite data are proven to be able to map for irrigated areas (Gao et al. 2018; Boken et al. 2004; Xiao et al. 2005). Boken et al. (2004) demonstrated the potential of Advanced Very High-Resolution Radiometer (NOAA-AVHRR) for estimating irrigated areas using NDVI and the Vegetation Health Index (VHI). Xiao et al. (2005) developed a paddy rice mapping algorithm that uses the time series of the Land Surface Water Index (LSWI), the Enhanced Vegetation Index (EVI), and the NDVI, derived from MODIS images. In addition, the availability of synthetic aperture radar (SAR) data offers new potential for irrigation monitoring by providing the ability to observe under any weather conditions. Shao et al. (2001) show that radar data can provide unique characteristics of irrigated rice fields. The radar remote sensing in soil analysis is sensitive to water content in the surface layer because of a noticeable increase in the soil dielectric constant with increasing water content (Baghdadi and Zribi 2017; Morvan et al. 2008; Zribi et al. 2011).

The accuracy of available DEM sources in order to produce a single DEM is an important issue and it is possible with merging all the available DEM data. A new range of recent DEM acquisition technologies includes airborne and ground-based light detection and ranging (LiDAR), and aerial photogrammetry based on images captured by unmanned aerial vehicles (UAV) can be applied for preparing raster-based DEM. The single composite DEM covers the location of interest with the highest possible resolution and accuracy (Leitão et al. 2016).

The accelerating rate of population growth and increasing demand for water have been considered two main challenges to achieve sustainable development goals for food security with zero hunger (Goals 1 and 2) in Bangladesh. The geomorphological conditions and climate change have further exacerbated freshwater availability and thus increases its demand, both temporally and spatially (Lemann et al. 2019). In Bangladesh, climate change alters hydrological responses and shifts in time of rainfall-runoff responses. Literature shows potential negative impacts of climate change on water supply for agriculture (Duran-Encalada et al. 2017; Obeysekera et al. 2011). However, the amount of water available for irrigation depends on the availability of freshwater resources (Huang et al. 2006; Keshta et al. 2019). Moreover, the groundwater is heavily contaminated with toxic levels of arsenic (Atkins et al. 2007; Hassan et al. 2005, 2021; Hassan and Atkins 2007) and the use of groundwater for irrigation shows the contamination of inorganic arsenic in many food staples in this country (Ciminelli et al. 2017; Awasthi et al. 2020; Bianucci et al. 2020; Flora 2015; Hassan 2018; Marmiroli 2020). Therefore, it is needed to migrate from the traditional agricultural practices depending on rainfall during the wet season towards surface water-based irrigation systems. Understanding the importance of agricultural development in the context of lack of available safe groundwater, the Government of Bangladesh undertook major efforts to provide surface water fed irrigation from some 700 and more rivers following over the country.

The agricultural sector plays an essential role in the economy of Bangladesh in terms of its contribution to gross domestic product (GDP) 15.59%, employment generation, livelihoods, and poverty alleviation (BBS 2015). The agricultural activities in Bangladesh are pursued intensively for the crop. Crop cultivation is a major contributor to the economy, and furthermore, paddy has a predominance with about 75% of crop planted area (FAO 2014).

The agricultural productivity in Bangladesh is directly influenced by soil properties, including those related to texture and drainage, which controls the water level in the soil. Terrain complexity of these regions determines the humidity, temperature, soil properties, soil moisture, and soil fertility that has direct effects on agricultural potential. Considering surface water resources in Bangladesh for irrigation in terms of its river and canal schemes as well as delivery of water to support the farming production is significant for handling water and food security concerns for the country. Agriculture, in Bangladesh, is mainly dependent on rainfall pattern. There is an irregular rainfall pattern during the wet season as the immediate effect of climate change in recent times which is different from our historical rainfall trends. Moreover, Bangladesh has been experiencing frequent atmospheric disasters in terms of floods, extreme temperatures, cyclones, and so on. Bangladesh is a land-hungry country, and the current agricultural production cannot ensure the food security of this country. Thus, agricultural development is needed with high cropping intensity with the utilization of DEM.

DEM, in the case of irrigation, can be used in simulations for estimating the area of land suitability. The criteria for land suitability for irrigation and its productivity is directly influenced by soil properties, including those related to texture and drainage, which controls available water holding capacity (AWHC). These "irrigation limiting factors" can be used for preparing irrigation suitability maps. Integrating DTM or DSM models with different types of soil properties, slope gradient, and sufficient water availability could contribute to sustainable agricultural development through irrigation. The paper aims to assess surface water availability during the dry season and evaluate its suitability for irrigation using DEM with different spatial resolutions. DEM can be used as an important instrument for landscape analysis for irrigation. It is noted that the prepared DEM will be helpful for detailed future irrigation design and management as well as agricultural development planning.

5.2 Agricultural Legislative Framework in Bangladesh

The guideline for Bangladesh "Vision 2021" and the associated "Perspective Plan of Bangladesh 2010–2021: Making Vision 2021 Reality" have set some solid development targets for Bangladesh by the end of 2021 (GOB 2012). To materialize this vision into a reality, the Ministry of Land has some specific targets to be achieved and those goals and targets have been reflected in the Sixth Five Year Plan (6FYP) and Seventh Five Year Plan (7FYP). In addition to national documents, several international documents [e.g., the Istanbul Program of Action (IPOA) and Sustainable Development Goals (SDG)] and their targets have also specified the activities and the targets to be achieved.

5.2.1 The Agricultural Development Corporation Ordinance 1961

This Agricultural Development Corporation Ordinance 1961 (Ordinance No. XXXVII of 1961) was convenient to establish an Agricultural Development Corporation for the purpose of increasing agricultural production in Bangladesh. According to Article 16 of this Ordinance, the Corporation may: "grant land vested in the Corporation to any person on any condition it thinks fit, and for this purpose issue a statement or statements of conditions on which the Corporation is willing to grant such land; provided that no land shall be granted to any person without the statement of conditions having been approved by the Government" (http://bdlaws.minlaw.gov.bd/act-320/section-16411.html). Provided further that no land shall ordinarily be leased for a period exceeding ten years, and where any lease for a longer period is given the provisions of the State Acquisition and Tenancy Act 1950, shall apply, and any other grant shall also be subject to the provisions of the said Act.

5.2.2 National Land Use Policy 2001

The existing land use policy was adopted in 2001 in order to minimize loss of cropland, stop indiscriminate use of land, prepare and use guidelines for different regions, rationalize land acquisition, and synchronization of land use with the natural environment (GOB 2001) It has given emphasis mainly on several aspects. They are:

(a) Protecting agricultural land to meet the additional food requirement for increased population
(b) Ensuring best utilization of land through "land zoning" for agriculture, fisheries, forestry, rural and urban settlement, industry, infrastructure, etc. through modern technology
(c) Rehabilitating landless people on newly reclaimed land (*char*-point bar land, coastal reclaimed land, *haor*-marshy land, etc.)
(d) Reserving *khas* (government land) land for future development projects
(e) Making land use environment-friendly
(f) Increasing the opportunity of income generation through proper utilization of land resources to reduce poverty and provide food security
(g) Protecting natural forest, river erosion, and hilly areas
(h) Protecting land from pollution
(i) Constructing multi-storied buildings for government, semi-government, and non-government organizations in limited land.

5.2.3 Agricultural Land Protection and Land Zoning Act 2010

The government first took steps to enact a new law titled "Agricultural Land Protection and Land Zoning Act" in 2010 to ban the construction of houses and industrial units on farmlands. Taking advantage of the absence of any arable land protection act, most of the multi-crop farmlands around different cities and district towns were filled up with sand for housing projects and industrial units by a group of people. This endangered the environment as well as food security.

5.2.4 Agricultural Land Protection and Land Use Act 2016

The Act has been prepared to protect the valuable fertile cropping land from land-grabbers in Bangladesh. The objective of this Act is "to ensure that the use of land and water areas and building activities on them create preconditions for a favourable living environment and promote ecologically, economically, socially and culturally sustainable development". There is a provision of this Act for three years of imprisonment, or BDT300,000 (US$3600) as fine, or both as punishment for those breaking the law. Brick kilns are also a major threat to arable land and the environment, as they mostly use topsoil—the most fertile part of the land—to produce bricks and burn a huge amount of fossil fuel. The land ministry had taken the initiative in 2001 to ban the construction of houses and industrial units on multi-crop farmlands but failed to do so due to the availability of appropriate law. It is recorded that at least 2.656

million acres of farmlands shifted to the non-agricultural sector between 2003 and 2013. “The nature of land cannot be changed. No one would be allowed to construct houses, industrial units, kilns, factories and other establishments by destroying farmlands”. It is also noted that neither the government nor private organizations can acquire arable land that yields three or two crops at any cost.

5.2.5 National Agriculture Policy 2018

The National Agriculture Policy 2018 has been prepared with a view to achieving sustainable food and nutrition security through efficient utilization of natural resources. In 1999, the Government of Bangladesh first formulated the national agriculture policy for making the crop production system profitable and then amended it in 2013 aiming to tackle challenges of climate change in the agricultural sector. This agriculture policy is addressing new issues and concerns for sustainable food production through the participation of women in production and marketing.

5.3 Materials and Methods

5.3.1 The Study Site

The study site has been selected within the Teesta Barrage Project (TBP) site to investigate the management of irrigation with DEM. Kishoreganj Upazila (subdistrict) is located in the northern part of Bangladesh in the Nilphamari district. The study area is located between 25°48′55″N and 25°59′17″N latitudes and between 88°56′59″E and 89°07′02″E longitudes covering about 205 km^2 of area (Fig. 5.1) with a total population of about 261,000 having the population density of about 1275 per sq. km. Kishoreganj was known for “Nil cultivation” while under British rule. The total cultivable land is about 16,363 ha (79.7%) in the study site of which 13,450 ha (82.2%) of land is under irrigation facilities with deep tubewells and shallow tubewells (http://kishorganj.nilphamari.gov.bd); while homesteads cover about 3840 ha (18.7%) and wetland about 312 ha (1.5%) in the study site. The main crops in Kishoreganj are paddy, potato, tobacco, ginger, and vegetables. The climatic characteristics are different from the other part of this country. The average maximum temperature during the dry season is about 35.7 °C and the minimum average is 12.6 °C. It is noted that during the winter and summer seasons, there is exceptionally low rain. In January, the average rainfall is calculated for 29 mm and in July the figure is 1574 mm (http://kishorganj.nilphamari.gov.bd). The rainfall is frequent during the wet season, but at present, it is very intermittent and sometimes severe that may cause local floods. During the dry season, it is needed sufficient water supply rabi (i.e., which are cultivated during the dry season) crops.

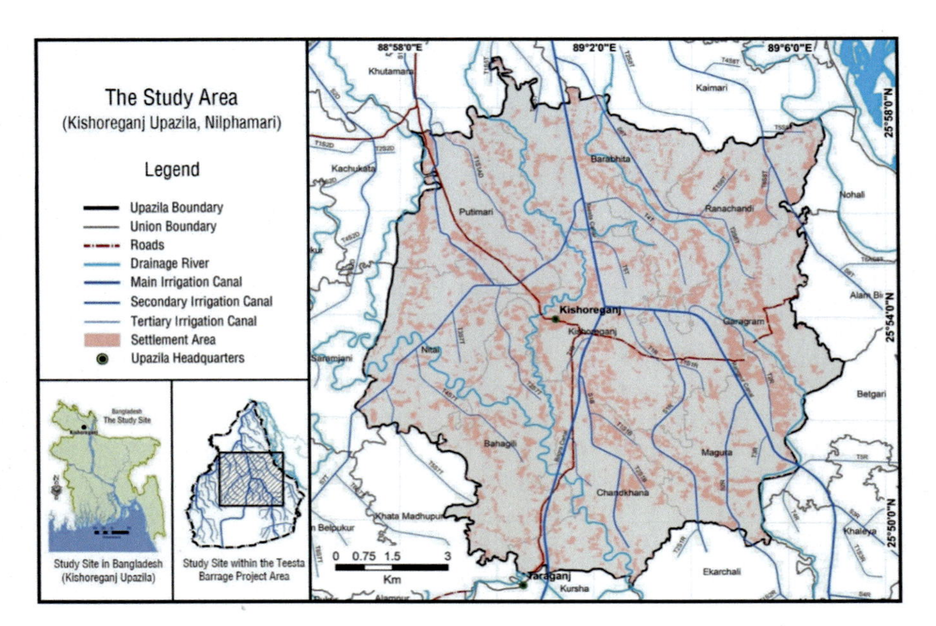

Fig. 5.1 The study area: Kishoreganj Upazila in Nilphamari district. This study site is within the Teesta Barrage project area

5.3.2 Data Collection

A number of methods were conducted for preparing DEM for the study site for irrigation management, and they were: (a) collected relevant maps, documents and records from different sources; (b) identified different types of existing natural and man-made structures; (c) conducted a topographic survey and GPS survey for important landmarks; (d) digitized different features like water bodies (perennial water bodies, seasonal water bodies, canals, rivers, etc.), settlement areas, roads, and so on from Google Earth; (e) dealt with digitized data with GIS technology; and (f) prepared DEM and boundary lineation for the cultivable area, water area, homestead area, etc. (Fig. 5.2).

In preparing DEM, it was needed to interpolate elevations (spot height and benchmark) on a topographic map in the study site. Accordingly, DEM for the study area was developed by digitizing the contours and all natural breaks (i.e., rivers, canals, roads, elevated homesteads, etc.) from the collected topographic maps (1:15,840) of 1960–1964 prepared by the Bangladesh Water Development Board (BWDB). To cover the whole study area, it was collected a total of nine topographic maps (Fig. 5.3). In addition, Google Earth Pro was utilized for this purpose. After generating the DEM, it was compared with the topographic survey data that was carried out by the authors in July 2016. Topographic survey data were vital for the preparation of a high precision DEM.

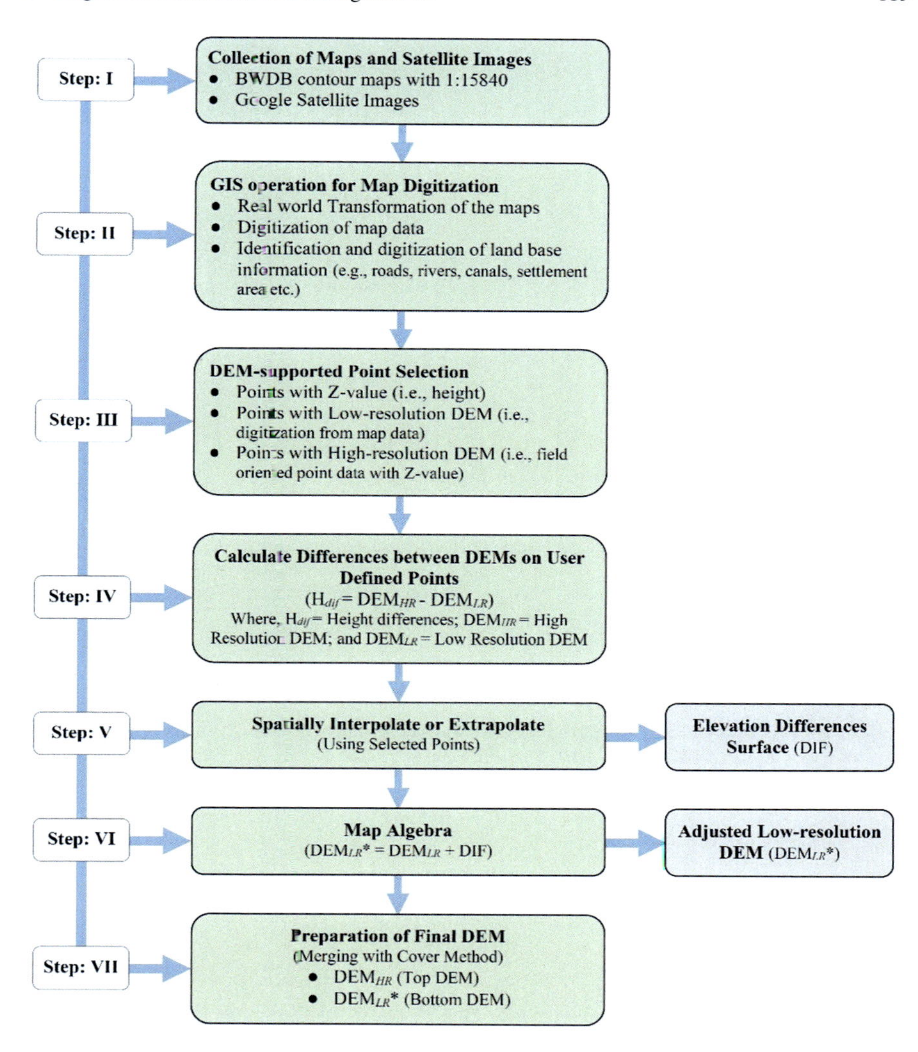

Fig. 5.2 Flow diagram for preparing DEM for irrigation management in the study site

The accuracy of this data was determined primarily by spatial resolution, but additional factors affecting accuracy were data type (integer or floating point) and the authentic sampling of the surface when creating DEM. Problems arose when the old database from BWDB contour maps of the study site was merged with patches of updated satellite data of streets and other fabric features. DEM generated by different interpolation techniques may have different characteristics having a spatial resolution, accuracy, geographic coordinate system, and acquisition dates. It is noted that several elevation values were considered from various available data sets for preparing DEM. Therefore, simple DEM merging methods increased inconsistencies

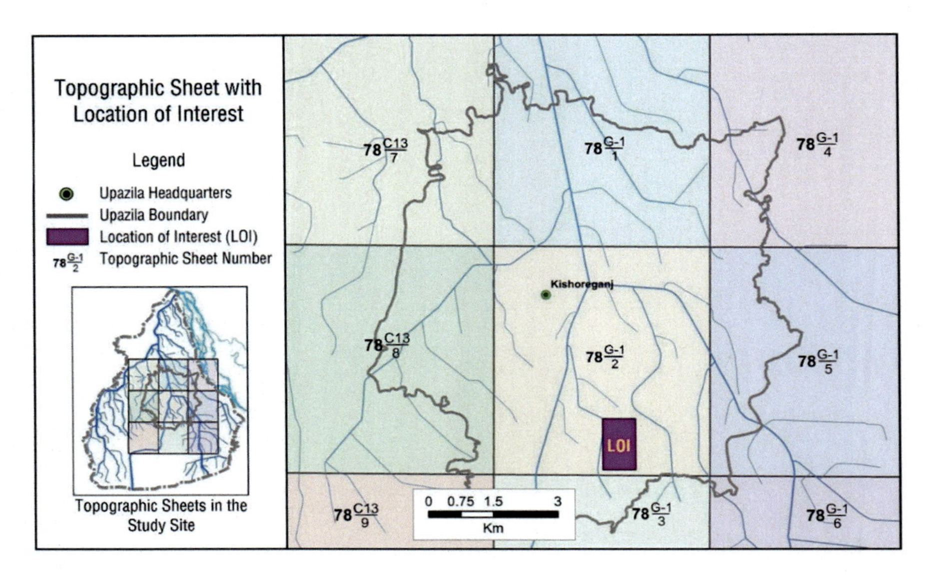

Fig. 5.3 Topographic sheets within the study site and selected location of interest (LOI) for the comprehensive fieldwork

and this, in turn, produced low accurate modelling results. Therefore, there was a need for novel methods that generated high accurate DEM. Such methods were able to extract all the accurate elevation data sets and retain the key features of the high accurate DEM.

5.3.3 Selection of Location of Interest (LOI)

LOI was selected in the study to cover the accuracy level of DEM. Figure 5.3 shows the location of LOI within our study site. The resolution and accuracy of DEM data sources are of utmost importance in modelling land-driven processes. In formulating spatial decision or spatial planning, it is of utmost importance to develop a DEM with high accuracy and precision. Irrigation, in terms of the amount of water needed, the pattern of water supply, and the pattern of overland flow can be hardly successful when the catchment area or parts of the catchment area can be prepared with low-resolution DEM. Therefore, high-resolution DEM is a prerequisite for a successful irrigation project. For the selection of LOI, a number of criteria were considered: (a) cover all the channel patterns; (b) areas where there was no information regarding contours and channels; and (c) cover the criteria for using a suitable methodology for adjusting low-resolution and high-resolution data. Accordingly, we selected Uttar Chandkhana Mouza (the lowest level administrative unit with jurisdiction in Bangladesh) within the study site. The positional corners of the LOI were:

(a) 25°51′15.80″ and 89°02′21.83″ in the Northwest corner; (b) 25°51′15.80″ and 89°03′07.28″ in the Northeast corner; (c) 25°50′08.37″ and 89°02′21.83″ in the Southwest corner; and (d) 25°50′08.37″ and 89°03′07.28″ in the Southeast corner (Fig. 5.3). The LOI covers about 90 ha of land and a canal (i.e., coded with T2S1B) is flowing over this LOI.

5.3.4 Field Investigations

Field investigations were carried out in the study site using a global positioning system (GPS), levelling devices (G2-32X), and Google Earth image. We used GPS (Garmin eTrex 30) device for relevant positional information. This handheld GPS boasts advanced tracking functionality with the capacity to lock on to 24 satellites and it is easy to use. This GPS has a high-sensitivity receiver with the facilities of a preloaded base map with topographic features. This device has the facilities for automatic routing with an electronic compass and barometric altimeter. The Garmin eTrex 30 GPS is capable of simultaneously tracking both GPS and Global Navigation Satellite System (GLONASS) so that it can obtain a desired position more efficiently than using GPS alone. This GPS provides impressive tracking features for high and low elevation points or store waypoints along a track to estimate the distance between points. We also utilized levelling devices (G2-32X) during our fieldwork. This levelling device has a standard deviation of $\leq$1.5 mm per kilometre double-run. The device has a working range of $\geq\pm15'$ with distance error of $\leq\pm0.4\%$, focusing error of $\leq$0.5 m, compensating error of $\leq\pm0.3''$, and levelling accuracy of $\leq\pm0.5''$.

The selected LOI was used for detailed elevation information. The main focus of the survey relates to the terrain and considers all elevated places such as homesteads, water bodies, roads, and so on. Several natural breaks (e.g., small canals, ponds, local unidentified roads etc.) were considered for field investigation. The roads and settlement areas were collected, and these features were considered to distinguish on the prepared DEM. The main limitation was that we were unable to collect the depth of any canals and water bodies. It is noted that we did not find any bathymetry data from the BWDB contour maps for small water bodies, but we have extracted a number of water body information from Google images, and we incorporated them with DEM. The digitized contours from the BWDB topographic maps were compared with the topographic survey data carried out by the authors of this paper.

5.3.5 Application of Geographical Information Systems

Geographical information systems (GIS) with geostatistical methods can be used for modelling DEM for spatial irrigation planning. The collected hardcopy BWDB topographic maps (1960–1964) were converted to digital form by scanning into "jpeg" format for digitization. The spatial data were geometrically corrected to the world geographic coordinate system (WGS1984) and shifting the "jpeg" image data for the study area and merging them with each other. DEM for the study area was prepared by digitizing the contours, spot height values, and natural breaks (rivers, canals, roads, elevated homesteads) with ArcGIS (version 10x). All these information were extracted from the topographic map (1:15,840) and the analysis in ArcGIS. The prepared DEM from the BWDB topographic maps were then compared with our topographic survey data. Topographic survey data were interconnected for preparing DEM with positional accuracy (Fig. 5.4). The main physiographic units of the region were identified by visual interpretation of DEM and guided by both GPS and Google Earth images (Table 5.1). The Google image is draped over the DEM and processed in ERDAS Imagine (version 9x) to define the different physiographic units.

A gridded array was used to prepare the DEM. The elevation data recorded on our grid were transported into ArcGIS as an ASCII file. These gridded survey data were used for preparing DEM and then compared with the vector DEM generated from the BWDB topographic maps with ArcGIS.

Isolines for contour maps were delineated using DEM at different elevation. DEM was used to classify the study site into different categories, namely arable lands, settlement areas, water bodies, etc. as per the requirement of irrigation management. The focus of the survey relates to the terrain and not elevated places such as settlement areas. Similarly, the low-lying areas of small water bodies were set to a constant level because of the unavailability of spot heights from the BWDB contour maps.

DEM was prepared by using 3D Analyst and Spatial Analyst for ArcMap. In addition, the inverse distance weighted (IDW) interpolation method was used for the DEM. Interpolation is the process of estimating the value of parameters at unsampled points from a surrounding set of measurements (Burrough and McDonnell 1998). When the local variance of sample values is controlled by the relative spatial distribution of these samples, geostatistics can be used for spatial interpolation (Oliver 1996) and point interpolation is significant in GIS operation (Hassan et al. 2021; Cinnirella et al. 2005). An IDW interpolator is a point estimation technique based on the weighting of a random function for a particular cell node of a grid. The IDW interpolator assumes that each input elevation point has a local influence that diminishes with distance (Hassan and Atkins 2011; Bianucci et al. 2020; Isaaks and Srivastava 1989). It weights the points closer to the processing cell greater than those further away, hence the name IDW interpolation or inverse squared distance (ISD) interpolation (Ashraf et al. 1997). In the IDW interpolation method, the maximum and minimum values in the interpolated surface can only occur at our collection points. All the investigated GPS land base points and all points within a specified radius were used to determine the output value for each location. In the IDW method, the surface is driven by local variation and calculation using the IDW method depends on power parameter and neighbourhood search strategy. We used the power value >1 for preparing our DEM.

We also applied an area-elevation-discharge curve to identify the level of elevation from MSL and the amount of surface water is required for irrigation during the dry months in the study site. The data for this curve was extracted from GIS-based DEM for this study. Water discharge capacity is the function of the water release which is

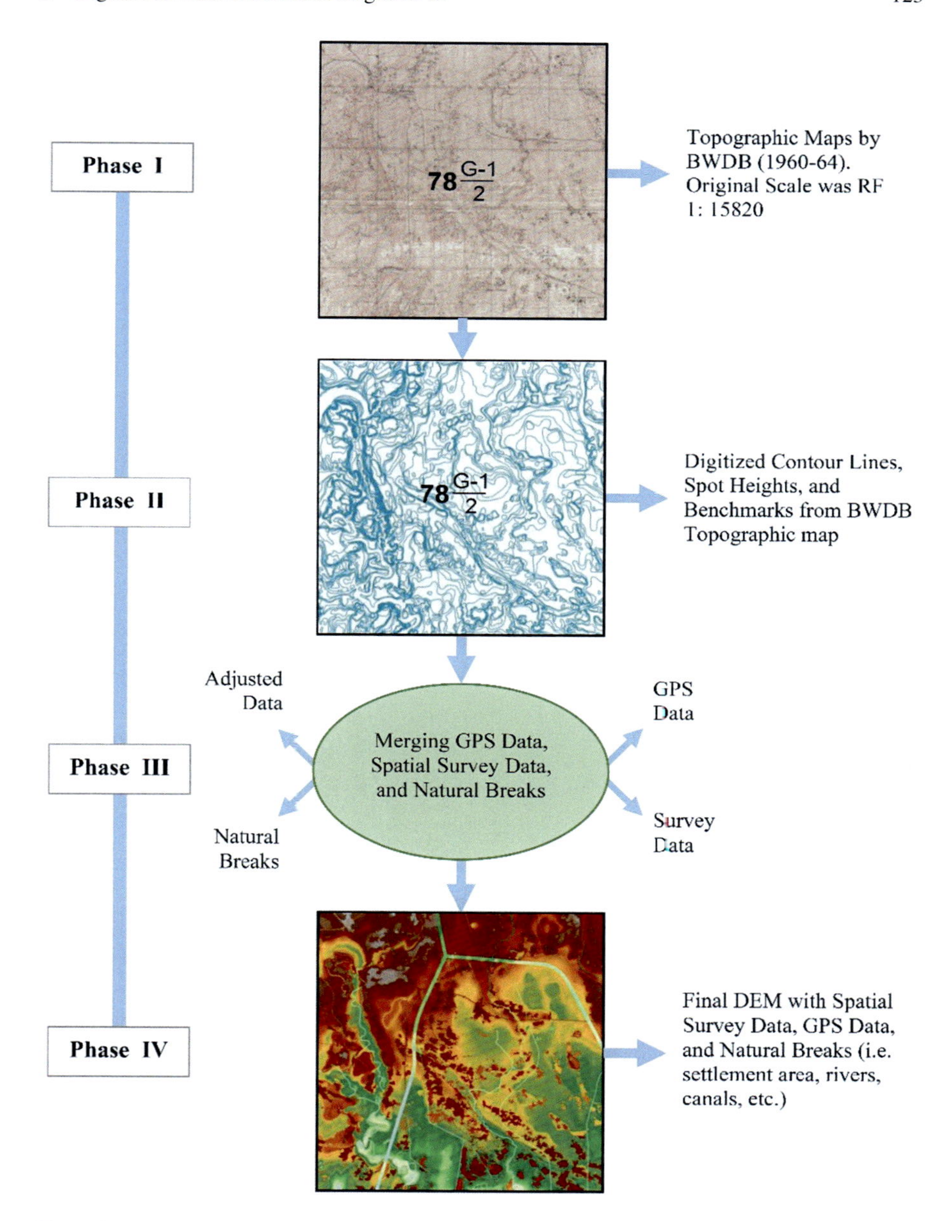

Fig. 5.4 DEM merging with different sources of information, e.g., satellite data, survey data, and GPS data as well as natural breaks

Table 5.1 Assumptions and interventions for preparing DEM in the study site

Major features	Features	Width with buffer	Assumed width (m)	Assumed height (m)	Assumed depth (m)
Irrigation canal	Primary canal	No	30	–	4.0
	Secondary canal	No	15	–	3.0
	Tertiary canal	No	10	–	2.0
Drainage canal	Primary canal	No	100	–	8.0
	Secondary canal	No	20	–	5.0
	Tertiary canal	No	10	–	3.5
Road/rail network	National road	Yes	25	2.0	–
	Regional road	Yes	20	2.0	–
	District road	Yes	15	2.0	–
	Local road	No	10	1.5	–
	Rail track with buffer	Yes	30	2.0	–
Wetlands	Polygons	No	Digitized from image data	–	–
Homesteads	Polygon	No	Digitized from image data	–	–
Field survey data	Canal	No	Aligned with survey data	–	–
	Grid	No	Aligned in 5 m grid	–	–

Source Fieldwork, 2020

also termed as "area and capacity curve" (Afshar and Takbiri 2012; Berglund et al. 2020; Elias et al. 2015; Jurík et al. 2015; Rasekh and Brumbelow 2014; Sayl et al. 2017). This curve describes the amount of water required for irrigation during the dry season in the form of a graph or table. Water surface area is calculated as the surface of the contour with the examined elevation and water discharge is usually calculated as the amount of water between two points within a channel.

5.4 Results and Discussion

5.4.1 Physiographic Characteristics for Irrigation

The prepared DEM can be used to classify the study site into different categories as per the requirement of irrigation. The elevation difference between survey data in selected LOI (Fig. 5.5) and the collected DEM was considered for preparing high-resolution DEM for the study site to meet the requirements of sufficient surface water for irrigation. Apart from irrigation planning, the prepared DEM can be used further for any development activities of the study site. It was delineated three different land features in the study site, for example, homesteads (3840 ha, 18.7%), cultivable area (16,363 ha, 79.7%), and wetland (312 ha, 1.5%) (Table 5.2). We tried to generate DEM with different spatial resolutions (e.g., 5, 50, 100, 150, and 200 m) for identifying the level of slope gradients within the study area for delineating the suitable areas for irrigation (Fig. 5.6).

The DEM map (Fig. 5.6), GPS data, and field data show different physiographic characteristics in the study site with different levels of elevation. The physical properties of land areas in the study site can be suitable for irrigation during the dry season as well as in formulating future irrigation planning.

The study site is characterized by two main characteristics. The first one is based on areas with an undulating surface. Areas within the elevation between 30.1 and 35.0 m and between 44.1 and 48.0 m from the MSL are in the first category. About 4152 ha (20.24%) of land cover this area. This area is mainly characterized by high soil fertility but is moderately suitable for agriculture. These areas are within the homestead areas having water bodies and roads. The second category is characterized by a gentle slope at most places. The area is located within the elevation between 37.1 and 44.0 m from the MSL. Some 16,363 ha (79.76%) of land cover this area. Soil properties are suitable for agriculture and productivity is high in this zone.

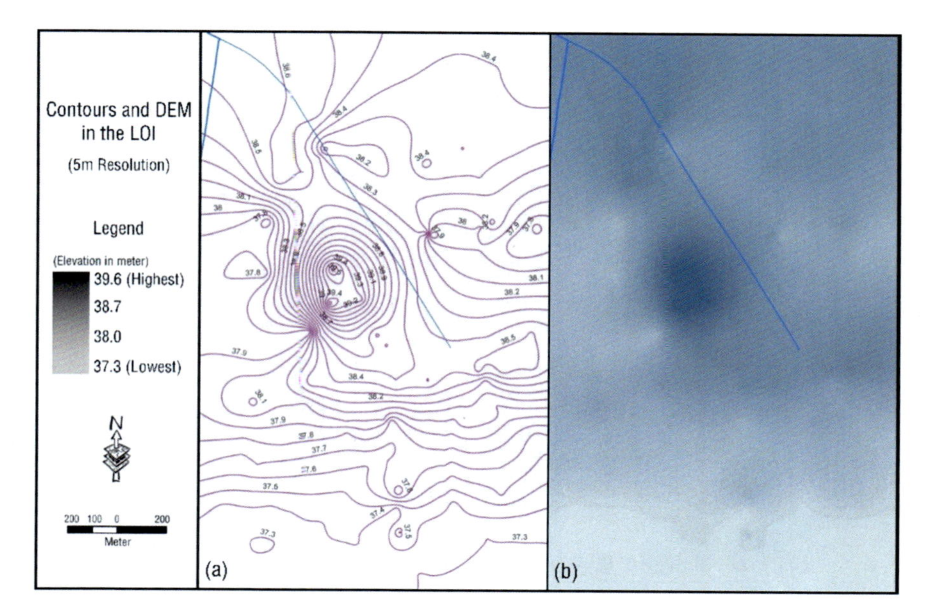

Fig. 5.5 Contours and DEM in the selected LOI: **a** contours at different elevations; and **b** DEM with 5 m spatial resolution

Table 5.2 Distribution of key land features and areas under irrigation in the study site

Union	Total area (ha)	Homesteads (ha)	Cultivable area (ha)	Wetland (ha)	Irrigable area (ha)
Bahagili	2346	412 (2.0)	1900 (9.3)	34 (0.16)	1330 (6.5)
Barabhita	2229	347 (1.7)	1850 (9.0)	32 (0.15)	1290 (6.3)
Chandkhana	2287	443 (2.2)	1810 (8.8)	34 (0.16)	1250 (6.1)
Garagram	2309	439 (2.1)	1835 (8.9)	35 (0.17)	1270 (6.2)
Kishoreganj	2408	501 (2.4)	1865 (9.1)	42 (0.20)	1330 (6.5)
Magura	2167	446 (2.2)	1688 (8.2)	33 (0.16)	1190 (5.8)
Nitai	2189	345 (1.7)	1814 (8.8)	30 (0.14)	1290 (6.3)
Putimari	2328	441 (2.2)	1849 (9.0)	38 (0.18)	1260 (6.1)
Ranachandi	2252	466 (2.3)	1752 (8.5)	34 (0.16)	1250 (6.1)
Study site total	**20,515**	**3840** (18.7)	**16,363** (79.7)	**312** (1.5)	**11,460** (55.9)

Note Figures in the parentheses show the area of net per cent of the total area in the study site

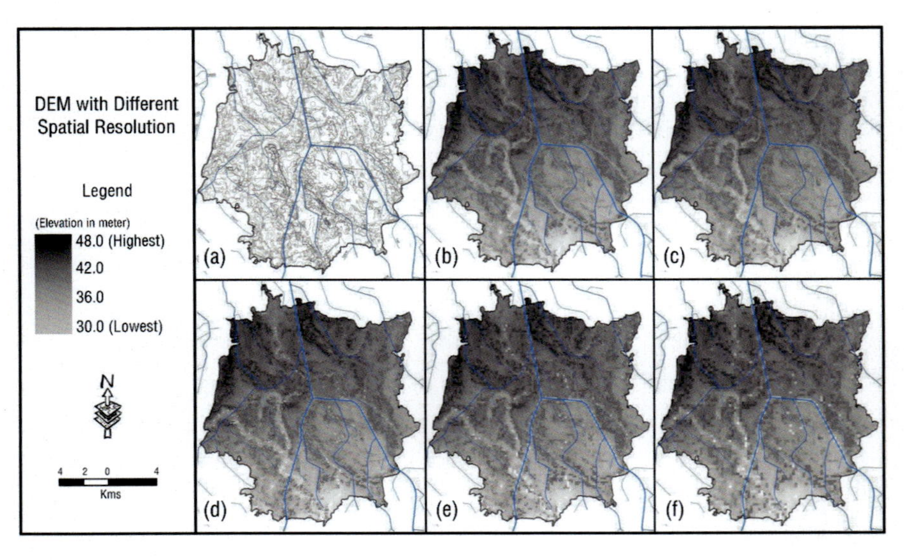

Fig. 5.6 Contours and DEM at different spatial resolutions in the study site: **a** contours at different elevations; **b** DEM with 5 m spatial resolution; **c** DEM with 50 m spatial resolution; **d** DEM with 100 m spatial resolution; **e** DEM with 150 m spatial resolution; and **f** DEM with 200 m spatial resolution

5.4.2 Potential Area for Irrigation

Soil type and terrain features are important components and prerequisites for a sustainable irrigation system. Physical and chemical properties of soil, as well as climatic data, are the factors that determine the land suitability of a given area. Since the soil structure is loamy and very fertile, we did not evaluate soil suitability criteria (physical and chemical properties of soil) to define the land suitability for irrigated agriculture. Nevertheless, we have classified the suitable area for irrigation based on DEM and surface water supply system. Based on the area-elevation-discharge curve, we can analyse the area for irrigation suitability (Table 5.2).

Based on water supply and demand during the dry season, the available water from the nearby perennial river can irrigate the available land that is suitable for agriculture. The result indicates that there is an excess amount of water resources available to satisfy the irrigation demand. The irrigation potential area was estimated as available water in nearby the Teesta River and the gross irrigation water requirements during the dry season for cropping.

The analysis of the prepared DEM gives useful insights into the suitable land resources development for irrigation at different slope levels. Overall, there is a large amount of land resources and potential irrigable area (11,460 ha, 55.9%) available for irrigation development if it is possible to provide sufficient water during the dry season (Table 5.2). The suitable irrigation land covers a very gentle slope and there is a remarkably high opportunity to increase the area of land suitability by developing new canals from nearby the Teesta River. In addition, there are some areas where due to the slope pattern and distance from the main irrigation canals, the irrigation potential is low. Our investigation shows that about 2482 ha of land (12.1%) in the study site were calculated as the land for low suitable irrigation since there is a high gradient of slope and it is not possible to utilize the provided water during the dry season. The area is in the middle and southwest part with a scattered distribution in the study site (Fig. 5.6). More new canals from the nearby perennial river could change the existing situation.

A vast amount of water is essential for Boro rice cultivation during the dry season in Bangladesh. It was calculated by Biswas and Mandal (1993) that about 11,500 cubic metres of water per hectare are required for Boro rice cultivation in Bangladesh. Geethalakshmi et al. (2011) reported that about 3000–5000 L of water is needed in producing one kilogram of rice from rice fields with different rice cultivation methods. Our investigation shows that about 120 cm of water is required for Boro rice cultivation in the study site. The study also shows that a huge amount of water is mismanaged for Boro rice cultivation during the dry season due to operation losses, evaporation, and seepage into the soil. Seepage is the most important in this aspect and this seepage loss in the irrigation canals accounts for water conveyance loss of 98.4% while approximately 0.3% of the total stream is lost due to evaporation (Akkuzu et al. 2006); while Sattar (2004) reported that about 30–40% water loss can be possible in the irrigation projects of Bangladesh through the earthen channels at the time of distribution.

The water conveyance loss can also be caused by improper design, poor management of the canal, insufficient freeboard, and social conflicts (Hossain et al. 2016) as well as the impact of climate change (Acharjeea et al. 2017, 2020). A major challenge for climate-oriented future agricultural production in Bangladesh would be sustainable water management in an environment with drought and high-temperature stress.

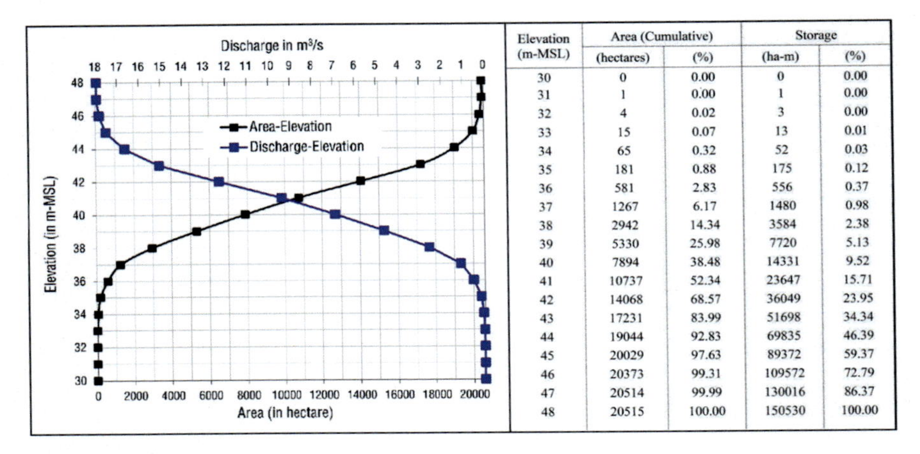

Elevation (m-MSL)	Area (Cumulative)		Storage	
	(hectares)	(%)	(ha-m)	(%)
30	0	0.00	0	0.00
31	1	0.00	1	0.00
32	4	0.02	3	0.00
33	15	0.07	13	0.01
34	65	0.32	52	0.03
35	181	0.88	175	0.12
36	581	2.83	556	0.37
37	1267	6.17	1480	0.98
38	2942	14.34	3584	2.38
39	5330	25.98	7720	5.13
40	7894	38.48	14331	9.52
41	10737	52.34	23647	15.71
42	14068	68.57	36049	23.95
43	17231	83.99	51698	34.34
44	19044	92.83	69835	46.39
45	20029	97.63	89372	59.37
46	20373	99.31	109572	72.79
47	20514	99.99	130016	86.37
48	20515	100.00	150530	100.00

Fig. 5.7 Area-elevation-discharge curve with slope gradient and area covered for irrigation. This figure is based on the information for the maximum requirement of Boro rice in Bangladesh

Considering the requirement of sufficient water for irrigation of Boro rice during the dry season in the study site, we implied the area-elevation-discharge curve that shows the irrigation potential with respect to the physical resources and irrigation water requirements for different slope classes (Fig. 5.7). It is estimated that about 11,460 ha (55.9%) of area are available for potential irrigation development in the study site if it is possible to provide a sum of about 10 m^3/s of water during the dry season (Fig. 5.7). This amount of required water can be minimized by almost half with the use of a plastic pipe distribution system in place of earthen channels (BRRI 2007; Sattar et al. 2009).

5.5 Discussion

There is a plethora of literature on DEM with computer simulation models. Utilization of available satellite data of different spatial resolutions is important regarding the behaviour of hydrologic characteristics for agricultural development and simulation models is important in this regard. We have utilized GIS with the IDW method to prepare DEM and we can find high-resolution DEM can be prepared with high-resolution satellite data and closed-spaced spot height information. Some use satellite-based Soil and Water Assessment Tool (SWAT) for agricultural development. SWAT is thought to be an effective and frequently used tool for implementing watershed simulation models. Al-Khafaji et al. (2020) investigated high-resolution DEM for agricultural development through hydrologic simulation models.

The prepared DEM and relevant literature from different countries show the accuracy of DEM and its effectiveness in agricultural development (Zhou et al. 2016; Al-Khafaji et al. 2020; Bai et al. 2015; Nobre et al. 2011). Chaubey et al. (2005) studied the Moore's Creek watershed in the USA using the SWAT model considering seven different DEM resolution scenarios (i.e., 30, 100, 150, 200, 300, 500, and 1000 m) and they found that spatial resolutions of DEM between 100 and 200 m cause less error (<10%) in streamflow. The accuracy of computed stream flow using the SWAT model depends on high-resolution DEM that could produce high stream flow results (Dixon 2009). However, Lin et al. (2013) examined DEM of SRTM 90 m using the SWAT and it was more accurate than the ASTER 30 m. Chaplot (2005) analysed DEM at spatial resolutions ranging between 20 and 500 m and indicated that the high spatial resolution of DEM has a high impact on stream flow. However, Zhang et al. (2014) considered the DEM resolution essentially does not affect the stream flow in analysing DEM with a spatial resolution of 30–1000 m.

It is noted that all these studies did not provide a conclusive response regarding the spatial resolution of DEM that provides accurate watershed modelling simulation results using SWAT. In addition, there is no interactive impact of land cover and DEM on computed stream flow (Al-Khafaji et al. 2020). Finally, a relationship between watershed characteristics and spatial resolution input data is important to understand watershed modelling. However, SWAT is an effective tool that is being used worldwide to describe different hydrologic components such as sediment, surface runoff, and pollution as well as the impact of climate change on watershed hydrology.

5.6 Conclusion

A number of methodological concerns (e.g., collection of hard copy BWDB map data, Google Image data, field investigation with GPS and levelling device, and application of GIS) were deployed for preparing the DEM for the study site. The prepared DEM shows the different physiographic unit and their properties based on land elevation as well as areas with irrigation suitability.

Suitable land for irrigation covers a very gentle slope and there is a remarkably high opportunity to increase the suitable land area by developing new canals from the nearby perennial river. Therefore, the prepared DEM can be utilized for future irrigation and agricultural development planning. Using the sophisticated analytical options with GIS methods, detected and delineated different terrain features can be utilized for potential irrigation.

The suitability map was classified based on only their land-level (elevation) and slope gradient. It has been mentioned that soil depth, soil drainage, soil infiltration rate, available water holding capacity (AWHC), etc. can be utilized for irrigation suitable areas and that can be achievable with a detailed investigation. Therefore, it is better in future to consider these factors for irrigation suitability. In addition, airborne LiDAR can be used for high accurate DEM. Besides its high resolution (1×1 m) and precision (<20 cm accuracy) in the altitude, LiDAR data offers to extract valuable topographic data where the terrain is covered with vegetation and buildings (Val et al. 2015).

This study indicates that there is enough water to expand irrigable areas and improve the livelihoods of the local communities through implementing an enhanced irrigation system and improving agricultural crop productivity. However, improvement in irrigation efficiency depends on the choice of irrigation type that plays a critical role in the water resource development and management of surface water (Shirazy et al. 2017). Therefore, to enhance the efficiency of irrigation water, it is recommended to develop small canals within arable land and adopt sprinkler/drip irrigation methods. This is expected to increase the irrigable command areas located on slopes by less than 5%. This study further noticed that the identified surface irrigation potential can assist in the decision-making process and implementation of irrigation development during the dry season.

Acknowledgements The authors would like to express our sincere thanks to Mr John D Prytherch (Team Leader) and Mr Mofazzal Ahmed (Deputy Team Leader) of the Irrigation Management Improvement Project (IMIP) of Mott MacDonald Ltd., Dhaka, Bangladesh for their kind support for relevant information.

Conflict of Interest The authors declare no conflict of interest.

References

Acharjee TK, Hellegers P, Ludwig F, van Halsema G, Mojid MA, van Scheltinga CT (2020) Prioritization of adaptation measures for improved agricultural water management in Northwest Bangladesh. Climatic Change 163(2):431–450. https://doi.org/10.1007/s10584-020-02852-w

Acharjeea TK, Ludwiga F, van Halsemab G, Hellegersb P, Supit I (2017) Future changes in water requirements of Boro rice in the face of climate change in North-West Bangladesh. Agric Water Manag 194:172–183. https://doi.org/10.1016/j.agwat.2017.09.008

Afshar A, Takbiri Z (2012) Fusegates selection and operation: simulation-optimization approach. J Hydroinformatics 14(2):464–477. https://doi.org/10.2166/hydro.2011.154

Akbari M, Mamanpoush A, Gieske A, Miranzadeh M, Torabi M, Salemi HR (2006) Crop and land cover classification in Iran using Landsat 7 imagery. Int J Remote Sens 27:4117–4135. https://doi.org/10.1080/01431160600784192

Akkuzu E, Unal HB, Karatas BS (2006) Determination of water conveyance loss in the Menemen open canal irrigation network. Turk J Agric for 31(1):11–22

Al-Khafaji M, Saeed FH, Al-Ansari N (2020) The interactive impact of land cover and DEM resolution on the accuracy of computed streamflow using the SWAT model. Water Air Soil Pollut 231:416. https://doi.org/10.1007/s11270-020-04770-0

Anwar AA, Ahmad W, Bhatti MT, Haq ZU (2016) The potential of precision surface irrigation in the Indus basin irrigation system. Irrig Sci 34:379–396. https://doi.org/10.1007/s00271-016-0509-5

Araya T, Nyssen J, Govaerts B, Deckers J, Cornelis WM (2015) Impacts of conservation agriculture-based farming systems on optimizing seasonal rainfall partitioning and productivity on vertisols in the Ethiopian drylands. Soil Tillage Res 148:1–13. https://doi.org/10.1016/j.still.2014.11.009

Ashraf M, Loftis JC, Hubbard KG (1997) Application of geostatistics to evaluate partial weather station networks. Agric for Meteorol 84(3–4):255–271. https://doi.org/10.1016/S0168-1923(96)02358-1

Atkins PJ, Hassan MM, Dunn CE (2006) Toxic torts: arsenic poisoning in Bangladesh and the legal geographies of responsibility. Trans Inst Br Geographers 31(3):272–285. https://doi.org/10.1111/j.1475-5661.2006.00209.x

Atkins PJ, Hassan MM, Dunn CE (2007) Environmental irony: summoning death in Bangladesh. Environ Plann A 39(11):2699–2714. https://doi.org/10.1068/a38123

Awasthi G, Singh T, Awasthi A, Awasthi KK (2020) Arsenic in mushrooms, fish, and animal products. In: Srivastava S (ed) Arsenic in drinking water and food. Springer Nature, Singapore, pp 307–323. https://doi.org/10.1007/978-981-13-8587-2

Baghdadi N, Zribi M (2017) Land surface remote sensing in continental hydrology. ISTE-Elsevier Press, Oxford. https://doi.org/10.1016/C2015-0-01226-6

Bai R, Li T, Huang Y, Li J, Wang G (2015) An efficient and comprehensive method for drainage network extraction from DEM with billions of pixels using a size-balanced binary search tree. Geomorphology 238:56–67. https://doi.org/10.1016/j.geomorph.2015.02.028

Bashir N, Saeed R, Afzaal M, Ahmad A, Muhammad N, Iqbal J, Khan A, Maqbool Y, Hameed S (2020) Water quality assessment of lower Jhelum canal in Pakistan by using geographic information system (GIS). Groundwater Sustain Dev 10:100357. https://doi.org/10.1016/j.gsd.2020.100357

BBS (2015) Statistical yearbook of Bangladesh. Statistics Division, Ministry of Planning, Government of the People's Republic of Bangladesh, Dhaka

Beltran CM, Belmonte AC (2001) Irrigated crop area estimation using Landsat TM imagery in La Mancha. Spain. Photogram Eng Remote Sens 67(10):1177–1184

Berglund EZ, Pesantez JE, Rasekh A, Shafiee E, Sela L, Haxton T (2020) Review of modeling methodologies for managing water distribution security. J Water Res Plann Manag 146(8):03120001. https://doi.org/10.1061/(ASCE)WR.1943-5452.0001265

Bianucci E, Peralta JM, Furlan A, Hernandez LE, Castro S (2020) Arsenic in wheat, maize, and other crops. In: Srivastava S (ed) Arsenic in drinking water and food. Springer Nature, Singapore, pp 279–306. https://doi.org/10.1007/978-981-13-8587-2

Biggs TW, Thenkabail PS, Gumma MK, Scott CA, Parthasaradhi GR, Turral HN (2006) Irrigated area mapping in heterogeneous landscapes with MODIS time series, ground truth and census data, Krishna Basin, India. Int J Remote Sens 27:4245–4266. https://doi.org/10.1080/01431160600851801

Biswas MR, Mandal MAS (1993) Irrigation management for crop diversification in Bangladesh. University Press Limited, Dhaka

Boken VK, Hoogenboom G, Kogan FN, Hook JE, Thomas DL, Harrison KA (2004) Potential of using NOAA-AVHRR data for estimating irrigated area to help solve an inter-state water dispute. Int J Remote Sens 25(12):2277–2286. https://doi.org/10.1080/01431160310001618077

Boulton SJ, Stokes M (2018) Which DEM is best for analyzing fluvial landscape development in mountainous terrains? Geomorphology 310:168–187. https://doi.org/10.1016/j.geomorph.2018.03.002

BRRI (2007) Annual report 2006–07. Bangladesh Rice Research Institute, Dhaka

Burrough PA, McDonnell RA (1998) Principles of geographical information systems. Oxford University Press, New York

Chaplot V (2005) Impact of DEM mesh size and soil map scale on SWAT runoff, sediment, and NO3-N loads predictions. J Hydrol 312:207–222. https://doi.org/10.1016/j.jhydrol.2005.02.017

Chaubey I, Cotter AS, Costello TA, Soerens TS (2005) Effect of DEM data resolution on SWAT output uncertainty. Hydrol Process 19:621–628. https://doi.org/10.1002/hyp.5607

Ciminelli V, Gasparon M, Ng JC, Silva GC, Caldeira C (2017) Dietary arsenic exposure in Brazil: the contribution of rice and beans. Chemosphere 168:996–1003. https://doi.org/10.1016/j.chemosphere.2016.10.111

Cinnirella S, Buttafuoco G, Pirrone N (2005) Stochastic analysis to assess the spatial distribution of groundwater nitrate concentrations in the Po catchment (Italy). Environ Pollut 133(3):569–580. https://doi.org/10.1016/j.envpol.2004.06.020

Croneborg L, Saito K, Matera M, McKeown D, van Aardt J (2015) Digital elevation models. International Bank for Reconstruction and Development, Washington DC
Dawit M, Olika BD, Muluneh FB, Leta OT, Dinka MO (2020) Assessment of surface irrigation potential of the Dhidhessa River basin. Ethiopia. Hydrology 7(3):68. https://doi.org/10.3390/hydrology7030068
Dehghani MH, Changani F (2020) Asghari FB (2020) Application of irrigation and drinking water quality indices to monitor ground water quality using geographic information system: a case study of the basins around Urmia Lake. Iran. Int J Environ Anal Chem 10(1080/03067319):1830982
del Val M, Iriarte E, Arriolabengoa M, Aranburu A (2015) An automated method to extract fluvial terraces from LIDAR based high resolution digital elevation models: the Oiartzun valley, a case study in the Cantabrian Margin. Quat Int 364:35–43. https://doi.org/10.1016/j.quaint.2014.10.030
Dixon B (2009) Resample or not?! Effects of resolution of DEMs in watershed modelling. Hydrol Process 23:1714–1724. https://doi.org/10.1002/hyp.7306
Duran-Encalada JA, Paucar-Caceres A, Bandala E, Wright G (2017) The impact of global climate change on water quantity and quality: a system dynamics approach to the US-Mexican transborder region. Eur J Oper Res 256(2):567–581. https://doi.org/10.1016/j.ejor.2016.06.016
Elias IE, Al-Ansari NA, Knutsson S (2015) Area-storage capacity curves for Mosul Dam, Iraq using empirical and semi-empirical approaches. In: Conference paper, Stavanger, Norway, 15–16 June 2015. https://www.ich.no/Opplastet/Dokumenter/Hydropower15/Issa_sweden.pdf
FAO (2014) Report on in-depth capacity assessment of Bangladesh to produce agricultural and rural statistics. Food and Agriculture Organization Representation in Bangladesh, Dhaka
Fenton GA, McLean A, Nadim F, Griffiths DV (2013) Landslide hazard assessment using digital elevation models. Can Geotech J 50(6). https://doi.org/10.1139/cgj-2011-0342
Flora SJS (2015) Arsenic: chemistry, occurrence, and exposure. In: Flora SJS (ed) Handbook of arsenic toxicity. Elsevier and Academic Press, Amsterdam, pp 1–49. https://doi.org/10.1016/B978-0-12-418688-0.00001-0
Gabr ME, Soussa H, Fattouh E (2020) Groundwater quality evaluation for drinking and irrigation uses in Dayrout city Upper Egypt. Ain Shams Eng J. https://doi.org/10.1016/j.asej.2020.05.01
Gallardo M, Elia A, Thompson RB (2020) Decision support systems and models for aiding irrigation and nutrient management of vegetable crops. Agric Water Manag 240:106209. https://doi.org/10.1016/j.agwat.2020.106209
Gao Q, Zribi M, Escorihuela MJ, Baghdadi N, Segui PQ (2018) Irrigation mapping using Sentinel-1 time series at field scale. Remote Sens 10(9):1495. https://doi.org/10.3390/rs10091495
Geethalakshmi V, Ramesh T, Palamuthirsolai A, Lakshmanan A (2011) Agronomic evaluation of rice cultivation system for water and grain productivity. Arch Agron Soil Sci 57(2):159–166. https://doi.org/10.1080/03650340903286422
Gillies MH, Smith RJ (2015) SISCO: surface irrigation simulation, calibration and optimisation. Irrig Sci 33:339–355. https://doi.org/10.1007/s00271-015-0470-8
Githui F, Hussain A, Morris M (2020) Incorporating infiltration in the two-dimensional ANUGA model for surface irrigation simulation. Irrig Sci 38:373–387. https://doi.org/10.1007/s00271-020-00679-y
GOB (2001) National land use policy 2001. Planning Commission, Government of the People's Republic of Bangladesh, Dhaka
GOB (2012) Perspective plan of Bangladesh 2010–2021: making vision 2021 reality. Planning Commission, Government of the People's Republic of Bangladesh, Dhaka
Gowing J, Walker D, Parkin G, Forsythe N, Haile AT, Ayenew DA (2020) Can shallow groundwater sustain small-scale irrigated agriculture in sub-Saharan Africa? Evidence from N-W Ethiopia. Groundwater Sustain Dev 10:100290. https://doi.org/10.1016/j.gsd.2019.100290
Grogan DS, Zhang F, Prusevich A, Lammers RB, Wisser D, Glidden S, Li C, Frolking S (2015) Quantifying the link between crop production and mined groundwater irrigation in China. Sci Total Environ 511:161–175. https://doi.org/10.1016/j.scitotenv.2014.11.076
Hassan MM (2018) Arsenic in groundwater: poisoning and risk assessment. CRC Press, Boca Raton. https://doi.org/10.1201/9781315117034

Hassan MM, Atkins PJ (2007) Arsenic risk mapping in Bangladesh: a simulation technique of Cokriging estimation from regional count data. J Environ Sci Health Part A 42(12):1719–1728. https://doi.org/10.1080/10934520701564210
Hassan MM, Atkins PJ (2011) Application of geostatistics with indicator kriging for analyzing spatial variability of groundwater arsenic concentrations in Southwest Bangladesh. J Environ Sci Health Part A 46(11):1185–1196. https://doi.org/10.1080/10934529.2011.598771
Hassan MM, Atkins PJ, Dunn CE (2005) Social implications of arsenic poisoning in Bangladesh. Soc Sci Med 61(10):2201–2211. https://doi.org/10.1016/j.socscimed.2005.04.021
Hassan MM, Shaha A, Ahamed R (2021) Water scarcity in coastal Bangladesh: search for arsenic-safe aquifer with geostatistics. In: Jana NC, Singh RB (eds) Climate, environment and disaster in developing countries. Springer Nature, Singapore (accepted)
Hossain MB, Roy D, Paul PLC, Islam MT (2016) Water productivity improvement using water saving technologies in Boro rice cultivation. Bangladesh Rice J 20(1):17–22. https://doi.org/10.3329/brj.v20i1.30625
Hosseinzadeh SR (2011) Drainage network analysis, comparis of digital elevation model (DEM) from ASTER with high resolution satellite image and areal photographs. Int J Environ Sci Dev 2(3):194–198. https://doi.org/10.7763/IJESD.2011.V2.123
Huang M, Gallichand J, Wang Z, Goulet M (2006) A modification to the soil conservation service curve number method for steep slopes in the Loess Plateau of China. Hydrol Process 20(3):579–589. https://doi.org/10.1002/hyp.5925
Imran RM, Rehman A (2019) Delineation of drainage network and estimation of total discharge using digital elevation model (DEM). Int J Innovations Sci Technol 1(2):50–61. https://doi.org/10.33411/IJIST/2019010201
Isaaks EH, Srivastava RM (1989) An introduction to applied geostatistics. Oxford University Press, New York
Jin N, Tao B, Ren W, Feng M, Sun R, He L, Zhuang W, Yu Q (2016) Mapping irrigated and rainfed wheat areas using multi-temporal satellite data. Remote Sens 8(3):207. https://doi.org/10.3390/rs8030207
Jurík L, Húska D, Halászová K, Bandlerová A (2015) Small water reservoirs—sources of water or problems? J Ecol Eng 16(4):22–28. https://doi.org/10.12911/22998993/59343
Kamthonkiat D, Honda K, Turral H, Tripathi NK, Wuwongse V (2005) Discrimination of irrigated and rainfed rice in a tropical agricultural system using spot vegetation NDVI and rainfall data. Int J Remote Sens 26(12):2527–2547. https://doi.org/10.1080/01431160500104335
Keshta E, Gad MA, Amin D (2019) A long-term response-based rainfall-runoff hydrologic model: case study of The Upper Blue Nile. Hydrology 6(3):69. https://doi.org/10.3390/hydrology6030069
Kharrou M, Page ML, Chehbouni A, Simonneaux V, Er-Raki S, Jarlan L, Ouzine L, Khabba S, Chehbouni G (2013) Assessment of equity and adequacy of water delivery in irrigation systems using remote sensing-based indicators in semi-arid region, Morocco. Water Resour Manag 27:4697–4714. https://doi.org/10.1007/s11269-013-0438-5
Leitão JP, Prodanović D, Maksimović Č (2016) Improving merge methods for grid-based digital elevation models. Comput Geosci 88:115–131. https://doi.org/10.1016/j.cageo.2016.01.001
Lemann T, Roth V, Zeleke G, Subhatu A, Kassawmar T, Hurni H (2019) Spatial and temporal variability in hydrological responses of the Upper Blue Nile basin, Ethiopia. Water 11(1):21. https://doi.org/10.3390/w11010021
Lin S, Jing C, Coles NA, Chaplot V, Moore NJ, Wu J (2013) Evaluating DEM source and resolution uncertainties in the soil and water assessment tool. Stoch Environ Res Risk Assess 27(1):209–221. https://doi.org/10.1007/s00477-012-0577-x
Lohani S, Baffaut C, Thompson AL, Aryal N, Bingner RL, Bjorneberg DL, Bosch DD, Bryant RB, Buda A, Dabney SM, Davis AR, Duriancik LF, James DE, King KW, Kleinman PJA, Locke M, McCarty GW, Pease LA, Reba ML, Smith DR, Tomer MD, Veith TL, Williams MR, Yasarer LMW (2020) Performance of the soil vulnerability index with respect to slope, digital elevation

model resolution, and hydrologic soil group. J Soil Water Conserv 75(1):12–27. https://doi.org/10.2489/jswc.75.1.12
Manaswi C, Thawait A (2014) Application of soil and water assessment tool for runoff modeling of Karam river basin in Madhya Pradesh. Int J Eng Sci Technol 3(5):529–532
Marmiroli M (2020) A brief status report on arsenic in edible vegetable species. In: Srivastava S (ed) Arsenic in drinking water and food. Springer Nature, Singapore, pp 325–331. https://doi.org/10.1007/978-981-13-8587-2
Mihajlov I, Mozumder MRH, Bostick BC, Stute M, Mailloux BJ, Knappett PSK, Choudhury I, Ahmed KM, Schlosser P, van Geen A (2020) Arsenic contamination of Bangladesh aquifers exacerbated by clay layers. Nat Commun 11:2244. https://doi.org/10.1038/s41467-020-16104-z
Morvan AL, Zribi M, Baghdadi N, Chanzy A (2008) Soil moisture profile effect on radar signal measurement. Sensors 8(1):256–270. https://doi.org/10.3390/s8010256
Nobre AD, Cuartas LA, Hodnett M, Rennó CD, Rodrigues G, Silveira A, Waterloo M, Saleska S (2011) Height above the nearest drainage—a hydrologically relevant new terrain model. J Hydrol 404(1–2):13–29. https://doi.org/10.1016/j.jhydrol.2011.03.051
Obeysekera J, Irizarry M, Park J, Barnes J, Dessalegne T (2011) Climate change and its implications for water resources management in south Florida. Stoch Environ Res Risk Assess 25:495–516. https://doi.org/10.1007/s00477-010-0418-8
Oh H-J, Kadavi PR, Lee C-W, Lee S (2018) Evaluation of landslide susceptibility mapping by evidential belief function, logistic regression and support vector machine models. Geomatics Nat Hazards Risk 9(1):1053–1070. https://doi.org/10.1080/19475705.2018.1481147
Oliver MA (1996) Geostatistics, rare disease and the environment. In: Spatial analytical perspectives on GIS. Taylor & Francis, London
Rahman MS, Saha N, Islam ARMT, Shen S, Bodrud-Doza M (2017) Evaluation of water quality for sustainable agriculture in Bangladesh. Water Air Soil Pollut 228:385. https://doi.org/10.1007/s11270-017-3543-x
Rasekh A, Brumbelow K (2014) A dynamic simulation-optimization model for adaptive management of urban water distribution system contamination threats. Appl Soft Comput 32:59–71. https://doi.org/10.1016/j.asoc.2015.03.021
Sattar MA (2004) Irrigation principles and on-farm water management. Nandita Prokash, Banglabazar, Dhaka
Sattar MA, Hasan N, Roy D (2009) Annual progress report on dissemination of water management technologies in the farmer's field for increasing water productivity in rice based crop cultivation. BARC-BRRI, Dhaka
Sayl KN, Muhammad NS, El-Shafie A (2017) Optimization of area-volume-elevation curve using GIS-SRTM method for rainwater harvesting in arid areas. Environ Earth Sci 76:368. https://doi.org/10.1007/s12665-017-6699-1
Shao Y, Fan X, Liu H, Xiao J, Ross S, Brisco B, Brown R, Staples G (2001) Rice monitoring and production estimation using multitemporal RADARSAT. Remote Sens Environ 76(3):310–325. https://doi.org/10.1016/S0034-4257(00)00212-1
Shirazy BJ, Islam ABMJ, Dewan MMR, Shahidullah SM (2017) Crops and cropping systems in Dinajpur region. Bangladesh Rice J 21(2):143–156
Thenkabail PS, Schull M, Turral H (2004) Ganges and Indus river basin land use/land cover (LULC) and irrigated area mapping using continuous streams of MODIS data. Remote Sens Environ 95(3):317–341. https://doi.org/10.1016/j.rse.2004.12.018
Vergine P, Salerno C, Libutti A, Beneduce L, Gatta G, Berardi G, Pollice A (2017) Closing the water cycle in the agro-industrial sector by reusing treated wastewater for irrigation. J Cleaner Prod 164:587–596. https://doi.org/10.1016/j.jclepro.2017.06.239
Wilson JP, Gallant JC (2000) Digital terrain analysis. In: Wilson JP, Gallant JC (eds) Terrain analysis: principles and applications. Wiley, New York, pp 29–50
Wu T, Li J, Li T, Sivakumar B, Zhang G, Wang G (2019) High-efficient extraction of drainage networks from digital elevation models constrained by enhanced flow enforcement from known river maps. Geomorphology 340:184–201. https://doi.org/10.1016/j.geomorph.2019.04.022

Xiao X, Boles S, Liu J, Zhuang D, Frolking S, Li C, Salas W, Moore B III (2005) Mapping paddy rice agriculture in southern China using multi-temporal MODIS images. Remote Sens Environ 95(4):480–492. https://doi.org/10.1016/j.rse.2004.12.009

Zhang P, Liu R, Bao Y, Wang J, Yu W, Shen Z (2014) Uncertainty of SWAT model at different DEM resolutions in a large mountainous watershed. Water Res 53:132–144. https://doi.org/10.1016/j.watres.2014.01.018

Zhou G, Sun Z, Fu S (2016) An efficient variant of the priority-flood algorithm for filling depressions in raster digital elevation models. Comput Geosci 90:87–96. https://doi.org/10.1016/j.cageo.2016.02.021

Zribi M, Chahbi A, Shabou M, Lili-Chabaane Z, Duchemin B, Baghdadi N, Amri R, Chehbouni A (2011) Soil surface moisture estimation over a semi-arid region using ENVISAT ASAR radar data for soil evaporation evaluation. Hydrol Earth Syst Sci 15:345–358. https://doi.org/10.5194/hess-15-345-2011

Chapter 6
Smallholder Tea Farming in West Bengal, India: An Exploratory Insight

Chinmoyee Mallik

Abstract India, like most of the developing countries, is dominated by smallholder farmers. While these small farms were typically of subsistence type, recently a considerable proportion of them have massively shifted in favour of cash crops in many parts of the country. This is very intriguing because escalating economic vulnerability of the small farmers due to erosion of state support from the farm sector in the neo-liberal policy context and concomitant monetization of smallholder economy are self-contradictory. Although South-East Asian countries have already experienced such a phenomenon few decades back with respect to the rubber production, the Indian tea production, particularly in Assam and West Bengal, is following a similar trajectory. This paper is mainly based on Agricultural Census of India, National Sample Survey unit level data and an exploratory field work in the tea producing district of Jalpaiguri in West Bengal undertaken in 2019. The fieldwork consists of an exploratory quantitative survey as well as in-depth interviews of few small tea growers to understand the recent trends and patterns of restructuring of the pre-existing agricultural system in the region. This paper seeks to draw insights from cropping pattern shift away from food crop towards cash crops and the socio-political and economic environment associated with this recent phenomenon. It emerges that the small farmers have shifted cropping away from food crops to cash crops and that the small farmers who have adopted tea farming have mostly replaced paddy cultivation.

Keywords Commercial agriculture · Food security · Cropping pattern shift · Small tea growers

6.1 Introduction

The recent Agricultural Census of India 2015–16 reveals that 86% of all farm holdings are of marginal category although area under these holdings is 47% of all operated

C. Mallik (✉)
Department of Rural Studies, West Bengal State University, Barasat, Kolkata 700126, India

N. C. Jana et al. (eds.), *Livelihood Enhancement Through Agriculture, Tourism and Health*, Advances in Geographical and Environmental Sciences,
https://doi.org/10.1007/978-981-16-7310-8_6

area. Traditionally smallholder farming has been looked upon as subsistence production units who are non-responsive to innovation and market. The Chayanovian thesis, while harps upon self-exploitation as means of survival in the context of adversities of the market economy, the Marxian constructions profess 'death of the peasantry' as a route to capitalist transition. Contrary to the traditional theoretical positions taken, the recent studies have indicated that smallholder agriculture in India is not only predominating but also gaining ground all the more (Mahendra Dev 2012). Their contribution to food security, poverty reduction as well as agricultural diversification through high-value crop production and cash cropping has been steadily gaining ground (Mahendra Dev 2012). Although the emergence of supermarkets and retail trade in the farm products offer opportunities to the small farmers in general, in the Asian context, the cultural inhibitions and policy drawbacks dampen the prospects considerably (Reardon et al. 2012). Notwithstanding the challenges, smallholder farming has gained strong foothold not only in the high-value cash crop production, but also in some of the plantation crops like tea, coffee, rubber, oil palm in different parts of the Global South (Bissonnette and Koninck 2017). Most of tropical Asia and particularly Southeast Asia a transition from plantations to smallholdings for an important number of cash crops is well under way by organized smallholder often under contract farming agreement.

The origin of the small growers draws its lease of life from the crisis of production and management of the large plantations. China, Kenya, Indonesia and Sri Lanka are forerunners in this aspect as more than seventy percent of the plantation crop output of the respective countries originates from the small grower farms. It was in fact a well-strategized scheme to encourage the growth of small farms in and around the peripheries of the nucleus plantation and attach the former with the latter functionally. Commonly the smallholder would depend upon the nucleus plantation for technical and marketing support while labour management would be negotiable (Hannan 2006). Several countries have institutions like the Rubber Industry Smallholders Development Authority (RISDA) in Malaysia and the Kenya Tea Development Authority (KTDA) in Kenya to organize the process. However, there are several instances where contradictory outcomes have been noted. In Indonesia, Li (2011) notes that smallholder contract schemes associated with the attached large plantations have drawn the land out of the subsistence economy and within the capitalist circuit while the labour that was attached to that land is rendered redundant. It also underlines how, in the long run, attaching a small farm with the enclave-like (Hall et al. 2017) plantation, loses the labour perspective and risks the independence and autonomy of the smallholders. A contradictory scenario is observed in Africa by Hall et al. (2017) where commercial farming areas and contract farming produced considerable local economic linkages and the plantations/estates created more jobs, although of low quality and mostly casual. Hence, the coexistence of the small farms alongside large plantations deliver diverse landscapes depending upon the local and pre-existing linkages as well as the national policies that govern the relative importance of the two sectors (Bissonnette and Koninck 2017).

Following the serious crises in the tea plantations in India and their reluctance to increase the area under tea, the small plantations and the small tea growers have been promoted to address the demands of the market (Bhowmik 1991; Kumar et al. 2008). The Tea Board has recognized the small tea growers as a vital component of the Indian tea industry during the 8th Plan Period. It is as recent as the 12th Plan that the small tea growers in India are incorporated in the policies to address the vulnerabilities of tea producers. It consists of the land holdings up to 10.12-ha land. The 59th Annual Report (2012–13) further brings forth some of the initiatives pledged for the small tea growers in general and the North Bengal region in particular (Roy 2011). Two types of small tea grower models are prevailing: (a) a small grower located along the periphery of a large estate and functionally linked with the nuclear plantation, (b) independent small grower who sells his green leaf independently to the middle men that subsequently is sold to the bought leaf factories (BLFs).

A recent study shows that more than 8000 small and marginal farmers have shifted in favour of tea cultivation since 2005–06 (Mallik 2016). It is commonly seen as a means of self-employment (Hazarika and Borah 2013). The unemployed youth of Assam started participating in the small-scale tea cultivation as a means to engage with meaningful economic activity (Roy 2011). Although the small growers are believed to have tremendous potential to 'prop-up' tea production (Bhowmik 1991), this emerging sector has several bottlenecks (Hannan, 2013; 2016; Hazarika and Borah 2013) related to the price of green leaf, land ownership problem, paucity of capital for investment and marketing. Borah and Das (2015) notes how most of the tea growing countries like China, Kenya and Sri Lanka presently depend upon the small growers for more than 70% of the output, whereas India offers the opposite scenario. Within India, Tamil Nadu and Assam have progressed considerably in promoting the small growers while in North Bengal it is still depressed.

This paper seeks to undertake a cursory look into the emergence of the small tea growers in India with special focus on the Bengal scenario. This paper attempts to contribute to the emerging questions of agricultural transformation in favour of cash cropping systems among the smaller land size categories that raise critical questions about food security and vulnerability on one hand and also seeks to deconstruct the contradiction of withdrawal of state support from the agricultural sector and recent policies to encourage cash cropping among the smaller peasants. This phenomenon seems to be very critical because given the pre-existing vulnerabilities of the smaller peasants in terms of the rising costs of cultivation along with systematic roll-back of public investment and support, indebtedness and marketing channel challenges the inclination towards commercialization clearly may make the households more insecure. This paper seeks to explore the broad issues of such cropping pattern change using mainly secondary data and a very brief fieldwork to explore the issues.

6.2 Database and Methodology

This paper uses two secondary data sources, namely unit level NSS 70th Round Situation Assessment of Farmers and Agricultural Census and a very brief field visit undertaken in 2019. Data on the small tea growers (STGs) is not readily available.

Using the crop code for tea (1501), visit 1 and visit 2 have been combined to arrive at the estimates from the NSS 70th Round. Appropriate weights have been applied to obtain the state-level aggregates. Farm size-wise cropping data, available with Agricultural Census, is used to understand the land size-wise dynamics. As the Tea Board considers less than 10.12-ha land size as STG, the Agricultural census data has been aggregated for marginal (less than 1 ha), small (1.0–2.0 ha), semi-medium (2.0–4.0 ha) and medium (4.0–10.0 ha) size classes to arrive at the STG size class. North Bengal is considered by aggregating data for the four tea producing districts of Coochbehar, Darjeeling, Dinajpur North and Jalpaiguri.

A very brief exploratory field visit was undertaken in 2019 in Jalpaiguri District around the Lataguri region in Jharmatiali Mouza located in the Mal Sub-division to gain exploratory insights in response to the issues raised by the secondary data insights. The region has seen a remarkable rise in the number of small tea growers in the recent times. The fieldwork consisted of household survey of about 33 small tea growers and a self-help group run tea factory. The samples were selected using snow balling method.

6.3 Analysis

6.3.1 All India Estimates from National Sample Survey 70th Round (2013)

Estimated from the NSS 70th Round Situation Assessment of Farm Households the total number of small tea growers figure 187,866 with Assam accounting for more than 80% followed by Tamil Nadu (17%) and West Bengal (2%) (Fig. 6.1). The small tea growers account for about 46% of total tea produced in India in 2017–18. While monthly per capita consumption expenditure (MPCE) of the STGs is the highest in Assam and lowest in West Bengal (Fig. 6.2), the productivity is highest in

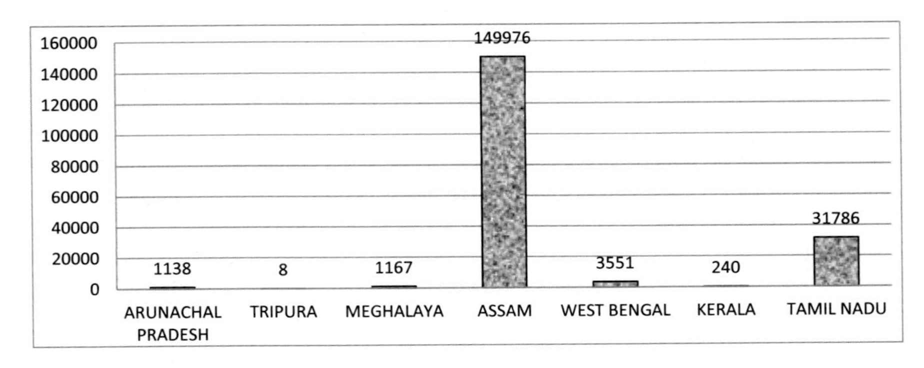

Fig. 6.1 Number of small tea growers, 2013

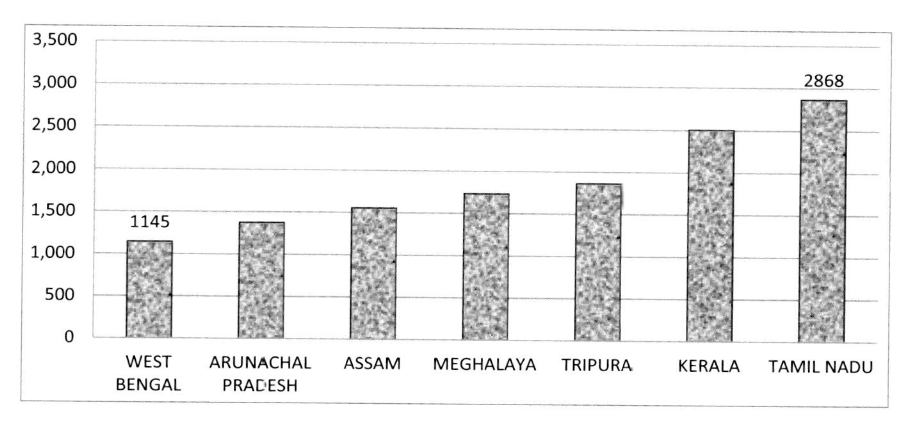

Fig. 6.2 Monthly per capita consumption expenditure of the small tea growers (Rs.)

Table 6.1 Productivity (kg/ha)

State	*N*	Mean
Assam	150,574	3520
West Bengal	3551	3322
Tamil Nadu	31,785	2326
Tripura	8	2227
Arunachal Pradesh	1138	1481
Kerala	240	892
Meghalaya	1168	757

Assam followed by West Bengal (Table 6.1). It is interesting to note that the small growers in West Bengal, inspite of registering lowest MPCE, have the second-highest productivity of tea.

Further, farm size disaggregation shows that tea acreage as well as productivity is highest among the marginal farmers within the STGs (Tables 6.2 and 6.3). It is indicative of intensified commercialization of the smallest land holdings. Given the current orientation of the agricultural policies that are adverse for the tiny land holdings, raising cash crops instead of subsistence agriculture is implicative and calls for deeper probe.

Table 6.2 ANOVA: productivity

Size class	*N*	Mean	Homogeneous subset for alpha = 0.05
Semi-medium	3827	1114	1
Small	95,791	2967	2
Marginal	88,246	3709	3

Table 6.3 Tea acreage

Size class	N	Mean	Homogeneous subset for alpha = 0.05
Marginal	88,247	74.7	3
Small	95,792	61.7	2
Semi-medium	3828	53.6	1
Total	187,867	67.6	

6.3.2 *Estimates from Agricultural Census: West Bengal (1995–96 to 2010–11)*

This section focuses upon the growth of tea farming in the small and marginal farms in West Bengal region based on Agricultural Census. Figure 6.3 shows that the number of small tea growers has increased steadily since 2000–01 clearly highlighting that cultivation of the plantation crop has emerged lucrative for the farm size less than 2 ha primarily. Excluding the tea plantations, area under tea in the STG size class has increased from zero hectares in 1995–96 Census to 15,315 ha in 2010–11 Census, this phenomenon being most prominent in the less than 2 ha category of farms (Fig. 6.4). This phenomenon raises two basic questions: (a) what has been the reason for this sudden spurt in the tea cultivation among the smaller land size classes, and (b) what is the broad implication of such cropping pattern transformation.

From Fig. 6.5, it is evident that area under food crops has declined considerably for the STG size classes taken together (see also Appendix). With decline in the acreage of paddy, which is the principal food crop, a nearly matching increase in acreage of tea may be noted. It may be indicative of reallocation of land from different food crop cultivation, specifically paddy, to tea acreage in the North Bengal region as a whole. However, within this region itself, a closer look reveals that Coochbehar does not conform to the regional trend where the smaller peasants have registered

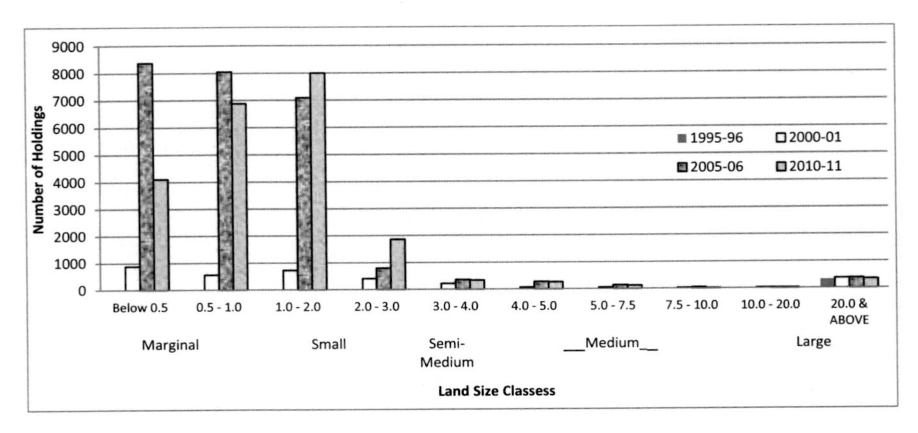

Fig. 6.3 Number of tea growers, West Bengal 1995–96 to 2010–11

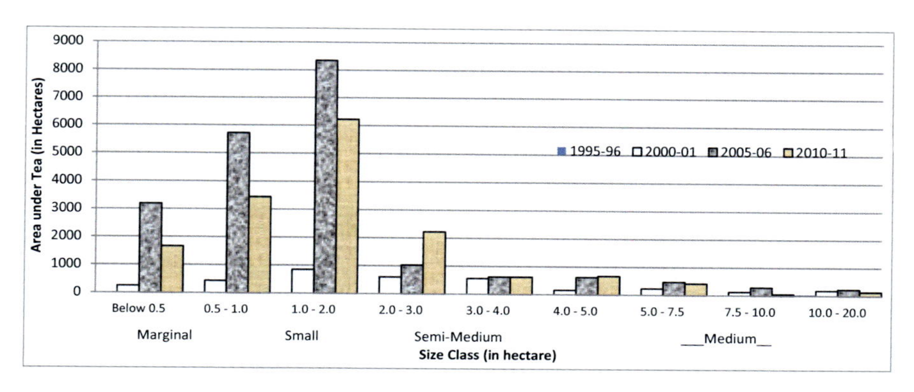

Fig. 6.4 Area Under Tea, West Bengal 1995–96 to 2010–11. *Note* The area under the largest size class, i.e., above 20 ha is not displayed in the figure to highlight the lower size classes. The corresponding areas are 177,126, 121,986, 118,812 and 126,270 ha respectively across the different years

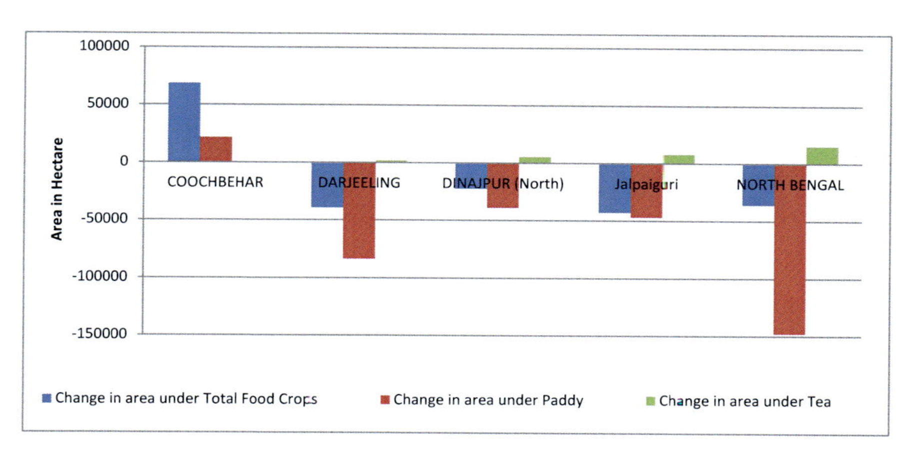

Fig. 6.5 Change in area 1995–96 to 2010–11: STG size class. *Source* Tables 6.7, 6.8 and 6.9 in Appendix

increased acreage of both total food crop and paddy along with some increase in tea acreage. This district being located in the geomorphologically plain segment of the North Bengal Districts, tea has not emerged outwitting food crops and that it has simply been another cash crop. But, the hilly districts of Darjeeling, rolling foothills of Jalpaiguri and Dinajpur (North), seminal decline in both food crop and paddy acreage and gains in tea acreage is notable.

The larger land size categories reveal a slightly different scenario (Fig. 6.6). For all the North Bengal Districts taken together, there is a net increase in the area under total food crop although acreage of paddy has reduced sharply by 200,855 hectare between 1995–96 and 2010–11 in the land holding above 10.0 ha size class. More

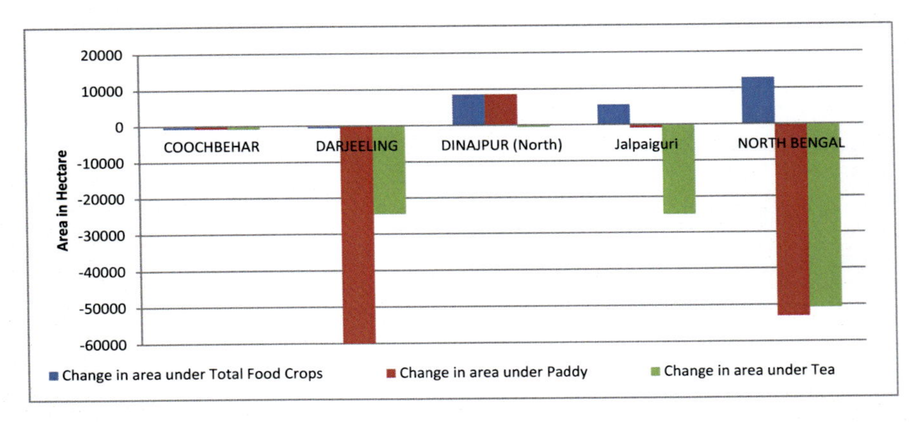

Fig. 6.6 Change in area 1995–96 to 2010–11: large size class. *Source* Tables 6.7, 6.8 and 6.9 in Appendix

Table 6.4 Acreage of tea

	N	% of households
Less than 50% of operational holding	8	24.2
50–99% of operational holding	12	36.4
100% of operational holding	13	39.4
Total	33	100.0

Source Field Work (2019)

critical is the reduction in tea acreage in this category of land holding which typically includes the tea estates that are so very vital to the national and global tea industry. Except for North Dinajpur where there is gain in food crop as well as paddy acreage, the rest of the three North Bengal districts have seen outflow from food cropping in addition to loss of tea acreage in the larger land sizes.

At the national scale where there are clear indications of outflows from net sown area, it is possible that North Bengal may also be experiencing a similar situation. Critical is the fact that outflows are being taking place from food crop production acreage (Tables 6.7, 6.8 and 6.9). This is all the more crucial for the smaller land size classes, specifically the less than 10 ha category (STG size class) as this phenomenon is taking place along with parallel trend of increasing cash cropping (Table 6.4).

6.3.3 *Insights from Fieldwork: Jalpaiguri in West Bengal (2019)*

The exploratory fieldwork in Jalpaiguri confirmed the trend that was observed at the secondary data analysis level. Among the 33 STGs interviewed, two-third of

Table 6.5 Crops that have been replaced by Tea

	N	% of households
Brinjal	1	3.0
Jute	1	3.0
Paddy	31	93.9
Total	33	100.0

Source Field Work (2019)

Table 6.6 Reason for cropping pattern shift in favour Tea cultivation

	N	% of households
Irrigation related issues	17	51.5
Low profitability of previous cropping system	10	30.3
Influenced by others	2	6.1
Elephant invasion	3	9.1
Labour problem	1	3.0
Total	33	100.0

Source Field Work (2019)

them had allocated nearly 100% of land to tea in their farms during the last decade. They further clarified that out of the surveyed farmers, 32 of them had replaced food crop chiefly paddy with tea (Table 6.5). Of the reasons cited by them (Table 6.6), most seminal was concerning problems with the paddy-based subsistence cropping system. That the foothills with porous soils raise the cost of paddy farming by increased irrigation costs and low profitability of the traditional crops which was incapable of sustaining livelihoods and the relatively assured and regular income streams from tea have been encouraging the cropping pattern shift. The respondents reported that in spite of problems of low green tea leaf prices received from the local agents and ill-developed marketing channels that are heavily dependent upon private agents, the exploitative tea leave prices received are still more remunerative and assured compared to the pre-existing paddy-based cropping systems. The informal discussions have also underlined the man-nature conflict in the ecologically distinctive region. Elephant invasion, especially during the paddy ripening seasons, is also a persistent threat to the paddy crop.

6.4 Conclusion

This paper brings forth the rapid emergence of the small tea growers within a span of about fifteen years in the North Bengal tracts although Assam had been a forerunner. The Tea Board has offered several schemes to encourage the small farmers to adopt tea farming as a strategy to maintain tea production levels because the larger estates are suffering from several management issues and labour unrests that has resulted in a dip in tea production. But, what remains a concern is the phenomenal shifting out from food crop by the peasantry whose bedrock of survival in the sector traditionally has been their near self-sufficiency atleast in terms of food grain production. But, the cropping pattern shift in favour of cash crops, given the agricultural policy environment and the agrarian crisis evident in the western and central Indian states, raises valid questions about the sustainability of the emerging commercial farming system centering smallholder tea farming.

Appendix

See Tables 6.7, 6.8 and 6.9.

Table 6.7 Change in area under total food crops (1995–96 to 2010–11)

Size class (HA)	Coochbehar	Darjeeling	Dinajpur (North)	Jalpaiguri	North Bengal
Marginal (<1)	80,416	−2615	31,476	5913	115,190
Small (1.0–2.0)	22,617	−14,793	3667	99	11,590
Semi-medium (2.0–4.0)	−6363	−17,566	−35,039	−31,249	−90,217
Medium (4.0–10.0)	−28,011	−4309	−22,585	−17,707	−72,612
STG size class	**68,659**	**−39,283**	**−22,481**	**−42,944**	**−36,049**
Large (>10)	−716	−532	8542	5526	12,820
All classes	67,940	−39,815	−13,942	−37,420	−23,237

Source Agricultural Census

Table 6.8 Change in area under Paddy (1995–96 to 2010–11)

Size class (ha)	Coochbehar	Darjeeling	Dinajpur (North)	Jalpaiguri	North Bengal
Marginal (<1)	46,862	−25,536	11,935	−1865	31,396
Small (1.0–2.0)	8591	−27,906	−5462	−5966	−30,743
Semi-medium (2.0–3.0)	−11,190	−24,278	−29,440	−24,829	−89,737
Medium (4.0–10.0)	−22,561	−5834	−16,054	−14,262	−58,711
STG size class	**21,702**	**−83,554**	**−39,021**	**−46,922**	**−147,795**
Large (>10)	−745	−59,945	8560	−922	−53,052
All classes	20,956	−143,499	−30,468	−47,844	−200,855

Source Agricultural Census

Table 6.9 Change in area under Tea (1995–96 to 2010–11)

Size class (ha)	Coochbehar	Darjeeling	Dinajpur (North)	Jalpaiguri	North Bengal
Marginal (<1)	42	408	1587	3065	5102
Small (1.0–2.0)	0	416	2549	3257	6222
Semi-medium (2.0–3.0)	0	552	1081	1216	2849
Medium (4.0–10.0)	0	483	234	425	1142
STG size class	**42**	**1859**	**5451**	**7963**	**15,315**
Large (>10)	−843	−24,461	−511	−24,906	−50,721
All classes	−801	−22,601	4940	−16,944	−35,406

Source Agricultural Census

References

Bhowmik SK (1991) Small growers to prop up large plantations, economic and political weekly, July 26–August 1, vol 26, no 30, pp 1789–1790

Bissonnette J-F, De Koninck R (2017) The return of the plantation? Historical and contemporary trends in the relation between plantations and smallholdings in Southeast Asia. J Peasant Stud 1–21. https://doi.org/10.1080/03066150.2017.1311867

Borah K, Das AK (2015) Growth of small tea cultivation and economic independence of the indigenous people of Assam. Int J Res Soc Sci Humanit 5(I):82–93

Hall R, Scoones I, Tsikata D (2017) Plantations, outgrowers and commercial farming in Africa: agricultural commercialisation and implications for agrarian change. J Peasant Stud 44(3):515–537

Hannan A (2006) Employment conditions in the small tea plantations (STPs) and their impact on the household economy: a case study of Islampur subdivision of North Bengal. Unpublished PhD thesis submitted to Jawaharlal Nehru University, New Delhi

Hannan A (2013) Organizational innovations and small tea growers (STGs) in India, NRPPD discussion paper no. 25, Centre for Development Studies (CDS), Trivandrum

Hannan A (2016) Livelihoods, labour market and skill development in small tea growers (STGs) gardens in India with special reference to India's North East, transactions, vol 38, no 2, pp 91–104

Hazarika K, Borah K (2013) Small tea cultivation in the process of self employment: a study on the indigenous people of Assam (India). Int J Latest Trends Financ Econ Sci 3(2)

Kumar P, Badal PS, Singh NP, Singh RP (2008) Tea industry in India: problems and prospects. Indian J Agric Econ 63(1):84–96

Li TM (2011) Centering labor in the land grab debate. J Peasant Stud 38(2):281–298

Mahendra Dev S (2012) Small farmers in India: challenges and opportunities, WP-2012–014, Indira Gandhi Institute of Development Research, Mumbai, June 2012. http://www.igidr.ac.in/pdf/publication/WP-2012-014.pdf

Mallik C (2016) Small holders and Bengal's Tea industry. Geogr You 50–52

Reardon T, Timmer CP, Minten B (2012) Supermarket revolution in Asia and emerging development strategies to include small farmers, PNAS, vol 109, no 31, July 2012. Accessed from https://doi.org/10.1073/pnas.1003160108 on 1/11/2019

Roy S (2011) Historical review of growth of tea industries in India: a study of Assam Tea. In: 2011 international conference on social science and humanity, IPEDR vol 5. IACSIT Press, Singapore
Tea Board of India: 59th Annual Report of Tea Board: 2012–2013

Chapter 7
The Transition of Traditional Agriculture in Nagaland, India: A Case Study of Shifting (*Jhum*) Cultivation

Devpriya Sarkar

Abstract *Jhum* cultivation is a practice of clearing the vast forest land for cultivating crops, where the land is left fallow after one or two growing cycles. It is the dominant form of traditional agricultural practice and continues to be the significant component of the livelihood of the state Nagaland's village communities. However, this has undergone many changes over time through various evolving practices and policy interventions. Further, the field survey found out that these self-sufficient communities are also looking towards modern economic development and activities and aim to earn more for a 'better' life. Such a changing attitude is showing reflection in their everyday lives, while state policies and innovations find ways to penetrate these traditional systems.

Keywords Traditional agriculture · *Jhum* cultivation · Agrarian transition · State policies

7.1 Introduction

7.1.1 Nagaland: A Geospatial Understanding

Nestled among the foothills of Northeast India, Nagaland forms an integral part of that region's political and geospatial space. It lies between 25°6'-27°4'N and 93°20'-95°15'E and has an area of 16,527 sq. Km. It is bordered on the north by Arunachal Pradesh, west and north-west by Assam, on the east by Burma and the south by Manipur. The altitude ranges here from 194 to 3,048 m, and the topography is severe with many ranges of spurs and ridges. The climatic conditions vary here from place to place because of the different altitudes present. The state experiences warm, subtropical types in the foothills, moderate and sub-moderate in the mid-slopes and lower range of the western part, and cold and temperate in the high hills. The rainfall is varied and has an average range of 200–250 cm with maximum receiving from

D. Sarkar (✉)
Jawaharlal Nehru University, New Delhi, India

N. C. Jana et al. (eds.), *Livelihood Enhancement Through Agriculture, Tourism and Health*, Advances in Geographical and Environmental Sciences,
https://doi.org/10.1007/978-981-16-7310-8_7

the southwest monsoon. The temperature varies from 5–25 and 12–32 °C in the hills and foothills, respectively. Flora and fauna are pretty rich here and vary from evergreen and semi-evergreen, deciduous, sub-tropical pine forest, and temperate types. Though depletion in natural forests is noticed because of shifting cultivation (locally known as *Jhum* cultivation) in this region but still a rich diversity can be seen (Nagaland State Profile, Government of Nagaland 2016).

7.1.2 Shifting (Jhum) Cultivation in Nagaland

Jhum cultivation is a traditional agricultural practice in the state. In this form of agriculture, the land is cultivated for 1–2 years and then left abandoned for a minimum of 15 years to rejuvenate its fertility. The indigenous people engaged in *Jhum* cultivation uses their traditional knowledge to till the land and grow crops for subsistence. Further, no chemical fertilizers are used to increase productivity and degrade the soil. Instead, they use physical barriers to restrict the movement of the insect's pests-pathogens. Shifting Cultivation, according to Pelzer (1958), is defined as 'an agricultural system characterized by rotation of fields rather than of crops, a short period of cropping (1 to 3 years), alternate fallow seasons (10 to 15 years) and clearing using slash and burn' (cited by Sachchidananda 1989: 3).

Further, Das (2006) in his work 'Demystifying the Myth of Shifting Cultivation: Agronomy in the North-East' argued that 'shifting cultivation has a production of food crops 'by default' organic because of the use of primitive tools (hoes, daos, digging stick) and human labour'. Although by the nineteenth century, the use of modern tools came into existence, the people of Nagaland continued to use primitive tools and their traditional knowledge of cultivation. Apart from agriculture, the people of Nagaland also have practices like weaving, blacksmithy, carpentry and other handicrafts but on a tiny scale to meet their daily needs (Bose 1991: 42–51).

Agriculture is the primary source of livelihood for the people of Nagaland. It contributed 21 per cent to the Net State Domestic Product in 2008–09, followed by industry, real estate, transport and communication. Here 68 per cent of the total workforce is engaged in agriculture, which is higher than the national average of 47 per cent. Thus, maximum people are involved in agriculture and are also the primary source of livelihood for the population of Nagaland. The food-grain production has shown an upward trend during 1961–62 to 2012–13. In 1961–62, the food-grain production was only 63,530 metric tonnes which increased to 598,960 metric tonnes in 2012–13 and showed an increase of 10.6 per cent (Statistical Handbook 2013). Though an increase is noticed in productivity, this sector's achievement is not very satisfactory and is unable to meet the needs of the existing population.

Traditionally, cultivation was done on the hilltops, and *Jhum* cultivation was the dominant form practised in Nagaland. Eventually, the increasing population created pressure on the agricultural sector and demand for food-grain production increased, to meet the demands of the rising population, a different form of cultivation was adopted, known as terrace cultivation. This form of agriculture made use of chemical fertilizers and pesticides, and modern technology to boost the production rate. Nevertheless, the substantial cost of implementing terrace cultivation makes it difficult for the small and marginal farmers to practice it on a large scale.

7.2 Research Questions of the Study

The current study aims to locate the small and hilly state of Nagaland, which still practices shifting cultivation in a pre-dominant form. Given the recent changes and issues related to such practices, the study has the following research questions.

1. Is shifting cultivation a viable and economically sustainable activity in the contemporary situation of Nagaland?
2. Is there any transition taking place in the traditional practice of *Jhum* cultivation? If any, how the state of Nagaland is responding to preserve the traditional indigenous knowledge?

7.3 Objectives of the Study

Within these research questions, the main objectives of the study are.

1. To study the traditional farming of *Jhum* cultivation and its related ecology, economy and cultural dynamics.
2. To understand the relevance and scope of *Jhum* cultivation in Nagaland and its transition from tradition to modernization and the state initiatives to preserve such traditional methods.

7.4 Data Sources and Methodology

The study is based on the information collected from both secondary and primary sources. The secondary data includes information from various government sources like Statistical Handbook of Nagaland, Basic Statistics of North-Eastern Region, District Census Handbook, Forest Department, Agriculture Department and other published and unpublished records in various departments. In the case of primary data, a field survey was carried out in two villages, namely Chuchuyimlang and Mongsenyimti of Mokokchung district. Both the villages are dominated by the Ao community, which practices shifting cultivation traditionally. The study covers the Ao community, their different agricultural land use patterns, and related environmental issues to understand the problems of shifting cultivation and the social implications of the farming communities. Further, an in-depth study has been carried out to understand the government policies implemented in the district and their effect on the livelihoods of such indigenous community.

The study has adopted a mixed method where both qualitative and quantitative methods to understand the phase of transition in the traditional methods of agriculture. Thus, various interviews, discussions and interaction with different groups of people during the fieldwork are used to lay the perspective of the study. In contrast, the information from the questionnaire and data from various secondary data sources have been used to quantify and bring empirical understanding to this study.

7.5 *Jhum* Cultivation: Traditional Methods Among the Ethnic Communities

Jhum cultivation is a practice of clearing the vast forest land for cultivating crops, where the land is left fallow after one or two growing cycles. It is the dominant form of traditional agricultural practice found among the tribes of Nagaland, including the Ao community (GOI-UNDP Report, Nagaland 2009; NEPED and IIRR 1999). These Ao communities pre-dominantly inhabit the Mokokchung district of Nagaland. The pattern and practice of *Jhum* cultivation differ from place to place amongst various tribes while having broad structural similarities. Thus, various ethnic communities like Ao, Angami, Sumi, Lotha, Rengma, Konyak, Sangtam, Phom, Chang, Yimchunger, Khiamniungan, Chakesang, Zeliang and Pochury practice *Jhum* cultivation in Nagaland within such broad structural frame. The clearing of forests usually begins in January and February, which is left for drying and finally burnt in March and April. Sowing of seeds takes place with the first rain, and the harvest begins in August (Report on Development of North East Region 1981). Paddy is the major crop grown on all the *Jhum* fields of Nagaland, while mixed cropping is also carried out by the *Jhummias* (people are practising *Jhum* cultivation), which include crops like maize, millet, beans, turmeric, ginger, tobacco, chillies, sesamum, tapioca and leafy vegetables.

Jhummias have a practice of rotating fields, but their place of residence is generally permanent, and every village has its demarcated region for the selection of *Jhum* plots. The Village Head or Village Council plays a vital role in the distribution of *Jhum* lands, and no villager is barred from owning land. Preferably, the households are provided land area according to the number of family members (Sachchidananda 1989: 37–38). According to the land use department records, the net area under *Jhum* cultivation in Nagaland was 99 345 hectares in the year 1985–86, which decreased to 78, 000 hectares in 1995–96 and then showed an increase to 86, 000 hectares in 2000–01 to 93, 000 hectares in 2010–11 (Senotsu and Kinny 2016: 123).

In the colonial and post-colonial period, *Jhum* cultivation in India, in general, has been under debate because of its harmful effects on the environment. The colonial state debated that the indigenous system of *Jhum* cultivation had no positive effects on the environment and is a driver of soil erosion, air pollution and causes biodiversity loss. Hence, the capitalist colonial state was not interested in the practices of *Jhum* cultivation, which primarily serves the purpose of a subsistence economy. Also, it did not put enough efforts into such traditional practices and, therefore, did not produce sufficient data and information on this type of cultivation. It has created a gigantic knowledge vacuum about the practice of such traditional cultivation in the post-colonial period as well. In the post-colonial period, the same scenario prevailed, and the practice of *Jhum* cultivation was hardly encouraged and acknowledged. The Indian Council of Agricultural Research (ICAR), the State Institution which was created in the year 1929 for promoting modern agricultural practices, has been deeply critical about *Jhum* cultivation in Northeast India. Eventually, one of the ICAR reports mentioned that *Jhum* cultivation is a significant driver of soil erosion in Northeast India and proposed various policies on soil conservation and soil erosion, forest and biodiversity conservation, and alternative agricultural methods to counter such indigenous methods of agricultural practices (Report on Development of NER 1981: 13).

Over a while, many alternative schemes were introduced by the Indian state and were implemented in various regions of India and *Jhum* cultivation was systematically marginalized by the state machinery. For example, in the Siang River Catchment of Arunachal Pradesh, a policy to control *Jhum* cultivation and Water Shed Management was implemented in the year 1974–75. The primary objective of the programme was to show the benefits of having settled agricultural practices like Wet Rice Cultivation, Terrace Cultivation, and Tanugiya over shifting cultivation (Report on Development of NER 1981). The policies were implemented by various Union Territories and State Governments, which were later aided by the Central government.

Northeast India has a unique and different form of land tenure system from the rest of the country, except for the plains of Assam, Tripura and valley areas of Manipur. Nagaland has its local customary laws to administer the land tenure systems, which are not only state-specific but sometimes even tribe or village specific (Sachchidananda 1989: 38–39; Erni 2014). The Village Head or Village Council generally does the conduct of these traditional laws and regulations. In the Ao village, as pointed out by Darlong (2004), 'they have their unique customary laws regulating ownership and access to land and resources'. There are differences in practices of such customary laws among various tribes in Nagaland, and they generally do not have rivalries and antagonism with neighbouring villages. In the case of the Mokokchung area, every village has its demarcated region for *Jhum* fields, homestead lands, permanent farms and clan forest, along with their personnel laws enacting within the villages. The Ao community owns both 'community land' and 'private land', which are governed by the Village Head. In the Mokokchung area, the land data is usually dependent on the reports provided by the village council, and therefore no extensive survey is found so far by the state machinery to understand their customary laws and regulations on such land systems. In the Mokokchung district, there are various practices of land use system; the dominant of them is again *Jhum* cultivation. According to the reports of District Agricultural Office of Mokokchung, Government of Nagaland, 70 per cent of the farmers are identified in the district as the practitioner of *Jhum* cultivation. The total area under such cultivation in Mokokchung is estimated to be 108,554 ha (Field Survey, District Agricultural Office, Mokokchung). The *Jhum* land in Mokokchung has a cropping period of 1–2 years along with a fallow period of 10–12 years, though such periods vary in different villages.

The interaction with various Village Head during the field survey revealed that the Village Council is the most significant body of governance in both the villages, namely Mongsenyimti and Chuchuyimlang. The Village Head holds the villagers always accept a prominent position in the society and their decisions. The Village Head is also responsible for all kinds of socio-economic and cultural practices and their overall well-being like health, education, land, market and economy, water and electricity.

The Village Council also plays a vital role in the events related to *Jhum* cultivation and decides everything throughout the year. For example, every year, the Village Council selects a plot of forest land for cultivation by the villagers, and subsequently, the dates are fixed for the sequence of events like slashing, burning and sowing. Also, the Village Council keeps a check on the *Jhum* fields during the burning process, and the villagers are advised to keep control over the fire by creating firebreaks. If a fire breaks out in the forest and the neighbouring areas and causes vast destruction, the farmer is liable to pay a compensation cost. This mechanism shows that the villagers have a secure system of protecting land and the environment in their traditional activities of cultivation and have a method to preserve and conserve the ecology.

The interactions in the field further revealed that among the Ao community, the *Jhum* land, including fallow, is divided into several blocks, which varies from village to village. Large villages have more than fifty blocks, whereas small villages comprise minimum fifteen to twenty blocks. The number of blocks in a village helps in determining the fallow cycle (land left for regeneration). A similar finding of block distribution is also present in the work of Jamir (2015). These *Jhum* blocks (JB) are generally located within a distance of 4 to 5 kms from the residential areas of these villages (Fig. 7.1). Amongst the two villages, Mongsenyimti has a total of 111 plots, and Chuchuyimlang has 126 plots. Each plot of land accounts for 50 hectares approximately, which is divided among the farmers for cultivation (Field Survey 2018). Thus, the field survey shows that these two villages have a good fallow cycle present in their respective villages, which are environment-friendly. The farmers cultivate paddy on a particular block for not more than two years. After that, the land is left for fallowing, vegetables, plantation and agroforestry is practised on the *Jhum* fields for an additional income and livelihood.

The practice of *Jhum* cultivation in Nagaland in general and Mokokchung, in particular, has shown in many ways an ingrained knowledge of the surroundings, their ecology and their operations (Erni 2015). Hence, the villagers, while practising *Jhum* cultivation, also protect and care about Mother Earth and ecology. The *Jhummias* also have a tradition of using the old primitive and indigenous tools, which is friendly to soil and causes less harm. Also, they have different cropping patterns to maintain soil fertility and derive a variety of crops from the same field. The various indigenous tools that are typically used in the *Jhum* land by the traditional *Jhummias* are given in the table below (Table 7.1):

Thus, *Jhum* cultivation has various methods and modes of practice which has evolved over centuries and represents a close link with the environment and its

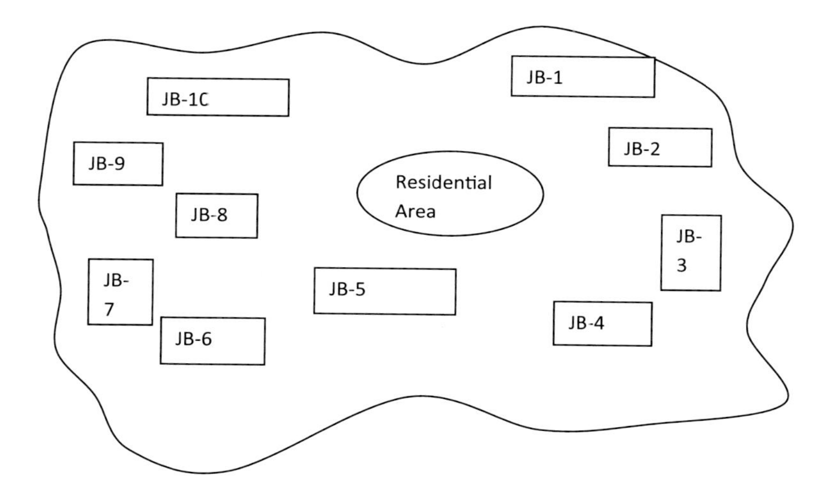

Fig. 7.1 *Jhum* block distribution in Ao Village: A Sketch. The figure is drawn by the researcher based on the field visit

Table 7.1 Indigenous Tools used in *Jhum* Land in Nagaland

Serial No	Name of the tool	Uses
1	Noklang (Axe)	Helps in clearing the forest area for cultivation
2	Dao	They are used for clearing bushes
3	Digging Stick (Bamboo, Iron)	It helps in digging holes for burying the seeds
4	Sickle	They are used for clearing weeds and reeds
5	Bamboo Basket	They are used for carrying the harvest
6	Tins	Used for measuring the harvest (1 tin = 6 kgs approx.)
7	Stone Grinders	They are For grinding maize and millets

Source Compiled from Husain (1988), Sachchidanda (1989) and Field Survey (2017–18)

components. Also, the indigenous tools and cropping patterns present among the indigenous communities of Nagaland and Mokokchung has traditional indigenous methods to overcome the drawbacks of *Jhum* cultivation. Such methods are listed below.

- To reduce the effect of soil erosion, the cultivators take out the central part of the crop, i.e. paddy panicle, maize cob. It helps in holding the topsoil, which, after decomposition, makes the soil fertile for the next period.
- Mixed cropping, intercropping and crop rotation is done to balance the nitrogen demand of the crops. More than 10 to 20 crops are grown together, with paddy as the main crop in the same plot of land. It provides other vegetables and food crops to meet the nutritional demands.

- Non-leguminous trees are also grown to meet the other nutrients of the soil that are beneficial for crop production.
- Apart from nutrient management, for pest and weed control, salt is spread on the field. Salt is majorly used to remove the broad-leaved weeds and control their growth on the field.
- Also, Dhatura (Datura stramonium) is used to overcome the problem of stem borer infection in the paddy. Dead frogs or crabs are used to control bugs.
- Use of ash on the field before sowing seeds helps in reducing the chances of drying up crops due to unfavourable conditions.
- The Alder base and Kolar-based fallow management system in Nagaland has benefitted in soil conservation and nutrient fixation (UNDP and IeLo Report 2015; Field Survey 2017–18).

These powerful ways of protecting soil and the environment by traditional agriculture prove their substantive knowledge over land ecology and environment, which has made the existence of *Jhum* cultivation possible to date. However, many innovations have taken place in the practice of *Jhum* cultivation in the recent decade because of the increased demand for food for the existing population and the need for cash to fulfil other desires of a healthy and comfortable life.

Such techniques and use of tools indicate that the cultivators or the *Jhummias* are deeply connected to the land and have sound indigenous knowledge to protect their land and environment. Therefore, the usual predicament of cultivators as primitive, ignorant and unaware about the environment may not be justifiable. They have a better knowledge of the surroundings and nature and have a necessary measure to protect the same. They have also given the topographic and climatic conditions, various modern technologies and equipment for cultivation in such land are not necessarily suitable. Also, various studies by Mishra & Jamir (2017), Mishra and Singh (2012), Erni (2015) and reports of Soil Survey Department have found that the soil structure of such hilly region has a weak layer typically and gets easily eroded with the use of ploughs and any other modern tools.

7.6 Transition and Transformation in Traditional Agriculture

As human civilization perpetually moves in progression, various methods, techniques and innovation are therefore become essential for them to fulfil their goals. Scholar like E. Rogers (1995) says that innovation is one of the first applications to the idea of progression, which can be either incremental (continuous) or radical (discontinuous). The path of innovation usually takes a linear path which begins with the components of invention and diffusion through markets. In this path to civilization, the human being always tends to improve and innovate to meet his needs of everyday life (Herbig 1994). Thus, scholars have always identified incremental innovation as a better form. These innovations have a direct effect on the livelihoods of the people, which also can improve various economic activities. Agricultural practices thus can also be improved through various useful innovations and can bring substantive changes in the lives of various communities who are dependent on them. Thus, the various small isolated tribal communities of Nagaland, like Ao Angami, Rengma, Sema, Koynak, who are currently facing the challenge of low productivity and low returns in their traditional agricultural practices, also can explore some innovative methods. Such explorations are expected to bring various transitions and transformations in their existing traditional agricultural practices. They could make a change from subsistence to the surplus agricultural economy through modern techniques and equipment of farming, intensification of markets and large-scale investments. The study made by Byres (1986) stated that in a developing country, the absence of agricultural transition has led to economic backwardness. He suggested that country like India, where the maximum of its human population engaged in agriculture, needed a capitalist mode of agricultural transition for economic betterment. Thus, the transition and transformation of agricultural practice is a natural phenomenon, came up with the changing time, increasing human population, and the capitalist system, which brought massive changes in land tenure, organization of production on farms, techniques and types of equipment on the farms, intensification of markets, and capital investment of small-holding farmers.

Nagaland already has experienced some such changes in their traditional fields. Even the colonial capitalist also have experimented with specific policies of innovation in the traditional agricultural fields of Nagaland. During the colonial period in India, a series of attempts were made to eradicate *Jhum* cultivation and inculcate the practice of settled agriculture among the tribal communities. The Indian Forest Act of 1865, 1878, 1927, *Jhum* Land Act (1947) were laid down to control, restrict and abolish the practice of shifting cultivation. The main focus of these acts was to preserve the forest resource and extract revenues from the same. Also, settled forms of agriculture like *Taungiya* and terrace cultivation were introduced, and allowance was provided to the cultivators by the colonial government to practice the same. A similar trend was being followed in the post-colonial period, when different state governments provided grants for a settled form of agriculture like Terrace Rice Cultivation, Valley Cultivation, Wet Rice Cultivation, Horticulture, *Taungiya* and a three-tier system (Report on Development of NEI 1981; Malik 2003; Maithani 2005). In 1987, the project *Jhum* Control Scheme was launched by the Government of India. This scheme was a part of the Fifth-Five Year Plan, and its main target was to wean away *Jhum* cultivation from the livelihoods of indigenous people. In 1994, the Watershed Development Project in Shifting Cultivation areas was launched (Tripathi and Barik 2003). A recent development in the policies of *Jhum* Cultivation took place in Mizoram in the year 2008, namely the New Land Use Policy (NLUP). This policy introduced a new pattern of land use, commercial use of the domestic resource and marketing of agricultural products. The common aim of these policies and schemes in Northeast India in general and Nagaland, in particular, was to introduce the cultivation of cash crops like rubber, cashew, tea, oranges, cardamom and provide *Jhummias* with a better economy via settled cultivation (Maithani 2005).

Further policies like the National Environment Policy (2006), Biological Diversity Act (2002) were enacted to preserve the environment and its components. Although limited attention was given to the interests of indigenous communities, forceful imposition of settled agriculture was done. However, these settled forms of agriculture received resistance from the cultivators, as these policies never provided a space for their cultures and customary laws. However, the Union Nations Intergovernmental Panel on Climate Change (5th Assessment Report 2015) identified that traditional agricultural practices like *Jhum* cultivation have the potential to flourish and adapt according to climate change without causing much harm. Thus, traditional agriculture gained importance in the policy frameworks in the recent period, and initiatives have been taken to uplift the production from traditional agriculture via minimum technological and scientific development. Such incremental innovation in the state has taken concrete shape in the forms of organic farming.

Organic farming is a practice of cultivating nutritious food without any use of chemical supplements, thus having a close relation to traditional agricultural practices like *Jhum* cultivation. As already discussed, *Jhum* fields have negligible use of chemical supplements. Nagaland adopted the Mission Organic. One of the examples of incremental innovation in Nagaland is Organic Farming, which aims to convert the 'Organic by default' lands to 'Organic by design' lands by 2025 (Vision 2025 Report, Government of Nagaland 2012).

However, though Nagaland has seen a series of policy interventions since the colonial period towards transition and transformation of its traditional agricultural practices, these external interventions have withstood over time. It is because these traditional practices are secure in their foundation and are naturally connected to the environment. In recent times, these traditional practices like *Jhum* cultivation have undergone some scientific-technological innovations and have shown a phase of transition in Nagaland. In this context in the recent decade, the villages of Mokokchung district, namely Mongsenyimti and Chuchuyimlang, also have seen some phase of transition in land use and agricultural practices. It has affected their existing traditional, social, cultural and economic lives. Thus, Nagaland is experiencing a phase of social and cultural transition, where agricultural innovations are playing a significant role. Some of the other factors responsible for the transition in the case study villages are discussed below.

7.6.1 Reduction in Land Cover Under Food Crops

The two case study villages, namely Mongsenyimti and Chuchuyimlang, consist of small-holding[1] farmers are occupying land between 1 to 2.5 acre. Both the case study villages show that every household has a minimum of 1 to 1.5 acre of land for the *Jhum* cultivation, which is used for the cultivation of both paddy and cash crops. Vast fields are hardly noticed, and cash crops are generally grown collectively on *Jhum* plots

[1] Farmers, pastoralists, forest keepers, fishers managing area less than 2.0 ha (FAO 2002).

during the fallow periods. The study reveals that in both the villages, a minimum of 1.5-acre land per household is always under *Jhum* practices and has shown the highest frequency. Paddy and millet is the primary crop grown in both the villages. However, leafy vegetables, ginger, garlic are also grown in the *Jhum* fields. In Mongsenyimti, the plantation and cash crop cultivation are minimal, whereas Chuchuyimlang has a higher share of land under the plantation crop. The main reason for this situation is the advanced practices such as the use of bio-pests for pest management, small equipment for digging the land, mini-pumps for water harvesting and the funds from the government programmes enacting in the Chuchuyimlang village, which enables them to practice plantation and horticulture on the fallow *Jhum* lands. At the same time, the Mongsenyimti village is still practising traditional mixed farming, i.e. *Jhum* cultivation, to meet their daily needs.

7.6.2 The Issue of Food Security

Conversation with the Village Heads in the field revealed that horticulture and plantation are practised in both these villages. Mongsenyimti produces cash crops like ginger, garlic, ramie beans, yam and tapioca, whereas the other village has additional cash crops, namely turmeric, apple, litchi, cardamom, banana, oranges, mango, pear, guava, pineapple, gooseberry, jackfruit, passion fruit, litchi and peach. The horticultural crops have benefitted the economic conditions of the *Jhummias* as these crops help them to fetch an extra income from the market as the price of cash crops is relatively higher than that of food crops. Also, cash crops have much more surplus than food crops in Nagaland. Paddy is only cultivated for self-consumption. However, the villages have not attained self-sufficiency in food-grain production.

Data collected from the Village Council of Chuchuyimlang village show that a household with six members consumes 4 kg of rice daily. The village has a total number of 905 households, which shows that the village needs approximately 3,620 kg of rice per day. Thus, to attain Food Security, the village needs a yearly production of 220,095 tins (1 tin = 6 kg approx.) but has only reached halfway with a production of 110,047 tins in 2016. Mongsenyimti has also not attained Food Security as informed by the Village Head, but they have a relatively high production of *Jhum* paddy. For attaining food security, this village needs a yearly production of 133,225 tins but has only reached a production of 79,935 tins in 2016, i.e. almost sixty per cent of the total consumption. Thus, the villages have the potential to meet the problems of food security, but one of the significant limitations is the increased involvement of the farmers in cultivating cash crops. It has shown a reduction in the share of land. Previously, one family used to cultivate on the entire 1.5-acre land, but with the introduction of cash crops, a portion of land (ranges between 0.5 to 1 acre) is always growing cash crops. Thus, the farmers get harvests throughout the year from such 'horticultural plots' (Field Survey 2018).

7.6.3 Labour Mobility Towards Other Sectors

The Economic Survey of Nagaland (2016–17) shows that the Gross State Value Added (GSVA) from the agricultural sector was 30.9 per cent in the year 2011–12 and is further reduced to 29.4 per cent in 2015–16. Although 71 per cent of the population in the state is dependent on agriculture, a gradual shift from agriculture to the industrial sector also can be noticed. The primary reason behind this shift is the low income of the farmers from the existing agricultural lands and practices. There is a persistent rise in aspiration amongst the people of the state, and therefore people are making efforts to provide formal education to their children, which they think can bring a better life to them. A woman from the Mongsenyimti village stated that:

> "We (narrator and her husband) are less educated, but we are earning hard to educate our children so that they can lead a better life and can work in the offices or industries."

Conversation with another family of 10 members revealed that eight of them have already migrated to the nearby or far-off towns for a better life and job opportunities. Thus, such a trend of mobility from farming lands to urban areas has created a shortage of labours in the agricultural lands. The hard labour and various changing climatic conditions are additionally creating limitations for the villagers to continue their practice in their traditional *Jhum* fields. Thus another farmer of Chuchuyimlang village narrated that:

> "Jhum cultivation is a hard labour work and does not produce much of surplus resulting in low income whereas plantation and horticulture crops fetch better value in the market (produces surplus) and also the State Government provides funding for the same and not for Jhum cultivation (Field Survey 2018)."

The other conversation with the Council member of Chuchuyimlang stated that:

> "Jhum cultivation is a hard labour work and needs more working hands. However, the new generation is not interested in agriculture and prefer to move to urban areas in search of better jobs. Also, the changing climatic conditions and uneven rainfall has affected the cultivation process. Thus, they have requested the District Agricultural Office that they want to practice Terrace Cultivation and want some help for the same."

However, in Mongsenyimti village, the field survey revealed that approximately 54 per cent of the total people interviewed during the field survey is engaged in traditional agricultural practices and its allied sectors. The other petty occupations that the village has are shopkeeping (5 per cent), car driving (4 per cent), teaching (3 per cent), and also a large number of them are the student (27 per cent), and some are Not Working (8 per cent). Whereas the Chuchuyimlang Village has approximately 55 per cent of the interviewed population, who are working in agricultural fields, followed by the student group (25 per cent), shopkeepers (7 per cent), teachers (5 per cent), private workers (2 per cent) and rest 5 per cent are not working. Thus, in both the villages, the largest concentration of the population is in the agricultural field followed by the younger generation who are students and aspired to have a 'better' life and job opportunity in future and they do not want to work in the fields. The

Jhummias in these villages also has some subsidiary occupations such as blacksmithy, ornament making, carpentry, bead making, weaving and livestock rearing. Livestock rearing is an essential occupation amongst the tribes, and every household has at least one pig and one chicken as their livestock. Broadly, the *Jhummias* are gradually becoming more dependent on horticultural and other such subsidiary occupations to gain economic surplus, and thus such transitions are becoming visible in the everyday lives of the cultivators.

7.6.4 Adaptive Farming

Climate change is a global challenge, and its widespread adverse effects have a direct impact on agro-ecosystems, agricultural production and human well-being. It has a worse effect in the region, where people are mostly dependent on agriculture and its allied sectors (Sugam et al. 2016). Nagaland's economy is mostly dependent on the agricultural sector, and change in climatic conditions has posed a threat to the development of agriculture. Thus, the state government, along with Central support, launched few schemes on adaptive farming, which focussed on developing agricultural practices and cultivators via local resources. The farmers of Mongsenyimti and Chuchuyimlang village are mostly small-holding farmers and have a low capital investment. The Central Government policies like Agriculture Technology Management Agency (ATMA) 2010, National Food Security Mission (NFSM) 2007, National Mission on Oilseeds and Oil palm (NMOOP) 2014–15, *Rashtriya Krishi Vikas Yojana* (RKVY) 2007–08 is providing funds for the cultivation of cereals, oilseeds, vegetables and cash crops. Also, the market price of plantation crops like rubber, tea, tapioca in the region has shifted the interest of the cultivators towards plantation crops. Thus, there is a sustained decline in the production of food grains from the *Jhum* fields.

7.6.5 Market-Driven Farming

As discussed earlier, due to incremental transition (linear to market), the crops grown in the Mokokchung district are highly influenced by the market. The major crops grown here to increase capital and productivity are ginger, ramie beans, gourd, oranges, banana, cardamom, tea which are mostly cash crops. Nevertheless, sometimes this market-driven farming affects the farmer's economic status. For instance, the weekly marketplace in Mokokchung needs proper registration of the farmers and pays an annual rent for the shop they hold. Additionally, the limited markets in the region give them less scope to sell out their products and are mostly dependent on the road-side *Mahila Goshtis*.[2] Also, the produce goes to waste sometimes because of the unavailability of the storage area. As told by the farmers of Mongsenyimti Village:

> "Last year (2016-17), the production of ginger was increased as the Government officials came to the village and distributed seeds and other necessary items. Nevertheless, they have limited access to the market, and the unavailability of storage places made the produce go to waste. Further, the village is located far from the urban areas, and lack of transportation creates a problem for them to take their produces to the marketplace."

Thus, the diversification of government support towards transitions for the market economy in the existing structure is not helping the farmers in a significant way. Hence, it will not be incorrect to say that the *Jhummias* have gradually shown a shift from the traditional agricultural practices towards a commercial one. On the other hand, the Village Head of Mongsenyimti states that:

> "Initiatives have been taken to improve the productivity of the Jhum fields, and now they are trying to gain benefits from the Organic Farming scheme, which is both economically and environmentally viable. Further applications have been put up so that the new policies cover their village also."

Thus, it can be easily seen that the villages covered under the schemes (Chuchuy-imlang Village) of Central and State government has shown uplift in their living standards whereas, the other has still many problems to overcome.

7.6.6 Cultural and Social Transition

The famous book 'Primitive Culture' of E. B. Taylor states that 'Culture and civilization taken in its ethnographic sense, is that complex whole which includes knowledge, beliefs, art, morals, law, custom and any other capabilities and habits acquired by man as a member of the society (1971). Further, Radcliffe, in his work, defined culture as a process of learning and acquiring knowledge from contact with other people, books, art, knowledge, skills (1952). Thus, in simpler words, culture is a product of social learning and not biological heredity, which understands human reactions and responses to existence. The case study villages have shown a transition as well as modifications in their cultural dimension because of the changing agricultural practices, as the new alternative practices do not provide proper time for celebration as different crops have different harvesting time. On the other hand, *Jhum* fields have a single harvesting period, and after the harvest, offerings are made to the 'supreme being' and celebrations were done. The villages of Nagaland have a very close link with the social norms and regulation, where the Head of the Villages decides for the celebration of different festivals on different occasions. Moatsu is the main festival

---

[2] *Mahila Goshtis* are road-side stalls appointed to the Women of a household, where she can sell agricultural products. This initiative was taken to improve the status of Women in agricultural practices in Nagaland.

of the two case study villages. This festival is generally celebrated in the first week of May after weeding is done in the agricultural fields. However, this social practice is weaning away slowly because of mechanized weeding or bio-weeding. A resident of Mongsenyimti village state that:

> "Previously different festivals were there during sowing, weeding and harvest but now these practices have been reduced."

Apart from the transition in festivals, a major transition has been noticed in the administrative setup of the village. The villages of Nagaland are monitored and regulated by the Village Head, but nowadays a change has been noticed, and a group of council members are elected from the villages to monitor different sectors. In Chuchuyimlang village, there are 108 council members present who look after different sectors such as health facilities, land use, crop selection, market availability, whereas Mongsenyimti has one Village head and three other council members to monitor the same departments. Thus, progressive villages have shown a move towards a more organized administrative setup of villages, whereas villages with limited resource have the same old ways of leading livelihood.

Thus, in Nagaland, the conversion of the farmers from traditional to modern was not a planned strategy and has caused severe threats to the environment and livelihood of the indigenous people. However, the government realized the importance of traditional agriculture and indigenous crops, which led to the adoption of the Mission Organic Value Chain Development (MoVCD) programme in 2016. This policy took the initiative to promote the traditional agricultural practices and indigenous crops but ended up being an agricultural practice focussing on the horticulture sector widely. Despite the fact, this programme has shown a positive development in the production of food-grains also.

7.7 Concluding Observations

Jhum cultivation is subsistence and does not have any scope to create a surplus. It, therefore, is not a surplus-producing system and cannot meet the increasing demands for food in Nagaland. Having this rising concern, the state of Nagaland sometimes aimed to subvert such practices on the ground of its harmful effects and unsustainable method. However, the indigenous communities of Nagaland always resisted such State agenda and tried to retain their traditional practices and living. The dominant tribe of Mokokchung, namely the Ao community, are situated in the inaccessible hill ranges, and their principal source of livelihood is *Jhum* cultivation. This practice is being supplemented by hunting, fishing, and livestock rearing. In the two villages, namely Mongsenyimti and Chuchuyimlang has chosen for the survey, it was found that almost all households were practising *Jhum* cultivation and are commonly known as *Jhummias*. Such practice of shifting cultivation has a close relation to religious beliefs and customary laws.

Jhum cultivation is the way of life amongst the Ao tribe in these two villages, where both male and female play a vital role in the practice. Males are generally engaged in the agricultural fields during the forest cutting and burning period, protection of crops and harvesting period, whereas females are engaged in dibbling and sowing of seeds and for the removal of weeds. It is a family-based practice, where each member of the household is engaged directly or indirectly in the agricultural fields. Children between the ages 8 and 12 follow the elders to the fields and learn about the practice though in the contemporary period more focus is paid on education, as good educational qualifications provide 'better' job opportunities. The *Jhummias* of these villages has individual ownership over their houses and homestead land, whereas the *Jhum* fields are a joint property of the village headed by the Village Head. Thus, shifting cultivation is more of a social activity than nearly an economic pursuit for such a community. The study categorically has found that *Jhum* cultivation needs low capital investment and is economically more viable than any other form of modern agricultural practices in terms of both economy and environment. The new modes of agriculture need high-yielding variety seeds, chemical supplements such as fertilizers, pesticides and insecticides, and irrigation facilities for a good harvest. On the other hand, *Jhummias* use indigenous seeds saved during the harvest period or obtain them from the neighbouring villages via exchange. Use of manures is negligible and the equipments used are primitive and are usually made up of wood and bamboo though iron is used to make daos (used for chopping forest cover).

The field report and various interviews have found out that the indigenous communities are deeply connected to such traditional practices, and this cannot be removed from their existing cultural economies. There have been persistent state efforts to motivate and teach the traditional farmers through regular classes (monthly and quarterly), incorporating some of the new methods in their traditional agricultural practices. Similarly, for the conservation of the environment, these traditional farmers are assisted by the state government. Thus, the idea of organic farming has already been introduced in this area within the *Jhum* land to enhance productivity while retaining environmental concerns. The *Jhummias* are also currently focusing on cash crop like horticulture to earn marketable surplus and to add to the formal economy of Nagaland. With State effort, the market for such products has started to expand in the neighbouring areas of Nagaland through the means of trade. For increasing organic farming in *Jhum* land, various forms of water harvesting are also introduced in these villages. Organic farming is now considered to have immense potential for the economy of Nagaland while protecting the environment. Nagaland possibly can achieve the status of the organic state as the communities already are enriched with several such knowledge bases. In this regard, the state initiative may remain fruitful in Nagaland shortly, which can help to attain a better living. Thus, the government schemes, together with the efforts of different organizations, made it possible to make the farmers keener about organic products and their benefits. It helped them to attain a 'better' livelihood and a 'better' economic condition to the state. In this process, a gradual subversion of *Jhum* cultivation is taking place in the state, while through policies, the agriculture is targeted to be modified and modernized over time.

Thus, the practice of *Jhum* cultivation is going under indirect suppression where the cultivators are now practising the modified forms of agriculture as these provide self-sustainability and economic development to them.

References

Annual Administrative Report (2009–10, 2011–12, 2014–15, 2015–16, 2016–17) Department of Agriculture. Government of Nagaland. Kohima, N. V. Press

Annual Report: National Agricultural Scenario (2016–17) Department of Agriculture Research and Education, Indian Council of Agricultural Research, Government of India, New Delhi

Borthakur DN (2002) Shifting cultivation in North-East India: an approach towards control. In: Bimal J Deb (ed) Development Priorities in North East India. New Delhi, Cocept Publishing Company

Bose Saradindu (July 1991) Shifting cultivation in India (ed.), Anthropological Survey of India. Kolkata, Government of India, pp. 42–51

Census of India, Office of the Registrar General and Census Commissioner, Ministry of Home Affairs, Government of India, 2001 and 2011

Darlong T Vincent (2004) To Jhum or not to Jhum, policy perspective on shifting cultivation. Society for Environment and Conservation, India

Das Debojyoti (2006). Demystifying the Myth of shifting cultivation: agronomy in the North-East. Econ Polit Week 41(47), Nov. 25- Dec. 1, 4912–4917

District Agriculture Handbook (2016–17) Indian Council of Agricultural Research, Ministry of Agriculture and Farmers Welfare, Government of India

District Census Handbook (2011) Office of the Registrar General and Census Comissioner, Ministry of Home Affairs, Government of India. http://censusindia.gov.in/2011census/dchb/DCHB_A/13/1302_PART_A_DCHB_MOKOKCHUNG.pdf. (Accessed on March 2018)

Erni Christian (Editor) (2015) Shifting cultivation, livelihood and food security; new and old challenges for indigenous peoples in Asia, The Food and Agriculture Organisation of the United Nations and International Work Group for Indegenous Affairs (IWGIA) and Asia Indigenous Peoples Pac (AIPP), Bangkok, 2014

Handbook on Farming, Published by Agriculture Technology Management Agency, Mokokchung, Nagaland, 2012

Jamir Amba (2015) Shifting options: a case study of shifting cultivation in Mokokchung District of Nagaland, India. In Erni Christian (ed) Shifting cultivation livelihood and food security. Bangkok, FAO, IWGIA and AIPP, pp. 159–202

Maithani BP (2005) Shifting cultivation in Northeast India (Policy, Issues and Options). New Delhi, Mittal Publication

Report on Development of North eastern Region (1981) Planning commission, government of India, November

Report of the Task Force Nagaland, UNDP and IeLO (2015) Department of Agriculture, Niti Ayog, Government of India, pp. 1–47. http://niti.gov.in/writereaddata/files/Nagaland_Report_0.pdf (Accessed on May and June 2018)

Sacchidananda (1989) Shifting cultivation in India. New Delhi, Concept Publishing Company

Senotsu M, Kinny A (2016) Shifting cultivation in Nagaland: prospects and challenges; ENVIS Bulletin Himalayan Ecology, vol. 24

Statistical Handbook of Nagaland, Directorate of Economics and Statistics, Department of Agriculture, Cooperation and Farmers Welfare, Ministry of Agriculture and Farmers Welfare, Government of India, 2006 to 2016

Sustainable Land and Ecosystem Management in Shifting Cultivation areas of Nagaland for Ecological and Livlihood Security (2009) United Nations Development Programme India (UNDP), Annual Work Plan
United Nations development Programme India, Annual Work Plan (2009) Sustainable land and ecosystem management in shifting cultivation areas of Nagaland for ecological and livelihood security. Government of Nagaland, India

Chapter 8
Land Use–Land Cover Dynamics in Baku Micro-watershed Area of Ausgram Block—II, Purba Bardhaman District, West Bengal, India

Raj Kumar Samanta and Narayan Chandra Jana

Abstract Micro-watershed management is the process of formulating action for the manipulation of natural, agricultural and human resources of the watershed area without affecting water, soil or land resources. Land use refers to various activities of human being over land area, whereas land cover indicates the biophysical state of the earth's surface. These are very dynamic. The changes in land use–land cover involve spatial and temporal aspects. Every watershed management project has a great impact on land use–land cover changes. Not only that but for proper utilization of land and other resources, there needs proper land evaluation and land use planning. Baku Micro-watershed project has a positive impact on land use–land cover aspects like increase of double-crop area, plantation activities, etc. It indicates a good mark of watershed development. In this paper, such kinds of spatio-temporal changes of land use–land cover patterns have been analysed. Relation of physiography and soil with land use has been identified. Here, low lands covering fertile soil with much agriculture and uplands are the areas of laterites and forest cover. The maximum portion of soils are sandy and acidic. Low lands indicate good quality land with a gentle slope and good quality soil. Land assessment and planning has been carried out for further better land utilization. The paper is based on extensive field works. Simple statistical and cartographic techniques have been applied to show various results.

Keywords Land use · Land cover · Watershed management · Micro-watershed · Land use planning

8.1 Introduction

Micro-watersheds lie at the lowest level of the hierarchical river system. It is administratively as well as operationally the most meaningful planning unit. The average

R. K. Samanta · N. C. Jana (✉)
Department of Geography, The University of Burdwan, Burdwan, India

N. C. Jana et al. (eds.), *Livelihood Enhancement Through Agriculture, Tourism and Health*, Advances in Geographical and Environmental Sciences,
https://doi.org/10.1007/978-981-16-7310-8_8

area as per Soil and Land Use Survey of India (SLUSI) is 1000 hector. A watershed is the area from where all rainwater flows to a single stream (Gurjar and Jat 2008). Watershed management is an integrated approach for the overall development of abiotic and biotic components of a watershed which includes rational and optimum utilization of natural resources and their conservation (Singh 2015). Land use simply means the use of land such as agricultural land and settlement. It is man's various activities and uses over the land surface. Land cover is the cover of the surface of the earth such as water bodies, grassland, forest cover and wasteland. Land cover is the biophysical state of the earth's surface. Land use–land cover is dynamic (Prakasam 2010). Land use is controlled by some factors such as geology, relief, drainage, climate, soil, vegetation, irrigation and government policies (De and Jana 1997). In this paper, all sorts of land use–land cover-related characteristics like spatio-temporal changes in LULC, the relation of soil and physiography with LULC of Baku micro-watershed project have been analysed.

8.2 Materials and Method

8.2.1 Study Area and the Rationale Behind the Selection of Study Area

Baku micro-watershed is the combination of parts of two mouzas—'Ba' for Babuisol and 'ku' for Kudiha of Bhalki Panchayet in Ausgram—II block of Purba Bardhaman district, West Bengal, India (Fig. 8.1). It is located between 23°28′80″ N to 23°29′30″ N latitude and 87°33′00″ E to 87°34′30″ E longitude. Baku micro-watershed was financed by National Bank for Agriculture and Rural Development (NABARD). The total area is 710 ha. Ridge is located on the east side, and the valley is located in the west. So, the general slope is from the east to the west and from the south to the north. In the east, there is a forested lateritic upland zone and in the west, there is drainage line and agricultural plain with the alluvial surface. The average height of this watershed is 57.5 m. The forest is mixed with Sal plants. Now after the watershed development programme post-monsoon average water level of ponds is 14 ft and the pre-monsoon situation is 7 ft. The groundwater level of the tube well is 24 ft (BGL) in pre-monsoon and 18 ft (BGL) in post-monsoon. Now there is no failure of tube well in summer. At present, the total population of Kuldiha is 850 (male—436 and female—414) and Babuisol have 739 total population (male—395 and female—344). Total SC families are 128 and ST families are 212. Before watershed project development, the area faced many problems such as soil erosion (Plate 8.1a), the water crisis in summer due to drought conditions, land degradation, lack of irrigation, the lower rate of agricultural production, deforestation, poverty and unemployment. But after watershed development, there has been a change in the biophysical and socio-economic conditions of the study area. These positive changes

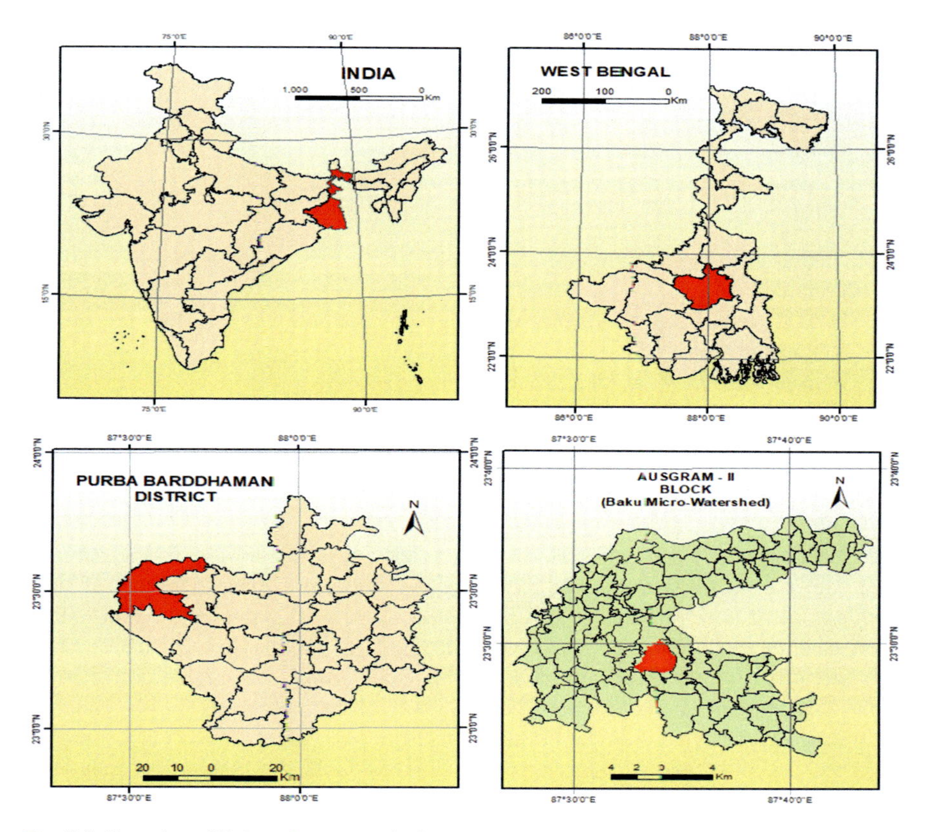

Fig. 8.1 Location of Baku micro-watershed

Plate 8.1 Pre-project land-related problems in Baku micro-watershed based on photographic survey—2013. **a** Severe soil erosion; **b** gully widening; **c** dry pond in summer; **d** wasteland formation; **e** degraded forest land

are well denoted by land use–land cover analysis. So the selection of this study area from the viewpoint of the watershed development is highly justified.

8.2.2 Objectives of the Study

1. To study the spatio-temporal changes of land use–land cover in the watershed area.
2. To analyse the relation between physiography and soil with land use–land cover.
3. To measure the land capability and quality.
4. To assess the existing land use for further planning.

8.2.3 Database and Methodology

The present paper is mainly based on an extensive field survey. GPS survey has been carried out to get locations and heights of different places. Survey of India Topographical Sheet—73M/11 and the Mouza maps of Babuisol and Kuldiha have been used to prepare land use–land cover map and other maps. Arc GIS 10.3 version software has been used for various mapping. Soil samples have been collected for the test to get their characteristics. Besides many cartographic and simple statistical techniques have been also used. Photographs have been taken to indicate pre-project problems and post-project improvements in land use–land cover situation. Land use–land cover changes have been explained by simple percentage analysis. Techniques of contour interpolation have been used to prepare contour map. Transect charts have been prepared to get the relation of LULC with physiography. Land Capability classification is based on the modified SLUSI method, 1970. Various problems have been identified by extensive field visits and question answering methods.

8.3 Results and Discussion

Integrated watershed management programmes are an important holistic approach for natural resource management. Whenever a watershed management programme is going on over the region, there happens an overall development from the physico-environmental and socio-economic point of view. Here in case of Baku micro-watershed project, there are various changes in biophysical aspects like improvement in soil and water conservation, reduction of soil erosion, increase in the groundwater table, changes in the cropping pattern and productivity, improvement in drinking water condition, water harvesting situation (Plate 8.2g), etc.

Plate 8.2 Post project situations of land in Baku micro-watershed-based photographic survey—2018. **a** Check dam cum reservoir; **b** new plantation; **c** earthen field drain; **d** mango forest; **e** inlet and outlet in pond; **f** contour bund; **g** excavated pond with water

8.3.1 Spatio-Temporal Changes in Land Use–Land Cover

One of the important objectives of watershed management is a better land use pattern. In case of land use–land cover a spectacular change has been detected before and after the Baku watershed development project (Table 8.1).

There are positive changes in the case of double crop land, ponds, mixed settlement, medicinal plant and plantation activities, etc. There are negative changes in the case of single crop land, fallow land, scrub forest land, etc. It carries a good mark of watershed development because the negative change in some parameters indicates success in watershed programme. So there is a reduction in single crop land and an increase in double crop land from 26.01 to 53.12 ha. Before the project, fallow and wasteland was 13.94 ha which has now become 6.68 ha after the project. Besides, before the project, there was no plantation or medicinal plants which now become 6.87 ha and 0.14 ha respectively (Table 8.1, Figs. 8.2, 8.3 and 8.4). Besides, there are some other activities which are the outcome of watershed management projects such as Kandar nala side plantation, earthen field drain, contour trenches

Table 8.1 Land use–land cover changes based on field survey 2013 and 2018

LULC classes	Pre-project area (ha)	Area in %	Post project area (ha)	Area in %	Land use change analysis	
					Difference in ha	Difference in %
Canal and field drain	1.13	0.16	3.13	0.44	2.0	0.28
Check dam	Nil	–	0.03	0.001	0.03	0.001
Crop land (double)	26.01	3.66	53.12	7.48	27.11	3.82
Crop land (single)	282.02	39.72	253.77	35.74	−28.25	−3.98
Fallow and wasteland	13.94	1.96	6.68	0.94	−7.26	−1.02
Water bodies and ponds	14.00	1.97	16.31	2.30	2.31	0.33
Mixed settlement	32.55	4.58	34.37	4.84	1.82	0.26
Playground	1.0	0.14	2.77	0.39	1.77	0.25
Forest	322.44	45.41	320.48	45.14	−1.96	−0.27
Scrub forest	15.39	2.17	10.81	1.52	−4.58	−0.65
Medicinal plants	Nil	–	0.14	0.02	0.14	0.02
Plantation	Nil	–	6.87	0.97	6.87	0.97
Miscellaneous	1.52	0.21	1.52	0.21	0	0
Total	710.00	100%	710.00	100%		

and bunds. These were constructed over the fallow land, scrub forest, forest areas to catch maximum rainwater. The average year-wise increase of double crop land is 5.42 ha, water bodies are 0.46 ha and plantation area is 1.37 ha. So, it is said that land use–land cover situations in Baku watershed have been changed over space and time.

8.3.2 Major Land Use–land Cover Units

8.3.2.1 Forest

Forest is a habitat of different types of plants, animals and other organisms. It helps to increase rainfall, prevent soil erosion, minimize surface runoff, increase recharge of groundwater, become a source of raw materials for industries, etc. Forest of Baku

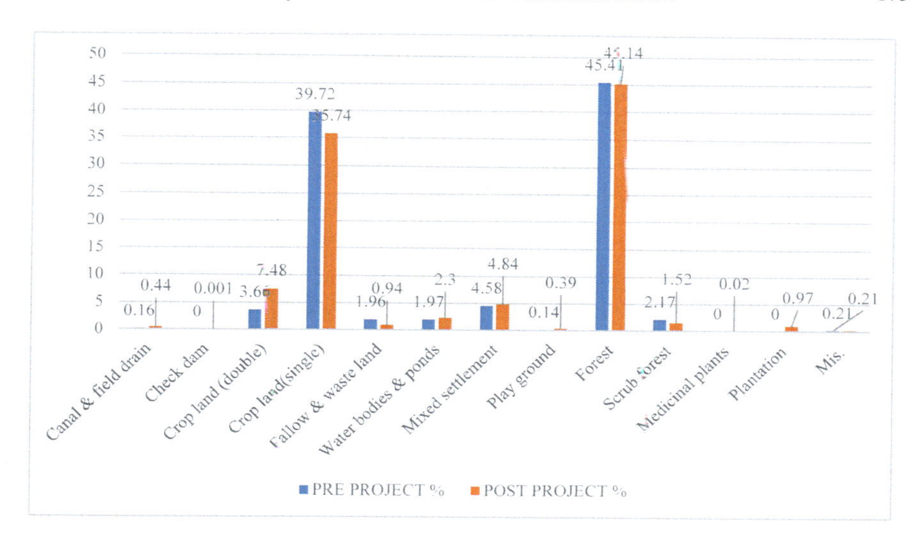

Fig. 8.2 Pre-project and post-project LULC percentages in Baku micro-watershed

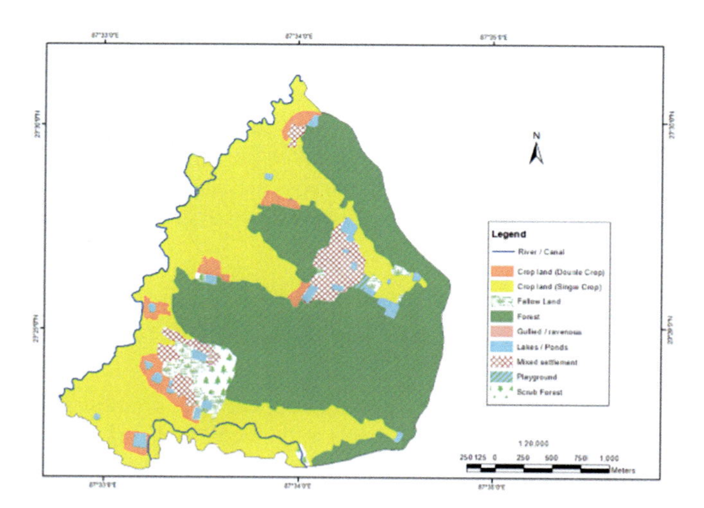

Fig. 8.3 Pre-project land use–land cover situation of Baku micro-watershed based on field survey—2013 and BL and LRO, Ausgram Block—II, Purba Bardhaman district

watershed is the open mixed type with maximum Sal plants. It is under the forest department of state government. It covers 45.14% of the total watershed area which is concentrated over lateritic and red soil tract. The entire eastern part is covered with forest land. The bulge of forest land is covered from east to west through the central part. Besides 0.65% of the scrub forest has been reduced after watershed development.

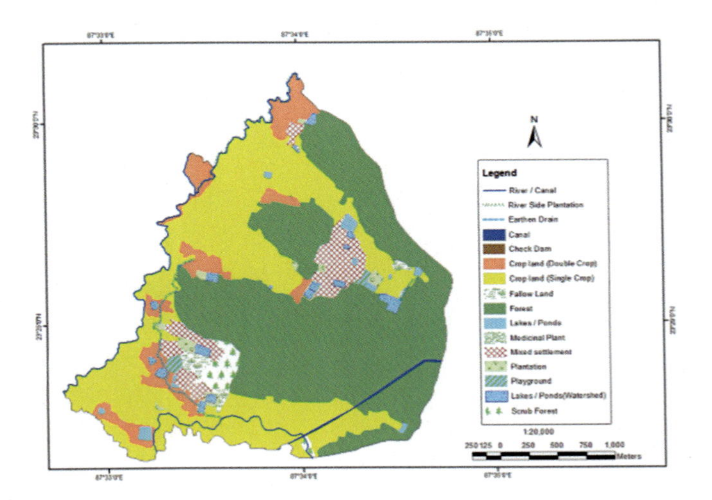

Fig. 8.4 Post project land use–land cover situation of Baku micro-watershed based on field survey—2018 and BL and LRO, Ausgram Block—II, Purba Bardhaman district

8.3.2.2 Fallow and Wasteland

Fallow land is such land that is not under cultivation at present but it was sown in the past. Wasteland is not a productive land in the present situation. It is in a sterile condition. But due to the improvement of the physical and socio-economic conditions, it could be changed. In Baku micro-watershed, these are found between Kuldiha village and forest land. After watershed development, 7.26 ha wasteland has been reduced.

8.3.2.3 Agricultural Land

Modern agricultural practices have given a fresh lease to agricultural land uses. To feed the increasing population, there needs to convert single crop land into double crop land. It is only possible by advanced agricultural technologies specially the improvement of the water harvesting system. In Baku watershed due to the improvement of water harvesting techniques 3.98% single crop land has been reduced and there is an increase of 3.82% double crop land. The double crop lands are mainly concentrated near the field ponds and tanks. This stored water is used in rabi (winter) cultivation mainly.

8.3.2.4 Water Bodies and Ponds

Water bodies are the most important ecosystem of the earth. Wetlands are called the kidney of the land surface. In Baku watershed, most of the ponds have been re-excavated and some new ponds have been excavated which are mainly situated outside the village and over agricultural fields. There is a 0.33% increase in water bodies than the pre-project situation.

8.3.2.5 Plantation Area

Afforestation or plantation is one of the important activities of each watershed development project. Here, new plantation of 6.87 ha have been done over vacant spaces, wastelands, scrub forest areas, pond embankments, etc. Besides medicinal plants have been planted over 0.14 ha land. So it is said that plant cover has been increased.

8.3.3 Relation Between Physiography and Land Use–Land Cover

Physiographically Baku watershed belongs to plain land. The maximum height is 65 m and the minimum height is 50 m (Fig. 8.5). The range of relief or relative relief is 15 m. The average height is 57.5 m. The height of the eastern side is much more than the west. Baku micro-watershed is divided into three physiographic units.

1. Below 50 m height zone: It is situated in the western margin along Kandar nala with fertile younger alluvial soil. It covers about 15% area of Baku micro-watershed. It is treated as low land.
2. 50–55 m height zone: It is situated in the middle portion with about 45% areal coverage. It is called midland or medium land with older alluvial soil and little undulations.
3. Above 55 m height zone: It is situated in the eastern part and extended towards the middle part. It is the upland part with laterite surfaces. It covers about 40% of the total area of Baku micro-watershed (Fig. 8.6).

The relationship between physiography and LULC in Baku micro-watershed has been easily shown by the transect chart. Transect chart shows the relationships between the physical and socio-cultural attributes of a region. In the west–east cross section west indicates low land where agricultural land is concentrated. In mid-portion, there is medium heighted land with human settlements, ponds, fallow land, scatter agricultural land and scrub forest (Fig. 8.7). In the eastern part, there is an upland zone which is covered by forest. Here, upland is the area of water discharge. Mid-land is the area of water harvesting, recharge and utilization. And low land is the area of water utilization and draining of excess water. The south–north profile also shows the relation between LULC and physiography. In the middle, there is upland which is covered with forest lands. In the extreme north, there is low land which is characterized by human settlement, ponds. In between human settlement and forest cover, there is agricultural land which is situated in a medium heighted zone (Fig. 8.8).

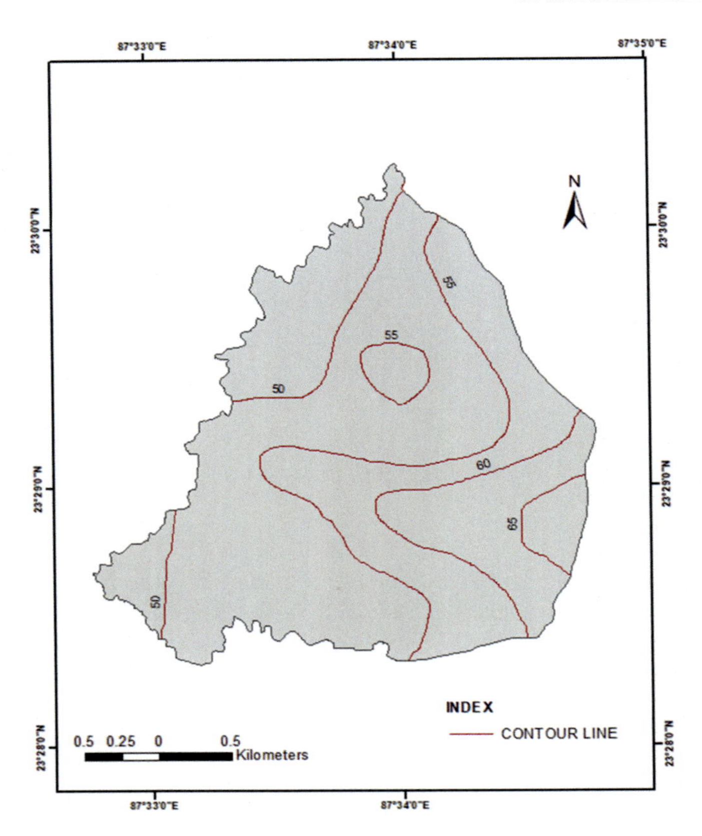

Fig. 8.5 Contours of Baku micro-watershed based on Topo. Sheet 73M/11 and field survey

8.3.4 Relationship Between Soil and LULC

Baku micro-watershed is characterized by basically two types of major soils—alluvial and laterite. About 60% area of this watershed has alluvial soil. In the extreme west, there is fertile younger alluvial soil. Southern and north-central parts have older alluvial soil with much clay concentration. The land use–land cover character of younger alluvial soil is fertile agricultural land, whereas in older alluvial soil, there are settlements, wasteland, playground, ponds, plantation areas, scrub forests, etc. Double crop lands are found much more in younger alluvial soil. Remaining 40% area of Baku micro-watershed is characterized by laterite soil (Fig. 8.9). It is unfertile. Mid-land covers this type of soil with open Sal forests and scrub forests. So, it is easily said that soil influences LULC pattern of Baku micro-watershed. The relation of soil characteristics with LULC classes is clearly shown in Table 8.2 in Baku micro-watershed.

Of the six sample sites, 66.67% area has above 50% sand content. About 33.33% area has silt content above 30%. Only 16.67% area has clay content of above 60%

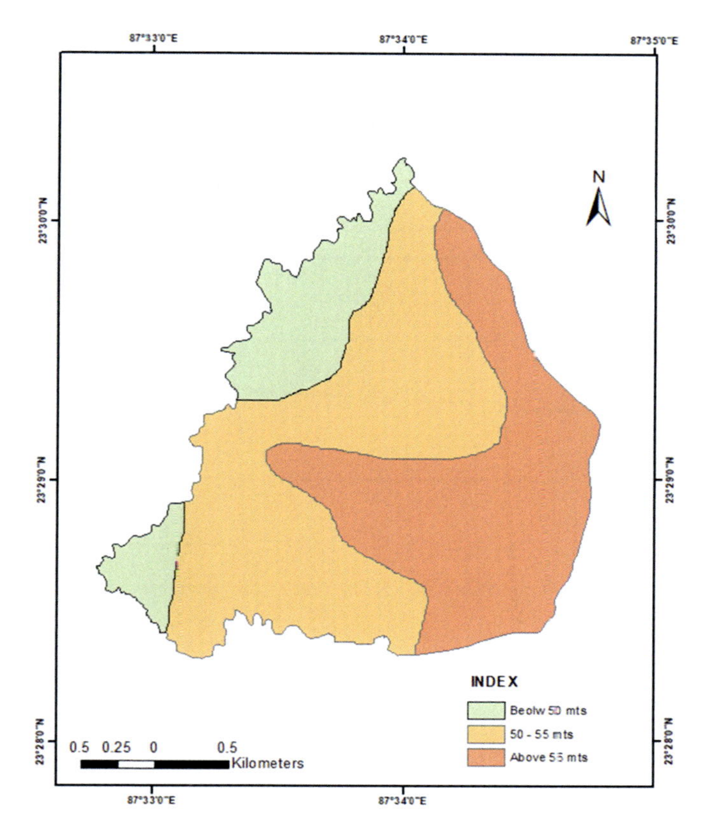

Fig. 8.6 Physiography of Baku micro-watershed based on Topo. Sheet 73M/11 and field survey

and organic carbon above 1%. Again 33.33% area of sample sites has organic matter above 1%. About 83.33% area has nitrogen below 0.10%. PH is acidic in nature. About 50% of the area has PH below 4.5. About 33.33% area have below 100 kg/ha, 33.33% have 100–200 kg/ha and the remaining 33.33% area have above 200 kg/ha average P_2O_5 content. 66.67% area has 200 kg/ha average K_2O content. Out of total sample sites, 50% have loamy texture and 16.67% of each has clay texture, silty clay loam and sandy loam texture. Of the sample sites, single crop lands have 33.33% share, forest/degraded forest land has 33.33% share, double crop land has 16.67% share and lateritic wasteland has 16.67% share (Table 8.2).

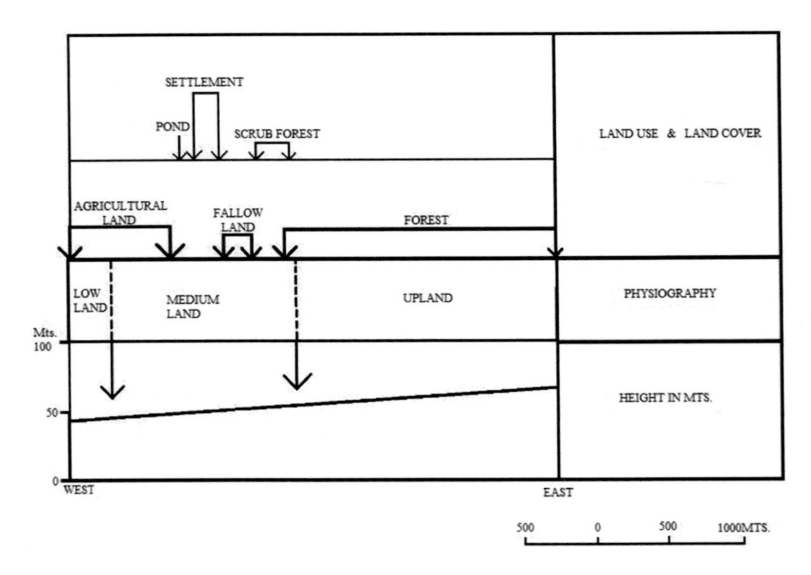

Fig. 8.7 West–east transect chart showing the relation among land use–land cover and physiography in Baku micro-watershed

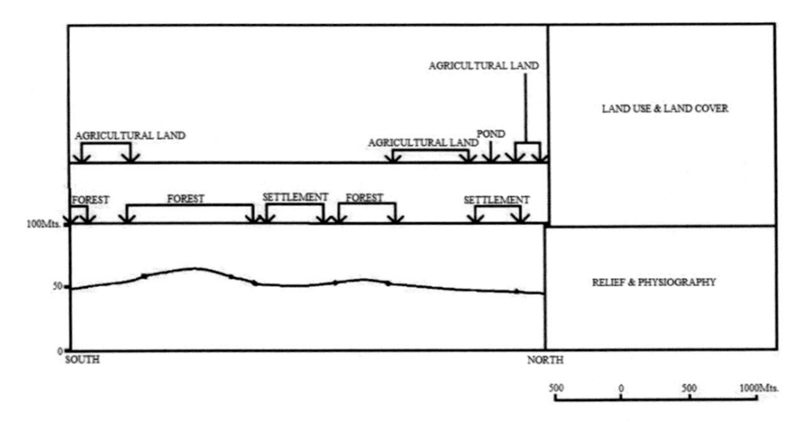

Fig. 8.8 South–north transect chart showing the relation among land use–land cover and physiography in Baku micro-watershed

8.3.5 Land Assessment and Capability Classification in Baku Micro-Watershed

Land assessment is the method of quality evaluation of a piece of land. Land assessment is the appraisal of land value under whole range of existing physical and socio-economic conditions. After proper assessment of land, it is classified on the basis

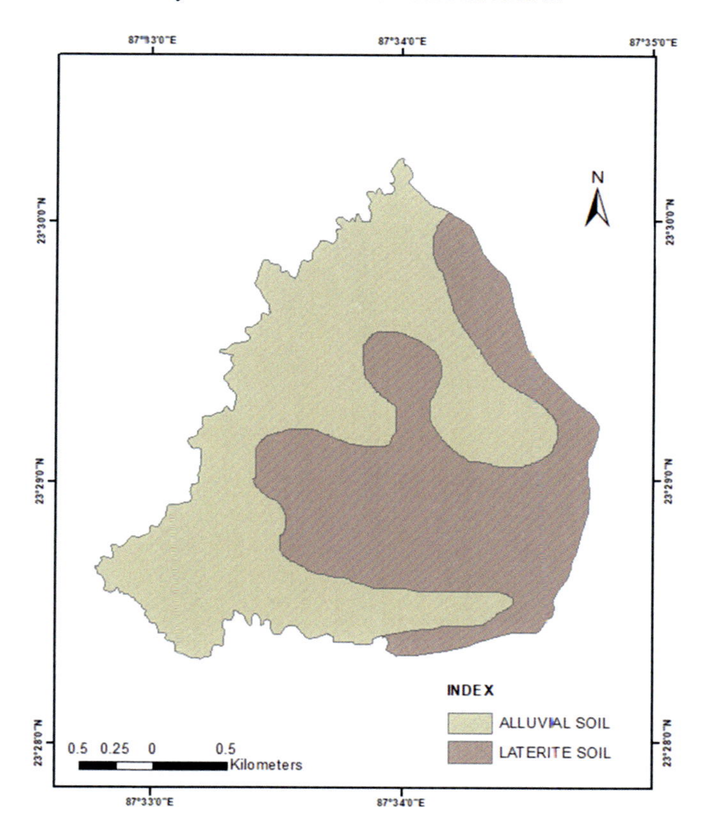

Fig. 8.9 Soils of Baku micro-watershed based on field survey

of its capabilities. Land capability or suitability indicates the potential capacity of a piece of land to support various types of land uses. Land capability classification is the method of interpretative grouping and grading of land according to its physical potentialities and limitations. The main objective of it is to realize capability, suitability and potentiality for optimum utilization of land. It is important because it gives scientific judgment for the conservation of land in a particular ecological set up (De and Jana 1997). Land capability is important in watershed management. The main objectives of soil and water conservation are to use the land as per its capability and need (Panhalkar et al. 2014).

There are mainly three types of land capability classes, i.e. II, III and IV. Class II capability has good quality land which is suitable for agriculture. Class III capability class has moderately good quality land with older alluvial. It is to be improved by watershed management. Class VI capability class has poor land quality which is undulating, lateritic and forested (Table 8.3 and Fig. 8.10). There need an immediate action plan by watershed development for its quality improvement.

Table 8.2 Relation of soil characteristics and LULC in Baku micro-watershed based on field survey and soil testing

Location of sample sites (latitude and longitude)	Sand (%)	Silt (%)	Clay (%)	Texture	PH	Organic carbon (%)	Organic matter (%)	Nitrogen (%)	Avg. P_2O_5 (kg/ha)	Avg. K_2O (kg/ha)	Present LULC type
23° 28′ 40″ N 87° 33′ 30″ E	60.6	15.4	22.0	Loamy	4.9	0.17	0.29	0.015	84	218	Lateritic wasteland
23° 30′ 00″ N 87° 34′ 00″ E	40.6	31.4	26.0	Silty clay loam	5.3	0.71	1.22	0.061	77	250	Double crop land
23° 28′ 42″ N 87° 33′ 00″ E	53.6	31.4	14.0	Loamy	4.2	0.28	0.52	0.026	208	107	Single crop land
23° 28′ 30″ N 87° 34′ 15″ E	14.6	26.4	60.0	Clay	4.3	1.20	2.10	0.105	256	410	Single crop land
23° 29′ 10″ N 87° 33′ 45″ E	60.6	23.4	14.0	Loamy	4.5	0.12	0.21	0.011	114	177	Degraded forest land
23° 29′ 15″ N 87° 34′ 50″ E	71.6	17.4	10.0	Sandy Loam	5.7	0.30	0.52	0.026	107	206	Forest land
Average	50.27	24.23	24.33		4.82	0.46	0.81	0.041	141	228	

Table 8.3 Land capability classification of Baku micro-watershed based on modified after SLUSI method, 1970

Major category	Capability class	Physiographic class	Soil type	Characteristics	Land quality
(A) Land suitable for cultivation	Class II	Low land	Younger alluvial	Gentle slope, good soil, good cultivable land	Good
	Class III	Medium land	Older alluvial	Gentle to moderate slope, comparatively good soil with subject to water and wind erosion, moderately good cultivable land	Moderately good
(B) Land not suitable for cultivation	Class VI	Upland	Laterite	Flat to gently sloping land with undulating surface, forest covered land, not suitable for farming	Poor

8.4 Risk Assessment and Recommendations

8.4.1 Various Problems Identified in the Study Area

Though micro-watershed management has been carried out in the area but still now the area faces some land-related problems.

(A) Problems found in the upland zone

- Rill and gully erosion
- Deforestation
- Overgrazing
- Illegal morrum extraction makes wasteland
- Acidic soil character due to high litter falling
- Loss of water due to maximum runoff during rainy season
- Objection of the forest department to cut pond and contour trench in the forested area.

(B) Problems found in the midland zone

- Poor water holding capacity of soil
- Concentration of fallow and wasteland in some places
- Fall of groundwater layer in summer

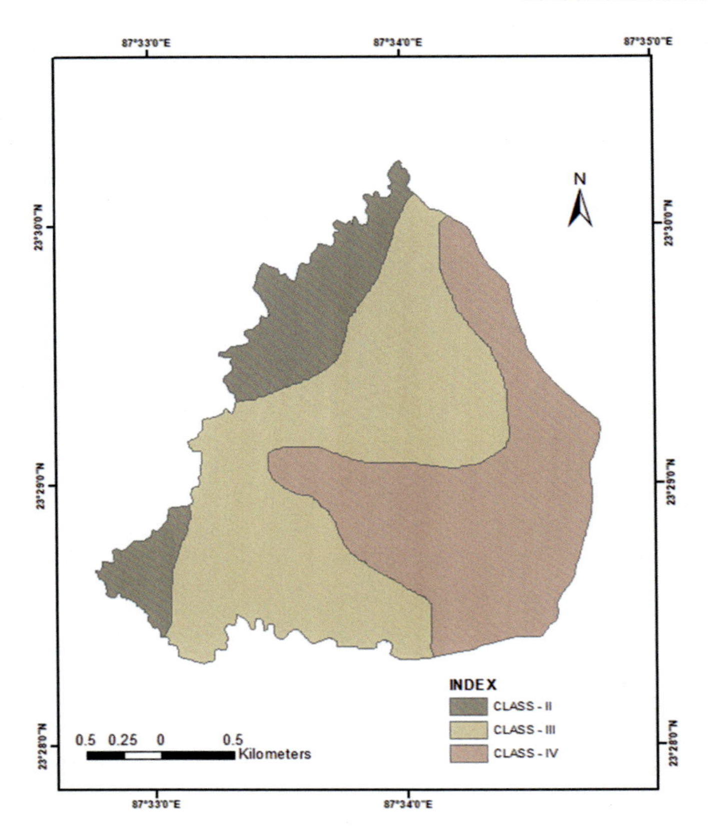

Fig. 8.10 Land capability classification of Baku micro-watershed based on modified after SLUSI method, 1970

- Illegal soil capturing by the brick field owner
- Lack of irrigation facility
- Soil erosion in rainy season.

(C) Problems found in the low land zone

- Decreasing productivity and soil fertility by increasing use of chemical fertilizer.
- Lack of water harvesting structures
- Cracks in clay soil during summer
- Silt deposition in field channels
- Reduction of crop productivity due to increasing soil acidity.

8.4.2 Management and Planning Measures

To overcome all such above problems there need proper management and sustainable land use planning measures in the study area.

(A) Planning and management measures for upland zone

- Afforestation in deforested lands.
- Encouragement of community forestry or Joint Forest Management.
- Conflict resolution with the forest department to construct water harvesting structures like contour bund, contour trench, etc. in the forested area.
- Gully plugging measures with natural ingredients.
- Proper regulation against illegal forest cutting and overgrazing.

(B) Planning and management measures for midland zone

- Fodder cultivation over wasteland.
- Social forestry over fallow land, wasteland, pond embankment, corners of playground, etc.
- Re-excavation of ponds and other water harvesting structures.
- Proper law and regulation against illegal soil capturing.
- Arrangement of more contour bunds, contour trenches and recharge wells, etc.
- Land levelling over undulating surface.

(C) Planning and management measures for low land zone

- Use of lime to reduce soil acidity
- Re-excavation of field drains and construction of new ponds.
- Proper sowing of land to increase its productivity and water holding capacity
- Arrangement of organic manure, vermin-compost and proper weeding.
- Introduction of zero tillage, drought farming, crop rotation, etc.
- Practice of double cropping, rabi (winter) cultivation and vegetable cultivation
- Arrangement of proper irrigation facilities.
- Emphasis on pluses, groundnut, til, mustard, wheat, maize and cashew cultivation.

8.5 Major Findings

- In Baku micro-watershed, there is an increase in double crop lands and water bodies, i.e. 3.82% and 0.33%, respectively.
- Below 50 m height relief zones are prone to fertile agricultural land and above 55 m height zones are mainly covered with forest land.
- Low land areas are characterized by fertile alluvial soil, whereas upland zones are covered with laterite soil
- Maximum portion of soils are sandy and acidic in nature.
- Low lands are characterized by good quality lands, whereas uplands are characterized by poor land capability.

- Lack of water harvesting and soil degradations are the main problems of the study area.
- To overcome all such problems emphasis is to be given on afforestation, construction of water harvesting structures, suitable land management and scientific cropping system.

8.6 Conclusion

Land Use–Land Cover studies relating to Baku micro-watershed project show a dynamic picture. It is important from the natural resource management and planning point of view. Spatio-temporal changes in LULC pattern have been clearly identified which indicates the fruitful effect of micro-watershed project. Relation of physiography and soil with LULC is also acute in Baku micro-watershed area. Though there are some problems but the above problems are to be eliminated with proper land use planning and management measures in a participative manner. Scientific land management measures are to be linked with Mahatma Gandhi National Rural Employment Guarantee Act (MGNREGA) programmes to get immediate results. In this context, lateritic wastelands and degraded forest lands are to be given more priority for their better treatment. Besides Joint Forest Management, social forestry, renovation of water harvesting structures, irrigation development, drought farming, proper regulation, etc. are to be popularized to get sustainable land use patterns in the study area.

Acknowledgements The authors are grateful to the Department of Geography, The University of Burdwan, India, for providing their necessary facilities to do this research work. The authors are also thankful to the villagers of the Baku micro-watershed area for their kind assistance and the persons who have given their effective co-operation. They are also grateful to the officers and staffs of various government and non-government offices who have provided their best help.

References

De NK, Jana NC (1997) The land—multifaceted appraisal and management. Sribhumi Publishing Company, Calcutta

Gurjar RK, Jat BC (2008) Geography of water resources. Rawat Publication, New Delhi and Jaipur

Panhalkar SS, Mali SP, Pawar CT (2014) Land capability classification in Hiranyakeshi Basin of Maharashtra (India): a Geoinformatics approach. Int J Eng Tech Res (IJETR) 2(6):18–21

Prakasam C (2010) Land use and land cover change detection through remote sensing approach: a case study of Kodaikanaltaluk, Tamil Nadu. Int J Geomat Geosci 1(2):150–158

Singh S (2015) Fundamentals of hydrology. Pravalika Publications, Allahabad

Chapter 9
Spatio-temporal Changes of Crop Combination in Selected C.D. Blocks of Purba Bardhaman District, West Bengal, India

Chanchal Kumar Dey and Tapas Mistri

Abstract Agriculture is the main driving force of the Indian economy contributing 18% country's GDP, and above 50% of the population has been engaged in agriculture and allied activities (Economic survey, 2017–18). Regional disparities are periodically surveyed in various parts of the world to monitor and measure by the several agricultural practices. Through the regionalization of agriculture, India is divided into some categories. Among a variety of methods, crop combination is the most significant technique for agricultural zoning and mapping purposes. Usually, crops are grown up in permutation and it is occasionally that a certain crop possesses a place of entire separation than additional crops in a given areal unit at a given point of time. Undivided Burdwan district was forever known as the "**Granary of Bengal**" but now, Purba Bardhaman district has occupied the greater parts of the agriculturally advanced area. To find out the agricultural zones of the district, crop combination is an important technique to understand in which areas how many crops are cultivated and to what extent. So present study is concentrated to find out the spatio-temporal changes of crop combination and their comparative analyses in selected blocks staying on both sides of the Damodar river in Purba Bardhaman district from 2000–01 to 2015–16. The present study is based on mainly secondary data and perception studies of the farmers. QGIS tools and Coppock's crop combination method have been used. Finally, changing trends have been analyzed that show two blocks are having unchanged status, while four blocks are gradually decreasing and two blocks have been showing the rising trend.

Keywords Agriculture · Crop combination · Changing trends and causes

---

C. K. Dey (✉)
Ph.D. Research Scholar, Department of Geography, The University of Burdwan, Golapbag, Purba Bardhaman, West Bengal, India

T. Mistri
Assistant Professor, Department of Geography, The University of Burdwan, Golapbag, Purba Bardhaman, West Bengal, India
e-mail: tmistri@geo.buruniv.ac.in

N. C. Jana et al. (eds.), *Livelihood Enhancement Through Agriculture, Tourism and Health*, Advances in Geographical and Environmental Sciences,
https://doi.org/10.1007/978-981-16-7310-8_9

9.1 Introduction

Agriculture has been the ground-breaking force in the headway of any nation since the antediluvian period but India has no exception. Arthur Keith, a Scottish anthropologist, says that "The discovery of agriculture was the first big step toward a civilization" (Saurabh 2019). Agriculture has silently and slowly laid down the cornerstone of Indian's development trajectory. On the basis of the NSSO survey (68th round), estimated nearly 49% of employments are coming from agriculture and related activities (Mishra 2019). Agriculture is the backbone of the Indian financial structure. Swaminathan observed that "If agriculture goes wrong, nothing will have a chance to go right in our country" (Mukherjee 2019). Regionalization of agriculture is a very significant perception in the discussion of agricultural science. In respect of homogeneity, incessant area and the superficial edge of the agricultural region, numerous concepts have been applied to conclude the regional diversities and disparities in agriculture, i.e., crop diversification, crop concentration, crop rotation, cropping intensity, cropping systems, cropping pattern and crop combination (Husain 1996). The concept of crop combination is a very significant idea to agricultural geography. Actually, crop combination plays an imperative role to take care of the soil and also improvement of soil fecundity and giving the maximum profits in farming. Crop combination refers to authentic proportions of total cropped area occupied by varieties of crops with concise theoretical distributions there in total cropped area is objectively distributed with in numerous crops (Singh and Dhillon 1984). Generally, crops are sown in combination and an individual crop acquires a position of entire separation from other crops in a particular areal component at a particular point of time (Husain 2015). Siddhartha and Mukherjee (2003) examined that crop combination is an art that is applied to set up the boundaries of agricultural regions based on statistical techniques measurement of the area (Siddhartha and Mukherjee 2003). Crop combination would finally trim down the scopes of oversimplified generalization of the spatial distribution of particular crops or combinations (Ali, 1878). Weaver (1954) was first introduced to find out the statistical technique for calculation crop combination (Weaver 1973). Coppock (1964) and Singh (1976) were suggested to modify and adopt least squares technique to find out the crop combination in a particular region (Coppock 1964). Many crops have been practiced in the Indian agriculture system but wheat and rice remain the dominant staple food. After the green revolution in India, a mono-crop combination has been developed that challenged the conventional crop rotation, crop succession and overall crops calendar (Thapar 1973). Swaminathan (2007) highlighted that wheat and rice are the most significant food crops those are increasing food grain production. Whereas, production of pulses and millets is decreased or becoming sluggish day by day (Swaminathan 2007). Union minister of agriculture, Shri Radha Mohan Singh, said that "India has been transformed from a food deficient country to a food exporting country about the last six decades". Shivay and Singh (2017) realized that wheat-rice crop combination is the backbone of Indian's food security and it contributes 80% of the total food basket in India (Shivay and Singh 2017). Besides, Sandhu and Chaturvedi (2018) observed that India is the maximum producer, consumer and importer of different types of pulses globally (Sandhu and Chaturvedi 2018). Consecutive research and agro-development attempts may be a fruitful improvement in these "nutri-rich food" or "healthy food" production and productivity during the past three successive plan periods (2002–2017). Besides, hostile or adverse crops succession of this system is deferment in productivity, deceleration of soil fertility and decline of ground-level water (Sindhu and Johl 2002; Singh 1997).

Jana (2017) has highlighted that crop combination and cropping intensity are co-related with the constructive result of infrastructural development of a certain area that produces different types of crops as well as a diversity of cropping intensity, and it is observed in Daspur-I C.D. block under Paschim Medinipur district (Jana 2017). Therefore, erstwhile Burdwan district was developed both in agriculture and industry. But now a greater portion of the agricultural areas are in Purba Bardhaman district and it is also well-known as "*Granary of West Bengal*". But it is observed that C.D. blockwise cropping intensity and crop combination are an uneven distribution in the whole district. Chakraborty and Mistri (2017) have observed that *Aman* (paddy)-*Boro* (paddy) crop combination is found in Burdwan, Raina-I and Galsi-I C.D. blocks, and this is the most significant crop combination that contributes 80% of *Rabi* and *Kharif* crops. Whereas Jamalpur block has observed rice-potato combination and *Aus*-*Aman*-potato-*Boro* crop combination has been recorded in Memari-I C.D. block. Therefore, the rice-rice-potato combination has been seen in the Raina-II C.D. block (Chakraborty and Mistri 2017). This paper tries to understand the spatio-temporal changes of crop combination in the seven C.D. blocks (lies on both banks of the Damodar river) among 23 C.D. blocks and the probable causes of such changes. Actually, proper crop rotation, crop combination and crop diversification support increasing soil fecundity and it would help for improved sustainability, ecological balance and socio-economic betterment of the study area.

9.2 Study Area

The study area covers some selected C.D. blocks, i.e., Bardhaman, Galsi-II, Memari-I, Jamalpur, Raina-I, Raina-II and Khandaghosh of Purba Bardhaman district in West Bengal. The study area lies on both banks of the Damodar River. The district has come into existence on 7th April 2017, after the division of erstwhile Burdwan district and its headquarter is Bardhaman. Purba Bardhaman is an agriculturally very prosperous (mainly rice and potato) district than other districts of West Bengal. This part of West Bengal is usually familiar as agriculturally advanced and is identified as the ***"Granary of the West Bengal" or "Rice bowl of West Bengal"***. The study area extended from 22° 54′ 08″ N to 23° 23′ 14″ N latitudes and 87° 36′ 33″ E to 88° 12′ 02″ E longitudes. It is bordered on the north by Bhatar, Ausgram-I, Monteswar and Purbasthali-I C.D. blocks in Purba Bardhaman district and Murshidabad district, on the east by Hooghly district and Kalna-II C.D. block under Purba Bardhaman district, on the south by Hooghly-Bankura district and west by the Paschim Bardhaman and Bankura district. Two important markets lie in the study area, i.e., Bardhaman and Memari market. According to the 2011 census, the total population is 1555064, the density of population is 828.75/ km^2 and contains an area of 1905.86 km^2 (735.85 sq miles) (Fig 9.1).

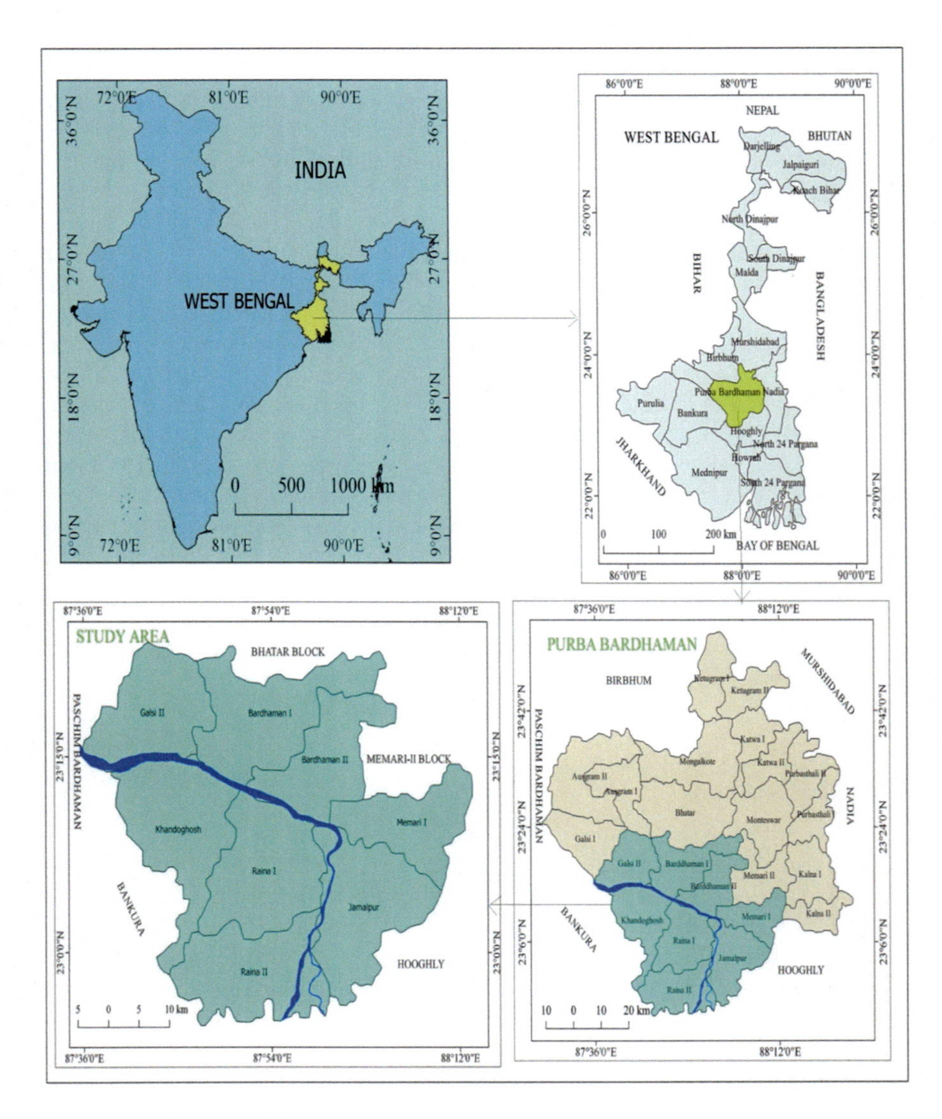

Fig 9.1 Location map of the study area

9.3 Objectives of the Study

The main objectives of the study are as follows:

- To find out the changing trends of crop combination in some selected C.D. blocks in Purba Bardhaman on a spatio-temporal scale.
- To explore the causes of such changes by making a comparative analysis of crop combinations in the study area.

9.4 Database

The present study is based on mainly secondary data. The data and information from 2000–01 to 2015–16 have had been collected from the Bureau of Applied Economics and Statistics (BAES), Government of West Bengal, and the office of the Deputy Director of Agriculture (DDA), Bardhaman. Therefore, data was compiled from different sources, and also supporting field observation has been done.

9.5 Methodology

Crop combination, an important concept in the regionalization of agriculture, depicts to find out the blueprint and processes of agricultural zoning Proper crop combination practices can save in farmer's agro and agri-sectors to combat the challenges of climate change in general and the vagaries of monsoon in particular. A crop distribution map is very imperative for planners, economists and scientists. So geographers have valid reasons to place emphasis to find out the best crop combination matching the ground reality. There are two universal methods for analyzing crop combinations as follows.

9.5.1 Optional Methods or Arbitrary Method

For example one crop area, two crops area or three crops area etc.

9.5.2 Scientific Method or Statistical Prescribed Method

J. C. Weavers, 1954 in his book "*Crop combination in the middle West*" first applied the statistical methods to calculate the crop combination in an area. Several geographers and economists have applied crop combination methods; they are Weavers

(1954), Coppock (1964), Doi (1959), Thomas (1963) and Rafiullah (1965), and many others modified this technique mainly with the help of a standard statistical algorithm. Weavers applied the standard deviation technique for the calculation of minimum deviation as follows:

$$SD = \sqrt{\frac{d^2}{n}} \tag{9.1}$$

where

SD Standard Deviation
d the difference between the actual percentage and theoretical percentage of areal units in an area
n zumber of crops given

Here, the authors followed **Coppock**'s method. Actually, the final modification by Coppock (1964) was applied as the Thomas technique (1963) to determine only the sum of the squared deviations and not to divide it by the total numbers.

$$\mathrm{SD} = \sqrt{d^2}$$

where

d the difference between the actual percentage and theoretical percentage of areal units in an area

The combination having the lowest or smallest sum of the squared deviations will be used as the conventional crop combination (Singh and Dhillon 1984).

9.6 Result

9.6.1 Trends Remain Unchanged of Crop Combination Zone

This scenario is found only in the Galsi-II C.D. block. (Tables 9.8, 9.9, 9.10 and 9.11 and Fig 9.3).

9.6.1.1 Galsi-II C.D. Block

Aman and Boro paddies are the most convenient crops in Galsi-II throughout study periods. A large number of *Boro* cultivators can be noticed due to high irrigation intensity mainly the Damodar canal irrigation (Dey and Mistri 2018a, b, c, d, e), but crop diversification is low (Dey and Mistri, 2017), where rice crop concentration

Table 9.1 Changing trends of crop combination in Galsi-II C.D. block

Sl. no.	Year	Values of crop combination	Types of crop combination
1	2000–01	Two	Aman, Boro
2	2005–06	Two	Boro, Aman
3	2010–11	Two	Aman, Boro
4	2015–16	Two	Aman, Boro

Compiled by authors

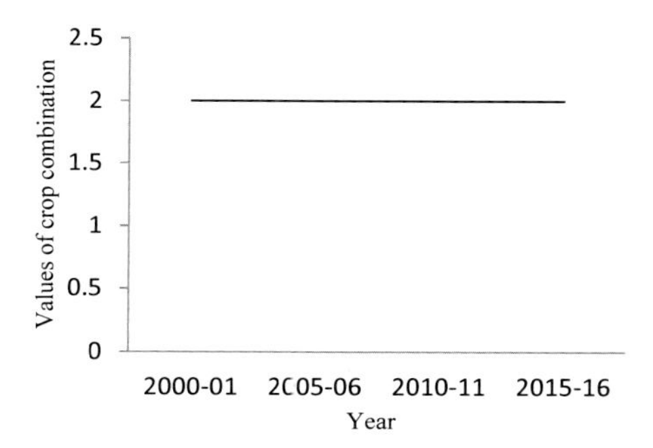

Fig. 9.2 Changing trends of crop combination in Galsi-II C.D. block

being high (Dey and Mistri 2018a, b, c, d, e). Besides, it is observed that *Aman* cultivation is increasing (i.e., in 2005, it is showing 40.9% whereas, in the cropping year 2015, it is 60.73%) steadily in entire crop production and *Boro* cultivation is decreasing for the reasons of water depletion, lack of Damodar's irrigation system and gradually decreasing fertility of land due to lake of scientific crop rotation system (Table 9.1; Figs. 9.2 and 9.3).

9.6.2 Increasing Trends of Crop Combination Regions

Only two C.D. blocks lie in this category. These are as follows (Tables 9.8, 9.9, 9.10 and 9.11 and Fig 9.3).

9.6.2.1 Khandaghosh C.D. Block

In 2000–01, there are mainly two crops combinations observed, i.e., *Aman* and *Boro* paddy. Khandaghosh is a C.D. block that has augmented crop concentration values of 0.19 in between 2005–06 and 2015–16 cropping years (Dey and Mistri 2018d). From the cropping year in 2005, there is a rising trend in potato cultivation and also crop combination is *Aman* paddy-*Boro* paddy-potato. In 2015, four crops combination practices, i.e., *Aman*-potato-*Boro*-mustard seed have appeared. Khandaghosh C.D.

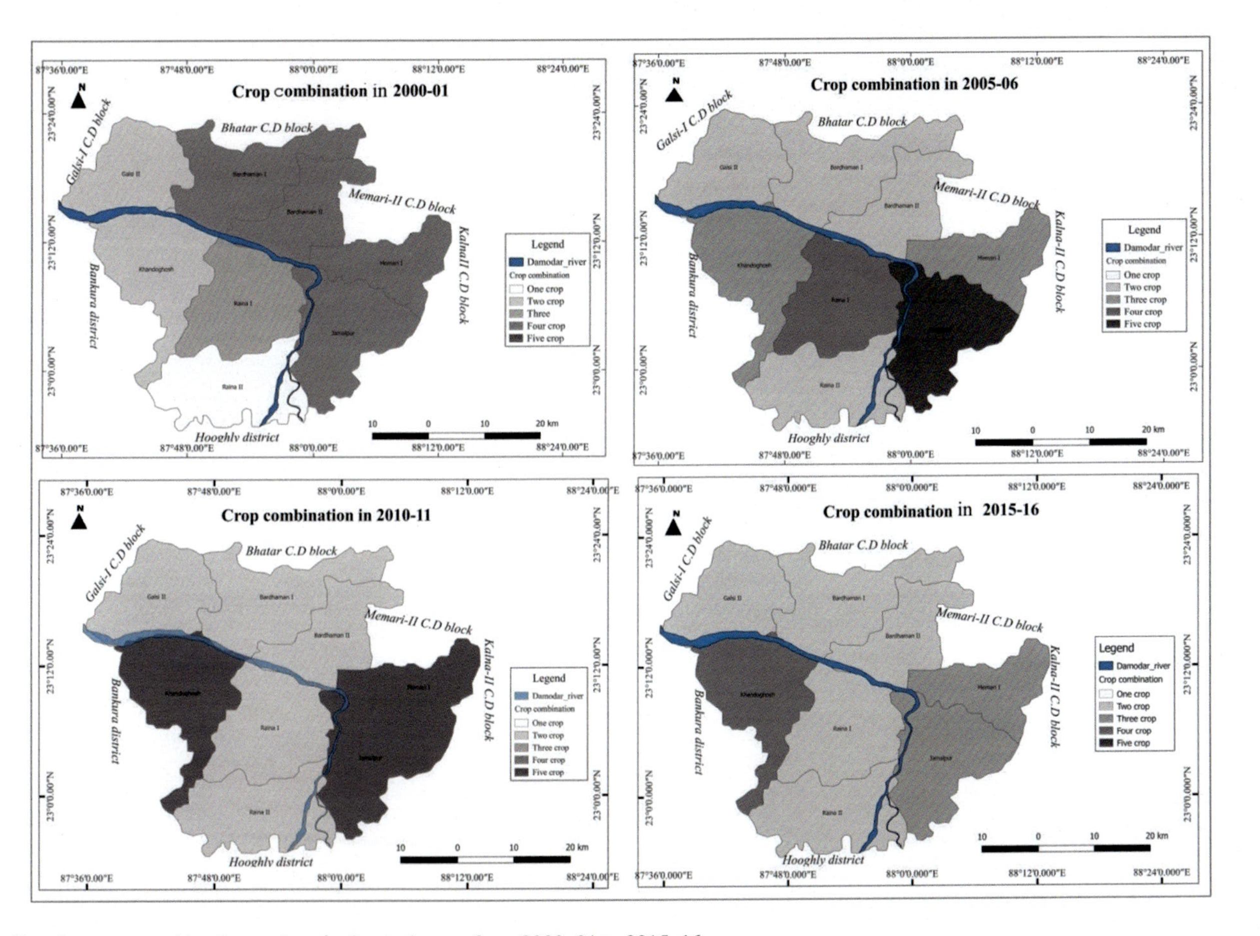

Fig 9.3 Showing crop combination regions in the study area from 2000–01 to 2015–16

Table 9.2 Changing trends of crop combination in Khandaghosh C.D. block

Sl. no.	Year	Values of crop combination	Types of crop combination
1	2000–01	Two	*Aman, Boro*
2	2005–06	Three	*Aman, Boro*, Potato
3	2010–11	Three	*Aman, Boro*, Potato
4	2015–16	Four	*Aman, Boro*, Potato, Mustard seed

Compiled by authors

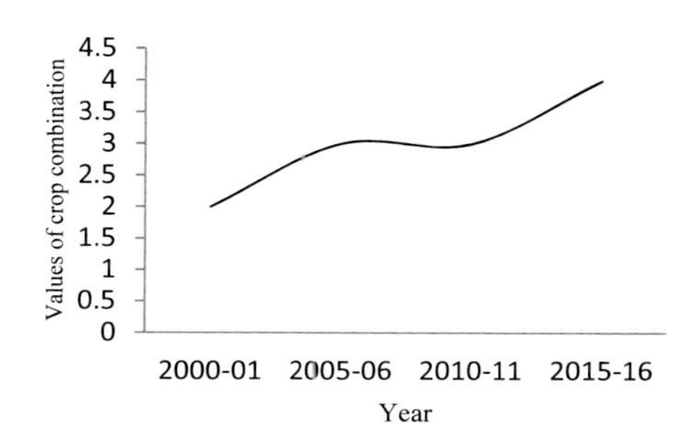

Fig. 9.4 Changing trends of crop combination in Khandaghosh C.D. block

block has been shifted from medium crop diversification to high crop diversification in between 2005–06 and 2015–16 cropping years due to positive agro-climatic condition, a large number of farmers being motivated crop diversification considering its multisided cropland by encroaching on fallow and grassland, farmers grow several crops to meet the family requirements (Dey and Mistri 2017) (Table 9.2; Figs. 9.3 and 9.4).

9.6.2.2 Raina-II C.D. Block

Mono-crop combination is found in Raina-II C.D. block in 2000–01 that is *Aman* paddy; it is concentrated in 66.81% of total cultivated land (Table 9.8). Since then, the number of crop combinations is going to be two. These are *Aman* and *Boro* paddy (Table 9.3). From 2000 to 2015, it is observed that *Aman* cultivation has occupied

Table 9.3 Changing trends of crop combination in Raina-II C.D. block

Sl. no.	Year	Values of crop combination	Types of crop combination
1	2000–01	One	*Aman*
2	2005–06	Two	*Aman, Boro*
3	2010–11	Two	*Aman, Boro*
4	2015–16	Two	*Aman, Boro*

Compiled by authors

Fig. 9.5 Changing trends of crop combination in Raina-II C.D. block

above 60%, and also *Boro* cultivation is showing an upward trend. It is said that paddy is a dominant crop in Raina-II C.D. block as Raina blocks lie on alluvial plains in between the Dwarakeswar or Dhalkisor river and the Damodar river, development of paddy marketing, rice milling concentration, traditional attitude of farmers, irrigation facilities, etc. (Table 9.3; Figs. 9.3 and 9.5).

9.6.3 Decreasing Trend of Crop Combination Regions

Four C.D. blocks are in this category. These are as follows (Tables 9.8, 9.9, 9.10 and 9.11 and Fig. 9.3).

9.6.3.1 Bardhaman C.D. Block

On the basis of 2000–01 data, four crops combinations are found in the Bardhaman C.D. block by applying the Coppock formula. Since then, there have been two crops combinations available, i.e., *Aman* and *Boro* paddy (Table 9.4). Bardhaman is laying in low crop diversification region in between 2005–06 and 2015–16 cropping year due to adverse effects of urbanization, acquisition of land for nonagricultural purpose and development of rice marketing system (Dey and Mistri 2017). Therefore, it is

Table 9.4 Changing trends of crop combination in Bardhaman C.D. block

Sl. no.	Year	Values of crop combination	Types of crop combination
1	2000–01	Four	*Aman, Boro*, Rabi.Veg., Mustard Seed
2	2005–06	Two	*Aman, Boro*
3	2010–11	Two	*Aman, Boro*
4	2015–16	Two	*Aman, Boro*

Compiled by authors

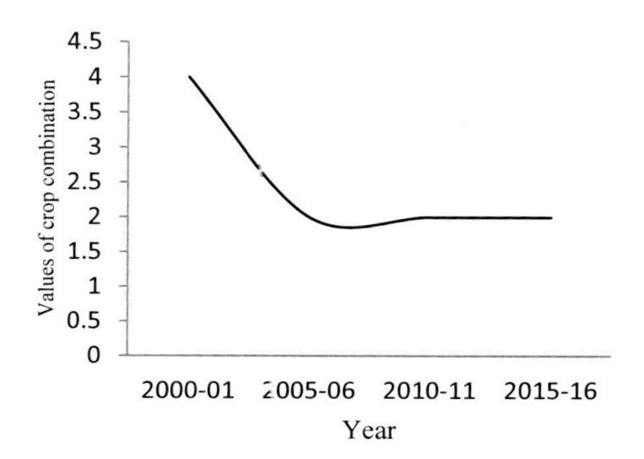

Fig. 9.6 Changing trends of crop combination in Bardhaman C.D. block

said that Bardhaman C.D. block is staying in a high rice crop concentration zone in between 2005 and 2015 due to low crop diversification practices, traditional attitude of farmers, good rice marketing system, high rice mill concentration, etc. (Dey and Mistri 2017; Dey and Mistri 2018a, b, c, d, e). It appears that the paddy cultivation area is increasing steadily in Bardhaman block mainly *Aman* cultivation that was 40.74% in 2001 and 56.62% in 2015 (Table 9.4, Figs. 9.3 and 9.6).

9.6.3.2 Jamalpur C.D. Block

In 2000, four crops combinations, i.e., *Aman*-potato-*Boro-Aus* are recorded, and in the 2005 cropping year, five crops combinations are observed that are *Aman*-potato-*Boro-Til-Aus*. Therefore, three crops combinations are fixed, i.e., *Aman*, potato-*Aus* (Table 9.5 and Fig. 9.7). In Jamalpur block, crop diversification remains always high without any major changes from 2005 to 2015 due to high quality alluvial soil, well irrigation facilities, suitable positive attitude, agricultural mechanization and after all storage facilities that lead to higher diversification (Dey and Mistri 2017) (Tables 9.8, 9.9 and 9.10). It can be found that *Aman* cultivation is decreasing and the area of potato cultivation is increasing. From 2000 to 2010 cropping year, increasing tendency in *Aus* cultivation is observed (Tables 9.8, 9.9, 9.10 and 9.11 and Fig. 9.3).

Table 9.5 Changing trends of crop combination in Jamalpur C.D. block

Sl. no.	Year	Values of crop combination	Types of crop combination
1	2000–01	Four	Aman, Potato, Boro, Aus
2	2005–06	Five	Aman, Potato, Boro, Til, Aus
3	2010–11	Three	Aman, Potato, Aus
4	2015–16	Three	Aman, Potato, Aus

Compiled by authors

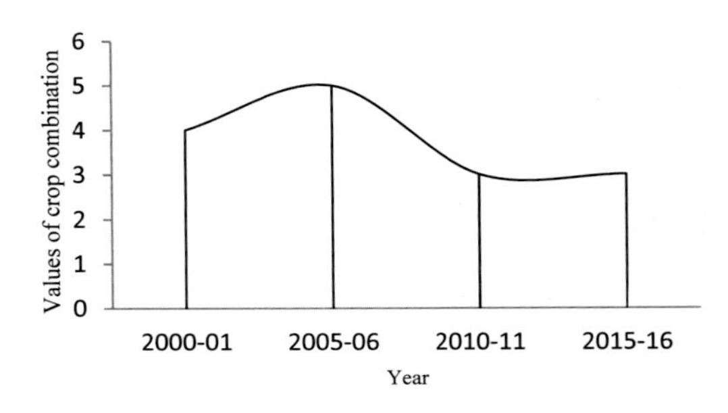

Fig. 9.7 Changing trends of crop combination in Jamalpur C.D. block

Table 9.6 Changing trends of crop combination in Memari-I C.D. block

Sl. no.	Year	Values of crop combination	Types of crop combination
1	2000–01	Four	*Aman*, Potato, *Boro, Aus*
2	2005–06	Three	*Aman*, Potato, *Boro, Aus*
3	2010–11	Three	*Aman, Aus*, Potato
4	2015–16	Three	*Aman*, Potato, *Aus*

Compiled by authors

9.6.3.3 Memari-I C.D. Block

Following Coppock's crop combination method, three crops combinations have been observed during the 2005–2015 cropping year (Table 9.6 and Fig. 9.3). Since 2010–11, there has been a combination of *Aus* paddy in place of *Boro* paddy. The gross cropped area of *Boro* is decreasing whereas *Aus* paddy is increasing as *Aus* paddy ripes quickly and so an increasing tendency toward potatoes harvesting can be noticed. As opposed to sesame cultivation is spread out as a crop of summer because post harvesting scorched soil is favorable, ground water deficiency, sometime rest of the cultivated land etc. (Table 9.6 and Fig. 9.8).

9.6.3.4 Raina-I C.D. Block

In 2000–01 databases, mainly three crops combinations are recorded. In the cropping year 2005, there are four crops combinations which were recorded, i.e., *Aman-Boro*-potato-mustard seed. Then decreasing trends of crop combination have been observed, i.e., *Aman* and potato (Table 9.7, Figs. 9.3 and 9.9). Mainly four C.D. blocks have recorded decreased agricultural efficiency (Ei) rate in between 2004–05 and 2014–15 cropping year. There are Bardhaman-I, Bardhaman-II, Galsi-I, Raina-I, etc. C.D. blocks due to the very high urbanization rate and shifting focus of the rural

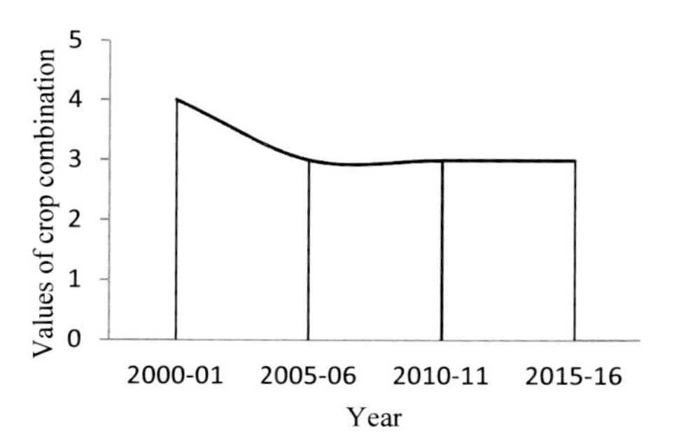

Fig. 9.8 Changing trends of crop combination in Memari-I C.D. block

Table 9.7 Changing trends of crop combination in Raina-I C.D. block

Sl. no.	Year	Values of crop combination	Types of crop combination
1	2000–01	Three	*Aman*, *Boro*, Potato
2	2005–06	Four	*Aman*, *Boro*, Potato, Mustard
3	2010–11	Two	*Aman*, Potato
4	2015–16	Two	*Aman*, Potato

Compiled by authors

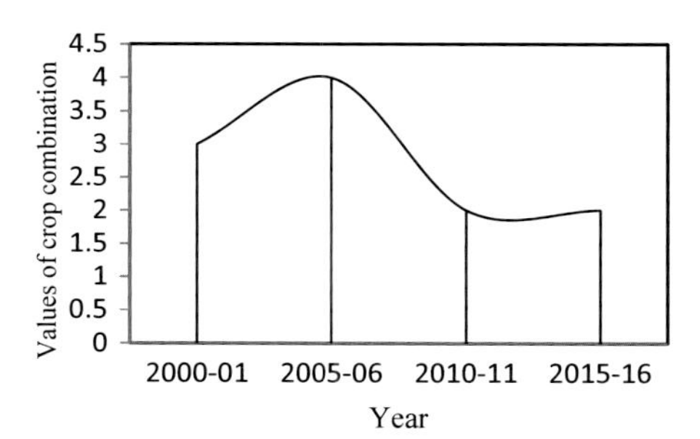

Fig. 9.9 Changing trends of crop combination in Raina-I C.D. block

economy from direct agriculture to secondary and other economic sectors mainly to agro-based industries (Dey and Mistri 2018c).

9.7 Conclusions

Formerly Bardhaman district was famous as the rice bowl of West Bengal. But now major parts of the agriculturally developed areas are now in the Purba Bardhaman district. The finding of Coppock's method exposes the sensible scenario of crop combination of the study area. Low and very low crop combination indicates high

Table 9.8 Area and percentage area of GCA of major crops in selected C.D. blocks in Purba Bardhaman district for the year 2000–01

Name of the block	CA/%	Aus	Aman	Boro	Wheat	Maize	Fiber	Sugar Cane	Pulses	Mustard	Til	Potato	Onion	Kharif Veg	Rabi Veg	Summer Veg	Chilly	Spices	Gross Cropped Area
Bardhaman	CA	1700	28,500	15,000	1000	10	11	10	149	7500	200	6000	65	850	8000	750	155	58	69,958
	%	2.43	40.74	21.44	1.43	0.014	0.014	0.014	0.21	10.72	0.28	8.58	0.09	1.22	11.44	1.07	0.22	0.08	N.A
Galsi-II	CA	1500	17,115	11,200	300	5	176	8	25	1555	60	1120	70	600	255	300	26	13	34,328
	%	4.37	49.86	32.63	0.87	0.01	0.51	0.02	0.7	4.53	0.17	3.26	0.2	1.75	0.74	0.87	0.07	0.03	N.A
Jamalpur	CA	3500	16,550	5500	75	0	105	17	51	1800	500	9600	100	600	1500	600	115	15	40,628
	%	8.61	40.74	13.54	0.18	Nil	0.26	0.04	0.12	4.43	1.23	23.63	0.25	1.48	3.69	1.48	0.28	0.03	N.A
Khandaghosh	CA	650	17,750	6500	875	10	33	5	158	1850	250	3500	175	550	1400	600	70	16	34,392
	%	1.89	51.61	18.9	2.54	0.02	0.09	0.01	0.45	5.38	0.73	10.18	0.51	1.6	4.07	1.74	0.2	0.05	N.A
Memari-I	CA	3100	13,550	5800	300	0	6	2	48	1950	900	6000	90	650	900	700	85	45	34,126
	%	9.08	39.71	17	0.87	Nil	0.01	0.01	0.14	5.71	2.64	17.58	0.26	1.9	2.64	2.05	0.25	0.13	N.A
Raina-I	CA	300	21,500	5500	1100	2	41	5	75	3700	1500	5000	90	800	900	800	90	55	41,458
	%	0.72	51.86	13.27	2.65	0.01	0.1	0.01	0.18	8.92	3.62	12.06	0.21	1.93	2.17	1.93	0.21	0.13	N.A
Raina-II	CA	950	18,100	2200	350	5	21	4	129	1400	150	2200	75	350	750	300	50	56	27,090
	%	3.51	66.81	8.12	1.29	0.01	0.07	0.01	0.48	5.17	0.55	8.12	0.27	1.29	2.77	1.11	0.18	0.18	N.A

Sources Office of the Deputy Director of Agriculture, Bardhaman and BASE
GCA Gross Cropped Area, *CA* Cultivated area, Area in hector
(Compiled by authors)

Table 9.9 Area and percentage area of GCA of major crops in selected C.D. blocks in Purba Bardhaman district for the year 2005–06

Name of the block	CA/%	Aus	Aman	Boro	Wheat	Maize	Fiber	Sugar Cane	Pulses	Mustard	Til	Potato	Onion	Kharif Veg	Rabi Veg	Summer Veg	Chilly	Spices	Gross Cropped Area
Bardhaman	CA	3000	28,780	23,920	400	12	2	5	71	2600	600	6000	32	700	2100	800	25	38	69,085
	%	4.34	41.66	34.62	0.58	0.01	0.01	0.01	0.1	3.76	0.87	8.68	0.04	1.01	3.04	1.16	0.04	0.05	N.A
Galsi-II	CA	1500	16,600	17,555	375	7	162	7	90	2200	175	1275	65	400	500	360	10	18	41,299
	%	3.63	40.19	42.51	0.91	0.01	0.39	0.01	0.22	5.33	0.42	3.09	0.15	0.97	1.21	0.87	0.02	0.04	N.A
Jamalpur	CA	4300	15,440	6000	100	7	80	0	41	1500	6000	13,800	50	600	500	1000	160	22	49,600
	%	8.67	31.13	12.1	0.2	0.01	0.16	Nil	0.08	3.02	12.1	27.82	0.1	1.21	1	2.02	0.32	0.04	N.A
Khandaghosh	CA	700	18,500	4000	700	27	25	5	585	2500	350	3400	100	600	1300	500	35	28	33,355
	%	2.1	55.46	11.99	2.1	0.08	0.07	0.01	1.75	7.5	1.05	10.19	0.29	1.8	3.9	1.5	0.1	0.08	N.A
Memari-I	CA	2000	14,485	5500	100	7	0	4	9	1700	600	7800	15	400	750	400	10	11	33,791
	%	5.92	42.87	16.28	0.3	0.02	0.01	0.01	0.02	5.03	1.78	23.08	0.04	1.18	2.22	1.18	0.02	0.03	N.A
Raina-I	CA	150	20,500	5315	1800	7	24	7	177	4300	1730	4980	105	355	1200	950	20	59	41,679
	%	0.36	49.19	12.75	4.32	0.01	0.05	0.01	0.42	10.32	4.15	11.95	0.25	0.85	2.88	2.28	0.04	0.14	N.A
Raina-II	CA	300	17,500	4500	200	7	50	0	112	1400	1100	2250	65	300	650	220	15	25	28,694
	%	1.05	60.99	15.68	0.7	0.02	0.17	Nil	0.39	4.88	3.83	7.84	0.22	1.04	2.27	0.77	0.05	0.08	N.A

Sources Office of the Deputy Director of Agriculture, Bardhaman and BASE
GCA Gross Cropped Area, *CA* Cultivated Area, Area in hector
(Compiled by authors)

Table 9.10 Area and percentage area of GCA of major crops in selected C.D. blocks in Purba Bardhaman district for the year 2010–11

Name of the block	CA/%	Aus	Aman	Boro	Wheat	Maize	Fiber	Sugar Cane	Pulses	Mustard	Til	Potato	Onion	Kharif Veg	Rabi Veg	Summer Veg	Chilly	Spices	Gross Cropped Area
Bardhaman	CA	2200	29,060	8500	35	2	3	4	46	3000	900	4700	40	850	2200	700	85	32	52,357
	%	4.2	55.5	16.23	0.06	0.01	0.01	0.01	0.08	5.73	1.72	8.98	0.07	1.62	420	1.34	0.16	0.06	N.A
Galsi-II	CA	2000	16,300	5600	200	0	0	2	134	1800	170	2500	35	290	400	200	40	6	29,677
	%	6.74	54.92	18.87	0.67	Nil	Nil	0.01	0.45	6.06	0.57	8.42	0.12	0.98	1.35	0.67	0.13	0.02	N.A
Jamalpur	CA	9000	11,200	1000	50	2	50	11	38	2500	3000	14,000	225	900	1100	600	115	20	43,811
	%	20.54	25.56	2.28	0.11	0.01	0.11	0.02	0.08	5.71	6.85	31.96	0.51	2.05	2.51	1.37	0.26	0.04	N.A
Khandaghosh	CA	400	18,700	3500	600	5	0	6	795	2700	600	3360	280	600	1700	600	135	20	34,001
	%	1.18	54.99	10.29	1.76	0.01	Nil	0.01	2.33	7.94	1.76	9.88	0.82	1.76	4.99	1.76	0.31	0.04	N.A
Memari-I	CA	7000	9600	1000	50	0	0	2	16	1700	650	8400	80	450	600	350	110	15	30,023
	%	23.32	31.98	3.33	0.17	Nil	Nil	0.01	0.05	5.66	0.22	27.98	0.26	1.5	2	1.16	0.37	0.04	N.A
Raina-I	CA	350	20,350	3200	480	0	0	3	232	1280	1400	4530	165	250	1020	525	140	39	33,964
	%	1.03	59.92	9.42	1.41	Nil	Nil	0.01	0.64	3.77	4.12	13.33	0.49	0.74	3	1.54	0.41	0.11	N.A
Raina-II	CA	750	17,450	5500	25	0	25	1	101	1000	630	2320	75	200	280	210	37	26	28,630
	%	2.62	60.95	19.21	0.08	Nil	0.08	0.01	0.35	3.49	2.2	8.1	0.26	0.69	0.98	0.73	0.13	0.09	N.A

Sources Office of the Deputy Director of Agriculture, Bardhaman and BASE
GCA Gross Cropped Area, *CA* Cultivated Area, Area in hector
(Compiled by authors)

Table 9.11 Area and percentage area of GCA of major crops in selected C.D. blocks in Purba Bardhaman district for the year 2015–16

Name of the block	CA/%	Aus	Aman	Boro	Wheat	Maize	Fiber	Sugar Cane	Pulses	Mustard	Til	Potato	Onion	Kharif Veg	Rabi Veg	Summer Veg	Chilly	Spices	Gross Cropped Area
Bardhaman	CA	2000	29,590	12,025	20	0	0	2	100	1000	600	4800	180	550	1000	300	60	30	52,257
	%	3.83	56.62	23.01	0.03	Nil	Nil	0.01	0.19	1.91	1.15	9.19	0.34	1.05	1.91	0.57	0.11	0.06	N.A
Galsi-II	CA	1200	16,835	3500	30	9	0	0	229	1200	1100	2800	65	190	380	120	43	20	27,721
	%	4.33	60.73	12.63	0.1	0.03	Nil	Nil	0.83	4.33	3.97	10.1	0.23	0.69	1.37	0.43	0.15	0.07	N.A
Jamalpur	CA	6000	14,500	800	4	0	80	10	175	1300	4000	14,000	75	600	1000	450	170	12	43,176
	%	13.9	33.58	1.85	0.01	Nil	0.18	0.02	0.41	3.01	9.26	32.43	0.17	1.39	2.31	1.04	0.39	0.02	N.A
Khandaghosh	CA	450	18,000	4000	5	0	0	0	2257	3000	2500	4700	55	700	650	450	85	10	36,862
	%	1.22	48.83	10.86	0.01	Nil	Nil	Nil	6.12	8.14	6.78	12.75	0.14	1.89	1.76	1.22	0.23	0.02	N.A
Memari-I	CA	6250	10,448	935	5	6	0	0	121	950	1410	9000	35	475	550	490	165	6	30,846
	%	20.26	33.87	3.03	0.01	0.01	Nil	Nil	0.39	3.07	4.57	29.18	0.11	1.54	1.78	1.59	0.53	0.01	N.A
Raina-I	CA	350	21,000	3500	0	0	0	1	1867	1200	1200	4500	10	325	300	180	130	10	34,573
	%	1.01	60.74	10.12	Nil	Nil	Nil	0.01	5.4	3.47	3.47	13.01	0.02	0.94	0.86	0.52	0.37	0.02	N.A
Raina-II	CA	650	17,600	5500	1	0	3	2	136	700	1200	2200	10	150	220	150	15	19	28,556
	%	2.27	61.63	19.26	0.01	Nil	0.01	0.01	0.47	2.45	4.2	7.7	0.03	0.52	0.77	0.52	0.05	0.07	N.A

Sources Office of the Deputy Director of Agriculture, Bardhaman and BASE
GCA Gross Cropped Area, *CA* Cultivated Area, Area in hector
(Compiled by authors)

to very high crop concentration and low to very low crop diversification in this district (Dey and Mistri 2017). From the above analysis and comparative study of seven C.D. blocks in Purba Bardhaman district from 2000 to 2015 cropping year, it is clear that two C.D. blocks, i.e., Khandaghosh and Raina-II have improved their crop combination due to improved crop diversification, the governmental initiative to cultivate minor crops, development of infrastructural facilities, motivation program of electronic media and the agricultural department, the development of agro-based industries and the overall positive attitude of farmers.

Only the Galsi-II C.D. block is lagging behind in terms of crop combination (Fig. 9.3) and four C.D. blocks that are Bardhaman, Memari-I, Raina-I and Jamalpur have recorded decreasing tendencies of crop combination values (Fig. 9.3). The probable causes of such decreasing tendencies are high rice crop concentration, development of rice milling industries, development of rice marketing system, sustained demands, traditional attitude of farmers, etc. It is hoped that if enhanced crop diversification, sprinkler-drip irrigation methods and increasing practices of microirrigation may be adopted with new and proper crop rotation in the future, these blocks will generate the new pathway of agricultural intensification and progress in the Purba Bardhaman district.

References

Ali M (1878) Studies in agricultural geography. Rajesh publication, New Delhi

Chakraborty K, Mistri B (2017) Irrigation system and pattern of crop combination, concentration and diversification in Bardhaman district, West Bengal. NEHU J XV(2):45–65

Coppock JT (1964) Crop livestock and enterprise combinations in England and Wales. Econ Geogr 40(a):65–81

Dey CK, Mistri T (2017) Changing patterns and causes of crop diversification in the Purba Bardhaman district West Bengal. Practising Geogr 21(2):109–135

Dey CK, Mistri T (2018a) Relationship between the infrastructural and agricultural development in the Purba Bardhaman district, West Bengal—a comparative analysis. Asian J Res Soc Sci Humanit 8(4):85–98

Dey CK, Mistri T (2018b) Causes of spatio-temporal variation of rice mill concentration in Purba Bardhaman district, West Bengal. Int J Sci Res Rev 7(6):877–886

Dey CK, Mistri T (2018c) Changing trends of agricultural efficiency (Ei) in Purba Bardhaman district, West Bengal, India. J Emerg Technol Innov Res 5(6):53–65

Dey CK, Mistri T (2018d) Changing trends of crop concentration in Purba Bardhaman district, West Bengal—a comparative analysis. Int J Appl Soc Sci 5(6):790–799

Dey CK, Mistri T (2018) Impact on irrigation intensity on cropping intensity in Purba Bardhaman district, West Bengal, India. Int J Manag Technol Eng 8(VIII):699–713

Doi K(1959) The industrial structure of Japanese perfecture proceding, I.G.U regional conference in Japan, pp. 310–316

Husain M (1996) Systematic agricultural geography. Rawat Publications, New Delhi

Husain M (2015) Systematic agricultural geography. Rawat Publications, New Delhi

Jana AK (2017) Analysis of crop combination and cropping intensity: a case study of Daspur–I block, Pachim Medinipur. Indian J Sci Res 16(02):33–40

Mishra JP (2019) Agro-industries to increase farmer's incomes, Kurukshetra 68(2):5–8

Mukherjee A (2019) Initiatives in agriculture sector. Kurukshetra 67(12):14–18

Rafiullah SM (1965) A new approach to functional classification of towns. The Geographer XII:40–53
Sandhu JS, Chaturvedi SK (2018) Increasing pulses production in India. Kurukshetra 66(4):54–59
Saurabh S (2019) Empowering farmers thought initiatives in agriculture. Kurukshetra 67(11):19–22
Shivay YS, Singh T (2017) Sustainable agriculture: aligning cropping patter with the availability of water. Kurukshetra 66(1):45–50
Siddhartha K, Mukherjee S (2003) A modern dictionary of geography. Kisalaya publication Pvt Ltd, New Delhi, p 117
Sidhu RS, Johl SS (2002) Three decades of intensive agriculture in Punjab: socio-economic and environment consequences. In: Johl SS, Roy SS (eds) Future of Punjab agriculture. Centre of Research in Rural and Industrial Development, Chandigarh
Singh J, Dhaliwal JGS, Randhawa NS (1997) Changing scenario of Punjab agriculture: an ecological perspective. Centre for Research in Rural and Industrial Development, Chandigarh
Singh J (1976) A new technique for measuring agricultural productivity in Haryana, India. The Geographer 19:17
Singh J, Dhillon S (1984) Agricultural region: concept and techniques. Agricultural geography. Tata Mc Graw-Hill Publication, Delhi, pp 175–207
Swaminathan MS (2007) Diversification of agriculture for human nutrition. Agriculture Cannot Wait, Academic foundation, New Delhi
Thapar R (1973) Backwash of green revolution. Econ Polit Weekly 34(8)
Thomas D (1963) Agriculture in wales during the napoleonic wars: a study in the geographical interpretation of the historian sources, Cardiff
Weaver R (1973) Crop combination regions for 1919 and 1929 in the Middle West, geographical review. Am Geogr Soc 44(2)
Weavers JC (1954) Crop combination regions in the middle west. Geograph Rev XLIV(2):173–200

Chapter 10
Development of Sericulture in Murshidabad with Special Reference to Women's Participation

Abhirupa Chatterjee

Abstract Sericulture, being an agro-based labour-intensive industry, includes both agricultural and industrial aspects and thus refers to the activities from the cultivation of silkworm food plants, rearing of silkworms and obtaining silk up to weaving. As this industry mainly depends on human power, it helps to provide an ample employment opportunities to the developing counties, likewise in India and considered as a remunerative cash crop; whereas, being retreated from the developed countries because of the increasing labour cost. Silk known as "Queen of Textiles" is an inseparable part of Indian ritual. India has secured the second position in raw silk production with more than 18% of the world's total production. Women play a vital role in this industry as 60% of the work has been done by them and simultaneously 80% of silk is consumed by them. In West Bengal, sericulture plays an important role in rural avocation by creating family employment round the year. The Murshidabad district of West Bengal is well equipped in the production as well as weaving of silk and so, as a matter of fact, the silk industry of West Bengal which is mainly confined around this state sometimes goes by the name of "Murshidabad Silk". This paper intends to analyse the active participation of women in the development and also the current status of sericulture as well as the silk industry of Murshidabad.

Keywords Development · Employment · Sericulture · Silk · Women

10.1 Introduction

Silk is the most elegant textile in the world and is something that can always add an extra elegance to any attire whether it is traditional or western. It is the most prestigious and natural fabric considered as the "Queen of textiles" (Thiripura Sundari and Rama Lakshmi 2015). The process of rearing silkworms for the production of raw silk is known as sericulture, which involves a few techniques starting from the plantation of trees to feed the silkworms, production of the cocoon, reeling and

A. Chatterjee (✉)
Nagaland University, Lumami, India

N. C. Jana et al. (eds.), *Livelihood Enhancement Through Agriculture, Tourism and Health*, Advances in Geographical and Environmental Sciences,
https://doi.org/10.1007/978-981-16-7310-8_10

spinning of cocoon for obtaining silk yarn and ultimately to processing and weaving, and thus, being a agro-based industry, it combines both the industrial and agricultural aspects (Banday 2001). This industry creates a linkage between the rich consumers and poor weavers. The money that comes from the consumers is directly distributed among the sericulturists, reelers, twisters, weavers and traders.

The Indian silk industry has played a significant role in the textile industry and is also famous among the oldest industries in India (Savithri et al. 2013). Though in the world, more than 30 countries are producing silk, India occupies the second position for producing silk and contributing to about 18% of the world's production (Thiripura Sundari and Rama Lakshmi 2015). According to the Annual Report of the Central Silk Board, the Indian Silk Industry has crossed the 30,000 MT mark with a production of 31,906 MT of total raw silk production in 2017–18 as the country is blessed to have a climate that is appropriate for all the varieties of silk called Mulberry, Tasar, Eri and Muga (Annual Report 2017–18). Mulberry Silk is the most renowned and popular form of silk in which West Bengal occupied third position (2351 MT in 2015–16) and secured fourth position (2391 MT in 2015–16) in total raw silk production in India (Sericulture-Statistical Yearbook India 2018). Sericulture plays an important tool for poverty alleviation in the rural areas of West Bengal by giving families employment round the year. More than 1.2 lakhs families and approximately 2.5 lakhs people of West Bengal are dependent on sericulture activity for their bread and butter. Murshidabad district of West Bengal is well equipped in both the production and weaving of mulberry silk and has occupied the second position in the production of silk (546.96 MT in 2016–17) after Maldah (Compendium on Seri-States 2019). Sometimes, the silk fabric of West Bengal goes by the name of "Murshidabad Silk" as the silk industry of this state mainly confines around this district (Directorate of Textile (Sericulture), Government of West Bengal 2012).

10.2 Women and Silk Industry

Women constitute more than half of the world's population and play a significant role in the Indian rural economy as a maximum of them are engaged in agricultural and allied activities. Though women are mainly considered as "home makers" in Indian society but the employment of women generates a superior impact on their children mainly through an increase in their nutrition and educational level and consider as a financial backbone to their entire family (Roy 2017; Roy and Roy Mukherjee 2015; Sandhya Rani 2006; Sarkar et al. 2017). Women are mostly found to be active in the silk sector as it has got some specific character suitable for them to adopt. Firstly, it is an indoor activity with less manual labour; secondly, it can be adjusted with the other household work as it has flexible working hours and thirdly, rearing of silkworms need motherly care instinct while reeling and spinning activities need nimble fingers (Directorate of Textile (Sericulture), Government of West Bengal 2012). This entire process of silkworm rearing and weaving needs skill and patience, which suits women well, and as a result, 60% of the work has been done by them. On other hand, 80% of silk consumption has been made by the women who conclude that they have a vital role in this industry both as a producer as well as a consumer (Bukhari et al. 2019).

10.3 Study Area

The district Murshidabad is located beside of the South of Ganga River and lies between 23°43′ and 24°52′ North latitudes and 87°49′ and 88°44′ East longitude (Fig. 10.1). It is one of the most populous districts of West Bengal with approximately 7 lakh inhabitants. Robert Clive remarked "the city of Murshidabad is as extensive and populous and rich as the city of London" (Guha 2005). The silk industry of Murshidabad district is well known for producing excellent quality of silk. This silk industry has had a long history that goes back to the early eighteenth century during the Mughal regime in India when the Nawab of Bengal, Murshid Kuli Khan, shifted his capital from Dhaka (presently in Bangladesh) to a town on the East of Bhagirathi river and named it Murshidabad and with that, the famous art of silk weaving started (Parinita 2019). At the same period of time, the East India Company, who had considerable trading interest in raw silk, considered Murshidabad as a remarkable place for trade and commerce since it was the capital of Bangla, Bihar and Orissa (Directorate of Textile (Sericulture), Government of West Bengal 2012). Though, with the flow of time, the capital has shifted with decreasing political importance, the domination of the silk weaving industry remains the same (Roy and Dey 2019). The fine quality "Murshidabad Silk" is the pride and carries the heritage of this district. To analyse the current status and the role of women in the silk industry of this district, two blocks Nabagram and Raninagar I have been selected for the study.

10.4 Objectives of the Study

- To emphasize the role of women in the production of silk in the district.
- To determine the condition of the silk industry in Murshidabad.
- To study the role of government in the up-gradation of this industry and its outcome.

10.5 Methodology

The present study will analyse the importance of women in the field of sericulture and the silk industry in Murshidabad. For this study, both primary and secondary data and also qualitative and quantitative data have been collected. Secondary data are collected from Central Sericultural Research and Training Institute (CSRTI) Berhampore; District office under Directorate of Textile (Sericulture) Berhampore,

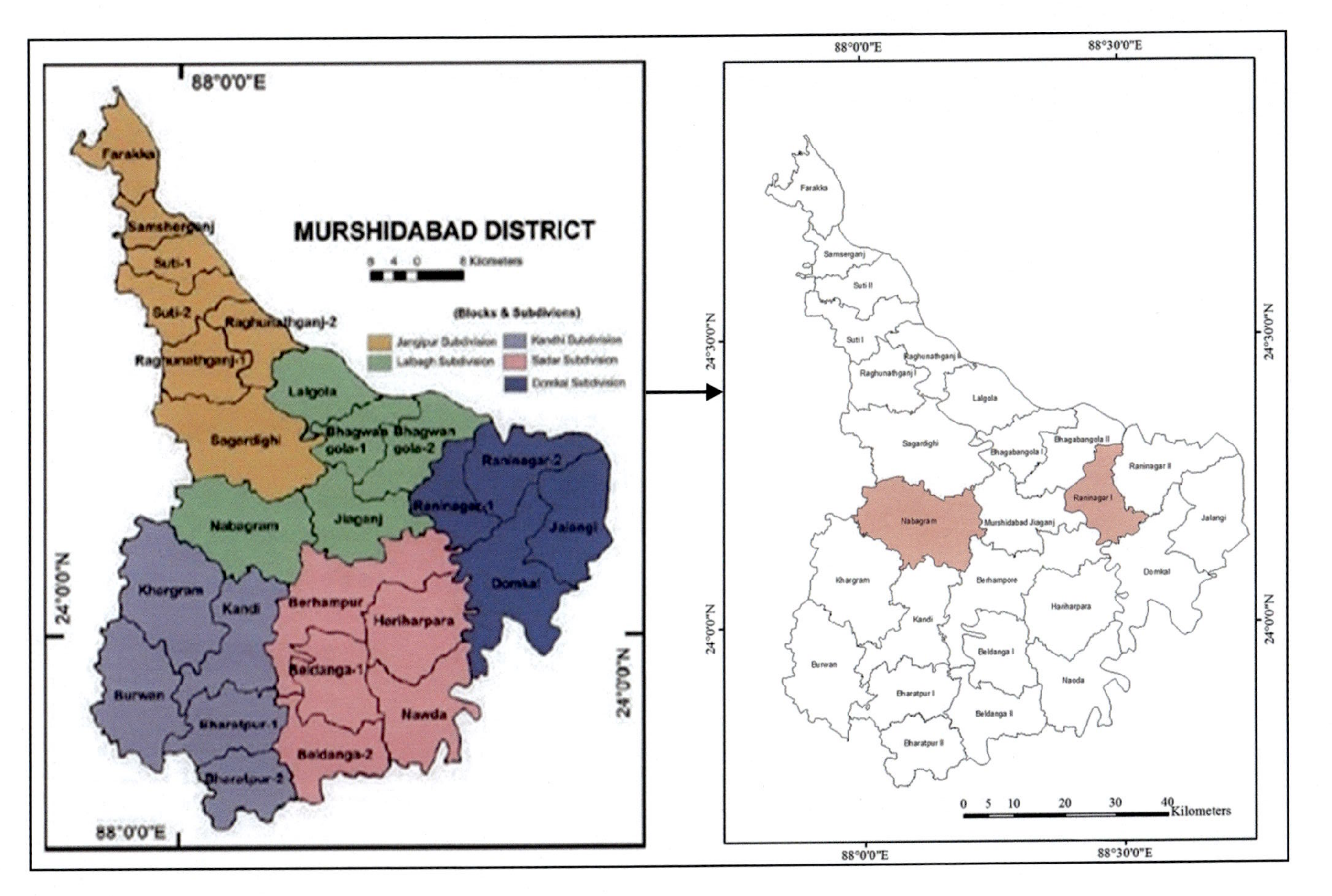

Fig. 10.1 Location map of study area

Murshidabad; Directorate of Textiles, Handlooms, Spinning Mills, Silk Weaving and Handloom based Handicrafts Division, Office of the Handloom Development Officer, Berhampore, Murshidabad and from the website of Ministry of Statistics and Programme Implementation, Government of India. Primary data are collected through a structured questionnaire survey. Multi-stage sampling procedure is used for collecting data. Nabagram block has been chosen for data collection as it is very famous and renowned for sericulture activities in the district. Data are collected from five villages of this block. These are Balaspur, Milkipalasi, Sahidpur, Bankipur and Derul. Total 75 houses are being taken and 185 female members have got. The data for weaving and reeling activities has been collected from Islampur and Harharia Chak town under Raninagar I for being famous in large-scale weaving activities. A total number of 30 houses and 79 female members are taken for the collection of data. Ultimately as a sample size, 264 numbers of females have been got. The collected data was classified and analysed in tabular forms and finally represented with a suitable cartographic technique.

10.6 Major Findings

It is clearly being seen (Table 10.1) that the production of mulberry silk is maximum in this district rather than other types and hence occupied the third position in India.

Location Quotient has been used to indicate the relative concentration or dispersion of block-wise sericulture adopted villages in Murshidabad district (Table 10.2). Location Quotient is a simple tool used to determine the spatial distribution (clustering/dispersal) of a phenomenon in an area compared to an entire region.

The formula of Location Quotient is

$$\text{LQ} = \frac{v_i/v}{V_{i/V}}$$

where

v_i number of sericulture adopted villages in block i,

Table 10.1 Raw silk production in West Bengal (2016–17)

Type	Raw silk production (MT)
Mulberry silk	2524
Vanya (non-mulberry silk)	
Tasar	37.10
Eri	3.8
Muga	0.2
Total	2565

Source Sericulture Statistical Year book India, 2018

Table 10.2 Location quotient showing concentration of sericulture adopted villages (2018–19)

Sl. No.	Name of the block	Total number of villages	Block-wise sericulture adopted villages	Percentage of sericulture adopted villages to total villages	L. Q. values
1	Khargram	155	60	38.71	6.47
2	Nabagram	118	63	53.39	8.93
3	Murshidabad Jiaganj	132	12	9.09	1.52
4	Hariharpara	63	06	9.52	1.59
5	Jalangi	53	05	9.43	1.58
6	Raninagar I	63	05	7.94	1.33
7	Raninagar II	36	05	13.89	2.32
8	Berhampore	138	05	3.62	0.61
9	Domkal	87	09	10.34	1.73
10	Bharatpur I	92	0	0	0
11	Bharatpur II	52	0	0	0
12	Naoda	39	01	2.56	0.43
13	Beldanga I	65	15	23.08	3.86
14	Beldanga II	71	02	2.82	0.47
15	Samserganj	44	0	0	0
16	Raghunathganj I	56	05	8.93	1.49
17	Raghunathganj II	77	08	10.39	1.74
18	Lalgola	94	08	8.51	1.42
19	Bhagabangola I	65	05	7.69	1.29
20	Bhagabangola II	70	05	7.14	1.19
21	Farakka	74	0	0	0
22	Sagardighi	197	09	4.57	0.76
23	Kandi	93	01	1.08	0.18
24	Burwan	160	02	1.25	0.21
25	Suti I	62	0	0	0
26	Suti II	48	0	0	0
	Total	2204	231	5.98	

v total number of sericulture adopted villages,
V_i number of villages in block i and
V total number of villages.

The values of Location Quotient (Table 10.2 and Fig. 10.2) show a higher concentration of sericulture adopted villages (>1.5) in Khargram (6.47), Nabagram (8.93), Murshidabad Jiaganj (1.52), Hariharpara (1.59), Raninagar II (2.32), Domkal (1.73),

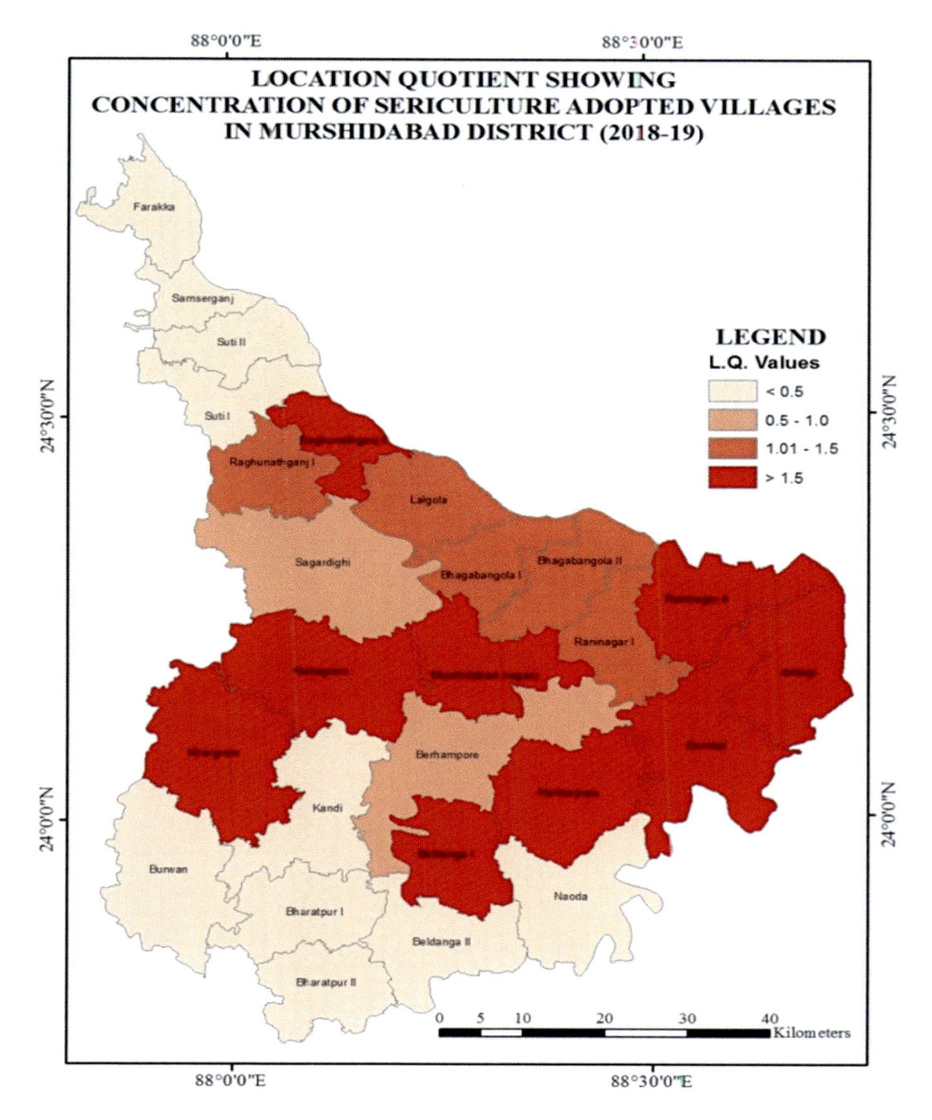

Fig. 10.2 Values of location quotient. *Source* District office under Directorate of Textile

Beldanga I (3.86) and Raghunathganj II (1.74) whereas moderate concentration (1.01–1.5) in Raninagar I (1.33), Bhagabangola I (1.29) and Bhagabangola II (1.19). The relative concentration of sericulture adopted villages is low (0.5–1) in Berhampore (0.61) and Sagardighi (0.76) and in the other blocks, the concentration is very low (<0.5).

The number of sericulturists (Table 10.3 and Figs. 10.3 and 10.4) is very high (1007–11,617) in Khargram (11,617) and Nabagram (7491) block whereas it is moderately high (363–1006) in Jalangi (1106) and low (122–362) in Beldanga I

Table 10.3 Block-wise variation of area under sericulture and number of sericulturists

Sl. No.	Name of the block	Block-wise number of sericulturists	Block-wise area under sericulture (acre)
1	Khargram	11,617	3495.09
2	Nabagram	7491	2297.22
3	Murshidabad Jiaganj	214	33.12
4	Hariharpara	362	100.32
5	Jalangi	1006	339.88
6	Raninagar I	272	94.75
7	Raninagar II	105	37.00
8	Berhampore	153	49.50
9	Domkal	156	55.08
10	Bharatpur I	0	0
11	Bharatpur II	0	0
12	Naoda	06	2.5
13	Beldanga I	305	178.10
14	Beldanga II	121	51.5
15	Samserganj	0	0
16	Raghunathganj I	68	30.0
17	Raghunathganj II	20	19.5
18	Lalgola	74	48.81
19	Bhagabangola I	45	25.97
20	Bhagabangola II	25	15.25
21	Farakka	0	0
22	Sagardighi	41	22
23	Kandi	10	4.66
24	Burwan	25	16.75
25	Suti I	0	0
26	Suti II	0	0
		22,116	6917

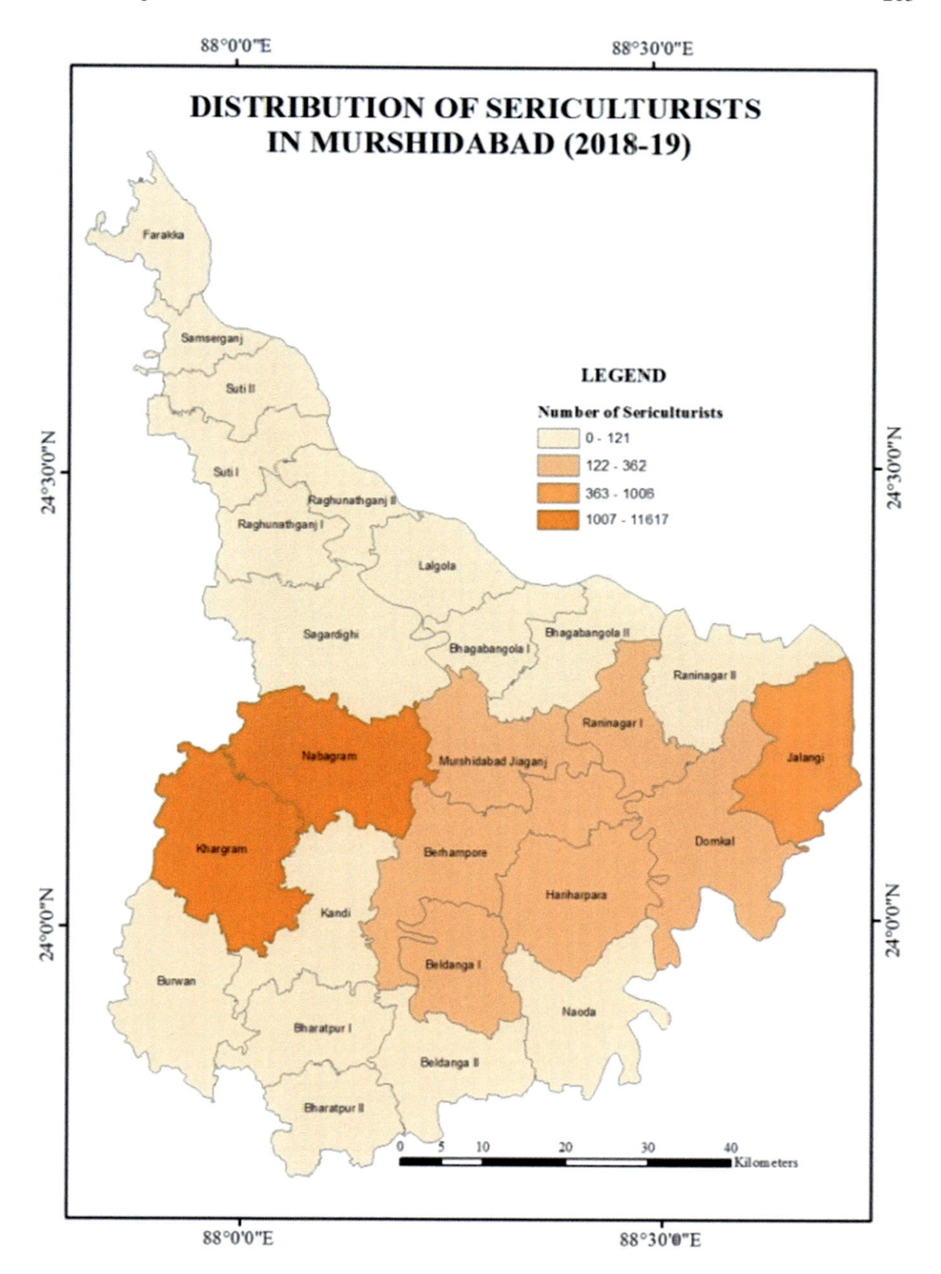

Fig. 10.3 Distribution of sericulturists in Murshidabad (2018–19). *Source* District office under Directorate of Textile

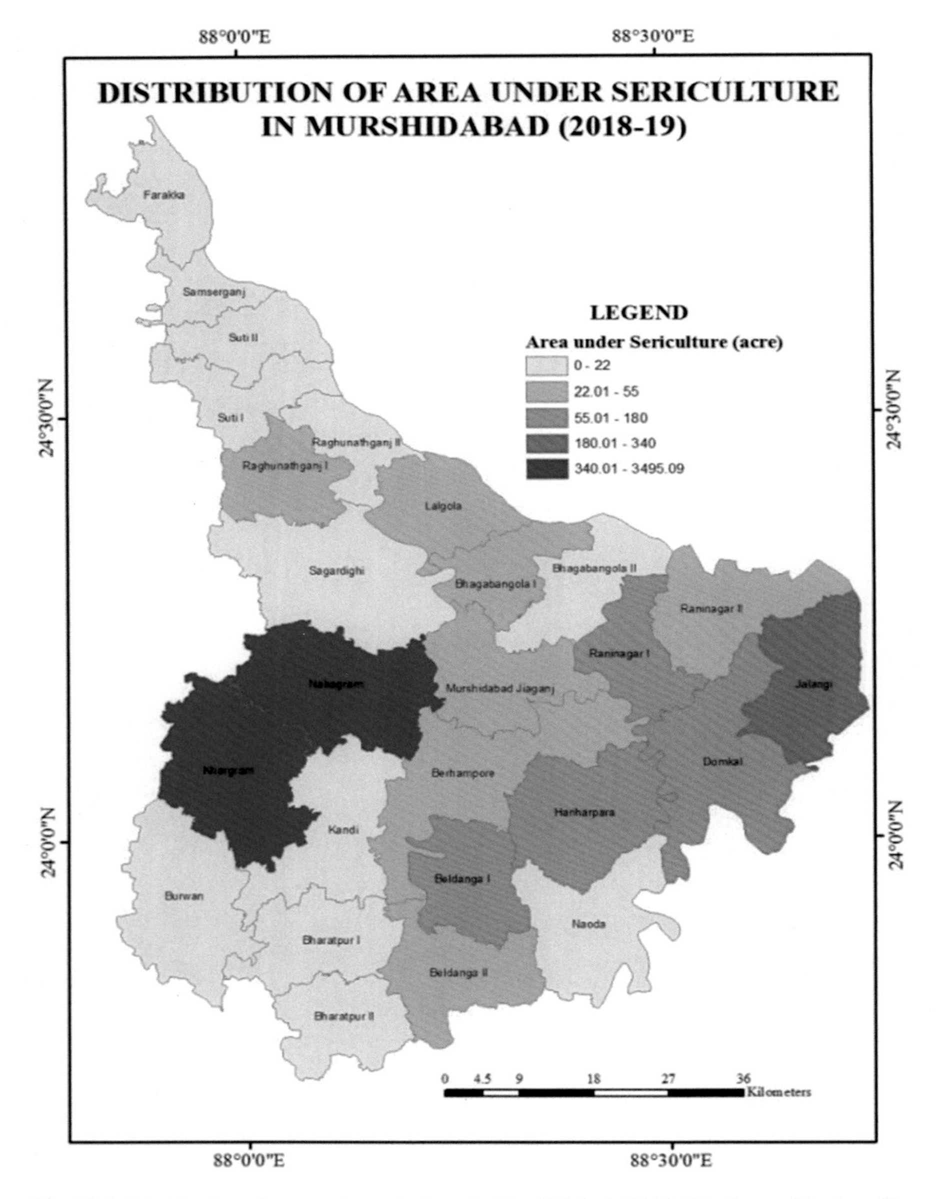

Fig. 10.4 Distribution of area under sericulture in Murshidabad (2018–19). *Source* District office under Directorate of Textile

Table 10.4 District profile on sericulture (as on March 2019)

Number of gram panchayats under sericulture	71 Nos.
Number of NGOs under sericulture	05 Nos.
Number of Self Help Groups (SHG) under sericulture	2450 Nos.
Number of silk reeling devices	3073 Nos.

Source District office under Directorate of Textile

Table 10.5 Work participation of women in sericulture of the selected area of Nabagram

Source of income	Number of working women	Percentage (%)
Sericulture	71	89.87
Other activities	8	10.13
Total	79	

(305), Raninagar I (272), Hariharpara (362), Domkal (156) and Murshidabad Jiaganj (214). In the other blocks, the number of sericulturists is found to be very low (0–121).

The area of mulberry plantation or sericulture (acre) (Table 10.3) is very high (340.01–3495.09) in Khargram (3495.09) and Nabagram (2297.22); high (180.01–340) in Jalangi (339.88) block and moderately high (55.01–180) in Raninagar I (94.75), Domkal (55.08), Beldanga I (178.10). Whereas low (22.01–55) in Bhagabangola I (25.97), Murshidabad Jiaganj (33.12), Raninagar II (37), Raghunathganj I (30) and Berhampore (49.50) and very low (0–22) in the rest of the blocks.

There are some blocks in Murshidabad with zero sericulture activity. The names of such blocks are Bharatpur I, Bharatpur II, Samserganj, Farakka, Suti I and Suti II. Table 10.4 represents the district profile on sericulture.

The percentage of work participation of women (Table 10.5 and Fig. 10.5) in the sericulture-related activities is very high (89.87%) in the selected villages of Nabagram block whereas the participation of women in the weaving and other allied

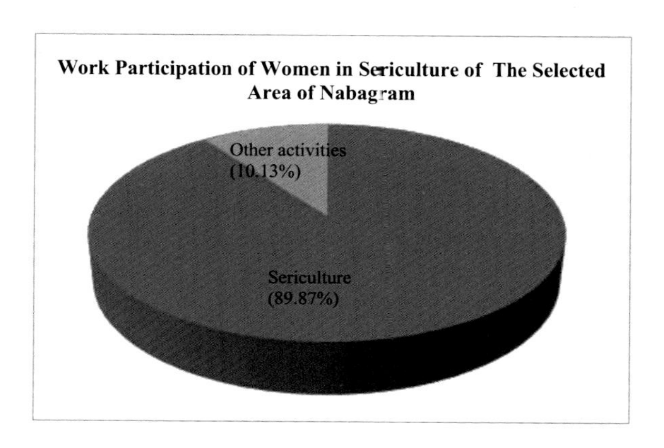

Fig. 10.5 Work participation of women in sericulture. *Source* Primary data

Table 10.6 Work participation of women in silk weaving industry of the selected area of Raninagar I

Source	Number of working women	Percentage (%)
Weaving and allied activities	61	95.31
Other activities	3	4.69
Total	54	

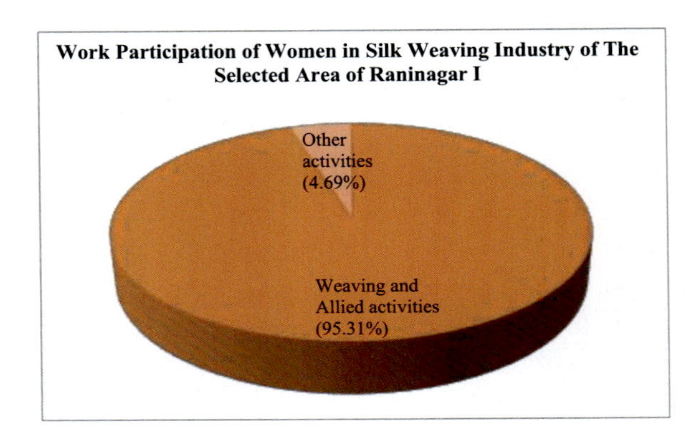

Fig. 10.6 Work participation of women in silk weaving industry. *Source* Primary data

activities is also very high (95.31%) in the selected towns of Raninagar I block (Table 10.6 and Fig. 10.6).

There is no such positive relationship between the educational level and the working status of the women silk workers (Table 10.7 and Fig. 10.7), as only 43.18% of the women engaged in the silk sector are under the educational level of class 8–class 12.

Through sericulture and silk industry women are contributing a remarkable economical support towards their families. As far as the rearing process is being concerned, all the effort goes to the women's court of a family engaged in sericulture activity (Table 10.8). The reeling process is considered as a preparatory weaving work and 50% of the work has been done by the female members of a family and so, the economic contribution in the same ratio. Finally, the weaving process is fully done by women in many cases, and if not, the yarn for wefting (varna) has been processed by the women and as a matter of fact, money that comes from the weaving activities is also a part of women's contribution. As maximum of the work of the silk

Table 10.7 Educational level of the women engaged in silk sector of the selected area

Level of education	Number of women	Percentage (%)
Up to Class 4	31	23.49
Class 5–Class 8	44	33.33
Class 8–Class 12	57	43.18
Total	132	

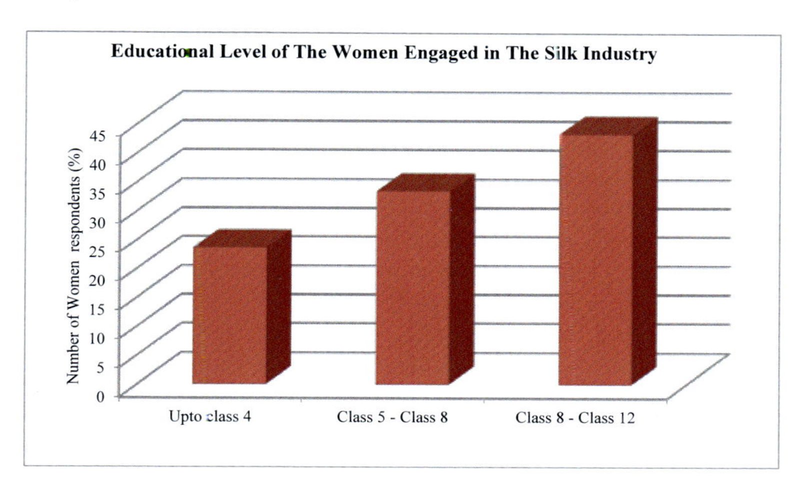

Fig. 10.7 Educational level of the women engaged in the silk industry. *Source* Primary data

Table 10.8 Monthly income from different silk-related activities

Activities	Income (Rupees)
Silkworm rearing	8000–10,000
Reeling	9000
Weaving	4500–7000

Source Primary Data

industry is being done by women, their participation is considered as a big factor for the success of this industry.

The decreasing price of raw silk affects the silk production (Table 10.9) directly as the silkworm cultivators are losing interest in this activity. According to the new budget, the price of 1 kg of raw silk is 1600–1700 rs per kg, which was previously 2500–3000 rs per kg.

Table 10.9 Sericulture at a glance (only for Mulberry)

Particulars	Unit	Financial year		
		2016–17	2017–18	2018–19
Total production of disease-free layings (dfls)	Lakhs	95.10	95.82	94.11
Reeling Cocoon produced	MT	5146	4843	4903
Total production of raw silk	MT	550	513	526
Total sale value of raw silk	Lakhs	11,907	12,055	10,520

Source District office under Directorate of Textile

10.7 Government Initiatives and Outcomes

Murshidabad silk has its own heritage and tradition, and to keep its prestigious position, the Government of West Bengal has created a separate Directorate in the year 1975 (Roy and Roy Mukherjee 2015). To strengthen the economical backbone of the district, the government put extra effort into this sector. Department of Sericulture (West Bengal) has fixed up an action plan for achieving the higher target by increasing the raw silk production with the help of expanding the area of the mulberry plantation as it is the primary key factor for silkworm rearing. For that Central Sericultural Research and Training Institute (CSRTI), one of the oldest training institutes of Directorate of Sericulture, introduced a new type of mulberry plant TR 23 which is suitable for the growth in the climatic condition of the Eastern part of India and also introduced a new type of Japanese silkworm, giving the sericulturists a new way of good hope and the successful rearing process of these silkworms gives 55–65 kg of cocoon from 100 dfls (disease-free layings). The mulberry plant TR 23 gives a yearly production of 24 tonne mulberry leaves in per hectare land which is qualitatively high and ideal to feed silkworms. Silkworm is very delicate and highly sensitive to environmental fluctuation. The optimum temperature for the normal growth of silkworm is 22–27 °C (Ram 2016). Climate change is a big factor nowaday, as a result, temperature rise and fall creates a huge problem in silk production. To increase the productivity in this condition, State Government is distributing airconditioner and room heaters to the sericulturists to keep the optimum temperature for the growth of silkworm. A yearly amount of 25,000 Rs. has been given to those sericulturists, who are cultivating 6 crops per year as a subsidy by the state government. With the help of the Handloom Development office, Berhampore 3000 pieces of looms in the year 2018 have been distributed by the State government throughout the district.

The concept of block-level cluster has been introduced by the initiation of the Central Government which will help the weavers to work under different projects and with these, they will get employment throughout the year, and also through this, they will get to know about the newly introduced schemes by the Government. A Common Facility Centre (CFC) mainly for giving training will be built for every cluster. A great initiative by the Central Silk Board about women empowerment through the promotion of sericulture came forward as the concept of gender budgeting was introduced by the Government of India during 10th five-year plan after the declaration of 1994 as a "Year of women in sericulture" and after that in the 11th five-year plan "Women Development Component" introduced mainly to emphasize the concern about the social security of women associated with silk industry (Sandhya Rani 2006). To extend the marketing of silk products, the Directorate of textile encourages and does the necessary for the weavers to participate in the different fares and expo in different parts of India. The district Murshidabad has been marked as a sericulture potential district by MSME-DI (Micro, Small and Medium Enterprises Development Institute), Government of India (2016) as the district is famous for both sericulture and silk weaving, and through this, the concept of Seri-tourism has been developed where sericulture and tourism intersect which further develop the importance of

Murshidabad as a silk weaving centre (Seri-Tourism in Murshidabad District: Some Potential Locations 2019).

10.8 Problems and Suggestions

- There are a total of 2450 Self Help Groups (SHG) (2018–19) governed by the co-operative society which have been formed on basis of sericulture sector and an approx 25,000 women are working under these groups but it seems many constraints are being faced by them as most of the houses are made up of mud and one storied, so it is difficult for them to rear silkworms in the summer season; secondly, many of them don't have their own reeling equipments. So it is necessary to give a common work shed to each SHG where they will get all the facilities to continue their rearing process without any blockage.
- For the farmers and the reelers who are producing cocoon and silk yarn, maximum time have been bound to sell their products in the open market and so, facing a huge loss because of the interruption of the middleman as they are forcing them to sell the products at a cheaper rate. The construction of the organized government authorized market is the only solution to this marketing issue so that Government can fix a market price of those products.
- A large amount of China Silk has been imported to Murshidabad and is being used for warping as there is no such domestic requirement of silk in China. The fine white china silk is very much eye catcher to the Mahajans (Middleman) and, for that reason, the market of Murshidabad silk has started deteriorating.
- Though the state government has taken initiative to distribute looms to the weavers, it seems that the quality of the looms is not good enough for the weaving of silk and can only be fitted for cotton weaving. So, it is important to make the looms with the proper technical guidance of the silk weavers.
- Government is giving money for making individual work shed for the weavers where only 25% of the total cost is to be paid by them, as they are running looms at their own living room. But these weavers have to be member of the co-operative society with a weaver's identity card and there are many weavers who are not registered under co-operative society as they are working for master weavers or under NGO's where female weavers are getting employment opportunities and also taking the training. So, it is necessary to take initiation from the Government so that all the weavers will be registered under them and can get their identity card as a weaver. This progress will help the female weavers greatly as they are not at all a part of Weaver's Co-operative Society.
- Lack of educational knowhow leads the weavers to their own loss by thinking that working under Mahajan is a great profit for them as they are getting the wages directly to their hands rather than working under Khadi Committee but the real scenario is totally different. The Khadi Committees are providing them different privileges like bonus (10% of the total annual income of each weaver). Even a part of the Khadi Commission's quarterly profit is being given to the weavers as per

their production and the most important is they are getting Provident Fund also. As they do not understand the actual importance of those government schemes, weavers are not interested to work under the Khadi Committee. To disclose the trapping process of these Mahajans, it is necessary to give proper guidance through root level training to the weavers regarding Government schemes.

10.9 Conclusion

There has been a continuous decline in the share of agriculture and allied sectors in the GDP over the last two decades because of the losing interest of the farmers (Savithri et al. 2013). Low and stagnant productivity, low return from the farm sector and also the erratic nature of monsoon are some of the main reasons behind this issue (Bhattacharyya and Kumar 2016). To support their own financial condition, farmers need a gainful employment opportunity, which can be absorbed by them easily. To eradicate this stress, the sericulture and silk industry play a vital role to reconstruct the rural economy by providing gainful employment as it is a labour-intensive activity. The farmers with a small land area can also start sericulture with ¾th of the land and would support three members in the family without hiring someone for the rearing process (Reddy 2019).

Sericulture and silk industry is also an eco-friendly activity (Soi 2019). The plantation of mulberry trees helps in soil conservation as it has good foliage and provides green coverage. The wastes from silkworm rearing are recycled to be as inputs of the garden.

Silk is the only remunerative cash crop that plays a significant role in shaping the destiny of rural people as it gives attractive returns with a minimum investment and low gestation period (Savithri et al. 2013). Sericulture and silk weaving industry bring employment to the people without judging their caste, religion and gender, and in a family, all the members irrespective of age can contribute to the sericulture, and hence, this sector is considered as a home-based industry (Kasi 2013). Both unskilled farm labourers and skilled artisans are part of this industry. Women have a crucial role in the sericulture activity and create opportunities by making them strong economically. In many cases, it shows that women are carrying both the family burden as well as the economic burden, which affects their health adversely, and to keep this problem in mind, sericulture is a perfect choice for them to be employed as it is an indoor activity (Goswami and Bhattacharya 2013). Silk is indispensable in ceremonies and religious rituals in India and so there is a strong domestic market for silk in India which is expected to continue for a long period and hence it can be taken as a positive part of the growth of this industry.

References

Annual Report 2017–18 (2017–18) Retrieved 9 July 2019 from Central Silk Board, Ministry of Textiles, Government of India: http://csb.gov.in/wp-content/uploads/2019/01/annual-report-english-2017-18.pdf

Banday S (2001) Sericulture its development and future prospect. Vinod Publishers and Distributors, Jammu

Bhattacharyya S, Kumar D (2016, September–October) Mulberry sericulture: an alternative, farm-based livelihood for small farmers. News Res

Bukhari R, Kour H, Aziz A (2019) Women and the Indian sericulture industry. Int J Curr Microbiol Appl Sci 857–871. https://doi.org/10.20546/ijcmas.2019.805.101

Compendium on Seri-States (2019) Retrieved 6 July 2019 from Central Silk Board, Ministry of Textiles, Government of India: http://csb.gov.in/wp-content/uploads/2019/02/Seri-States-Profiles-2019.pdf

Directorate of Textile (Sericulture), Government of West Bengal (2012) Retrieved 5 August 2019 from http://www.seriwbgov.org/gen_mandate1.aspx

Goswami C, Bhattacharya M (2013) Contribution of sericulture to women's income in Assam—a case study in Goalpara district of Assam. India. Int J Sci Res Publ 3(3):1–6

Guha SC (2005) Silk industry of Murshidabad from 1660 to 1883. Retrieved 11 December 2019 from https://shodhganga.inflibnet.ac.in/handle/10603/137105

Kasi E (2013) Role of women in sericulture and community development: a study from a South Indian village. SAGE Open 1–11. https://doi.org/10.1177/2158244013502984

Parinita (2019) Retrieved 5 June 2019 from https://www.parinita.co.in/pages/murshidabad-silks

Ram RL (2016, June) Impact of climate change on sustainable sericultural development in India. Research Gate

Reddy J (2019, January 28) Sericulture project report, cost profit, economics. Agri Farming

Roy C (2017) The Artisanal silk industry of West Bengal: a study of its history, performance and current problems. Turk Econ Rev KSP J 4(3)

Roy C, Roy Mukherjee S (2015) Issues of productivity, employment and exploitation in Artisanal silk industry of West Bengal. Indian J Soc Nat Sci 49–68

Roy C, Dey A (2019) Murshidabad silk industry of West Bengal: a study of its glorious past and present crisis. SSRN 2–18

Sandhya Rani G (2006) Women in sericulture. Discovery Publishing House, New Delhi

Sarkar K, Majumdar M, Ghosh A (2017) Critical analysis on role of women In sericulture. Int J Soc Sci 211–222. https://doi.org/10.5958/2321-5771.2017.00024.2

Savithri G, Sujathamma P, Neeraja P (2013) Indian sericulture industry for sustainable rural economy. Int J Commer Res 3(2):73–78

Sericulture-Statistical Yearbook India (2018) Retrieved July 3, 2019, from Government of India Ministry of Statistics and Programme Implementation: http://mospi.nic.in/statistical-year-book-india/2017/180

Seri-tourism in Murshidabad district: some potential locations. Retrieved 22 September 2019 from http://www.csrtiber.res.in/Seri_tourisim.htm

Soi S (2019, June 15) Sericulture—an introduction to silk cultivation and production in India along with its policy initiatives, Krishijagaran

Thiripura Sundari K, Rama Lakshmi P (2015) An analysis of silk production in India. Int J Bus Manage 3(3):151–162

Chapter 11
Opportunities and the Challenges of Tourism Industry in Bangladesh

K. M. Rezaul Karim

Abstract Tourism is a growing industry all around the world. Though Bangladesh has a huge prospective to improve tourism due to its natural scenery and enriched heritage, the industry flops to the extent of its end because of its different challenges. The *Sundarban* mangrove forest, *Shatgombuj* Mosque, and *Paharpur* Buddhist Vihara are the three world heritage sites in Bangladesh. The total contribution of tourism to GDP is 4.4%, but the global contribution is 10.4% in 2018. Attracted by the natural beauty of the country, a significant number of domestic and overseas tourists visit its different tourist sites. Despite its immense potentials, the sector is facing different challenges like inadequate infrastructure and backward communication system, deficiency of accommodation facilities, lack of safety and security, scarcity of professionalism, lengthy visa processing, and political instability, which discouraged both international and internal tourists to visit the attractive places in the country. Tourism also brings socio-economic and environmental benefits for the country, albeit mass tourism is also associated with negative effects on the social environment. The opportunities for tourism of the country are religious tourism, sports tourism, eco-tourism, educational tourism, spa tourism, rural tourism, cultural tourism, etc. But there is a lack of research and plan to explore the development of the tourism industry of Bangladesh. The paper tries to focus on the important and attractive tourist spots and the impact of tourism on the economy of the country. The paper also explores the challenges and opportunities of the tourism sector of Bangladesh.

Keywords Tourism · Potential · Challenge · Opportunity · Bangladesh

---

K. M. R. Karim (✉)
Department of Sociology, Government M. M. College, Jashore 7400, Bangladesh
e-mail: rezakarim.km@gmail.com

N. C. Jana et al. (eds.), *Livelihood Enhancement Through Agriculture, Tourism and Health*, Advances in Geographical and Environmental Sciences,
https://doi.org/10.1007/978-981-16-7310-8_11

11.1 Introduction

Tourism is regarded as a promising industry in the world. As a tourist country, Bangladesh is located in South Asia between Myanmar and India. It is the eighth-most populous country in the world, with a population exceeding 161 million people in the area of 147,570 square kilometers (Nazem 2017). The country is a land of diversified culture, history, heritage, and full of natural beauty. There are sea beaches, archeological places, holy places, mountains, waterfalls, tea gardens, safari parks, eco-parks, etc. here. Historians rightly observe that the natural beauty of Bangladesh has always attracted a significant number of priests, traders, and wanderers from different parts of the universe (Howlader 2013). The *Sundarban mangrove forest*, *Shatgombuj* Mosque, and *Paharpur* Buddhist Vihara are the three world heritage sites here. In recent years, a number of new exclusive tourist spots have been discovered and have become popular to tourists. Bangladesh Parjaton Corporation looks after the tourism part here.

To observe the beauty of nature, several internal and external tourists visit the country and its attraction sites. It is estimated that about 9–9.5 million visitors travel to different tourist spots within the country per year and the majority of them (around 70%) prefer to visit Cox's Bazar, the world largest sea beach, and Chittagong hill tracts followed by *Sundarban* mangrove forest and Sylhet tea garden (Antara 2019). On the other hand, many internal tourists visit overseas countries for recreation due to insufficient facilities in the home. It is reported that about 2.25 million people from Bangladesh visited neighboring country India, 0.7 million Thailand, and several million visited Thailand, Malaysia, Singapore, and Indonesia a year for tourism in 2019 (Afroz and Hasanuzzaman 2012). According to BBS statistics, around 1.5–1.6 million people from Bangladesh visited overseas countries in 2016, while only 0.1–0.2 million overseas tourists visited the beauty of Bangladesh. In the year 2017, Bangladesh generated Tk 18.8 billion from overseas tourists, and the direct contribution of the tourism industry is $5.4 billion which is approximately 2.2% of the GDP of the country (Uddin 2020). Besides, travel and tourism generated about 1.2 million jobs directly which is 1.8% of total employment across the country (Siddique 2020).

11.2 Objectives of the Study

The key objective of the article is to emphasize on challenges and opportunities for the tourism sector of Bangladesh and the contribution of tourism to the Bangladesh economy also. The specific objectives are as follows:

(a) To find the present scenario of Bangladesh tourism industry is currently facing;
(b) to explore the opportunities and challenges for the tourism industry of Bangladesh; and
(c) to develop some recommendations for the improvement of the present tourism industry of Bangladesh.

11.3 Methodology

The study is basically exploratory in nature. Content analysis has been applied as a method in the study. The study is mainly based on secondary information. Tourism-related studies have been reviewed and relevant data have been gathered from different books, edited books, national and international journals, magazines, bulletins, newsletters, different published research works, different reports of government and non-government organizations, Bangladesh Parjatan Corporation (BPC), World Travel and Tourism Council (WTTC), Bangladesh Bureau of Statistics (BBS), daily newspapers, different websites, etc. After processing and analyzing the collected data, it has been interpreted in descriptive and simple tabular form.

11.4 Statement of the Problem

In the twenty-first century, tourism emerges, as a significant income-generating sector as well as a weapon for development and survival for various sectors (Sultana 2016). It is argued that Bangladesh's share in the total visitor arrivals in the south Asian region is too small (Jahangir 1998). Besides, Islam and Islam claimed that Bangladesh has high potentiality for tourism (Islam and Islam 2006). In past, Bangladesh was a pretty place for tourists, but presently its location is not weighty in the light of international tourism (Sultana 2016). It is noted that many issues are concerned with socio-economic aspects of the tourism industry. For the last couple of years, Bangladesh has been underlined as a striking place for tourists. At present, Bangladesh is the 120th ranked nation among 140 in terms of availability of tourist-friendly facilities such as air transport, accommodation, security, culture, and stable travel opportunities for visitors (Bureau of Tourism Research 1995). Besides, *Sundarban* and Cox's Bazar have been included in the Worldwide New Wonders of Nature Campaign among the 440 candidate locations from 220 countries (Travel and Tourism Competitive Report 2019). The total contribution of the tourism sector to GDP is 4.4. It is indicated that the possibilities of the tourism sector of Bangladesh are so high (Sakib and Chowdhury 2016). It is found that Bangladeshi visitors spent at least Tk 60 billion abroad, whereas Bangladesh got only Tk 13 billion from overseas tourists (Reviving tourism sector of Bangladesh 2018). It is mentionable that overseas tourists are the key components of the direct contribution of travel and tourism of Bangladesh because they spend a good sum of money during their visit in the country. In 1984, about 100,000 tourists visited the country and Bangladesh earned foreign currency of Tk 771.81 million, which increased substantially over time and reached up to Tk 1.2273 billion in 2014 (BBS 2016). In 2018, around 0.27 million tourists arrive in Bangladesh to visit and enjoy its beauty. Another source revealed that some 0.84 million foreigners visit Bangladesh annually, but the number of pure tourists is limited. In contrast, 13.7 million internal visitors visit the same spots annually (Sazzad 2020). Table 11.1

Table 11.1 Year-wise number of overseas tourists and income (2008–2018)

Serial no	Year	Number of overseas tourists (million)	Job in travel and tourism	Income (in Tk) (million)
01	2008	0.46	–	–
02	2009	–	–	5762.20
03	2010	0.30	–	5562.90
04	2011	–	2,880,500	6201.60
05	2012	0.12	–	8254.00
06	2013	–	1,328,500	8495.60
07	2014	0.16	1,984,000	12,273.0
08	2015	0.14	234,600	–
09	2016	0.20	2,401,000	–
10	2017	0.26	2,432,000	1369.10
11	2018	0.27	–	8073.20

Source Bangladesh Parjaton Corporation, 2019, Roy and Roy (2015)

displays the year-wise number of overseas tourists and income from the Bangladesh tourism industry (2008–2018).

The tourism sector of Bangladesh is now an effective partner of the country's journey on the development highway. According to the WTTC, at present, 1.5 million people are working directly (in the hotel, motel, travel agencies, leisure industries, airlines, and other transport) and 2.3 million indirectly in the tourism sector of Bangladesh. According to the WTTC, tourism creates 12% of global GNP and it occupies about 200 million people in the world. It is hoped that by 2028, Bangladesh tourism industry will contribute to the job market of Bangladesh with a total of about 3.24 million jobs (Ujhellye 2019).

11.5 Results and Discussion

Tourism encourages the movement of tourists to the region in the world with certain natural features marked to leisure and rest (Rahman and Chakma 2019). In fact, Bangladesh with its superb natural beauty and archeological sites has enormous opportunities to be a site for international tourists for getting overseas currencies. Tourism is the world's largest justifiable service industry that is generating huge revenue for tourist countries. It is observed that Bangladesh requires a vast amount of money for reaching Sustainable Development Goals (SDGs) by 2030. Currently, the chief bases of earning money are export earnings of garments sector and the remittance provided by the Bangladeshi Diasporas working overseas country and minor other raw and finished commodities exported overseas country. These are not enough to meet the development requirements of the country. As a result, the country has to rely on foreign aids which will stand in the present fiscal year at Tk 680.16 billion and about Tk 1.45 trillion including aids taken from different internal sources (Roy and Roy 2015). Under these situations, it is better for the country to search for alternative ways for earning money to break the upsurge of the burden and reliance on external aids. In this respect, the Bangladesh government emphasizes viable improvement of tourism in the country. The tourism board recorded 778,143 inbound tourists last year and the country made revenue of $97.05 million from overseas tourists. To facilitate the incoming of Buddhist tourists, Bangladesh stressed the importance of regional cooperation in the development of strategic tourism resources of the country such as age-old rich Buddhist heritage sites, *Paharpur*, *Shalban Bihar*, *Mahasthangarh*, *Wari-Bateshwar*, *Varotvaina*, and others. Bangladesh has a big number of tourist sites with fanciful natural sights and beauties. As for tourists demand, the peak season in Bangladesh is the period from December to February, while the off-pick season is between June–August. It has been argued that such seasonality is caused by the disaster patterns in the country to some extent (Ali 2018). Major tourist spots of Bangladesh are displayed in Table 11.2.

Bangladesh necessitates to sufficiently develop these potential tourists' spots and their approach roads together with tourists' resorts. Such measures can facilitate the incoming of tourists to Bangladesh from all over the world including western countries like Europe and America. If adequate advertisements are published, incidentally the western tourists spend more than their Asian counterparts.

11.6 Impact of Tourism Industry

Tourism is the emerging economic sector in many countries around the world. It has social and environmental importance, discussed that tourism can bring many economic, social, and environmental profits, particularly in developing countries, although mass tourism is also associated with negative effects. It is proved that the variety of tourist spots and tourism policy can contribute to the economy of a country. It contributes to the creation of jobs, minimizing social inequalities, and conservation of the environment (Wilks et al. 1996). It will be viable if it is cautiously managing the probable impacts on the host community and the atmosphere is not allowed to balance the economic profits. The socio-economic, cultural, and environmental key benefits of tourism are discussed below.

11.6.1 Economic Benefit

Tourism is funding ominously to the economy of many countries. As an important driver of economic growth, it can earn a huge amount of foreign currency and support the balance of payment of a country. It can also generate employment opportunities. Indirect employment is also generated through other sectors like agriculture, food

Table 11.2 Major tourist spots of Bangladesh: at a glance

Serial no	Division	Location in Bangladesh	Places to visit
01	Chittagong	Southeast	Patenga sea beach, Chittagong seaport, Cox's Bazar sea beach, Inanisea beach, Parki beach, St. Martins island, Himchori falls, War Cemetery, Lalmai hill, Moinamoti and Shalbon Bihar, Chittagong hill tracks, Kaptai lake, Foy's lake, etc
02	Dhaka	Middle	Shaheed Minar, Jatiyo Smriti soudho, Lalbagh fort, Ahsan manzil, Jatiya Sangshad Bhavan, Sonargoan, wari-bateshwar, etc
03	Khulna	Southwest	Sundarban mangrove forest, Kotka beach, Hiron point, Mongla seaport, Sixty dome mosque, Sagardari, Rabindrakutibari, *Dakhindihi,* Khanjahanali bridge, etc
04	Rajshahi	Northwest	Paharpur Buddist Vihara, Mahasthangarh, Gokul meth, Varendra museum, Bangabandhu bridge, Harding bridge, etc
05	Barisal	Southwest	Kuakata sea beach, Pairadeep seaport, etc
06	Rangpur	Northern	Kantagir temple, Tajhat rajbari, Sayedpur railway junction
07	Sylhet	Northeast	Jaflong, Ratargul, Bisanakandi, Madhabkunda waterfall, Satcharinational park, Srimongal, Lawachararain forest, Tea garden, etc

Source Bangladesh Parjatan Corporation, 2018, *The Daily Star*, 20 November 2019

production, and retail. Besides, infrastructure development and visitors' expenditure generates revenue for the country which helps to poverty alleviation.

11.6.2 Social Benefit

Tourism is the industry of peace and helps to promote mutual understanding between different cultures and people. In addition to the revenue, tourism can introduce to the country. It also permits them to look at their society, history, and heritage which help the countrymen to maintain their history, tradition, and culture.

11.6.3 Environmental Benefit

Tourism has close relation with the environment. Although Environment delivers the raw material for tourism, it executes impacts on the environment over the manufacture of different by-products (Rasekhi and Mohammad 2015). It affords financial aid for the protection of ecosystems and natural resource management, making the sport more attractive for tourists.

11.7 Challenges for Tourism in Bangladesh

The tourism industry of Bangladesh is facing manifold challenges to construct the industry as internationally renowned. There are numerous reasons that stand as a hindrance to the development of the tourism sector of Bangladesh. In order to achieve sustainable development of the tourism industry, there is a range of challenges that need to be addressed.

11.7.1 Inadequate Infrastructure Facilities

Tourism aids a country in constructing requisite infrastructures which will facilitate the tourists and local people (Johannesburg 2002). Tourism-friendly infrastructure is required for smooth and free movement of tourists of all ages and even for the physically challenged tourists, but the tourism infrastructure is not far too improved as well in the country.

11.7.2 Backward Communication

The role of a well-organized transport and communication system is essential for the socio-economic development of a country. Lack of decent public transports and smooth and safe roads are challenges for the internal and overseas visitors in Bangladesh.

11.7.3 Lack of Accommodation

Most of the towns and cities of Bangladesh lack decent accommodation facilities: standard hotels, motels, and resorts especially for overseas tourist that discourages them to visit Bangladesh. Actually, the country has yet to offer hotels, motels, and resorts of quality.

11.7.4 Lack of Safety and Security

Safety and security are influential factors of domestic and foreign tourists (Henderson 2011). To feel unsafe in Bangladesh is a great barrier that demotivates not only international but also local tourists. In the country, the lack of well-equipped and modern hospitals; access to safe water; and hygienic foods at hotels, motels, resorts, and tourist spots are barriers to the tourists.

11.7.5 Lack of Professionalism

The success of any tourism industry depends on the professionalism of its workforce. But in the country, most of the employers of hotels, motels, resorts, tourist spots, guides, and drivers are not efficient enough to deal with the overseas and internal tourists. As a result, the tourist's especially overseas tourists do not feel comfortable visiting Bangladesh further.

11.7.6 Political Instability

Political stability is also one of the crucial factors to attract domestic and foreign tourists, especially in south Asian countries. The tourism sector of Bangladesh experienced several political crises like strikes, *hortal*s, and conflicts. Political instability produces lawlessness and terrorism and it also damages socio-economic development and impedes infrastructural development, and in both cases, overseas visitors notice the spot negatively (Hai and Chik 2011). It is, therefore, becoming more and more important to analyze negative events from every angle, to systematically identify critical success factors, to integrate them, and to take them into account when considering strategic corporate orientation (Ujhellye 2019).

11.7.7 Lack of Recreation and Tourist Facilities

The country has yet to offer well-equipped recreational facilities attractive to traveling tourists. The other attractive entertainment facilities such as quality shopping malls, cineplex's, theme parks, heritage parks, and museums, are seen as poor here, which decreases the interest of tourists to visit Bangladesh.

11.7.8 Inadequate Allocation in National Budget

In comparison with other tourist destination countries in south Asia, the Bangladesh government allocates a poor amount of finance for the tourism industry sector per year. With this poor allocation, it is not possible to reconstruct the existing tourist spots or build newly tourist spots that make a barrier to flourish the tourism industry of the country.

11.7.9 Lack of Human Resources

Human resource is the most significant capital which has competitive advantages over the other resources for any economic entity. The lack of well-trained tourist personnel to guide the tourists is a problem for the tourism development of the country.

11.7.10 Lengthy Visa Processing

Visa processing system in Bangladesh is complex and time-consuming. As a result, the overseas tourists feel disinterest to visit here.

Besides, these many other challenges are behind underdevelopment of the tourism industry in Bangladesh, such as lack of foreign direct investment and inefficient marketing policy.

11.8 Opportunities for Tourism in Bangladesh

Tourism is defined as an industry that is involved in leisure and travel (Crossan et al. 2015) and has turned into one of the fastest-growing sectors of the economy within the past few years. It has huge potentials to earn a substantial amount of foreign currencies from this sector. But Bangladesh tourism industry is yet to be developed to its desired level despite the immense potential for its improvement (Nazem 2017). After the independence, the Government of Bangladesh established Bangladesh Parjatan Corporation with a view to developing tourism as an important industry. The corporation was restructured in 1973 and formulates policy and strategy to improve the tourism industry in the country (Islam and Hossain 2015). This sector might result in a multiplier effect on the country's economy by not only earning overseas currencies but also creating job opportunities for the huge unemployed population.

11.8.1 Religious Tourism

Bangladesh is a country of lots of spiritual spots like mosques, shrines, tombs, and temples. These places can be promoted to overseas tourists. There are huge opportunities to improve religious tourism here.

11.8.2 Sports Tourism

Bangladesh has huge possibilities to promote sports tourism by developing some venues. Meanwhile, few cricket and football venues have been made in different cities of the country. The Government can seek both local and foreign investment for the improvement of sports tourism. The government also can help develop many backward linkage sectors in the country which will help generate informal jobs like tea vendors, food corners, betel shops, hawkers, etc.

11.8.3 Ecotourism

Pollution-free environment is necessary for the visitors to get refreshed mentally, physically, and spiritually. Ecotourism provides incentives to local communities, entrepreneurs, and the government.

11.8.4 Educational Tourism

Educational tourism was established due to the popularity of teaching and learning of knowledge and enhances technical knowledge out of the classroom. The focus of educational tourism is to visit the overseas country by a tourist to learn the culture and heritage. In this way, a student can exchange programs, organize special lectures of eminent personalities, and visit for field study.

11.8.5 Rural Tourism

Recently, a number of tourist sites have been developed in the rural area of Bangladesh which is eco-friendly for the tourist. To develop rural tourism, it is necessary to organize roundtable meetings, workshops, and seminars with citizens for promoting the development of the rural tourism; arrange and equip the area for tourist; encourage the enrichment of live-stock fund in size and variety; and establish programs.

11.8.6 Cultural Tourism

Culture is one of the most important factors, which attract tourists to a destination. Cultural tourism gives insight to away of people's life of a distant land, its dressing, jewelry, dance, music as well as architecture, customs and traditions, fairs, and festivals.

11.9 Summary of the Findings

It is indicating that the natural and cultural heritage of regions represents the excellent basis for the tourism development of Bangladesh. Bangladesh's tourism industry has some strengths, weaknesses, opportunities, and threats also. Here, SWOT analysis has been applied for the interpretation of major findings of the study related to the tourism industry in Bangladesh (Table 11.3).

Table 11.3 SWOT analysis of tourism sector in Bangladesh

Strength	Weakness	Opportunity	Threat
> Bangladesh is unique for Natural beauty and Archeological site > Cost-effective transportation facility > Rich history and heritage > Well known as a struggle for the mother tongue > Liberal behavior of local people toward tourists > Nation famous for hospitality	> Backward communication and transportation system > Unavailability of proper mode of communication > Insufficient number of Restaurants > Lack of accommodation (resorts, hotel, and motels) > Length booking system > Many attractive locations remain unexplored > Lack of safety and security > Lack of professionalism > Lack of investment	> Creating awareness among tourist, policy-maker, and local people > Arranging of Training program to local community > Campaign make a positive attitude toward tourism > Building positive attitude toward the country > To flourish handicraft and locally made organic food > Founding better Transportation system (luxury bus, luxury boat, helipad, etc.) > Creating cost-effective accommodation facility > Ensuring well security system for the tourist > Creating Job	> Political instability > Fear of abduction > Poor coordination among tourist, local community, and policy-makers > Fear of local communities for loss of land > Misinterpretation or misconception about tourism in loca lcommunities > Non-sustainable behavior of Bengali and tourist > Destructing nature > Conservative social and religious system

11.9.1 Recommendations

To flourish the prospects of tourism, the government of Bangladesh formulated a new Tourism Policy in 2010. This policy emphasizes the expansion of eco-tourism, community tourism, rural tourism, pilgrimage tourism, riverine tourism, archeological tourism, and other tourisms in the context of Bangladesh's cultures and traditions. Only 0.3 million overseas tourists came to Bangladesh in 2010, in which more than 70% came for business and official purposes. Observing the opportunities in the tourism sector of Bangladesh, we need to think more about earning more foreign currencies. Moderate tourist facilities have also been developed but more standard facilities are required to keep up with trends and tastes. The following recommendations for the future development of the tourism industry in Bangladesh are given. Road communications and transport facilities must be improved, so that tourists can reach their desired and attractive sites without hassle.

(a) Infrastructures around the tourism spots should be developed and maintained on regular basis.
(b) To ensure uninterrupted power supply especially around remote and rural tourist spots, continuous access to electricity must be ensured.
(c) It is essential to develop human resources in the sector of tourism.
(d) To preserve the sites of natural beauty for the attraction of tourists.
(e) For the growth of the economy of the country, it is essential to improve the marketing strategy of the government.
(f) To make proper plans and policies of the government for ensuring sustainable tourism.
(g) Political stability should be maintained strictly by the government for the attraction of overseas tourists.
(h) Infrastructural facilities should be improved with target-oriented planning.
(i) To make a plan to preserve environmental sustainability especially in the tourist spots.
(j) In Bangladesh, the majority of the renowned tourist spots are organized by the government sector. As a result, the attraction of the tourists' spot is deteriorating day by day. So, tourism sites should be run by the private sector.
(k) Government should take necessary steps about the security system at the tourist spots. So, the safety and security of the tourists should be given the utmost priority.
(l) Need to upgrade tourism policies in Bangladesh and it should be compared to the world tourism market.

The tourism sector of Bangladesh can add to acquire the country's vision for 2021 in different ways. However, it needs several short, mid, and long-term projects with enough budgetary allocation. The sector also needs to be highlighted in the national development plans and policy programs.

11.9.2 Conclusion

Tourism is a promising sector with an increasing contribution to GDP. Presently, tourism is one of the major sources of income for many countries. Forty years have elapsed of Bangladesh's tourism industry, yet we still see it in a nascent position in comparison to our neighboring countries. Despite having all the potential, Bangladesh tourism is growing slowly. Mere some infrastructural development can make the sector sustainable. Though Bangladesh has lots of attractive tourism spots, most of them are unexplored. As a result, the tourism industry could not develop satisfactorily in Bangladesh. This article will facilitate the policy- and decision-makers to assess the passion of the challenges and to take measures for the improvement of Bangladesh tourism industry which might contribute a large share in the GDP of Bangladesh.

References

Afroz NN, Hasanuzzaman M (2012) Problem and prospects of tourism in Bangladesh: bandarban districtcase. Glob J Manag Bus Res 12(23):234–241

Ali B (2018) Influence of destination factors on touristssatisfaction: the moderating role of securitysystem. J IBS 41:91–102

Antara NF (2019) World tourism day. The Dhaka Tribune, Dhaka, 20 November, p 7

BBS (2016) Statistical yearbook of Bangladesh 2015. Bangladesh Bureau of Statistics, Govt. of Bangladesh, Dhaka, p 19

Bureau of Tourism Research (1995) p 22

Crossan M, Cunha MPE, Cunha J (2015) Time and organizational improsation. Acad Manag Rev 30(1):54–61

Hai A, Chik AR (2011) Political instability: country image for tourism industry in Bangladesh. In: Proceedings of the international conference on science, economics and art 2011, Hotel Equatorial. Bangi-Putrajaya, Malaysia, 14–15 January, pp 131–39

Henderson JC (2011) Tourism and politics development, The Philippines. Tourismos 6(2):159–173

Howlader ZH (2013) Thegreat potential of tourism. The Daily Star, Dhaka, 17 March, p 4

Islam Z, Hossain S (2015) Challenges and opportunities of human resource management practices evidence from private commercial banks in Pakistan 5(1):23–34

Islam F, Islam N (2006) Tourism in Bangladesh: an analysis of foreign tourist arrivals. http://stad.adu.edu.tr/TURKCE/makaleler/stadbah2004/makale040103.asp

Jahangir MJ (1998) Facilitation of international travelin Bangladesh. Bangladesh Parjatan Corporation, Dhaka, p 17

Summit J (2002). World Health Organization, 26 Aug–4 Sept 2002, p 13

Nazem NI (2017) National atlas of Bangladesh. Asiatic Society of Bangladesh, Dhaka, p 166

Rahman S, Chakma J (2019) Tourism booming with economy. The Daily Star, Dhaka, 20 November, p 5

Rasekhi S, Mohammad S (2015) The relationship between tourism and environmental performance. Iran J Econ Stud 4(2):51–81

Reviving tourism sector of Bangladesh, 2018.Reviving tourism sector of Bangladesh (2018) The Daily Sun, Dhaka, 5 Aug 2018, p 4

Roy SC, Roy M (2015) Tourism in Bangladesh: present and future prospects Int J Manag Bus Adm 1(8):53–61

Sakib BA, Chowdhury ST (2016) Enlivening the tourism industry in Bangladesh. The Daily Star, Dhaka, p 5

Sazzad A (2020) Tourism contributes 4.4 PC to GDP. The Daily Star, Dhaka, 22 May, p 6

Siddique R (2020) Contribution of neglected tourism sector to Bangladesh economy is raising. The Bangladesh Monitor, 1 January, p 11

Sultana S (2016a) Economic contribution of tourism industry in Bangladesh. J Tour Hosp Sport. 22:45–54. www.iiste.org

Travel and Tourism Competitive Report (2019) Travel and tourism competitive report-2019 prepared by the world economic forum

Uddin S (2020) Taping tourism potentials crucial to Bangladesh economy. The Financial Express, Dhaka, 6 June, p 3

Ujhellye I (2019) Challenges and opportunities in tourism. Parliament, 23 October, pp 21–25

Wilks J, Pendergast D, Service M (1996) Newspaper reporting of tourism topics. Aust Stud Journal 8:240–255

Chapter 12
Socio-economic Development Through Tourism: An Investigative Study for the Himalayan State Sikkim, India

Debasish Batabyal and Dillip Kumar Das

Abstract Having been accorded the status of the largest service industry in India, tourism is an instrument for economic development and employment generation, particularly in remote and backward areas. Tourism in the Indian Himalayan Region has shown a perpetual and increasing trend over many decades despite several disasters and crises. Sikkim is one of the peaceful and hospitable small Indian states that boasts of rich ecological and cultural diversities. Tourism in almost all the alpine regions of the Indian sub-continent is conceptualized to have been a means of spending from disposable and discretionary income, mostly for non-essential activities. This old and stagnant idea has been changing drastically as tourism either energies through a total long-lasting experience or an essential mean of the present time. On the supply side aspect, it is imperative to provide new avenues for income and employment in the destination. This article has dealt with the modern socio-economic environment of tourism in the backdrop of its essential sustainable development indicators. More specifically, this article has shown how the tourism phenomenon is influencing the ecology and community benefits with important tourism marketing and destination supply trends. The study is based on primary data collection of tourists and local communities in Sikkim, and the statistical tool used herein is Kendall's Coefficient of Concordance.

Keywords COVID-19 · Destination · Development · Contingent valuation method (CVM) · Kendall's coefficient of concordance (Kendall's W) · Marketing · Strategy · Sustainability

JEL Classification P28 · Q5 · R1 · R4

D. Batabyal
Amity University, Kolkata, India

D. K. Das (✉)
University of Burdwan, Burdwan, West Bengal, India

N. C. Jana et al. (eds.), *Livelihood Enhancement Through Agriculture, Tourism and Health*, Advances in Geographical and Environmental Sciences,
https://doi.org/10.1007/978-981-16-7310-8_12

12.1 Introduction

Globally, mountain areas cover almost 27% of the land surface and contribute to the existence and habitation of more than 720 million people. Over half of the human population extensively depends on mountains for water, food, and clean energy. Climate change, avalanche, landslides and slips, flush flood, overpopulation, illegitimate construction activities, and natural disasters are with potentially far-reaching and devastating adverse impacts and consequences, both for hosts and guests in upland areas and downstream populations. It is estimated that one out of three persons in developing countries is vulnerable to food crisis and live in poverty and socio-economic isolation. While there is great scope for tourism growth in Indian Himalayan Region (IHR), it must be managed so that it is dignified, inclusive, and sustainable (ensuring jobs, promotion of local culture and tourism products) and contributes to achieving the Sustainable Development Goals (SDGs), more specifically, 1st, 8th, 12th, 13th, and 15th goals, respectively. Had not this COVID-19 pandemic occurred, the region would have been able to grow the number of tourists at an average annual rate of 7.9% from 2013 to 2023 in continuation with its previous positive growth rate. The 12th Five-Year Plan of the government of India clearly recognized pro-poor tourism for inclusive growth that was expected to reduce economic marginalization resulting in socio-economic disparity.

In Table 12.1, domestic and international tourist arrivals are showing an overall increasing growth rate. The table also depicts a handsome increase in foreign tourist arrivals in the present decade, and out of 1,33,388 foreign tourists in 2019, a total of 60,542 were from Bangladesh and 56,781 from Nepal only. Overall increasing domestic tourist arrival and continuously increasing emergence of tourism as a source of powerful industry and employment brought about a new orientation for the socio-economic growth and development through tourism in the study area.

COVID-19 outbreak has a more adverse consequence on economic and social setup than any other form. Drastic reduction in physical accessibility and actual mobility is found to have been an area of uncertainty and insecurity for the new

Table 12.1 Domestic and foreign tourist arrivals in Sikkim

Year	Domestic tourists arrivals	Foreign tourist arrivals
2011	552,453	23,945
2012	558,538	26,489
2013	576,749	31,698
2014	562,418	49,175
2015	705,023	38,479
2016	740,763	66,012
2017	1,375,854	49,111
2018	1,426,127	71,172
2019	1,615,000	133,388

Source Sikkim Tourism Development Corporation (2020)

normal travel industry. The recent outbreak has brought about a loss of employment, reduction in income, and more life risk associated with travelling and overnight accommodation, and such Indians are expected to spend less on tourism as a noticeable trend in coming years as never before. Travel and tourism phenomenon is seemed to be moving towards the unorganized, non-institutional, and unprofitable market segments from Visiting Friends and Relatives (VFR) to other small-scale forms of tourism. With the loss of capital and provision for health and safety measures, many travel houses, hotels, and other principal suppliers in the region have been operating at a break-even level. Again, maintaining such levels for a longer time span is very difficult for travel principal suppliers with huge fixed costs while commission agents have already got their working capital blocked during the earlier stage of lockdown which occurred during the pandemic process. At this point of time, an assessment of socio-economic analysis is imperative to reintroduce tourism for fighting against poverty alleviation and social inclusiveness.

12.2 Brief Overview of the Study Area

Sikkim is surrounded by the stretches of Tibetan plateau in the North, the Chumbi Valley and Bhutan in the East, Nepal in the West, and Darjeeling (West Bengal) in the South, respectively. Sikkim is famous for scenic valleys, rich vegetation patterns, snow-clad mountains, magnificent monasteries, Buddhist heritage and the most hospitable people (Pandey et al. 2011). This small land has surprisingly a presence of diverse ethnic communities. It is also a land of Arya-Mongoloid confluence with the Hindu-Buddhist coexistence. When Sikkim was an independent state, it experienced many invasions by its neighbouring countries, and as such the then king sought protection from British India and, later, gifted Darjeeling and adjacent areas to British India. This 22nd Indian State has over eighty-one per cent of the total geographical area, which is under the administrative and managerial control of the Ministry of Environment and Forest, Government of India. More than forty-five per cent of the whole geographical area of the state has forest coverage and about thirty-four per cent of the geographical area is set aside as a protected area network as a national park, wildlife sanctuary, etc. So, Sikkim deserves more environmental consciousness, alternative and eco-friendly approaches, and destination management with sustainable principals. The following is the environmental performance index for all the Himalayan states and Sikkim for a relevant compression (Table 12.2).

Sikkim's rank in overall environmental performance is two with an alarming air pollution and waste management. Both are assumed to be more than their normal rate during peak season when tourists are visiting (NITI Aayog 2018).

Sikkim conserves its rich biological diversity that includes coexistence and protection of over 5000 species of angiosperm which is one-third of the total national angiosperms, respectively. Among the multi-ethnic communities, the Bhutias came from Tibet while the Lepchas are the aboriginal local population of Sikkim. The Nepalese came from Nepal. When Sikkim was an independent state, it experienced

Table 12.2 Environmental performance index (EPI) of Indian Himalayan States

State	Air pollution		Water		Forests		Waste management		Climate change		Final environmental performance	
	Score	Rank	Score	Rank	Score	Rank	Score	Rank	Score	Rank	Score	Rank
Arunachal Pradesh	0.3333	33	0.3333	32	1	1			0.4885	6	0.4310	31
Assam	0.929	12	0.6536	19	0.4993	18	0.7643	12	0.3658	18	0.6426	21
Himachal Pradesh	0.8939	15	0.9843	1	0.6531	10	0.8550	5	0.3208	20	0.7414	3
Jammu and Kashmir	0.8571	20	0.6758	11	0.5783	13	0.4161	28	0.2139	26	0.5483	22
Manipur	0.9048	12	,6667	12	0.4601	23			0.3740	15	0.4811	28
Meghalaya	0.8939	15	0.6544	18	0.4355	25	0.8718	4	0.4061	14	0.6524	10
Mizoram	1	1	0.6667	12	0.5071	17	0.4220	27	0.6280	2	0.6448	12
Nagaland	0.9608	7	0.6458	22	0.3677	28	0.4679	23	0.0378	31	0.4960	26
Sikkim	**1**	**1**	**0.6399**	**9**	**0.6230**	**11**	**0.9333**	**2**	**0.4892**	**7**	**0.7478**	**2**
Tripura	0.5881	31	0.6667	12	0.7851	8	0.4008	29	0.3713	17	0.5624	20
Uttarakhand	0.7850	25	0.5948	28	0.8280	5	0.4283	26	0.4351	12	0.6142	16
West Bengal	0.7425	29	0.5739	30	0.4009	26	0.4567	24	0.4909	5	0.5330	25

Source Planning Commission of India

many invasions by its neighbouring countries and the king sought protection from the then British India (Rahman 2006). Later on, this state became a protectorate state of India. Now this twenty-second state of India has over 81% of the total land area under the administrative managerial control of the Ministry of Environment and Forest, whereas nearly 34% of the total area is declared as a protected area in the form of the national park and wildlife sanctuary. The temperature is varying from 28 °C in summer to 0 °C in winter. The temperatures in the valleys (up to 600 m) situated at lower elevations, particularly during summer, are like the monsoon type of climate. The temperature decreases from 600 to 2000 m above sea level with cool temperature climatic conditions and further up till 3000 m enjoys a cold temperate climate, and above 5000 m is arctic type, respectively. The state has a variety of deposited mineral resources such as coal, limestone, iron ore, graphite, and pyrite.

Increased accessibility by roadways and air transport, fast social and economic development, and competitive advantage exhibited through geographical proximity availability of tourist generating states contributed to the development of tourism in Sikkim. Considering the ever-increasing tourist arrivals, accommodation units were set up initially in and around Gangtok, and later on, in a few towns mostly by outsiders without proper land use planning and architectural design (Batabyal 2018).

12.3 Literature Review

NITI Aayog (2018), in coordination with some crucial national institutions and ICIMOD International Centre for Integrated Mountain Development (ICIMOD), set up an Action Agenda for "Sustainable Development of Mountains of Indian Himalayan Region (IHR)", in which non-mass tourism in Indian Himalayan Region (IHR) gained ultimate attention for the world's common future. The report incorporated a review of current tourism and key cross-cutting policies and plans in the Indian Himalaya, identified gaps and best initiatives (policy and practice) related to sustainable tourism, convened state-level dialogues (multi-sector and multi-stakeholder), and revisited integrated sustainable tourism policies, financial and institutional efforts, and regulatory framework of all the states at the local and regional level. Though context specific, there are some promising cases from the Indian Himalayan States (IHS) on solid waste management, controlling visitor flow to Rohtang Pass in Himachal Pradesh, controlled number of pilgrims at Vaishno Devi in Jammu, community-based tourism, and tourism policies embracing sustainability principles in Sikkim (NITI Aayog 2018).

Tourism planning put forward goals, strategies, and objectives for the tourist destinations that require tourism development marketing, institutionalized setup, awareness, and other support services and activities while tourism development plan normally provides overall guidelines for development, outlines broad development concepts, and identifies individual development opportunities worthy of in-depth analysis through feasibility studies and/or cost–benefit analyses (Mill and Morrison

1986). While moving with tourism development, the destination area would establish overall development guidelines in conformity with the area's economic, environmental, social, and cultural policies, goals, and strategies (Saarinen 2003). The most well-accepted destination development plans are worked out with an unanimous multiple participation and consultation of all possible private, public, and nonprofit sectors (Müristaja 2003). Organizing and coordinating are found to be among the core activities in tourism planning and strategy formulation as well. Another recommended approach is the development through a holding enterprise with the participation and consultation of all stakeholders including local groups (Moulin et al. 2001).

In 1997, the Tourism Policy of the Government of India exhibited a belief that the future prospects of tourism would be quantified by the yields of the hosts and was primarily an industry of spare time not considering the interest and monitoring of the policymakers and destination managing authority. According to this Policy, this belief was based on an information gap that chronologically undermined and constrained the development of tourism in India over the years. National Action Plan for Tourism (1992) suggested to bring off different types of tourism products, the continual growth of tourism infrastructure, effective marketing and promotional efforts in the overseas markets, and mitigating measures of all impediments of tourism but could not set out the infrastructural requirements and the investments needed to meet the targets and the sources of funding for the same.

Maharana et al. (2000a, b) showed the application and relevance of the Contingent Valuation Method (CVM) to measure the Willing to Pay (WTP) that is an important part of Polluter Pays Principles (PPP). They found WTP was strongly influenced by age, education, and income. While Kumari et al. (2010) advocated factor indices incorporating wildlife, ecology, ecotourism value, ecotourism attractivity, ecotourism diversity, and environmental resiliency to identify and prioritize the potential alternative or ecotourism sites in Sikkim.

The tourism strategy for a region generally revolved around five key principles, viz. Strategic allies, total quality assurance, resilience, competitive performance, and ease of access. It is obvious that a formulated strategy addresses a holistic theme within which tourism can explore and grow with dignity for all stakeholders (Porter 1980). Also, tourism strategy creates and evolves more strategies, e.g., Country Cultural Strategy, Community Strategy, etc. (Cambridge Tourism Strategy 2001–2006).

In his article "Why Is the State of Sikkim India's COVID-1v9 Exception?", Bhattacharyya (2020) mentioned that the first step was the closure of tourist destinations on fourth March along with hotels, restaurants, cinema halls, schools, and colleges, all of which were shut down after two weeks to stop the large crowd. Consequently, all important check cum entry posts along the state's borders with the three countries were closed which included Nathu La. Only two gates at Rangpo and Melli remained functional as entry and exit points with West Bengal as required. Further, initiatives were taken to bring back the residents of Sikkim with all necessary arrangements for supplying foods, medicines, etc. with residents and volunteers. At present, the government wants to unveil a plan to boost the rural economy during the lockdown period. The economy is found steadily depressive due to almost a ban on tourism, which contributes around 8% to the state's gross domestic product.

12.4 Objectives and Methodology

In a holistic sense of established proposition, tourism provides the local mountain people with valuable economic and business opportunities and jobs, and for state governments and private entrepreneurs, it brings revenues and profits (Tapper 2001). But it was evident that the principal supplies are being owned and managed by the entrepreneurs or business houses of other states in India. The following are the objectives of the study.

- To measure the distribution pattern of income from tourism among local people in the study area.
- To find out the supply trends for post-COVID-19 tourism in Sikkim.

Based on the objectives above, the following are the hypotheses taken into consideration in the study.

H01: Tourism in the study area is for the upper layer of the local people as well as hosts and benefits don't accrue to the local people to enhance social inclusion.
H02: Due to COVID-19 pandemic, supplies of travel products will reduce and entrepreneurs will diversify themselves.

The time period for collecting data was the last quarter of the year 2019 and the effects of COVID-19 pandemic were not considered with regard to future visits.

The number of respondents in the study was 400, and the overall response rate was 50.25% (201 completed, usable questionnaires). A total of 500 local people of the different areas of Sikkim participated in a community survey and the overall response rate was 19.80% (99 completed, usable questionnaires). Though, an online opinion survey was conducted for the local entrepreneurs and travel associates during the pandemic to assess the present situation and future direction of the tourism sector in Sikkim.

Kendall's Coefficient of Concordance was considered for analysis and interpretation of the primary data as it directs the degree of association among several (k) sets of ranking of N objects or individuals. This descriptive measure of the consensus has special implications in providing a standard method of ordering objects according to the agreement when the data set does not have an objective order of the objects. The basis of this test is to conceptualize how the given data would shape if there were no agreement among the several sets of rankings, and then to visualize how it would look if there were perfect agreement among the sets available for the study. Apart from this, the collected data was analysed by using Excel and SPSS 19.0, respectively.

12.5 Data Analysis and Discussion

One of the major factors in the destination population studies in respect of monthly income from all available occupations and monthly income only from tourism in Sikkim is taken into note of in this study. In this study, the changing mean of monthly average income and the mean of monthly average income absolutely from tourism have been studied and found relevant.

Thirty-eight per cent of all participants are found to be associated with tourism business in Sikkim and their range of monthly average income is less than nine thousand five hundred to above thirty-four thousand five hundred, justifying the need for tourism development for the poor vulnerable section to lower middle-class people in the state.

During the study year 2019, it is found that the lower is the earning of the local population, the higher is the mean of monthly average income from tourism. It signifies tourism as an emerging socio-economic activity among the local population for maintaining a livelihood. A non-parametric Kendall W rank test is conducted to test the consistency of ranks. These ranks are found consistent in this study with the suitable P value (>0.10). Table 12.3 showing the importance of different variables in this study.

Table 12.3 recommends basic infrastructural support for tourists along with the moderate fulfilment of their associated derived demand for tourism facilities and services. Price-conscious domestic tourists are the majority and recently increasing foreign tourists put added scope for future income as the second important criteria for a retained visit while the same ranks fourth for the foreign nationals. Foreign nationals visit Sikkim mostly for different niches of tourism and travel, with alternative tourism predominating. They recommend proximity and more frequency of vehicles, fair, uniform, and cheap price, information and communication aids, and outlets. But after the pandemic, the safety support system with its overall rank of 6th in Table 12.3 would come first or much ahead.

12.6 Conclusion

The widely accepted earlier concept of tourism for the rich and elite is found to be the perception of hosts as usually the tourists are richer than the common local people. This study has contributed to further scope for tourism development and promotion in a distinct socio-economic condition of the Sikkim Himalaya Except for price consciousness, almost all desire for moderate infrastructural supports. Local entrepreneurs and associates are of the opinion that owing to the effects of COVID-19 pandemic, the supply of travel products incorporating hotel rooms, seats in vehicles, etc. will reduce, and a need for tourism capacity building will increase. They also indicated a massive grass route redirection of travel trade oriented to community-based, responsible, and alternative tourism in Sikkim. Informal and subsidiary sectors

Table 12.3 Determinants of repeat visitation of tourists (through mean of ranks)

Criteria	All respondents		National and internal respondents		Inbound 1 respondents	
	Arithmetic mean of ranks	Ranks	Arithmetic mean of ranks	Ranks	Arithmetic mean of ranks	Ranks
Near future better scope for earning	3.16	1	3.38	1	4.38	4
Tourism products (economical)	3.67	2	3.72	2	3.84	2
Changing attitude of hosts towards guests	8.59	10	8.32	10	9.72	12
Skilled staff and their friendly treatment	6.82	7	6.80	7	7.66	9
Awareness/information about equipments at tourist spots	6.05	5	5.951	5	6.58	6
Disaster management measure	6.64	6	6.66	6	6.97	7
Proximity, frequency of vehicles for travel	4.03	3	3.93	3	3.59	1
Pathways at attraction premise	7.66	9	7.60	9	7.56	8
Display board, etc. at major tourist spots	9.27	11	9.33	12	8.50	10
Installed elevators or construction of ramp at tourist spots	7.20	8	7.32	8	6.31	5
Effective information communication support	5.190	4	5.370	4	3.88	3
Tour operators and guides	9.720	12	9.63	11	9.02	11.00

Note Statistical analysis has been made using SPSS 19.0 statistical package

are expected to be stronger. High fixed costs of operation, waning of capital, risks associated with large occupancy, increasing costs of safety measures, and rebranding for safe tourism practice are found to be the reasons for this. Local entrepreneurs are also excited for offering more homestays with more safe means with the support of the local government. The government officials and local entrepreneurs want a zero base new initiation with more inclusion and new avenues for tourism. Owing to this pandemic, local government representatives and businessmen want a dilution of tourist traffic and a greater number of alternative routes and packages in continuation with the earlier tourism development plan. Tourism carrying capacity for each destination is more important for safe tourism and social distancing. The opinion survey also highlighted that the local people are disciplined, and the rural lifestyle helped them in mentioning lockdown strictly. They did not consider inaccessibility as a reason for the successful implementation of lockdown.

References

Batabyal D (2018) Assessing sustainable tourism and crisis situations: an investigative study of a Himalayan state, Sikkim (India). In: Batabyal D (Eds) Managing sustainable tourism resources, IGI Global, pp 87–110. https://doi.org/10.4018/978-1-5225-5772-2

Bhattacharyya R (2020) Why is the state of Sikkim India's COVID–19 exception? The Diplomat. https://thediplomat.com/2020/05/why-is-the-state-of-sikkim-indias-covid-19-exception

Kumari S, Behera MD, Tiwari HR (2010) Identification of potential ecotourism sites in West District. Sikkim Using Geospatial Tools. Trop Ecol 51(1):75–85

Maharana I, Rai SC, Sharma E (2000a) Environmental economics of the Khangchendzonga National Park in the Sikkim Himalaya, India. Geo J 50:329–337. Retrieved from http://search.proquest.com/docviews on 15 Mar 2011

Maharana I, Rai SC, Sharma E (2000b) Valuing ecotourism in a sacred lake of the Sikkim Himalaya, India. Environ Conserv 27(3):269–277. http://search.proquest.com/docviews

Mill RC, Morrison AM (1986) The tourism system: an introductory text. Prentice Hall Publication, Englewood Cliffs

Moulin C, Boniface P (2001) Routeing heritage for tourism: making heritage and cultural tourism networks for socio-economic development. Int J Heritage Stud 7(3):237–248. https://doi.org/10.1080/13527250120079411

Müristaja H (2003) Development trends and the association of stakeholders in tourism development process in the case of Pärnu County (Estonia). University of Tartu, Pärnu college, Estonia

National Action Plan for Tourism (1992) Ministry of civil aviation and tourism, government of India. https://tourism.gov.in/sites/default/files/2019/10/National%20Action%20Plan%20For%20Tourism%201992compressed.pdf

NITI Aayog (2018) Sustainable tourism in the Indian Himalayan region, report of working group II. NITI Aayog. https://www.niti.gov.in/writereaddata/files/document_publication/Doc2.pdf

Pandey H, Pandey PR (2011) Socio-economic development through agro-tourism: a case study of Bhaktapur, Nepal. J Agric Environ 12

Porter ME (1980) Competitive strategy: techniques for analyzing industries and Competitors. Free Press, New York

Rahman SA (ed) (2006) The beautiful India-Sikkim. Reference Press, New Delhi

Report of working group II sustainable tourism in the Indian Himalayan region (2018). NITI Aayog, Government of India Publication

Saarinen J (2003) The Regional economics of tourism in northern Finland: the socio-economic implications of recent tourism development and future possibilities for regional development. Scand J Hospitality Tourism 3(2):91–113. https://doi.org/10.1080/15022250310001927

Sikkim Tourism Development Corporation (2020) Sikkim, the land of peace and tranquility. India Brand Equity Foundation. https://www.ibef.org/download/Sikkim-June-2020.pdf

Tapper R (2001) Tourism and socio-economic development: UK tour operators' business approaches in the context of the new international agenda. Int J Tourism Res 3(5):351–366. https://doi.org/10.1002/jtr.348

Chapter 13
Tourism Potentials of Fossil Parks as Geoheritage Sites: A Study in Western and South Western Region of West Bengal, India

Rahul Mandal and Premangshu Chakrabarty

Abstract Fossils are paleontological treasures manifesting preserved remains, impression, or traces of organisms that existed in past geological ages. Fossil parks are one of the major geotourism attractions especially when they attain the status like world heritage sites. From geotourism promotion perspective, the fossil parks have exceptional heritage and scientific values as admired by UNESCO. Geosite and geomorphosite tourism is still at its juvenile stage in India despite of the positive efforts of Geological Survey of India who recognized 26 geological sites as National Geological Monuments including the seven fossil parks. With increasing interest on fossils as geoheritage, geopark network of the country has been strengthened with inclusion of new sites. In the year 2006, angiosperm wood fossils have been discovered from Illambazar forest of Birbhum District in and around a tribal village named Amkhoi. A fossil park is inaugurated in the year 2018, and its success encourages its extension and further planning for geotourism. This paper is an attempt to apply SWOT-AHP analysis on Amkhoi Fossil Park, the first fossil park of West Bengal which is youngest among the fossil parks of the country in order to evaluate its sustainability aspects.

Keywords Paleontological · Geosite · Heritage · Geotourism · Sustainability

13.1 Introduction

The word 'fossil' is derived from a Latin word 'fossilis' which means something obtained from digging. Usually found in sedimentary rocks, fossils are paleontological treasures manifesting preserved remains, impression, or trace of organism that existed in past geological ages (Chakrabarty and Mandal 2019a). The process by which the remains of organism gradually transformed into fossil is known as fossilization which occurs rarely since in normal circumstances, the decomposers of the environment quickly decompose the organic body. Followed by a rapid burial

R. Mandal · P. Chakrabarty (✉)
Department of Geography, Visva-Bharati University, Santiniketan, West Bengal, India

N. C. Jana et al. (eds.), *Livelihood Enhancement Through Agriculture, Tourism and Health*, Advances in Geographical and Environmental Sciences,
https://doi.org/10.1007/978-981-16-7310-8_13

under sediment cover if an environmental condition is suitable to prevent the action of the decomposers, the process of petrification may operate in which the organic remains are replaced by minerals. The minerals which are involved in this process are silica, calcite, pyrite etc.

Fossils are admired as the data sources on paleo-environmental history (Chakrabarty and Mandal 2019b). They are non-renewable earth heritages imparting the scientific knowledge on the past climate, past depositional environment of sediments, past geographical conditions, relative age determinations and past ecology (Dietz et al. 1987). A research gap exists on analysing of its tourism potentials, and this paper is an attempt in this context with reference to a case study on the youngest wood fossil park of India located at Amkhoi (Fig. 13.1) in West Bengal.

The fossil parks distributed worldwide have been recognized as the cradles of modern geotourism. Fossil parks are one of the major geotourism attractions especially when they attain the status like world heritage sites. There are numbers of such fossil rich sites worldwide such as Australian Fossil Mammal Sites (Riversleigh/Naracoorte), Australia; Joggins Fossil Cliffs, Canada; Dinosaur Provincial Park, Canada; Chengjiang Fossil Site, China; and Messel Pit Fossil Site, Germany, which have already obtained the status of World Heritage Sites from UNESCO. As the fossil parks have exceptional heritage and scientific values as admired by UNESCO with its recognition under the Global GeoPark Network, they can play a crucial role in geotourism promotion and geoheritage conservation perspective (Császár et al. 2009). At present a number of geoparks in the world are primarily based on fossils. Geological Reserve of Haute Provence (France), Petrified Forest of Lesvos (Greece), Nature Park Terra Vita European Geopark (Germany), Abberley and

Fig. 13.1 Glimpse of Amkhoi Fossil Park (23° 37′ 25″ N, 87° 34′ 56″ E) (*Source* Authors, 2020)

Malvern Hills (UK), Geopark Schwabian Alb (Germany), Hateg Country Dinosaur Geopark (Romania), and Forest Fawr Geopark—Wales (UK) are among the popular geoparks in Global Network of Geoparks assisted by UNESCO where fossils are the main attraction (Turner 2006).

In India however, none of the geological sites have ever been nominated for World Heritage recognition. Geosite and geomorphosite tourism is still at its juvenile stage in India despite of the positive efforts of Geological Survey of India who recognized 26 geological sites as National Geological Monuments (Anantharamu et al. 2001) for the visitors among which there are seven fossil parks namely Siwalik Fossil Park, Himachal Pradesh; Marine Gondwana Fossil Park, Chhattisgarh; National Fossil Wood Park, Thiruvakkarai, Tamil Nadu; National Fossil Wood Park, Sathanur, Tamil Nadu; Akal Wood Fossil Park, Rajasthan; Jhamarkotra Stromatolite Park, Rajasthan; and Bhojunda Stromatolite Park, Rajasthan. Apart from these, there are a number of fossil parks scattered in the different part of the country viz, Indroda Dinosaur and Fossil Park, Gujarat; Dinosaur National Park Bagh, Madhya Pradesh; Mandla Plant Fossil Park, Madhya Pradesh; Ghugha National Fossil Park, Madhya Pradesh; Salkhan Fossil Park, Uttar Pradesh; Wadadham Fossil Park, Maharashtra; Dinosaur Fossil Park, Balasinor, Gujarat; Kutch Fossil Park, Gujarat; and Amkhoi Wood Fossil Park, West Bengal which could be classified based on four main fossil groups—invertebrate fossil, vertebrate fossil, wood fossil and stromatolite fossil (Fig. 13.2).

The primary aim in establishing a fossil park is to conserve the paleontological resources, which is followed by objectives like to educate the public about these fossils and promote geopark tourism for the purpose of sustainable socio-economic development of the local community. When a fossil park becomes a new tourist destination, it may offer new opportunities of employment and create many new economic activities to benefit the local people (UNESCO 2010; Farsani et al. 2011). The chief functions and activities of the geopark network include protection and conservation of fossils, scientific interpretation of the fossil resources, establishing geomuseum, organizing fossil oriented scientific and cultural events to educate the visitors further (Zouros 2010). How such activities could further explore the geotourism potentials is the research question that has been dealt in this study.

13.2 The Study Area

In West Bengal, rich wood fossils and vertebrate fossils sites are situated in the Himalayan foot-hills of Darjeeling District as well as in some discrete lateritic patches scattered in the western and south western region of West Bengal (Table 13.1).

Figure 13.3 reveals the relation of the location of fossil sites with the rivers flowing from north west to south east in the region. Among the major sites, the Directorate of Forests, Government of West Bengal concentrated on Amkhoi for developing a fossil park in an organized manner that has been inaugurated on 3rd January, 2018, for the geotourists.

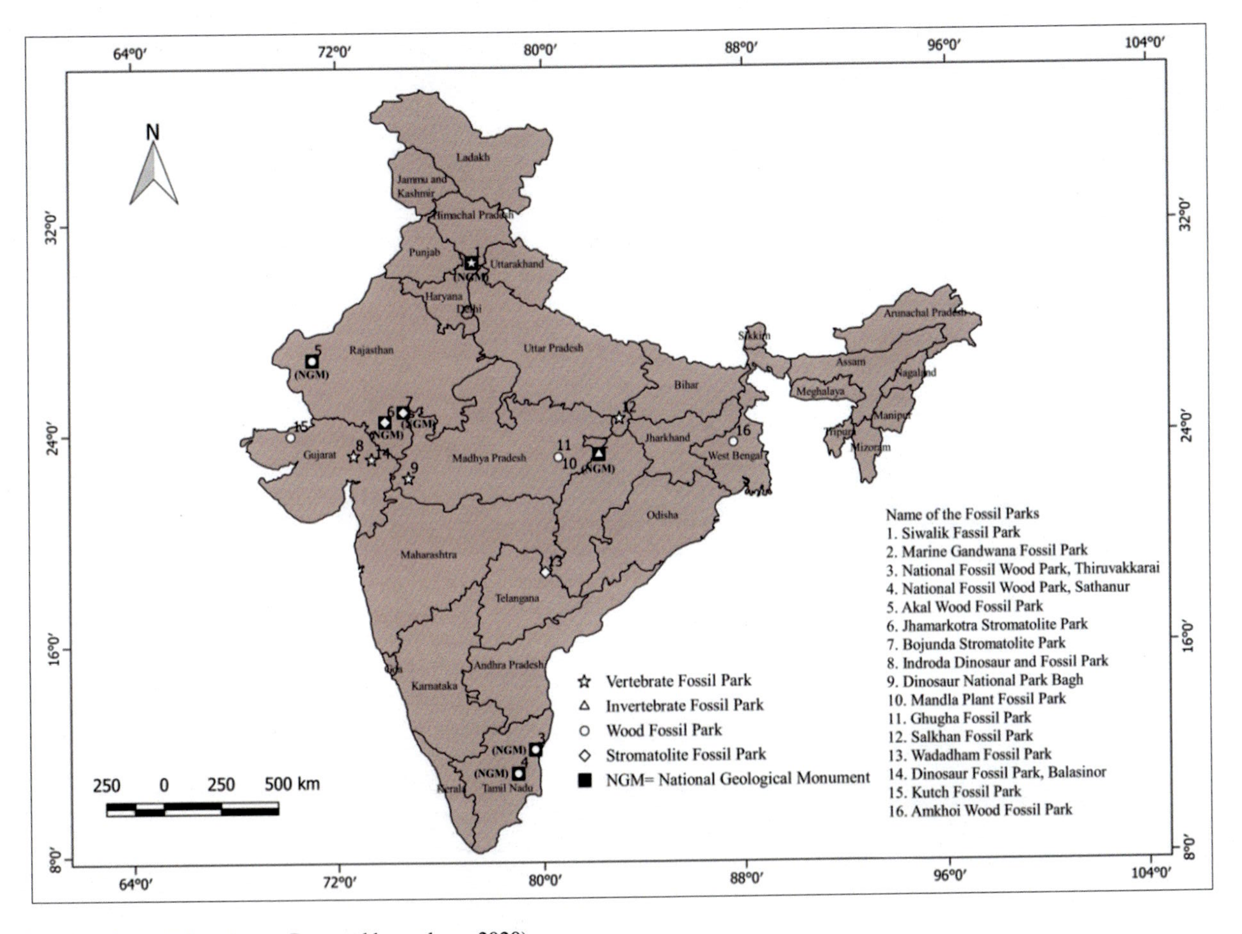

Fig. 13.2 Fossil parks in India (*Source* Prepared by authors, 2020)

Table 13.1 Major fossil sites in the western and south western region of West Bengal

Fossil sites	Geological and palaentological heritages and attractions
Susunia hills (23°23′ 4″ N, 86° 58′ 34″ E)	Susunia hills and surrounding Gandheswari river valley are very rich with vertebrate fossils. Susunia hills and its adjoining area are considered to be one of most significant paleontological sites with abundant of mammalian fossils and pre-historic artefacts. Due to this attraction, this area had been investigated, excavated and explored by various eminent geologists, archaeologists and palaeontologists. A large number of fossils mostly vertebrate animal fossils and bone tools have been discovered from the Susunia hills and its surroundings e.g. Bamundiha (south of Dhankora Jor), Biribari (near the confluence of Gandheswari and Jhikuria) and Suabasa and two caves namely, Jamthol (southwest of Susunia) and Bhaluksoda (in the northeast of Susunia hill) (Dassarma et al. 1982). The age of these fossils dates back to Upper Pleistocene to Early Holocene period. The geological formation of Upper Pleistocene sediments deposited in Bankura District is termed as Baltora Formation (Dassarma and Biswas 1978). A large number of vertebrate fossils have been found from it. Above the Baltora formation, there exist old alluvial terrace sediments of Early Holocene time known as Badladanga-Bamundiha Formation (Sastry et al. 1976) from where numerous mammalian fossils have been discovered During the excavation in 1959–66, Directorate of Archaeology, Govt. of West Bengal under the supervision of P.C. Dasgupta collected a large number of vertebrate fossils along with more than two thousand artifacts ranging from Chello-Acheulian to Neolithic types in Susunia hill complex (Dasgupta 1967). It has been assumed that the starting of this culture might be in later half or the end of Pleistocene. The artefacts which have been collected in and around Susunia are showing high level development of Acheulian industrial technology in Bengal (Sen et al. 1963). In 1966, Geological Survey of India (GSI) recovered a large number of mammalian fossils of upper Pleistocene age viz, *Palaeoxodon namadicus*, *Bubalus palaendicus*, *Equus namadicus*, *Bos namadicus* etc. (Sastry 1968). C-14 dating on the fossilized remains of *Palaeaxadan namadicus* (extinct elephant species) revealed that it belonged to a period before 40,000 BC (beyond the limit of the sensitivity of the instrument, Jadavpur University) in this region (Dasgupta 1967). Further field investigation in this area by GSI discovered huge number of fossils of uppermost Pleiostocene to early Holocene time which includes both living forms as well as some extinct forms viz, *Bos cf namadicus*, *Bubalus bubalis var. palaeindicus*, *Boselaphus namadicus*, *Palaeoloxodon sp.*, *Crocuta sp.*, *Hystrix crassidens*, *Antilope cervicapra*, *Sus cristatus*, *Panthera leo*, *Batagur baska*, *Gavialis gangeticus* etc. (Dassarma et al. 1982). It is noteworthy to mention that Dutta (Dutta 1976) discussed the occurrence of fossil lion and spotted Hyena from the Pleistocene deposit of Susunia; Banerjee and Ghosh (Banerjee and Ghosh 1977, 1978) reported the occurrence of *Giraffa* and Saha; Banerjee and Talukder (Saha et al. 1984) reported occurrence of fossil *Panthera pardus*. The animals' fossils recovered from this area once upon a time roamed through the elevated tract of Susunia Palaeontological study of fossils and geostratigraphy excavations of Gandheswari river banks clearly ascertained the oscillation of climate and environment during Pleistocene and early Holocene period (Dutta 1976). This area experienced a sequence of wet and dry phas

(continued)

Table 13.1 (continued)

Fossil sites	Geological and palaeontological heritages and attractions
Tarafeni river valley (22° 38′ 25″ N, 86° 50′ 56″ E)	Numerous vertebrate fossils have been discovered on the right bank of Tarafeni river (a tributary of the Kansabati river) at Dhuliapur, Jhargram from the surface of calcrete which is actually a part of Sijua formation comprising of brown/grey colour sands and loams of fluvial origin (Basak et al. 1998). Most noteworthy among them are an assemblage of animal fossils of late Pleistocene age including *Bos namadicus* (extinct cattle), *Axis axis* (spotted deer), *Antilope cervicapra* (Black back), *Equus* sp. (Horse), fragments of Turtle carapaces etc. Detailed study on the paleontological account of the valley area indicates that in late Quaternary period, there prevailed sub-humid climatic condition along with savanna habitats, characterized by waterpool conditions and abundant grass growth in the foothill region (Badam 2000)

(continued)

Table 13.1 (continued)

Fossil sites	Geological and palaentological heritages and attractions
Gondwana fossil sites near Panchet area (23° 39′ 05″ N, 86° 51′ 07″ E)	The Panchet area of Domodar river valley is well recognized for its rich vertebrate fossils, particularly the presence of the dicynodont genus *Lystrosaurus* fauna (Gupta 2009). The *Lystrosaurus* fauna have been found from the Early Triassic Panchet Formation, which is underlain by Raniganj Formation and overlain by the Supra-Panchet Formation. Lithologically, the Panchet Formation is divided into two parts namely, Lower Panchet and Upper Panchet sediments (Dasgupta 1922; Tripathi 1962, 1969; Tripathi and Satsangi 1963). The Lower Panchet sediments are rich in *Lystrosaurus* fossils and most of the fish fossils have been collected from Upper Panchet sediments (Gupta 2009). Geological Survey of India (GSI) and Indian Statistical Institute (ISI) have been undertaken extensive field survey in Panchet series and collected a huge number of fossils from this region (Tripathi and Satsangi 1963). A diverse assemblage of fossil fishes is reported in the Lower Triassic Gondwana sediments of Damodar river valley (Gupta 2009; Tripathi 1969). The fish fossils recovered has a close similarity with the Early Triassic fish fossils found in different parts of Gondwanaland which actually reveals the evidence of the geographical unity of the Gondwana landmasses during the Triassic period and also giving stress to the concept that there was no such faunal provinciality at that time (Battail 1998; Bender and Hancox 2004; Lucas 1998)
Gangani badland (22° 51′ 24″ N, 87° 20′ 24″ E)	Numerous trace fossils and petrified woods have been recovered from spectacular Gangani badlands area, popularly known as the grand canyon of Bengal. The area is covered by 3.5 km^2 of Pleistocene lateritic upland and it is distinguished for amazing ravine development on the concave right bank of Silabati river (Bandyopadhyay 1988). Trace fossils are considered as one of the most important environmental indicators of the geological succession of Tertiary-Quaternary strata in western part of Bengal basin (Bera 1996). The outcrop sediments exposed at Gangani badland along the Silabati river contain abundant trace fossils. This area is known as the youngest trace fossil zone of Bengal basin. These trace fossils have been recovered from two separate lithostratigraphic units of this area i.e. (i) from Lower Lalgarh Formation of Lower Pleistocene period where the larger specimens have been recovered and (ii) from Holocene Sijua Formation, smaller and narrower specimens have been found (Bera 1996; Ghosh and Majumder 1981). The trace fossil assemblages (Trace of autochthonous burrow fossils and petrified woods) recovered, clearly indicate a tropical to sub tropical humid condition with near-shore shallow water environmental deposition of sediments during the Lower Pleistocene (Bera 1996; Sen 2008). The burrow fossil zone has been exposed from the lower horizons near Silabati river as a result of dissection of different horizons by gully erosion. These burrows are resting trace of *Beguerreia hemispheria.* The *Ophiomorpha nodasa, Cylindricum sp.* and large forms of *Thaalassinoids* are the dominant fossils identified and recovered in the lateritic horizons of Lalgarh formation of Gangani area (Bera and Banerjee 1989)

(continued)

Table 13.1 (continued)

Fossil sites	Geological and palaentological heritages and attractions
Amkhoi Fossil Park (23° 37′ 25″ N, 87° 34′ 56″ E)	Situated at Amkhoi village in Illambazar forest of Birbhum District in West Bengal, Amkhoi fossil park is considered as great geological and paleontological treasure for experts, researchers, students as well as tourists for its heritage: a million years old buried forest manifested through the presence of abundant number of angiosperm wood fossils. The trees which later transformed into wood fossils found in this area, were present in the uplands area of Rajmahal hills and Chotonagpur plateau in Late Miocene period. Due to natural calamities, they have been uprooted and carried by occasional floods. Under suitable conditions, they became petrified. Geologically, these angiosperm woods fossil lie in the Late Tertiary sedimentary sequence called as Santiniketan Formation which has been found in the western part of Bengal and now exposed in several discrete patches (Ganguli 1995). The fossils of this park have been showcased as open air museum in the park premises. It is regarded as one of the most appropriate destinations to know the geological and biological past of the country (Ghosh 2019)

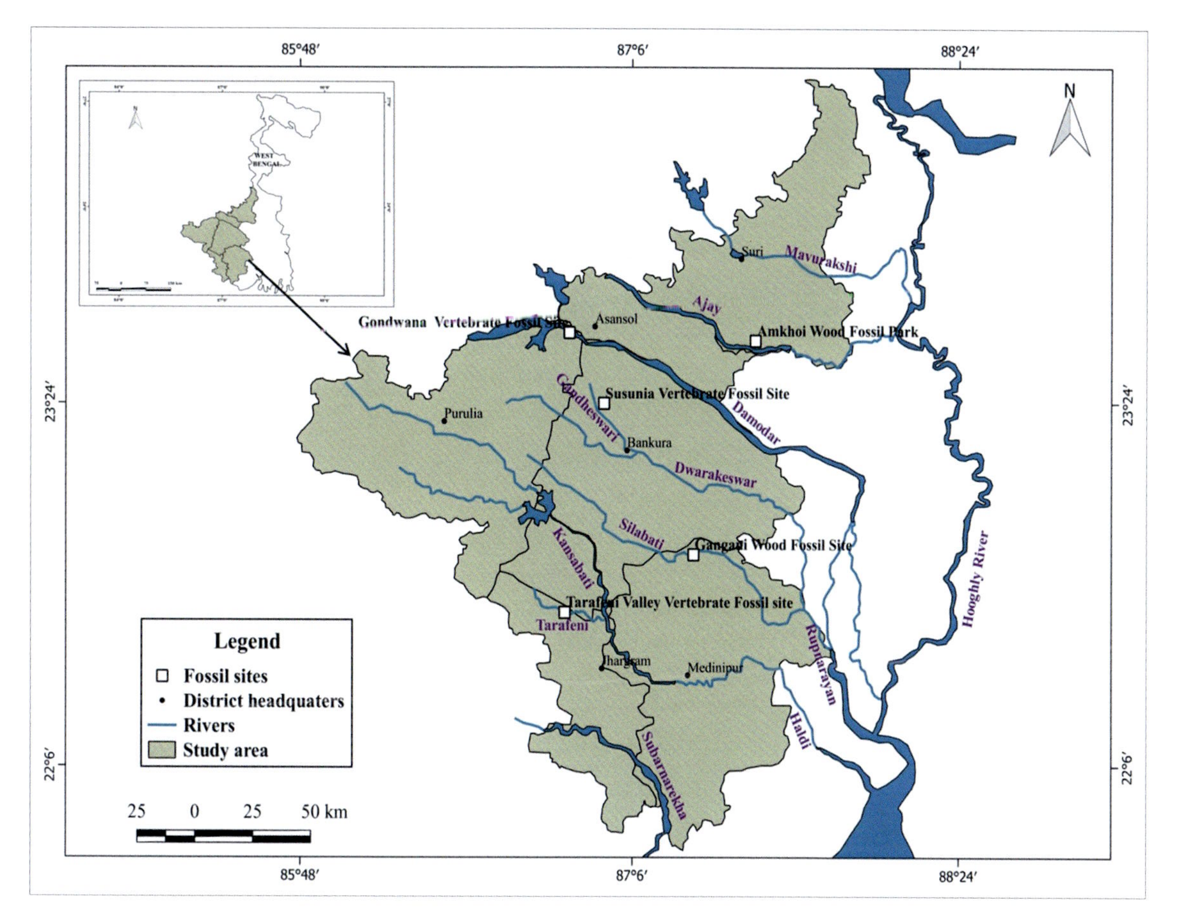

Fig. 13.3 Rich fossil sites in Western and South Western region of West Bengal (*Source* Prepared by authors, 2020)

Table 13.2 SWOT factors for Amkhoi fossil park

Strengths	Weaknesses
S1. Location in terms of accessibility and connectivity S2. Aesthetic value of the destination S3. Community bondage S4. Knowledge domain on geomorphosites	W1. Lack of promotion/advertisement W2. Infrastructural bottlenecks W3. Inadequate amenities and facilities for geotourism W4. Poor awareness of the stakeholders on geoconservation
Opportunities	**Threats**
O1. Worldwide expansion of geotourism market O2. Employment opportunities for the host population O3. Educative tourism opportunities for the guests O4. Rejuvenation of cultural properties associated with landscape	T1. Inadequate conservation and management of the geomorphosite T2. Anthropogenic pressure leading to unsustainable landuse conversion T3. Damage/loss of the geomorphological features T4. Less involvement of local people in geotourism and geoheritage conservation

13.3 Materials and Methods

GIS technology has been applied for thematic mapping. With the help of QGIS 3.12 Bucuresti and Google Earth software the maps have been prepared. A combined methodology of SWOT-AHP (SWOT—Strength, Weakness, Opportunity and Threat and AHP—Analytic Hierarchy Process) has been attempted to analyse the scope of geotourism potentials of Amkhoi fossil park. SWOT–AHP Model is a hybrid method of strategic planning where quantitative as well as qualitative; both the aspects could be taken into consideration. It has been used as an effective tool for situational assessment and to identify significant factors before recommending any strategic planning measure (Celik 2017). Initially the key SWOT (Strengths, weaknesses, opportunities and threats) factors (Table 13.2) to assess the scope of geotourism potential of Amkhoi fossil park have been identified based on the feedbacks on structured questionnaires and group discussions with sampled visitors. In the next stage, the Analytic Hierarchy Process (Saaty 1980) is undertaken, which involves the hierarchical construction of SWOT matrix. Each factor are subjected to pair—wise comparison based on the nine—point scale of relative importance represented by a square or reciprocal matrix, each of which are then normalized and prioritized. Consistency analysis is performed to test whether the judgements are consistent or not because if found inconsistent, reassessment is needed.

13.4 Discussion and Analysis

Among the fossil sites, Amkhoi is selected for SWOT-AHP analysis since it has been declared as the first fossil park of West Bengal dedicated to educational tourism on fossils. It is only 15 km drive from the Santiniketan, one of the major cultural tourism destinations of West Bengal. Since Santiniketan draws a good number of visitors whose motive is serious leisure; Amkhoi has been designed to satisfy them. The park is located only two kilometres from the main road (joining Bolpur-Santiniketan with Durgapur, an industrial hub with airport facility nearby) that penetrates through the Illambazar forest under the Birbhum Forest Division. Sal is the main floral species at present. During the digging of a pond in the year 2006, angiosperm wood fossils in considerable quantity have been discovered representing specimens preserved by natural process in the laterites of the area that prevailed 15–20 million years back (Late Miocene). It seems that there was a vast dry deciduous forest and trees carried by occasional flood transformed into petrified woods (petra is a Greek word meaning rock/stone) due to burial under fine sand and clay. The minerals of these soils gradually replace the original plant materials through a process called permineralization (also called petrification) in which the outer morphology of the parts of the tree and cellular structure has been maintained. Apart from petrification, cementation is another process in which the materials filled up the cavity and form cast fossils. A cast fossil appears like surface of original wood instead containing any wood material. Both cast fossils and petrified woods are the attractions of Amkhoi fossil park. It is regarded as one of the most important destinations to understand the geological and biological past of the lateritic belt of West Bengal and thereby emerging as a geotourism destination, the potentials of which is subjected to a SWOT-AHP analysis.

It is derived from the Pair wise comparison matrices for the identified SWOT factors (Table 13.3) that the worldwide expansion of geotourism market emerges as the prevailing opportunity while the greatest strength of the fossil park lies in is its knowledge domain. Each and every specimens of the open air museum is vividly described. Inadequacy of amenities and facilities however are still the dominant weakness even after two years of the inauguration of the geopark. The greatest threat emerges from the less involvement of local people. The access of them in the park enclosure is prohibited without permission. While responding in interviews, they expressed grief on their detachment with the park but at the same time expressed hope for the better future if they have been provided with the opportunity to cater the visitors. An integrated plan to involve them with the promotion of geotourism is therefore an immediate necessity.

13.5 Conclusion

As beautiful petrified wood have lapidary value and turned into items like jewellery, clock faces, sculptures etc., the area was under the scanner of fossil hunters who often employed the local youth to collect the fossil and sell to them in meagre price. Thus, this treasure land had been robbed by fossil hunters and antique thieves for years (Ghosh 2019). This is one of the reasons for prohibiting entry in the geopark without permission today. But it is essential to provide an alternative livelihood for

Table 13.3 Pair wise comparison matrices for SWOT factors

SWOT group	Strengths (S)	S1	S2	S3	S4	Priorities of SWOT factors/Local weight
Strengths	S1. Location in terms of accessibility and connectivity	**1**	2	1/3	1/5	0.110553
	S2. Aesthetic value of the destination		**1**	1/3	1/5	0.0784029
	S3. Community bondage			**1**	1/4	0.22583
	S4. Knowledge domain on geomorphosites				**1**	***0.585214***
Maximum Eigen Value = 4.15799 CI = 0.0526645 CR = 5.85161111						
SWOT group	Weaknesses (W)	W1	W2	W3	W4	
Weaknesses	W1. Lack of promotion/advertisement	**1**	1/4	1/4	3	0.139757
	W2. Infrastructural bottlenecks		**1**	1	3	0.377652
	W3. Inadequate amenities and facilities for geotourism			**1**	4	***0.397805***
	W4. Poor awareness of the stakeholders on geoconservation				**1**	0.0847857
Maximum Eigen Value = 4.20715 CI =0.0690515 CR = 7.67238889						
SWOT group	Opportunities (O)	O1	O2	O3	O4	
Opportunities	O1. Worldwide expansion of geotourism market	**1**	4	3	4	***0.53847***
	O2. Employment opportunities for the host population		**1**	1/2	1	0.120963
	O3. Educative tourism opportunities for the guests			**1**	2	0.219604
	O4. Rejuvenation of cultural properties associated with landscape				**1**	0.120963
Maximum Eigen Value = 4.02062 CI = 0.0068734 CR = 0.763711111						
SWOT group	Threats (T)	T1	T2	T3	T4	
Threats	T1. Inadequate conservation and management of the geomorphosite	**1**	3	1/2	1	0.245722
	T2. Anthropogenic pressure leading to unsustainable landuse conversion		**1**	1/3	1/4	0.0862654
	T3. Damage/loss of the geomorphological features			**1**	1/2	0.295376
	T4. Less involvement of local people in geotourism and geoheritage conservation				**1**	**0.372637**

(continued)

Table 13.3 (continued)

SWOT group	Threats (T)	T1	T2	T3	T4	
Maximum Eigen Value = 4.17074 CI = 0.0569132 CR = 6.32368889						

the local youth by promoting geotourism so that they may voluntarily turn into the protectors of the fossils of their own land. If it is understood that tourism development centring geopark may provide the opportunity to lead respectable life, fossil hunting could be combated by community involvement. Addressing the threats and weakness according to priority basis as derived from SWOT-AHP analysis is very much essential for sustainable utilization of fossils as geoheritage.

References

Anantharamu TR, Bellur D, Bhasker AA (2001) National geological monuments. Geol Surv India Special Publication 61(6):98

Badam GL (2000) Pleistocene vertebrate paleontology in India at the threshold of the millennium. J Palaeontol Soc India 45:1–24

Bandyopadhyay S (1988) Drainage evolution in badland terrain at Gangani in Medinipur District. West Benagl. Geogr Rev India 50(3):10–20

Banerjee S, Ghosh M (1977) On the occurrence of Giraffe, *Giraffe* cf. *Camelopardalis Brisson*, from the prehistoric site of Susunia, District Bankura, West Bengal. Sci Cult Calcutta 43:368–370

Banerjee S, Ghosh M (1978) Evidence of the occurrence of Giraffe *Giraffe* cf. *Camelopardalis Brisson*, from the prehistoric site of Susunia, in West Bengal. News Lett Zool Surv India 2:46–47

Basak B, Badam GL, Kshirsagar A, Rajaguru SN (1998) Late Quarternary environment, palaeontology and culture of Tarafeni valley, Midnapur District, West Bengal—a preliminary study. J Geol Soc India 51:731–740

Battail B (1998) Gondwanan and Laurasian amniote faunas compared, Late Permian and Early Triassic. J Afr Earth Sc 27:20–22

Bender PAH, Hancox PJ (2004) Newly discovered fish faunas from the Early Triassic, Karoo Basin, South Africa, and their correlative potential. Gondwana Res 7(1):185–192. https://doi.org/10.1016/S1342-937X(05)70317-8

Bera S (1996) Remarks of paleoenvironment from the trace fossils from subsurface & outcrop Tertiary-Quaternary sediments of the Western Part of Bengal Basin. J Geogr Environ 1:1–15

Bera S, Banerjee M (1989) Ichnofossils of shallow Marine Environment from the Lateritic Sediments of Western Bengal basin (Abstract) workshop on coastal processes & coastal quaternaries of Eastern India. Geol Surv India (Eastern Region):40

Celik N (2017) Strategy making with quantified swot approach: a case analysis on tourism industry in Black Sea Region of Turkey. J Lexet Scientia 14(1):167–176

Chakrabarty P, Mandal R (2019a) Geotourism development for fossil conservation: a study in Amkhoi Fossil Park of West Bengal in India. GeoJ Tour Geosit 27(4):1418–1428. https://doi.org/10.30892/gtg.27425-444

Chakrabarty P, Mandal R (2019b) Geoarchaeosites for geotourism: a spatial analysis for Rarh Bengal in India. GeoJ Tour Geosit 25(2):543–554. https://doi.org/10.30892/GTG.25221-379

Császár G, Kázmér M, Erdei B, Magyar I (2009) A possible Late Miocene fossil forest PaleoPark in Hungary. Notebooks on geology—Book 2009/03 (CG2009_B03), Chapter 11:130

Dasgupta, HC (1922) Notes on the Panchet reptile. Sir Asutosh Mukherjee Silver Jubilee Volumes, Science, Calcutta University 2:237–241

Dasgupta PC (1967) Pragoitihasik Susunia (in Bengali). Calcutta
Dassarma DC, Biswas S (1978) Study of the quaternary deposits of Bankura District, West Bengal. Geological Survey of India Unpublished Report for 1972–73
Dassarma DC, Biswas S, Nandi A (1982) Fossil vertebrates from the late quarternary deposits of Bankura, Burdwan and Puruliya Districts, West Bengal. Mem Geol Surv India (palaeontological Indica) 44:1–65
Dietz RS, Pewl TL, Woodhoush M (1987) Petrified wood (*Araucarioxylon Arizonicum*): proposed as Arizona's state fossil. J Arizona-Nevada Acad Sci 22(2):110
Dutta AK (1976) Occurrence of fossil lion and spotted hyena from Pleistocene deposits of Susunia, Bankura, West Bengal. J Geol Soc India 17(3):386–391
Farsani NT, Coelho C, Costa C (2011) Geotourism and geoparks as novel strategies for socio-economic development in rural areas. Int J Tour Res 13(1):68–81. https://doi.org/10.1002/jtr.800
Ganguli U (1995) A new lithostratigraphic unit at the Western fringe of West Bengal India. India J Geol 67(4):282–288
Ghosh D (2019) Amkhoi—youngest wood fossil park of India. Sci Rep NISCAIR-CSIR India 56(8):45–48
Ghosh RN, Majumder S (1981) Neogene-Quaternary sequence of Kasai Basin, West Bengal, India. In: Sastri et al (eds) Proceedings of Neogene-Quaternary boundary field conference, India, 1979. Geological Survey of India, pp 63–73
Gupta A (2009) Ichthyofauna of the lower Triassic Panchet formation, Damodar valley basin, West Bengal, and its implications. Indian J Geosci 63(3):275–286
Lucas SG (1998) Global tetrapod biostratigraphy and biochronology. Palaeogeogr Palaeoclimatol Palaeoecol 143:347–384. https://doi.org/10.1016/S0031-0182(98)00117-5
Saha KD, Banerjee S, Talukder B (1984) Occurrence of fossil *Panthera purdus Linn.* From the Pleistocene deposits of Susunia, Bankura, West Bengal. Bull Zool Surv India 6(1–3):257–259
Sastry MVA (1968) Pleistocene vertebrates from Susunia, Bankura District West Bengal. Indian Miner 20(2):195–197
Saaty TL (1980) The analytic hierarchy process. McGraw-Hill, New York
Sastry MVA, Dassarma DC, Biswas S (1976) Quaternary deposits of Bankura District, West Bengal. Rec Geol Surv India 107(2):176–185
Sen J (2008) Geomorphology of Garhbeta Badlands: West Medinipur District, West Bengal. (Unpublished Doctoral Dissertation) University of Calcutta, Kolkata, India
Sen D, Ghosh AK, Chatterjee M (1963) Palaeolithic industry of Bankura. Man India 43(2):100–113
Tripathi C (1962) On the remains of Lystrosaurus from the Panchets of Raniganj Coalfield. Recor Geol Surv India 89(2):407–419
Tripathi C (1969) Fossil Labyrinthodonts from Panchet series of Indian Gondwana. Palaeontol Indica (new Series) 38:1–45
Tripathi C, Satsangi PP (1963) Lystrosaurus Fauna from the Panchet Series of Raniganj Coalfield Palaeontological Indica (new Series) 37:1–53
Turner S (2006) Promoting UNESCO Global Geoparks for sustainable development in the Australian-Pacific region. Alcheringa, Special Issue 1:351–365. https://doi.org/10.1080/031155 10609506872
UNESCO (2010) Guidelines and Criteria for National Geoparks Seeking UNESCO's Assistance to Join the Global Geoparks Network (GGN). United Nations Educational, Scientific and Cultural Organization: 1–12
Zouros NC (2010) Lesvos petrified forest geopark, Greece: geoconservation, geotourism, and local development. George Wright Forum 27(1):19–28

Chapter 14
Classifying the Million-Plus Urban Agglomerations of India—Geographical Types and Quality of Life

Habil Zoltán Wilhelm, Róbert Kuszinger, and Nándor Zagyi

Abstract India is one the fastest growing and developing economies and also societies of the world. An evident consequence of this is urbanisation, which poses a huge challenge for the population and the political decision-makers of the country and is also one of the most topical research trends of the social geographical researches concerning India. The paper first introduces the general urbanisation trends experienced in the sovereign India in the 1951–2011 period, in the framework of an analysis of statistical data recorded in censuses, indicating the volume and trends of urbanisation. This is followed by the demonstration of the structural features and diverse development paths of the million-plus agglomerations (i.e. agglomerations with at least a million inhabitants), connected to one of its main characteristics depicted by this introductory summary: metropolisation. Using the quantitative categories defined during their analysis, the authors classify the metropolises of India in urbanisation types, with the method of cluster analysis. In what follows, we sought to answer whether any correlation could be justified between these urbanization types and the complex quality of life indicators we generated for the central settlements of the agglomerations.

Keywords India · Urbanisation · Metropolisation · Agglomeration · Urban statistics · City clusters · Quality of life

---

The publication was supported by the University of Pécs, Szentágothai Research Centre, Research Centre of Historical and Political Geography and PADME Foundation.

H. Z. Wilhelm (✉)
Faculty of Sciences, University of Pécs, Pécs, Hungary
e-mail: wilhelm@gamma.ttk.pte.hu

R. Kuszinger
Doctoral School of Earth Sciences, University of Pécs, Pécs, Hungary

N. Zagyi
Szentágothai Research Centre, University of Pécs, Pécs, Hungary
e-mail: zana@gamma.ttk.pte.hu

N. C. Jana et al. (eds.), *Livelihood Enhancement Through Agriculture, Tourism and Health*, Advances in Geographical and Environmental Sciences,
https://doi.org/10.1007/978-981-16-7310-8_14

14.1 Introduction: General Urbanisation Trends in India in the Second Half of the 20th and the Beginning of the Twenty-First Century

Urbanisation in India, with almost four and a half millenniums of past (Ramachandran 2001; Tirtha 2002), had a radical change of direction after gaining the sovereignty of the country in 1947 (Wilhelm and Zagyi 2018; Wilhelm 2015; Wilhelm 2008). The new socio-economic factors of this time led to a growth in the number of cities and the proportion of urban inhabitants that had never been experienced before. In addition, a significant phenomenon of the period from the mid-twentieth century until now is the rapid strengthening of the weight of cities with hundreds of thousands or millions of population. At the time of the first census of the sovereign India, in 1951 just 45% of urban citizens lived in settlements with more than a hundred thousand inhabitants, while cities in this size this category gave home to more than 60% of the population of India, parallel to a six and half-fold increase in their population; also, their number grew from 76 to 423 by the time of the census in 2001. The growth in the number of million-plus cities (metropolises) was also dynamic: as opposed to five such cities in the mid-twentieth century, India had 35 of them by the dawn of the new millennium (Census India 2001).

The concentration of urban citizens in settlements with more than a hundred thousand inhabitants took a new momentum in the first decade of the twenty-first century. Their population further increased, due, on the one hand, to the increase of the number of cities now in this magnitude (Class-I category), and to the surplus population coming from natural increase and in-migration, on the other hand. In 2011, 70% of all urban citizens lived in such settlements, already.

An examination of the spatial concentration of urban population at the level of districts that are the lower administrative units of states and union territories reveals that urban agglomerations, sometimes grown into megapolises, are being born, with tight correlation with the SENTIENT index featuring their social, economic and infrastructure development level (Wilhelm et al. 2011, 2013, 2014; Wilhelm 2011). These formations were even more striking by 2011 than they had been ten years before.

The number of cities, accordingly the number and proportion of urban population, is further increased by the so-called census towns that are registered among the urban settlements only for the manipulation of the pace of urbanisation, sometimes by plastically interpreted urban definition, in the lack of real central functions and self-governments. The number of census towns at the time of the census of 2011 was 3,894, as opposed to 1,362 a decade earlier, while that of the functional towns increased from 3,799 to 4,041 Singh (2014), which means a more than one-and-a-half-fold growth in the number of urban settlements, from 5,161 to 7,935. According to our computation, a strong (Pearson method) correlation ($r = +\ 0.89$) exists between the change in number of census towns and in rates of urban population from 1961 until 2011 (Census India 2011; Census Newsletter 2001) verifying the definite role of census towns in Indian urbanisation.

Looking at the growth of the number of cities and urban agglomerations from 2001 to 2011 we can see that the number of small towns (small town areas) with less than 100,000 inhabitants grew one-and-a-half-fold, from 4,738 to 7,467, due to the sudden increase of the number of census towns; their share from the urban population, however, fell from 38 to 30%. The number of middle towns and agglomerations in the category of 100,000–1,000,000 population grew by 7%, from 388 to 416, while the increase in the number of their population was twice as much, 14%. The most dynamic growth could be observed on the whole at the million-plus cities and agglomerations, with an increase in their number from 35 to 52 in ten years. The number of their inhabitant showed a similar growth, approximately 50%, as a consequence of which it grew from 107.9 million to 160.7 million. As a result of these processes, with the exception of Odisha and the states with small territories and low population numbers, by 2011 there was at least one million-plus city in all federal states, in addition to the Delhi union territory. The metropolises, giving home to 43% of the urban population of India by 2011, already, showed a tremendous growth both as regards their number and their population, although the growth of the latter showed significant extremes between the last two censuses. Cities and agglomerations showing outstanding growth, with the exception of Ghaziabad, can all be found in Kerala state, far above average Indian urbanisation and socio-economic development level, with a 47.7% share of urban population, where the proportion of urban population grew by 98.2% in ten years, advancing all union territories and states with at least one million-plus city (agglomeration). A considerable part of this growth was due to the intensive, often tenfold population growth of cities grown into metropolises, being in the category of middle cities in the previous decade, except for Kochi. As a consequence of this, the first, approximately 500-km-long metropolitan axis of India was born in the south part of the Western Coast, in the Kerala territories of the Malabar Coast in the foreground of the Western Ghats, stretching from Kasaragod to Thiruvananthapuram.

In the approximately 4,500-year-long history of Indian urbanisation from the flourishing of the Harappa culture to the birth of the Kerala megalopolis,[1] the latter phenomenon is a significant milestone even if we consider the fact that it is largely the result of statistical manipulation. Of all 2,352 Indian settlements declared census towns in 2011, almost 13%, i.e. 320 settlements, can be found in the federal state that makes just over 1% of the territory and approximately 3% of the population of the country, with 33.4 million inhabitants in 2011 (Pradhan 2017), and the 8.8 million population surplus of the new Kerala census towns is more than 40% of the 16 million urban population of the state, and almost 90% of the 7.7% growth of urban population.

[1] In our interpretation megalopolises are extended and continuous urban areas having more than 10 million inhabitants.

14.2 Characteristic Features of the Indian Metropolisation on the Ground of Structural Features, Development Processes, and Quality of Life of Agglomerations

14.2.1 Survey Methods

In this chapter, the structural and growth characteristics of settlements involved effectively in cluster analysis are presented, on the basis of the authors' own database compiled from the data registered during the censuses for the population of the 52 cities with more than a million inhabitants, and the population of the agglomeration and the size of their administrative territories (Towns and urban agglomerations 2011). On this ground, an attempt is made for the identification of certain settlement types as well.

The database generated by the authors contains, in addition to the time series values of the numbers of population in the cities, data for the agglomerations: the number of inhabitants in the core settlements, the satellite towns, within this separately the population of the real towns and the census towns, the proportions of the centre and the census towns in relation to the total population of the agglomeration, and also the proportions of the census towns within the total population of all real and census towns in the hinterlands. In addition, the authors made separate calculations for the volume of the change of population in the sub-settlement categories (total of the agglomeration, centre, hinterland, real and census towns of the hinterland), by decades, for the period 1951–2011.

In a breakdown similar to this, aggregate data for the towns and agglomerations were collected: size of the administrative territory of the whole agglomeration, of the central settlement and of the hinterland; also, proportions of the territories of the centres and the adjoining towns within the agglomerations were calculated, and the volumes of the changes of these in time series. Finally, population density values were calculated for the total of the respective settlements, in the case of agglomerations separately for the central settlement and the hinterland, as were the data indicating the changes of these by decades.

Taking account of these population, territory and population density data, different examination goals were set for the time of the 2011 census and for the period from 1951 to 2011, and on the basis of these aspects and using the relevant data the metropolises were analysed, assessed and grouped as follows:

- Types of agglomerations as per the 2011 population rate of their central settlements;
- Types of agglomerations as per the 2011 population rate of census towns;
- Types of agglomerations as per the 2011 extension of their administrative areas;
- Types of agglomerations as per the 2011 territorial rate of their central settlements;
- Types of agglomerations as per their 2011 population density;
- Types of agglomerations as per the 2011 population density of their central settlements;

- Types of agglomerations as per the 2011 differences between population densities of their central settlements and hinterlands;
- Types of agglomerations as per their growth rate in the period of 1951–2011;
- Types of agglomerations as per population change of their census towns in the period of 2001–2011;
- Types of metropolises (cities and UA's) as per their population density change in the period of 1961–2011;
- Types of metropolises (cities and UA's) as per their population density change in the period of 2001–2011and
- Types of agglomerations as per differences of their central settlements' and hinterlands' population density change in the period of 2001–2011.

In order to identify the urbanisation categories of million-plus agglomerations, classified into static and dynamic (time series) categories (ordered to variables), after the analysis of the dataset mentioned above a cluster analysis was implemented, the results of which, as the abstraction of the urban statistics examinations of the authors, is described below.

During this analysis variables were created—in a statistical approach—each of them with a numerical value, as were the categories belonging to these, which are essentially the textual data (factors). Data were processed with the R programming language (R core team 2020). In the first step it had been analysed to what extent the respective variables support the cluster analysis, and how much they are independent of each other. The majority of the procedures applied require that each variable included in the survey has a value for all cases (locations). Thus, it is more reasonable to have variables in the data series of which there are less <NA> values.

At the one-by-one examination of variables, the plot of the estimated density function was visually evaluated. The R language integrated density() function was used with default parameters (Becker et al. 1988). It was also examined which factors had stronger ($|r| > 0.3$, Pearson method) correlations. By this, variables were evaluated on the ground of how many other variables they correlated with. This means that variables indicated with the larger correlation numbers derived this way (Table 14.1, column '$|r| > 0.3$ corr. number') can be neglected, if necessary. The definitive capacity of the respective variables for the sample were evaluated with K means clustering, where groups sum of squares (WSS) were calculated for 2–15 clusters. Demonstrating this by the number of clusters, with the *elbow* method frequently used in practice, the practical number of clusters for the given variable was defined (Table 14.1, column 'K-means WSS based number of clusters'). In order to counterbalance subjectivity, a model-based clustering was also implemented. The procedure chosen was model-based clustering based on parameterised finite Gaussian mixture models, where the number of clusters is defined by the algorithm itself. In the process, default settings of the system were used (Scrucca et al. 2016). Cluster numbers gained as a result are included in column 'Model-based clustering number of clusters' in Table 14.1.

As the respective quantifiable values had already been ordered into categories, the actual multi-variable clustering was practically continued with these categories

Table 14.1 Key characteristics of each variable. Those involved in cluster analyses are in bold

Variable description	Numerical variable name (category variable name)	Number of <NA> occurrences	Single variable k-means WSS based optimal number of clusters	Model-based clustering ideal number of clusters	Number of $\lvert r\rvert>0.3$ correlations computed against all other variables
Static					
Population density	popdens (popdens_c)	0	4	1	8
Population density of centre	centdens (centdens_c)	0	3	2	5
Centre and agglomeration population density ratio	centaggldensrat (centaggldensrat_c)	12	3	2	3
Size of administrative area	adminarea (adminarea_c)	0	2	2	4
Centre and agglomeration area size ratio	centagglarearat (centagglarearat_c)	12	3	3	5
Weight of census towns	cenzcitywei (cenzcitywei_c)	12	2	3	7

(continued)

Table 14.1 (continued)

Variable description	Numerical variable name (category variable name)	Number of <NA> occurrences	Single variable k-means WSS based optimal number of clusters	Model-based clustering ideal number of clusters	Number of |r|>0.3 correlations computed against all other variables
Centre and agglomeration population ratio	centagglpoprat (centagglpoprat_c)	12	3	2	6
Dynamic					
Population growth rate (1951–2011)	growth19512011 (growth19512011_c)	0	3	3	2
Population change of census towns (2001–2011)	cenzpopchange20012011 (cenzpopchange20012011_c)	26	3	3	0
Population density change of metropolises (1961–2011)	metropopchange19612011 (metropopchange19612011_c)	2	3	3	6
Population density change of metropolises (2001–2011)	metropopchange20012011 (metropopchange20012011_c)	0	4	3	5
Agglomeration population density change (2001–2011)	agglopopdenschange20012011 (agglopopchange20012011_c)	14	3	3	4

Table 14.2 Lists of involved variables in each cluster analysis

Cluster analysis	Involved variables
On static variables	Adminarea, centdens, centagglarearat, centaggldensrat, centagglpoprat
On dynamic variables	Popgrowth19512011, metropopchange20012011, metropopchange19612011

(according to the R terminology: factor) values. The shift from the set of real numbers to discrete factor values will definitely change the ideal number of clusters, but will still carry the definitive characteristics of the numerical variable used as the basis of the category.

Values of dissimilarities among the factors as "distances" were calculated with the R daisy function. This, taking factors into consideration, uses Gower's procedure (Gower 1971). For multi-variable clustering, agglomerative hierarchical clustering was chosen and implemented with the hclust function. Within this, ward.D2 had been selected as the agglomeration method (Murtagh and Legendre 2014), meanwhile, other options had been left at their defaults.

At this procedure too, the ideal number of clusters is not obvious, so it had been preliminarily evaluated what would give ideal results by several indices, from the static and dynamic variables. Results of Table 14.1 about the characteristics of the variables were also used for compiling the series of variables. As the calculation of optimum variable combinations is not a goal of this paper, what is given here is only the final set of variables used for the two analyses (Table 14.2).

For the definition of the cluster number, a routine that calculates multiple variables was used, run for 1–8 clusters. Cluster analysis was finalised with the ideal number of clusters defined this way. Values taken into consideration were the application of the elbow method also used at the single-variable analysis at the WSS chart, the search for the maximum of the silhouette values (Rousseeuw 1987) and dunn2 index (Halkidi et al. 2001). The results of clustering are presented in a dendrogram. For the examination of the regularities of spatial appearances, settlements ordered into clusters were also presented in a map, using the QGIS geographical information system. Helping the interpretation of the spatial distribution, maps also contain the standard deviational ellipses and their centres (Yuill 1971) for each respective cluster.

The basic aim of our study is to examine whether there is any correlation between the urbanization types of agglomerations based on the methodology described above and the quality of life of the settlements belonging to the latter. For this purpose, relying on the population enumeration as well as houselisting and housing databases of the 2011 census Census India (2011), we selected the following indicators, showing partly the human resource development of the population of each settlement and partly the infrastructural level of households, which in our opinion are suitable for forming separate clusters based on quality of life and a complex quality of life indicator:

- percentage of main workers to total population;
- percentage of literate main workers to total main workers;
- percentage of literate female main workers to total main workers;
- percentage of graduate and post-graduate main workers to total main workers;
- percentage of female graduate and post-graduate main workers to total main workers;
- percentage of main workers in quaternary sectors (information & communication, financial & insurance activities, real estate activities, professional, scientific & technical activities) to total main workers;
- percentage of female main workers in quaternary sector to total main workers;
- percentage of migrants immigrating with aim of business & education & moving with household to total migrants from outside of the state and abroad;
- percentage of population attending college to total population of age group 18–24;
- percentage of population living in normal households to population living in normal and slum households;
- percentage of households in good condition;
- percentage of households with tapwater from treated source;
- percentage of households with electric lighting;
- percentage of households with piped sewer system;
- percentage of households having bathroom;
- percentage of households with LPG/PNG & electric fuel used for cooking;
- percentage of households having kitchen inside;
- percentage of households availing banking services;
- percentage of households having computer/laptop with internet and
- percentage of households having car/jeep/van.

As no data on agglomerations are available in the database of population enumeration for the above mentioned categories, we used city-level data to assign the appropriate numerical values (percentages) to million-plus cities of and to central settlements of urban agglomerations, the size of which is directly proportional to the quality of life. We were able to rate 51 out of 52 metropolises as Kannur UA's naming settlement is not included in the Population Enumeration being a Class-II category town. It is important to note that for settlements that are distributed among several sub-districts; therefore, a single data for the whole city is not available, the indicators reflecting the infrastructural development of households are given by the average of the relevant values of the wards in the sub-districts. Based on the data set formed such a way, cluster analysis was performed as follows.

Data were processed with the R programming language (R core team 2020). Each dataset was treated as a single vector without further weighting or other modifications. All of them were incorporated into the cluster analysis process.

Having all vectors numerical the well-known K-means method was selected for the purpose of cluster analysis kmeans () function of the stats package, method: default (Hartigan and Wong 1979), nstart: 25. Groups sum of squares (WSS) were calculated for 2–15 clusters to support the estimation of cluster numbers. The *elbow* method showed 5 or 6 clusters to be ideal.

Instead of using other cluster number estimation processes both of the 5 and 6 cluster scenarios were run. Graph visualizations of the clusters (fviz_cluster() function of the factoextra package, axes are results of built-in principal component analysis) made it clear that 6 clusters provide a clearer separation (Fig. 14.6). Also the spatial distribution on map presented in the form of standard deviational ellipses (Yuill 1971) enforced the 6 cluster scenario to be chosen (Fig. 14.7).

This clustering was also compared to static and dynamic urbanization type cluster values of each city. The comparison was done using Pearson's method.

14.2.2 Types of Agglomerations Assigned to Static and Dynamic Variables of the Cluster Analyses

As mentioned above, Indian metropolises and agglomerations were examined and typified according to their number of population, administrative area and population density from different perspectives based on both the 2011 (static) state and the dynamic processes interpreted for the periods of 1951/1961–2011 and 2001–2011. For extensional reasons, we are obliged to omit the detailed analysis of these data, and only the types of agglomerations embodying the static and dynamic indicators actually included in the cluster analysis and the associated population, area and population density value intervals are reported below.

On the basis of the share of central settlements from the total population the following agglomeration types were distinguished:

- agglomerations with dominant centres, if the population ratio of the central city to the total population of the agglomeration is at least 60%;
- agglomerations with centres of balanced weight, if neither the centre nor the adjoining settlements have population shares below 40% of the total of the agglomeration, but this figure does not exceed 60%, either;
- agglomerations with centres of limited weight but great significance, if the central city has less than 40% but more than 15% of the total population (in this group the centre despite its relatively low share from the population, has a dominance over the other settlements, due to the fragmented hinterland and
- agglomerations with centres of negligible weight, if the eponymous settlement has less than 15% of the total population (in this case it is not more than a pole in a multi-polar cluster of cities, which, however, has an outstanding significance due to its role in culture, economy, or urban hierarchy).

As regards the administrative areas of agglomerations the settlement types were grouped as follows:

- giant urban settlements, if their territories are at least 1,000 km^2;
- large urban settlements, if their territories are between 500 and 999 km^2;
- medium sized urban settlements, if their territories are between 200 and 499 km^2 and

- small urban settlements, if their territories are less than 200 km^2.

Looking at million-plus agglomerations on the ground of the administrative territory ratios of the centre and the hinterland the following categories have been set up:

- agglomerations with centres of dominant territories, if the administrative area of the centre is at least 60% but does not reach 95%;
- agglomerations with centres of balanced weight, if the administrative area of both the eponymous central settlement and the towns or cities in its hinterland makes 40–59.9% of the territory of the whole agglomeration;
- agglomerations with centres of limited territories but great significance, if the territorial weight of the centre is 15–39.9%, still it has a decisive role and significance in comparison with the settlements in its definitely fragmented hinterland and
- agglomerations with centres of negligible territories, if the territorial weight of the eponymous centre is negligible (less than 15%), it is only one pole of outstanding significance due to its role in culture, economy or urban hierarchy.

Comparing population densities of metropolises consisting of one single city and eponymous settlements of agglomerations, the composition of them among size categories is as follows:

- congested metropolises, if their population density is more than 10,000 people/km^2;
- densely populated metropolises, if their population density is between 5,000 and 9,999 people/km^2;
- metropolises with medium population density (2,500–4,999 people/km^2) and
- metropolises with low population density (less than 2,500 people/km^2).

Regarding the differences between the relative positions of the centres' and the hinterlands' population density the agglomerations were organized into the following categories:

- agglomerations in which population density of the centre is smaller than that of the hinterland;
- agglomerations in which population density of the centre is not more than double of that of the hinterland;
- agglomerations in which population density of the centre is 2–4 times higher than that of the hinterland and
- agglomerations in which population density of the centre is more than 4 times higher than that of the hinterland.

As per population growth rate of metropolises between 1951 and 2011 the following categories were set up:

- regular settlements growing at a slower pace than average (789%) with growth rate less than 200% in each decade;

- regular settlements growing at a faster pace than average with growth rate less than 200% in each decade;
- irregular settlements following an extraordinary growth path with growth rate exceeding 200% in one or more decades.

Considering the direction and volume of change of population density for the period between 1961 and 2011 million-plus settlements were categorised as follows:

- metropolises whose population density more than doubled;
- metropolises whose population density did not grow more than twofold and
- metropolises whose population density decreased.

Relating to direction of change of population density for the period between 2001 and 2011 million-plus settlements were categorised in two groups:

- metropolises whose population density stagnated or grew and
- metropolises whose population density decreased.

14.2.3 Findings of the Cluster Analysis of Structural Features and Development Processes of Agglomerations

The cluster analysis done for the designation of urbanisation types was extended to those settlements that were defined as real agglomerations in the framework of the statistical data analysis above, i.e. those six metropolises were excluded that are registered as ***statutory towns*** (Jaipur, Visakhapatnam, Ludhiana, Faridabad, Vasai-Virar and Kota), as were cities that were considered by the authors as ***quasi*** (i.e. not real) ***agglomerations***[2] (Bengaluru, Aurangabad, Surat, Amritsar, Nagpur and Gwalior). It was done for technical reasons explained in the section on the methodology of cluster analysis, on the one hand, and because these settlement clusters are significantly different, due to their own classification features, from the real agglomeration, on the other hand.

As a result of the ***static analysis***, i.e. an analysis built on the data of the 2011 census, the 40 agglomerations included in the research can be best classified into five clusters (Fig. 14.1) Their classifying factors are given by different combinations of also five relevant variables indicated in Tables 14.1 and 14.2 (size of administrative area; population density of centre; centre and agglomeration area size ratio; centre and agglomeration population density ratio; centre and agglomeration population ratio). This makes it evident that the number of population of the agglomerations is not a clustering factor in the absolute sense.

Cluster 1, the second largest one consisting of 12 elements, is definitely heterogeneous as regards the size of administrative area but shows a relatively homogeneous

---

[2] Urban agglomerations where the population rate of the eponymous core city within the total agglomeration reaches 97.5% or at least 95%, and at the same time the administrative area of the hinterland makes less than 5% of the territory of the total urban area, are not taken as real agglomerations.

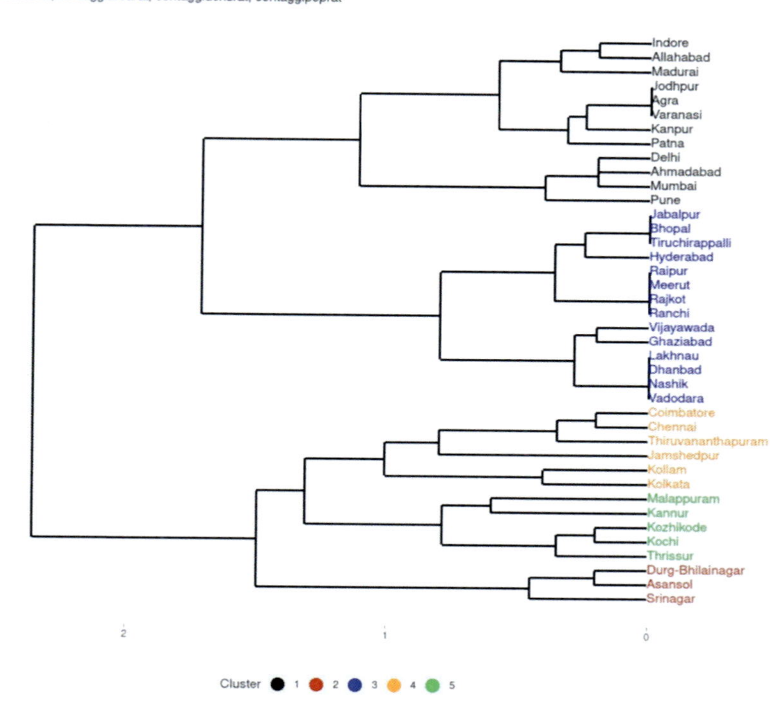

Fig. 14.1 Clusters of the static analysis. Edited by the authors

picture as regards the other categories. Agglomerations listed here are all densely populated, congested (>10,000 people/km^2), have central settlements that are mainly dominant (>60%)—with only one exception—and in the second place have balanced territories (40–59.%) in which the density of population is significantly—typically 2–4 times or more than 4 times—higher than that of the hinterland. All of the agglomerations ordered into this cluster have single-centred cores dominant in number of population (>60%). The research findings show that these settlement clusters make the type of ***agglomerations definitely prevailed by congested centre (top-heavy UAs)***. In a topographic sense these 12 settlements are partly scattered; half of them, however, are clearly concentrated in the Great Plains, along the rivers of the Yamuna (Delhi and Agra) and the Ganges (Kanpur, Allahabad, Varanasi and Patna) (Fig. 14.2).

In the definition of the largest cluster, *Cluster 3* with 14 members, the size of the territory of the settlements plays a role, already: with one exception, they are middle-sized (200–499 km^2), in the second place, small agglomerations (<200 km^2). The centres are densely populated (5,000–9,999 people/km^2) in each case. With no exception, they have one single centre dominant in territory and population whose population density, apart from one single case (Ghaziabad), exceeds that of the rest

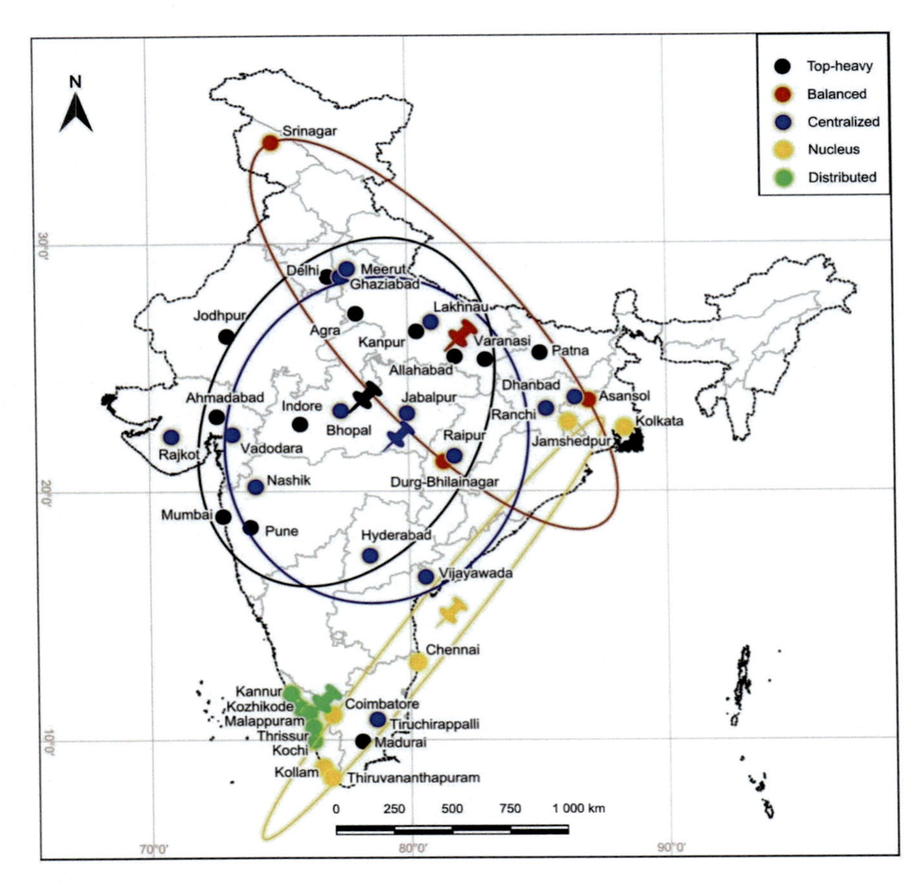

Fig. 14.2 Geographical distribution of static urban clusters featuring their standard deviational ellipses and topographical centres. Edited by the authors

of the settlements in the hinterland at least 2–4 times, in some case more than 4 times. Taking all these facts into consideration, settlements in this cluster are ***agglomerations with small-medium sized area definitely prevailed by densely populated centre (centralised UAs)***. Their geographical location shows no regularity at all: they can be found scattered all over the country, from Tiruchirappalli in Tamil Nadu to Ghaziabad adjacent to Delhi, from Rajkot in Gujarat to Dhanbad in Jharkhand.

Cluster 2 with its only 3 elements features settlements with medium-sized territories, with centres that have medium population density (2,500–4,999 people/km^2). The ratio of the territory of the centre and the hinterland proved to be irrelevant in this case. The position of the centre within the agglomeration, as regards population density and population number, is much less definite: population density of the central settlements is maximum 2 times that of the hinterland and its share from the number of population is even more balanced. On the ground of these facts, these three clusters of settlements are seen as ***homogeneous agglomerations with medium sized area and moderately populated centre (balanced UAs)***. Similarly to the settlements in Cluster 3, these three agglomerations are situated far from each other, with no topographical connection detectable among them.

In *Cluster 4*, consisting of 6 members, settlement size shows a significant standard deviation as in Cluster 1, i.e. it has no clustering role. Most of the centres are densely populated; in two agglomerations they are congested. As regards the territories of the centres, they are, with the exception of one agglomeration with balanced centre, smaller (15–39.9%) than the hinterlands, but with decisive significance in each case. Their population density with the same one exception (Jamshedpur) exceeds that of the settlements in the hinterland 2–4 times or more than 4 times. As regards population number in the central cities, agglomerations in this cluster have typically balanced, in a smaller part low (<40%) weight but dominant centres. Accordingly, settlements in this cluster are categorised as ***agglomerations with nucleus type centre drawing widespread and populous urban area (nucleus UAs)***. The cluster is hard to grasp also in a territorial sense. Four settlements (Chennai, Coimbatore, Kollam and Thiruvananthapuram) can be found in the Dravidian South, but no other closer geographical relation among them seems to exist. The other two agglomerations (Kolkata and Jamshedpur) are far from these in the northeast part of the country. Although their standard deviation seems to stand out along the south-eastern coast of the country.

Cluster 5, with 5 elements, consists of large territory agglomerations predominantly, but the population density of the centres show considerable differences, and thus has no clustering effect in this case, either. The sizes of the administrative area of the centres are negligible in proportion to the hinterlands, but their population density is significantly higher (in three out of the five cases 2–4 times higher than that of the adjoining settlements). The centres of the agglomerations in this cluster typically have low weight in the population but dominant roles otherwise, in a smaller number of cases; however, the share of the centre from the total population is negligible. This made authors identify these settlements as ***large-sized decentralised agglomerations (distributed UAs)***. As opposed to the clusters described above, members in these settlement clusters are in a visibly same geographical environment: each is part in the so-called Kerala Megalopolis, located next to each other from Kannur to Kochi in the northern part of the Malabar Coast (Fig. 14.2).

The set of ***dynamic cluster analysis***, based on the changes of urban statistical data of metropolises from the mid-twentieth century until the time of the latest census in 2011, was made by 38 agglomerations. On the ground of the reasons already mentioned above, analysis was not extended also in this case to the six cities with no hinterlands and the six quasi agglomerations, and Hyderabad and Rajkot were not included, either, due to the deficient data concerning their administrative areas. The best grouping of agglomerations resulted in five clusters again (Fig. 14.3), where clustering factors are defined by the combinations of the value ranges of the three relevant variables featured in Tables 14.1 and 14.2 (population growth rate 1951–2011, population density change of metropolises 1961–2011, population density change of metropolises 2001–2011).

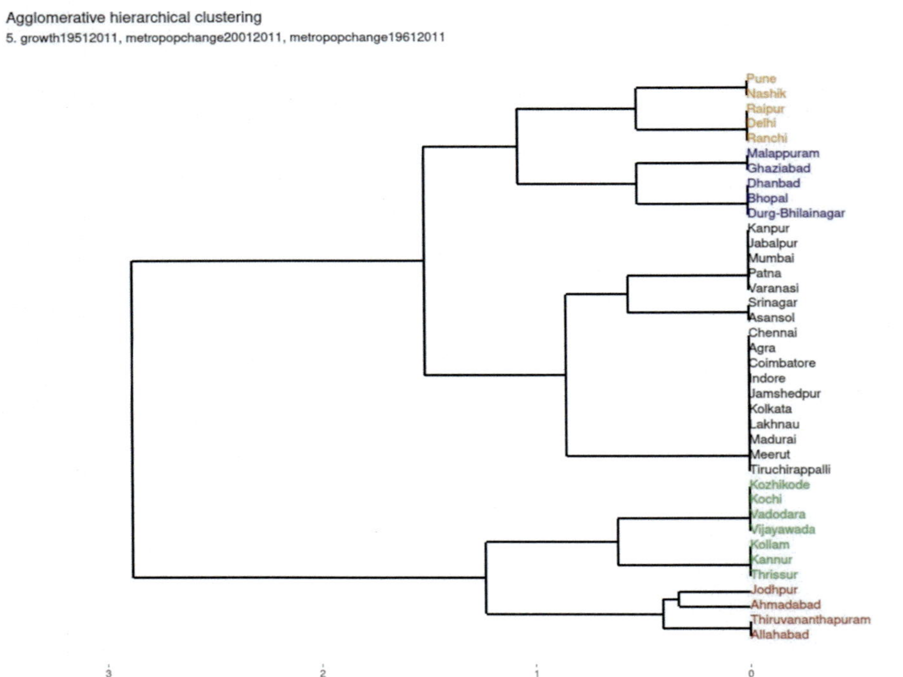

Fig. 14.3 Clusters of the dynamic analysis. Edited by the authors

The largest cluster (*Cluster 1*), consisting of 17 elements, concentrates almost half of all settlements examined. All of these are characterised by a slower population growth from 1951 to 2011 than the average (789%) and a regular growth rate by decades (not exceeding 200% in one decade). As a reminder: these settlements were the most privileged members of the Indian urban hierarchy already at the time of the census of 1951, all but a few with long and rich historical past, significant cultural and administrative functions and born as a result of organic development, with population numbers typically in excess of one hundred thousand in the middle of the last century, already. Population density in most settlements in this cluster maximum doubled from 1961 to 2011, in some of them more than doubled, in two of the 17 cases it decreased, whereas in the 2001–2011 period it also grew or stagnated (which means a few per cent increase). On the ground of this, these settlements are considered as ***moderately growing and densifying autochthonous (regular) agglomerations (conformist UAs)***. As regards the geographical distribution of these settlements, no clear-cut spatial concentration can be identified. The only connection of physical and regional geographical character seems to be justifiable among the six agglomerations in the plain of the Ganges and Yamuna rivers (Meerut, Agra, Kanpur, Lakhnau,

Varanasi and Patna) (Fig. 14.4). In addition, two other smaller settlement clusters, showing not much similarity apart from the relative geographical proximity, can be detected in the southern (Coimbatore, Tiruchirappalli and Madurai) and northeast part of the peninsula (Kolkata, Asansol and Jamshedpur).

The population of settlements in *Cluster 2*, with only four members, grew similarly as did that of the agglomerations in the previous urbanisation type, slower than the average and at a balanced pace from 1951 to 2011. The volume and direction of the change in their population density in the 1961–2011 period vary, so it has no clustering effect in this case, either, but the tendency from 2001 to 2011 is decline in each case. This latter fact, as it was mentioned already, indicates the growth in the administrative area exceeding that of the population number. Taking the above-said into consideration, these settlements are referred to as moderately ***growing, recently***

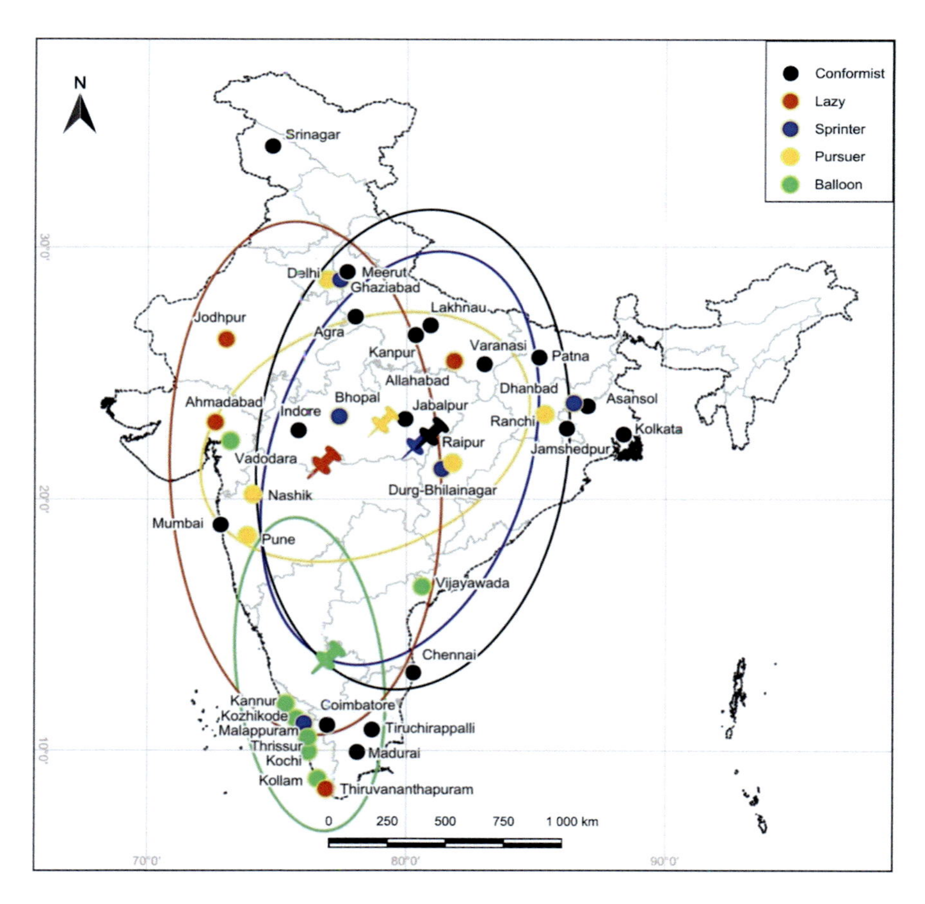

Fig. 14.4 Geographical distribution of dynamic urban clusters featuring their standard deviational ellipses and topographical centres. Edited by the authors

sparsifying autochthonous (regular) agglomerations (lazy UAs). They cannot be related to each other in a topographical sense.

The common feature of agglomerations in *Cluster 3*, with five elements, is that their population number in the 60 years until 2011 grew in an extraordinary way, in some decades in excess of 200% and definitely fast. Most of these have been described above as settlements among the latest creations of the urban network of India, with a short historical past in almost all cases; they are typically industrial centres and satellite towns with residential functions, founded in the twentieth century or becoming urban areas in the second half of the century. The increase of their population density from 1961 to 2011 was intensive—more than twofold in most cases, less then twofold in fewer cases—and continued to grow in the 2001–2011 period with no exception. These settlements, on the basis of their population trends outlined above, are ***intensively growing and densifying, newly developed (irregular) agglomerations (sprinter UAs)***. As regards their geographical situation, they are not related to each other.

The population of the five settlements in *Cluster 4* grew regularly from 1951 to 2011, but faster than the average. Population density in the half century before 2001 grew to maximum double in most of the cases and it is more than doubled in fewer cases; also, in the 2001–2011 period the tendency was growth. Accordingly, these agglomerations are seen as ***rapidly growing and densifying autochthonous (regular) agglomerations (pursuer UAs)***. No spatial relation can be traced among them.

The most considerable common feature of the seven-member *Cluster 5* is the decrease of the population density both in the period from 1961 to 2011 and in the last census decade (2001–2011), with no exception. The reason for this is the significant increase in their administrative territories from 2001 to 2011. This is in line with the fact that settlements of the Kerala megalopolis, significantly enlarged recently with census towns, are all in this group, with the exception of Malappuram and Thiruvananthapuram. The character of the growth of their population number, however, is mostly regular but faster than the average, in a fewer cases it is irregular and definitely intensive. On this ground these settlements are taken as ***rapidly growing but sparsifying, newly developed (irregular) agglomerations (balloon UAs)***. As it has been mentioned before, most of these settlements are in the so-called Kerala Megalopolis, so they are located next to each other along the Malabar Coast (Fig. 14.4).

For the joint visualisation of the clusters identified and described above, i.e. for the sake of the demonstrability of complex urbanisation types, static–dynamic pairs were generated by placing the agglomerations examined in a 5×5 matrix, in accordance with the number of clusters. It must be remarked that only those 38 agglomerations are visualised in the matrix that were parts of the domains of both the static and the dynamic cluster analysis; accordingly, Hyderabad and Rajkot were omitted. Linking the two dendrograms visualising the results of the cluster analyses, kind of overlapping them, it is seen how much a respective static cluster attracts a respective dynamic agglomeration type, whether settlements making the cluster pairs are concentrated into certain complex types or are located scattered in the most diverse combinations.

The results show the presence of a few complex types with relatively large numbers of elements; however, 11 of the possible 25 cluster pairings do not have one agglomeration. As regards urbanisation pairs containing at least 10% of the presentable 38 settlement clusters, i.e. a minimum of four members, the biggest one is the classic type with at least seven agglomerations: the ***top-heavy & conformist*** type. In addition, three more complex urbanisation types with four elements can be identified: ***centralised & conformist***, ***nucleus & conformist***, and ***distributed & balloon*** (Fig. 14.5).

Besides agglomerations that can be listed into complex types, agglomerations interpretable only in the static cluster, the quasi agglomerations and the city metropolises consisting of only one city are also summarised by urbanisation types in Table 14.3.

14.2.4 Findings of the Cluster Analysis of Quality of Life Values and Their Connections with Urban Clusters

Clustering based on quality of life data resulted in six well-separated groups of settlements (Fig. 14.6), as indicated above in the context of the methodological description of the analysis.

In order to grade the clusters separated on the basis of development level quality of life, the individual cities and the values of the development indicators assigned to them were ranked as follows. The values were averaged per clusters and development indicators and ranked from 1 to 6 so that the cluster with the highest average value for the given variable was ranked 1st, while the cluster with the lowest average was ranked 6th. The six clusters were ranked based on the sum of the products of the rankings and their frequency of occurrence (i.e., how many times indicators occur in total). According to this calculation method, the cluster with the lowest aggregate value is considered to be the highest, while the cluster with the highest value is considered to be the least developed (Table 14.4). For each city, we aggregated the values of quality of life indicators separately, which provided an opportunity to rank the 51 settlements examined independently of their cluster classification based on the values of their quality of life characteristics. However, it should also be noted that value ranges of the aggregate quality of life indicators of the cities belonging to each settlement group show some overlap between the clusters, but at trend level they basically indicate well the degrees of development between them (Table 14.5).

The metropolises that make up quality of life clusters are quite definite in terms of their geographical location. Although there are spatial matches between them, it is clearly visible that they are lined up in southwest-northeast direction arranged in deviational ellipses with northwest-southeast longitudinal axes. However, the spatial order among the topographic centres of the individual clusters cannot be clearly corresponded to the ranking of their relative levels of human and infrastructural development. The order of development and spatiality of clusters 2, 3 and 5, which

	(1) agglomerations definitely prevailed by congested centre (top-heavy)	(2) homogeneous agglomerations with medium sized area and moderately populated centre (balanced)	(3) agglomerations with small-medium sized area definitely prevailed by densely populated centre (centralized)	(4) agglomerations with nucleus type centre drawing widespread and populous urban area (nucleus)	(5) large-sized decentralized agglomerations (distributed)
(1) moderately growing and densifying autochthonous (regular) agglomerations (conformist)	[Agra] [Indore] [Kanpur] [Madurai] [Mumbai] [Patna] [Varanasi]	[Asansol] [Srinagar]	[Jabalpur] [Lakhnau] [Meerut] [Tiruchirappalli]	[Chennai] [Coimbatore] [Jamshedpur] [Kolkata]	
(2) moderately growing, recently sparsifying autochthonous (regular) agglomerations (lazy)	[Ahmadabad] [Allahabad] [Jodhpur]			[Thiruvananthapuram]	
(3) intensively growing and densifying, newly developped (irregular) agglomerations (sprinter)		[Durg-Bhilainagar]	[Bhopal] [Dhanbad] [Ghaziabad]		[Malappuram]
(4) rapidly growing and densifying autochthonous (regular) agglomerations (pursuer)	[Delhi] [Pune]		[Nashik] [Raipur] [Ranchi]		
(5) rapidly growing but sparsifying, newly developped (irregular) agglomerations (balloon)			[Vadodara] [Vijayawada]	[Kollam]	[Kannur] [Kochi] [Kozhikode] [Thrissur]

Fig. 14.5 Urban agglomerations ordered in complex (static–dynamic) urban types. Edited by the authors

Table 14.3 All million-plus cities and UAs of India by various urban types

Urban types of metropolises	Million-plus cities & UAs
City metropolises	Faridabad, Jaipur, Kota, Ludhiana, Vasai-Virar, Visakhapatnam
Quasiagglomerations	Amritsar, Aurangabad, Bengaluru, Gwalior, Nagpur, Surat
Top-heavy & conformistUAs	Agra, Indore, Kanpur, Madurai, Mumbai, Patna, Varanasi
Top-heavy & lazyUAs	Ahmedabad, Allahabad, Jodhpur
Top-heavy & pursuerUAs	Delhi, Pune
Balanced & conformistUAs	Asansol, Srinagar
Balanced & sprinter UAs	Durg-Bhilainagar
Centralised & conformistUAs	Jabalpur, Lakhnau, Meerut, Tiruchirappalli
Centralised & sprinter UAs	Bhopal, Dhanbad, Ghaziabad
Centralised & pursuerUAs	Nashik, Raipur, Ranchi
Centralised & balloonUAs	Vadodara, Vijayawada
Nucleus & conformistUAs	Chennai, Coimbatore, Jamshedpur, Kolkata
Nucleus & lazyUAs	Thiruvananthapuram
Nucleus & balloonUAs	Kollam
Distributed & sprinter UAs	Malappuram
Distributed & balloonUAs	Kannur, Kochi, Kozhikode, Thrissur
CentralisedUAs	Hyderabad, Rajkot

includes 37 out of 51 cities, i.e. almost three quarters of the examined settlements, is clearly in line with the north–south regional inequality characteristics of the country. Depending on the number of elements, the clusters cover a larger or smaller, but in each case a well-defined area, there is no trace of spatial variance other than the directions indicated above. The primary confine among the clusters is the northwest-southeast axis designated by the Jodhpur–Nagpur–Visakhapatnam line, of which northeast only cities in clusters 5, 6 and 1 and southwest only settlements of clusters 3, 2 and 4 are situated (Fig. 14.7).

Examination on interdependence among quality of life clusters formed by cities included in this study and the urbanization types of their agglomerations indicates that correlation values are found to be low (Table 14.6). Practically we can say that quality of life and urbanization types represent independent cluster sets, that is, the role of quality-of-life factors in the formation of different types of urbanization of agglomerations cannot be mathematically justified, they are influenced by other socio-economic factors that need further clarification.

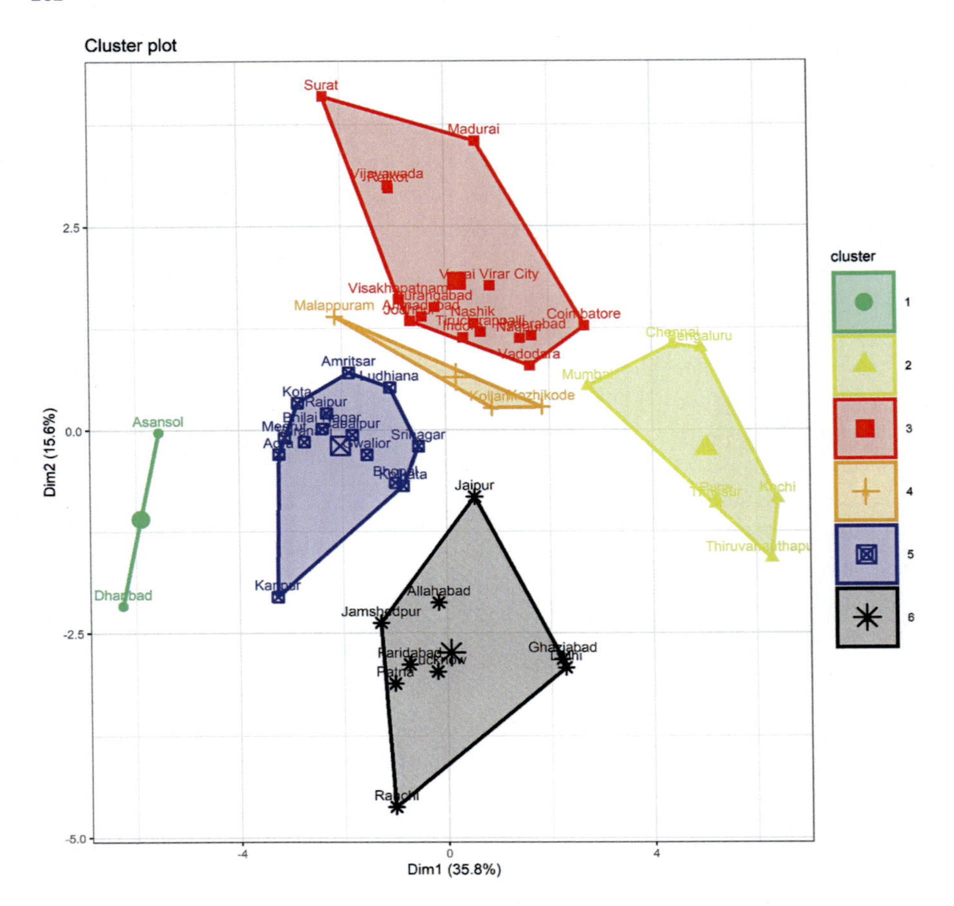

Fig. 14.6 Result of K-means cluster analysis for 6 clusters plotted against Dim1 and Dim2 of primary components. Edited by the authors

Table 14.4 Development ranks of quality of life clusters

Cluster	Number of occurrences of						Sum of
	Rank 1	Rank 2	Rank 3	Rank 4	Rank 5	Rank 6	Product of ranks and occurrences
2	11	7	0	2	0	0	33
6	3	7	1	7	2	0	58
3	3	3	7	3	3	1	63
4	3	2	5	3	4	3	72
5	0	0	4	5	9	2	89
1	0	1	3	0	2	14	105

Table 14.5 Metropolitan cities with individual ranks and value ranges of aggregate quality of life indicators per clusters

Cluster	Cities with rank among all the 51 settlements	Interval of cumulative values of quality of life indices
2	1. Kochi; 2. Pune; 3. Chennai; 4. Bengaluru; 7. Thiruvananthapuram; 10. Mumbai; 22. Thrissur	1061.6–938.9
6	6. Delhi; 9. Ghaziabad; 11. Jaipur; 13. Allahabad; 23. Lucknow; 26. Jamshedpur; 29. Jamshedpur; 36. Patna; 42. Ranchi	1032.7–848.5
3	5. Vadodara; 8. Nagpur; 12. Coimbatore; 14. Jodhpur; 15. Hyderabad; 16. Madurai; 17. Nashik; 18. Aurangabad; 19. Ahmadabad; 20. Vasai-Virar; 24. Rajkot; 25. Indore; 28. Tiruchirappalli; 31. Surat; 33. Vijayawada; 38. Visakhapatnam	1037.3–870.7
4	34. Kozhikode; 40. Kollam; 48. Malappuram	893.1–794.7
5	21. Ludhiana; 28. Srinagar; 30. Gwalior; 32. Kolkata; 35. Varanasi; 37. Bhopal; 39. Amritsar; 41. Bhilai Nagar; 43. Kanpur; 44. Kota; 45. Agra; 46. Meerut; 47. Jabalpur; 49. Raipur	942.2–788.7
1	50. Asansol; 51. Dhanbad	733–690.5

14.3 Summary

In our study, apropos of metropolisation, one of the main characteristics of Indian urbanization, i.e. the increasing prevalence of million-plus cities and agglomerations within urban areas, it was first examined that these settlements, especially urban agglomerations, how can be typified on the basis of differences in their structural characteristics—population, size of administrative area, population density—registered at the time of the 2011 census and on the grounds of divergences in change of these features taking place in the previous decades. We summarized our results in the form of five static and five dynamic settlement clusters constituting different types of urbanization. On the other hand, our study focused on how Indian metropolises can be grouped in terms of their quality of life characteristics. Based on the values of indicators marking the level of human development and household infrastructure of the affected settlements, core cities of million-plus urban agglomerations have been arranged into quality of life clusters, among which differences in development have been managed to prove. Finally, we sought to answer whether there is any relationship between quality of life clusters and types of urbanization. In this respect, it has been found that such a correlation cannot be mathematically justified, i.e. other socio-economic factors requiring further clarification play a role in shaping of characteristics of India's urban agglomerations.

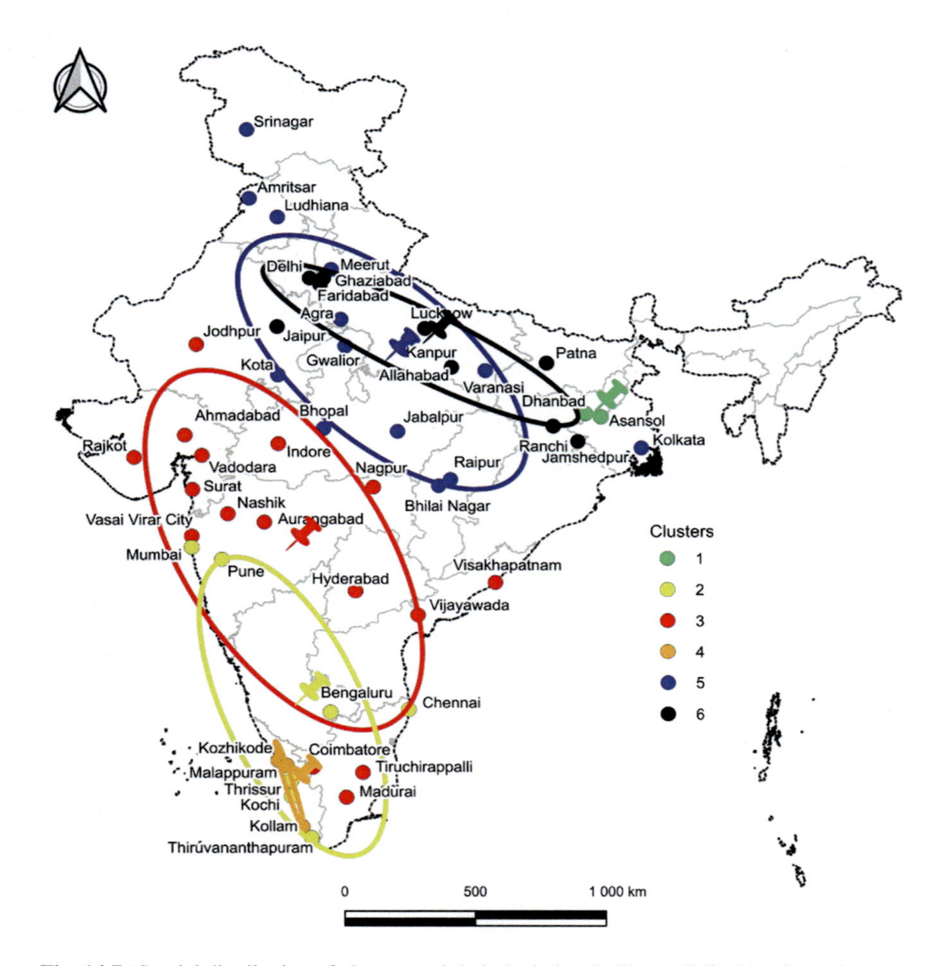

Fig. 14.7 Spatial distribution of clusters and their deviational ellipses. Edited by the authors

Table 14.6 Correlation values between quality of life and urbanization type clusters

Correlation values between cluster analyses	Static urbanization type cluster values	Dynamic urbanization type cluster values
Quality of life cluster values	−0.1662446	−0.1327364

References

Becker RA, Chambers JM, Wilks AR (1988) The new S language: a programming environment for data analysis and graphics, Wadsworth & Brooks/Cole, Monterey

Census India (2001) Census Data 2001. Office of the registrar general & census commissioner, India. https://censusindia.gov.in/2011-common/census_data_2001.html

Census India (2011) Census Data 2011. Office of the registrar general & census commissioner, India. https://censusindia.gov.in/2011-common/censusdata2011.html

Census Newsletter (2001) eCENSUS India: Issue Number 3: 2001. https://censusindia.gov.in/Census_Data_2001/Census_Newsletters/Newsletter_Links/eci_3.htm

Gower JC (1971) A general coefficient of similarity and some of its properties biometrics 27(4):857–874

Halkidi M, Batistakis Y, Vazirgiannis M (2001) On clustering validation techniques. J Intell Inf Syst (17):107–145.

Hartigan JA, Wong MA (1979) Algorithm AS 136: A K-means clustering algorithm. Appl Stat 28(1):100–108. https://doi.org/10.2307/2346830

Murtagh F, Legendre P (2014) Ward's hierarchical agglomerative clustering method: which algorithms implement Ward's criterion? J Classif 31:274–295. https://doi.org/10.1007/s00357-014-9161-z

Pradhan K (2017) Unacknowledged urbanisation: new census towns of India. In: Denis E, Zérah MH (eds): Subaltern urbanisation in India: an introduction to the dynamics of ordinary towns, pp 39–66 https://doi.org/10.1007/978-81-322-3616-0_2

R core team (2020) R: A language and environment for statistical computing. R Foundation for statistical computing, Vienna, Austria. https://www.r-project.org/

Ramachandran R (2001) Urbanization and urban systems of India Oxford University Press, New Delhi

Rousseeuw PJ (1987) Silhouettes: A graphical aid to the interpretation and validation of cluster analysis. J Comput Appl Math (20):53–65. https://doi.org/10.1016/0377-0427(87)90125-7

Scrucca L, Fop M, Murphy TB, Raftery AE (2016) mclust 5: Clustering, classification and density estimation using Gaussian finite mixture models. R J 8(1):289–317.

Singh N (2014) Mapping metropolises in India. Modern geográfia 2014 (3):77–94. http://www.moderngeografia.eu/?=44

Tirtha R (2002) Geography of India rawat publications, New Delhi

Towns and urban agglomerations (2011) Towns and urban agglomerations classified by population size class in 2011 With variation since 1901 office of the registrar general & census commissioner, India. https://censusindia.gov.in/2011census/PCA/A4.html

Wilhelm Z (2008) Adatok az indiai urbanizáció folyamatának vizsgálatához ModernGeográfia 2008 (2):1–57. http://www.moderngeografia.eu/?cat=22

Wilhelm Z (2011) The survey of spatial disparity in India with the application of the SENTIENT index Hungarian geographical. Bulletin 60(1):45–65

Wilhelm Z (2015) India regionális földrajza. Publikon Kiadó, Pécs

Wilhelm Z, Déri I, Szilágyi S, Nemes V, Zagyi N (2013) Urbanizáció Indiában a 2001-es, valamint a 2011-es népszámlálás előzetes eredményeinek tükrében. Településföldrajzi Tanulmányok 2 (1):60–74.

Wilhelm Z, Pete J, Nemes V, Zagyi N (2011) The survey of spatial disparity in India with the application of the sentient index with special focus on religious composition. Human innovation review 2 (1): 56–76.

Wilhelm Z, Zagyi N (2018) Történeti és jelenkori urbanizációs folyamatok Indiában. In: Pap N, Szalai G (eds) Táj geográfus ecsettel. Pécsi Tudományegyetem, Pécs, pp 119–143

Wilhelm Z, Zagyi N, Nemes V, Kiss K (2014) A gyermekkori nemi arányok átalakulásának területi összefüggései Településföldrajzi Tanulmányok 3 (1): 139–149.

Yuill RS (1971) The standard deviational ellipse: an updated tool for spatial description geografiska annaler series B. Hum Geogr 53(1):28–39

Zagyi N, Kuszinger R, Wilhelm Z (2021) Characteristics of recent urbanisation in India in light of the divergent development paths of metropolises. Reg Stat 11(3):60–94. https://doi.org/10.15196/RS110301

Chapter 15
Urban Rejuvenation and Social Sustainability in Smart City: An Empirical Study of Community Aspirations

Virendra Nagarale and Piyush Telang

Abstract Sustainable environment is the need of any society. Urban renewal in Smart Cities is the major concern for incorporation of sustainable environment in cities. Significant increase in population and lack of proper planning strategy has led to the series of problems of urban decay in metropolitan areas intimidating community wellbeing and security. To seize urban decay, rejuvenation is normally an adopted move toward regeneration of rundown areas. Rejuvenation often results negatively and may lead to bother existing social networks. The success of the renewal practices mostly depend on active participation of residents. The approach of sustainable development in urban rejuvenation should balance the interests of stakeholders in different socio-economic and demographic class. According to Ease of Living Index 2018, Pune ranks highest among 111 cities in India while in Smart city ranking by Ministry of Housing and Urban Affairs (MoHUA) in 2019 Pune stood 11th with 213.50 marks. For ranking purpose MoHUA considers a variety of factors like performance of civic institution, spent expenditure, and implementation of different projects in five years. On the one hand, Pune is considered as the most liveable city in contrast its ranking being a smart city shows negative effect. Pune is selected for the present study that is governed by Pune Municipal Corporation (PMC). PMC governs 331.26 km^2 area comprising 15 administrative and 144 electoral wards. The total population of PMC is 3,371,626 where, 452,240 are SCs and 37,630 are STs. The present study explores the preferences and aspirations of citizens in regard to urban renewal through smart city mission. The questionnaire survey is been used to collect the responses of citizen from Pune Municipal Corporation. This paper aims to assess how residents perceive the urban renewal strategy through smart city mission and to identify the responses in view of individual's socio-economic and demographic structure. The results show that how age, educational level, employment status, etc. changes the perspective of responses.

V. Nagarale (✉)
Department of Geography, SNDT Women's University, Pune Campus, Karve Road, Pune 411038, India

P. Telang
ICSSR Project, Department of Geography, SNDT Women's University, Pune Campus, Karve Road, Pune 411038, India

N. C. Jana et al. (eds.), *Livelihood Enhancement Through Agriculture, Tourism and Health*, Advances in Geographical and Environmental Sciences,
https://doi.org/10.1007/978-981-16-7310-8_15

Keywords Smart city · Social sustainability · Sustainability · Rejuvenation · Urbanization

15.1 Introduction

We have constantly taken a move toward developing our cities from traditional to modernized spaces and networks. An increased population and urbanization in major cities worldwide has created numerous opportunities and challenges to plan and manage resources. Many global cities are now facing different types of problems like congestion, slums, deterioration of resources, pressurizing local urban amenities, etc. To deal with these problems and maintain the resources in cities rejuvenation practices are often used. In the year 2014, Smart City Mission is announced by Government of India to redevelop 100 major cities in India. "The Smart City Mission needs to be equipped to provide solutions to India's urban challenges" (Bhattacharya and Rathi 2015). The demographic profile, size of household, education attainment, income level, and distribution of resources and services can be recognized to its crucial function for implementing any planning strategy. Urban renewal in Smart Cities is the major concern for incorporation of sustainable environment in cities. Significant increase in population and lack of proper planning strategy has led to a series of problems of urban decay in metropolitan areas intimidating community wellbeing and security. Pune is one of the cities selected for Smart City Mission. According to Ease of Living Index (2018), Pune ranks highest among 111 cities in India while in Smart city ranking by Ministry of Housing and Urban Affairs (MoHUA) in 2019 Pune stood 11th with 213.50 marks. For ranking purpose MoHUA considers a variety of factors like performance of civic institution, spent expenditure, and implementation of different projects in five years. On the one hand, Pune is considered as the most liveable city whereas, its ranking being a smart city shows negative effect. Considering this fact the present study explores the preferences and aspirations of citizens in regard to urban renewal through smart city mission.

15.2 Social Sustainability

Sustainable environment is the need of every society. Sustainability is often categorized in three major types dealing with environment, society, and economy. Social sustainability is a process rather than a concept which means maintaining social capital. Social capital includes love and peace, cohesion of communities, discipline, shared values and equal rights, equity in accessibility of civic amenities, safe and healthy environment, decent employment opportunities, minimal insecurities, etc. Maintaining balance in these social parameters can improve social well-being which is the key element for sustainable social environment. There are some causes which sometimes disturb the existing social networks. Rejuvenation and rehabilitation practices are the ones which also affect the social well-being and can deteriorate the social communications. Therefore, it makes essential to understand how the residents perceive the role of governmental policy framework and implication strategies. The study will also find out whether rejuvenation practices are more reliable for the development and can they maintain or disturb social sustainable networks.

15.3 Review of Literature

Social sustainability is being a major concern in recent years for the researchers in urban studies. Social sustainability is mostly seen as a concept and has been experimented by considering various aspects such as housing conditions, rejuvenation, and redevelopment practices, pedestrian space, urban connectivity, intelligent transportation, etc. Along similar lines, but with more emphasis on individuals' rights, Omann and Spangenberg (2002) argued that the basic focus of social sustainability should be on the personal assets like education, skills, experience, consumption, income, and employment and comprises every citizen's right to actively participate in his/her society as an essential element. Thus, in their analysis, access to societal resources is the major part of social sustainability. "Redevelopment often results in negative outcomes such as disturbances to existing social networks and burgeoning construction and demolition waste" (Ho et al. 2012). In his study, the author discovers the aspirations and preferences of stakeholders in relation to the Rehabilitation and Redevelopment through a structured survey. Dogu and Aras (2019), tried to develop a sample scale for measuring social sustainability in urban context using Measurement of the City from Social Aspects (MCSA) model. This model was examined empirically in the city of Güzelyurt. In the MCSA scale, dependent variables are considered to be "Sense of Belonging", "Social Capital", "Perceived Environment", "Social Interactions/Security", "Interaction with Space", "Satisfied with Space", and "Voice and Influence" and age, educational level, professional and region migration are considered to be the categorical (independent) variables. Analysis results indicate that the developed model has a good fit and it works and can be suitable to measure social sustainability in other cities. The OECD report by Martinez-Fernandez (2012) on Demographic Change and Local Development examines the cases from twenty different countries across the world. The report mainly indicates local strategies and initiatives for policy consideration and learning. The report also illustrates the problems of modern local development in relation with change in population and its structure. A chapter in book entitled Smart Cities in a Smart World Murgante and Borruso (2015) explains smart city concepts and suggests that the devices which smart city provide has to be used by citizens to monitor the management and highlight the positive and negative aspects for better management. It is therefore important that the citizens participation is important in making smart cities also their preferences for checking in and doing particular activities in certain places allow planners and scholars to better understand how cities shape themselves from a social—not only in the ICT way which can be better way to achieve sustainable social environment. Baffoe and Mutisya (2015) in their empirical research discussed indicators of social sustainability and applied them using composite index approach in Kibera in Kenya and found a medium level of social sustainability mechanism. The study together with the help of results and evidences put forward the policy support for many aspects of society like employment generation, better housing, safety and security, etc. for improving social sustainability in Kibera. It also advised that the validation of approaches to deal with social sustainability should be done by using various methods such as Analytic Hierarchy Process (AHP) or Fuzzy Logic. Severson and Vos (2018) worked to operationalize social sustainability in an affordable housing context at the level of the housing provider in the sector by presenting a possible measurement framework. In that they have identified four key dimensions for social sustainability in the context of social and affordable housing: housing standards; non-shelter needs; community integration and social inclusion; and capacity building and resiliency and chosen indicators for each of these dimensions. In this study, 49 different measures for 15 different indicators along the four different dimensions have been included. As this study has only considered one aspect and the others are neglected such as income level of residents, occupational structure and other demographic characteristics like total population and density of population which is also responsible for the dwelling and affordability of housing and deciding the overall quality of life. Andrea Colantonio (2007) in his research on Social Sustainability provides wide range summary about social sustainability concept and discovers the main intention of the concept. The author assesses various methods and tools that are deliberately used for measuring the social sustainability and examines the obstacles that restrict its practical implementation. The author argues that the measurement and assessment of concept of social sustainability are still in debate into holistic against reductionist approach. It is also highlighted in the study that there is scarcity of tools for implementing the concept of social sustainability. Janubova and Gress (2018), in the study, entitled 'Urbanization of Poverty and the Sustainable Development of Urban Areas in Chile assesses urbanization level in Chile and correlates it with the sustainable economic development'. It provides historical background of urbanization in Chile along with the housing policy deal with a phenomenon of urbanization and its impact not only on the sustainable economic development in Chile. The author used Urban Sustainable Index (USI) to assess the impact of Chile's housing policy on the sustainable development. The obtained results show that urban areas have a gradual improvement in sustainability from 2000 to 2013. At that time, average growth rate of sustainability increased by 6.15%. Križnik (2018) in his article argues that "declining social cohesion and a lack of citizen participation are a consequence of speculative urban development, in which urban regeneration and urban redevelopment are paved to attract investments, strengthen economic competitiveness, and improve the city's global appeal rather than address diverse local challenges" (Križnik 2018). The Consequences and perception of urban development has been surveyed using interview method and the results show that Wangsimni used to have a well-developed economic and communal life before the urban development started. This largely contrasts with the notion of a deprived urban area, which the local government used in public to legitimize the transformation of Poblenou and Wangsimni. The lack of citizen participation along with a decline in social cohesion—two major dimensions of a localized social sustainability agenda—has been clearly identified in this study by qualitative measures but there is a gap found in the present research that the lack of statistical measures. Social sustainability concept is multifaceted, and it has to be assessed with the help of multidimensional aspects. The demographics and spatial arrangement plays major role in this context.

15.4 Study Area

Pune Municipal Corporation (PMC) governs 331.26 km^2 area which is divided into 15 administrative wards (Fig. 15.1). PMC is a part of Pune District which is situated in the western region of Maharashtra state between 17°50′ North to 19°24′ North latitude and 73°19′ East to 75°10′ East longitudes. A significant increase in the population of PMC is observed from the year 1951 to 2011 where, in 1951 the population was 488,419 which is been increased to 3,124,458 in the year 2011. The great proportion of population residing in PMC limits makes Pune the second most populous city in the State of Maharashtra and eight largest city in India. In October 2018, PMC included 11 fringing areas (Villages) to its administrative limits adding 239,483 more population residing in 57,928 households. The total population of PMC after including 11 villages is 3,371,626. Total SC and ST population is 452,240 and 37,630 respectively. The main reasons for a significant change in the population size from 1950 to 2011 were increase in employment and business opportunities as well as educational facilities. From 1951 to 2011 various changes have taken place in PMC and these quick changes play a vital role to give a makeover of the city's old identity 'Pensioners Pune' to 'IT Pune' and now 'Smart Pune'.

15.5 Research Design

An empirical research is descriptive in nature. To fulfil the output of the present study the sound methodology has been adopted. The methodology is divided in the following three phases.

15.5.1 Pre-Field Work

Before starting the actual research, it is important to understand that whether there are any studies were conducted in relation with the present research work. It is therefore seen through literature survey, visiting different libraries, reading books, published

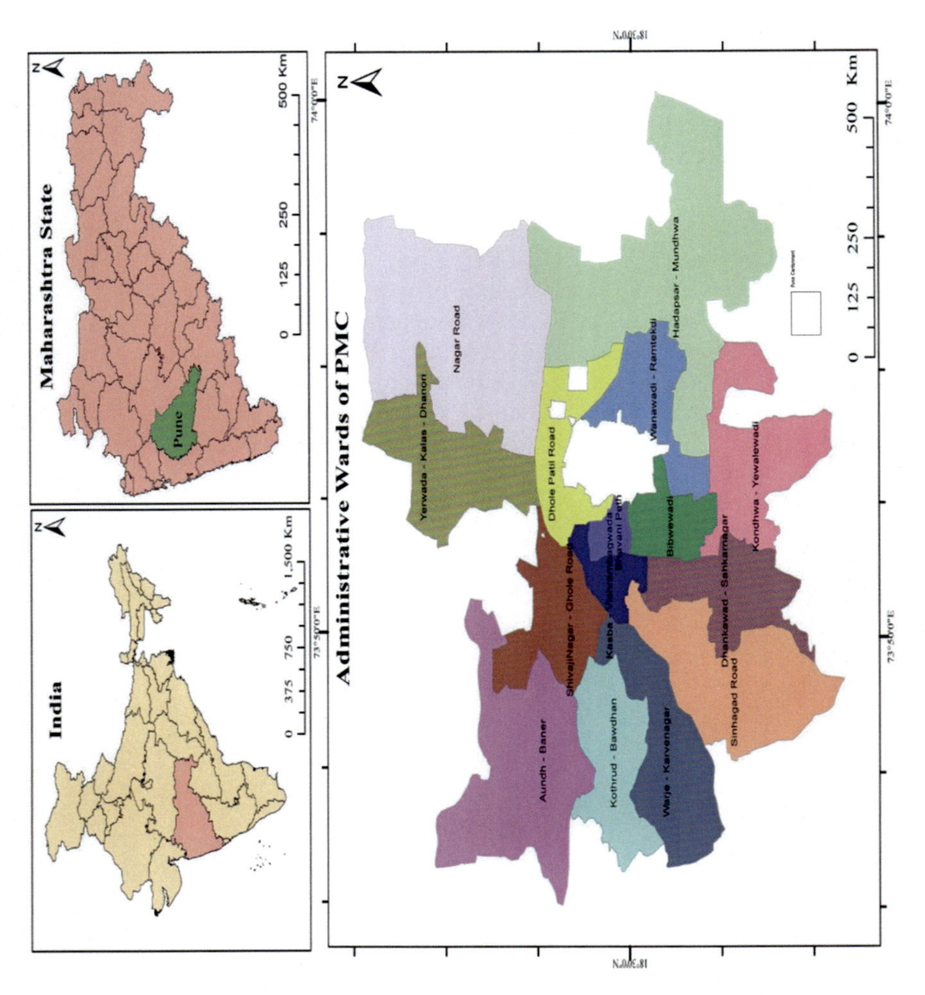

Fig. 15.1 Map of PMC

research works, etc. Along with that the observation through field visits and informal talks are made with residents to clear the idea and frame the present research question. The preparation of questionnaire was also done in this phase.

15.5.2 Field Work

In this phase the collection of datasets like population data of each ward, base maps, toposheets, etc. has been made. After that the questionnaire survey has been conducted visiting each administrative ward. Online available resources like household data and other relevant information are collected in this phase.

15.5.3 Post-field Work

The compilation of gathered data and its arrangement, conversion of data in digital format, is done in this phase. After the proper arrangement of all the datasets the analysis part has begun to find out the results and reach the conclusion.

15.6 Background

There are 15 administrative wards in PMC in which the questionnaire survey has been conducted using random sampling method. Thirty individual samples were collected from each administrative ward from different socio-economic and demographic background. The questionnaire comprised of 47 questions which includes Personal information, Family information, Social Status, Smart City, and Sustainability related questions. The combination of open ended and multiple choice based as well as Likert scale based questions were given variety of dimensions and were challenging for an empirical analysis. The responses are analyzed based on the social, economic, and demographic status of the individuals. The study aims to see that how the Socio-economic and demographic structure drives the choices and expectations of the stakeholders. The study also illustrates the level of awareness about a smart city and sustainability measures of the stakeholders. This study tries to assess the smart city progress by considering the progress of each component through the responses of city's stakeholders. The question is asked where four main choices were given to know how each individual perceives smart city in terms of their age group and where they are located. They are Equitable, Sustainable, Effective, and Efficient. The explanation of those is as follows.

15.6.1 Equitable

This aspect explains that each resident is getting equal and fair treatment as well as urban resources without compromising those with effect of any circumstances for example, on the basis of regional identity or social status.

15.6.2 Sustainable

It is the situation where the social networks are maintained in such a way that those shall not be disturbed and hindered by policy framework of redevelopment.

15.6.3 Effective

It explains how the desired strategy of redevelopment is being implemented and how it has been successful in producing the intended results.

15.6.4 Efficient

It explains that how the planned strategy of development has been produced maximum outcomes in case of employment, minimizing deprivation, maintaining well-being with minimal efforts and resources.

15.7 Results

The population of PMC for the year 2001 and 2011 is shown in Table 15.1. As per the data, it is seen that in 2001 the total population of PMC was approximately 2.38 million which has increased to 3.31 million in 2011. Total SC and ST population is higher in Sangamwadi–Yerwada ward and lower in Yevalewadi ward. The no. of households is much higher in Hadapsar and lower in Yevalewadi administrative ward. A considerable increase in population is seen in Hadapsar which is more than double from 2001 to 2011.

From the total 450 responses of questionnaire, 23 participants have not responded to the questions which are considered under this study but provided other information (Table 15.2). The maximum number of unanswered questionnaires is seen in Kasba–Vishrambagh administrative ward which is mostly a commercial area of the city. Very few responses are recorded from ST and NT category as the random sampling

Table 15.1 Population of PMC: 2001 and 2011

Sr. no	Ward name	Total no. of households	Total SC population	Total ST population	Population: 2001	Population: 2011
1	Aundh	45,032	27,910	3,349	175,755	181,124
2	Dhole Patil Road	35,208	32,773	2,089	154,919	155,413
3	Bhavani Peth	39,419	35,532	923	160,504	192,932
4	Bibvewadi	68,013	36,436	2,757	174,811	291,446
5	Dhankawadi	56,682	22,121	2,864	151,692	236,648
6	Ghole Road	39,001	21,525	2,242	198,286	171,678
7	Hadapsar	78,445	47,219	3,548	123,288	324,751
8	Kothrud	54,480	14,109	1,612	171,632	209,331
9	Kasba–Vishrambagh	43,138	6,923	1,170	211,388	178,484
10	Sahakarnagar	46,355	39,788	1,682	166,902	205,441
11	Sangamwadi–Yerwada	58,421	56,641	4,512	166,364	261,957
12	Sinhagad Road	60,387	29,877	2,261	175,112	242,290
13	Warje Karvenagar	58,977	17,594	2,754	140,250	233,399
14	NagarRoad	59,044	30,948	2,262	203,110	239,564
15	Yewalewadi	1,733	1,369	104	4,564	7,685
	Total[a]	744,335	420,765	34,129	2,378,577	3,132,143

Source City Population Census PMC (2001 and 2011)
[a]In 2018 Wanawadi administrative ward is newly created accounting total population 204,644 and then the total population PMC is 3,371,626 as per the PMC Election 2017 including 11 villages

method has been used to collect the responses. Although overwhelming responses are captured from open category followed by SC, Maratha and OBC category. The ward wise distribution of responses shows that out of thirty responses highest no. of responses for SCs are captured from Tilak road administrative ward and for STs it is gathered from Kasba–Vishrambagh ward.

Table 15.3 shows the employment status of the total collected respondent. Out of total 450 respondents high number of respondents is employed in the private sector and only 105 in government sector. From the total respondent, 101 are either students, housewives, or unemployed. This sample data can give an idea about the dependency level of population. Maximum number of respondents those are involved in private sector is seen in Kasba–Vishrambagh ward where mostly the commercial activity takes place and minimum is in Dhankawadi ward. As opposite to that, Dhanakawadi ward has high number of respondents who are involved in government sector and minimum is seen in Bhavani Peth ward which is adjacent to Kasba–Vishrambagh ward. From the total dependent respondent maximum no. are seen is Tilak Road and minimum in Kasba–Vishrambagh ward.

An attempt has been made to see the awareness level of the resident for Smart City Mission and its Pillars. It has been found that out of total responses 352 respondents are aware about the same and 98 are unaware (Table 15.4). Gender wise awareness

Table 15.2 Category wise responses for questionnaire survey

Sr. no	Ward	Open	Maratha	OBC	SC	ST	NT	Not answered	Total
1	Aundh	10	8	4	6	0	0	2	30
2	Dhole Patil Road	17	4	1	7	1	0	0	30
3	Bhavani Peth	20	1	5	4	0	0	0	30
4	Bibvewadi	17	2	8	3	0	0	0	30
5	Dhankawadi	16	2	1	6	1	0	4	30
6	Ghole Road	19	4	4	2	0	1	0	30
7	Hadapsar	15	0	2	8	0	0	5	30
8	Kothrud	14	13	2		0	0	1	30
9	Kasba–Vishrambagh	7	4	4	5	3	1	6	30
10	Wanawadi	22	7	0	1	0	0	0	30
11	Sangamwadi–Yerwada	14	5	1	7	1	0	2	30
12	Sinhagad Road	17	1	1	10	0	1	0	30
13	Warje Karvenagar	8	2	17	0	0	1	2	30
14	NagarRoad	24	2	1	2	0	0	1	30
15	Yewalewadi	25	2	2	1	0	0	0	30
	Total	245	57	53	62	6	4	23	450

Source Created by Authors on the basis of Questionnaire Survey

Table 15.3 Employment status

Sr. no	Ward name	Government	Private	Student/housewife/unemployed
1	Aundh	9	13	8
2	Dhole Patil Road	16	14	0
3	Bhavani Peth	0	20	10
4	Bibvewadi	4	14	12
5	Dhankawadi	18	11	1
6	Ghole Road	3	19	8
7	Hadapsar	7	14	9
8	Kothrud	4	18	8
9	Kasba–Vishrambagh	9	21	0
10	Wanawadi	11	18	1
11	Sangamwadi–Yerwada	10	16	4
12	Sinhagad Road	2	15	13
13	Warje Karvenagar	4	17	9
14	NagarRoad	5	13	12
15	Yewalewadi	3	21	6
	Total	105	244	101

Source Created by Authors on the basis of Questionnaire Survey

Table 15.4 Awareness about the concept of smart city and its pillars

Sr. no	Ward	Total awareness			Total unawareness		
		Total	Females	Males	Total	Females	Males
1	Aundh	22	4	18	8	4	4
2	Dhole Patil Road	21	4	17	9	0	9
3	Bhavani Peth	30	11	19	0	0	0
4	Bibvewadi	30	14	16	0	0	0
5	Dhankawadi	23	11	12	7	1	6
6	Ghole Road	14	3	11	16	8	8
7	Hadapsar	28	10	20	2	1	1
8	Kothrud	25	0	25	5	0	5
9	Kasba–Vishrambagh	19	2	17	11	0	11
10	Wanawadi	14	4	10	16	1	3
11	Sangamwadi–Yerwada	22	4	18	8	3	5
12	Sinhagad Road	21	0	21	9	0	9
13	Warje Karvenagar	25	10	15	5	4	1
14	NagarRoad	29	12	17	1	0	1
15	Yewalewadi	29	14	15	1	1	0
	Total	352	103	251	98	23	63

Source Created by Authors on the basis of Questionnaire Survey

level shows that total male awareness is much higher than awareness in females. Highest awareness level in male as well as in female is found in both Bhavani Peth and Bibwewadi administrative ward. Where, minimum awareness is recorded in Ghole Road ward. Other facts can also be seen in Table 15.4.

The awareness about the concept of Sustainability is also recorded and higher awareness is recorded again in Bhavani Peth and Bibwewadi ward (Table 15.5). Lowest awareness is recorded in Ghole Road ward. Out of total responses, 339 respondents are aware about the concept of sustainability and 111 are unaware. The awareness and unawareness level is higher in males from the totals.

To explore the preferences of residents for Smart City and to know their expectation from smart city mission a questionnaire was framed and the responses were recorded. Figure 15.2 shows the total no. of responses and their representation for each aspect which has been discussed earlier in this paper. High number of respondents have responded for sustainable and equitible smart city that accounts more than 80% of responces. Further, the responses are recorded for efficient accounts 80% responces. Though there are more number of respondents who does not responded they are categorised in not answered.

The understanding of respondents and their experiences mostly depends on the age group. Therefore it is important to understand how each age group perceive their interpretation for smart city redevelopment plan. Figures 15.3, 15.4, 15.5, 15.6, 15.7

Table 15.5 Awareness about the concept of Sustainability

Sr. no	Ward	Total awareness			Total unawareness		
		Total	Females	Males	Total	Females	Males
1	Aundh	22	4	18	8	4	4
2	Dhole Patil Road	21	4	17	9	0	9
3	Bhavani Peth	**30**	11	19	0	0	0
4	Bibvewadi	**30**	14	16	0	0	0
5	Dhankawadi	23	10	13	7	2	5
6	Ghole Road	14	3	11	16	8	8
7	Hadapsar	27	10	17	3	1	2
8	Kothrud	18	12	6	12	2	10
9	Kasba–Vishrambagh	21	5	16	9	2	7
10	Wanawadi	14	4	10	16	1	15
11	Sangamwadi–Yerwada	22	4	18	8	3	5
12	Sinhagad Road	20	6	14	10	2	8
13	Warje Karvenagar	19	9	10	11	5	6
14	NagarRoad	29	12	17	1	0	1
15	Yewalewadi	29	14	15	1	1	0
	Total	339	122	217	111	31	80

Source Created by Authors on the basis of Questionnaire Survey

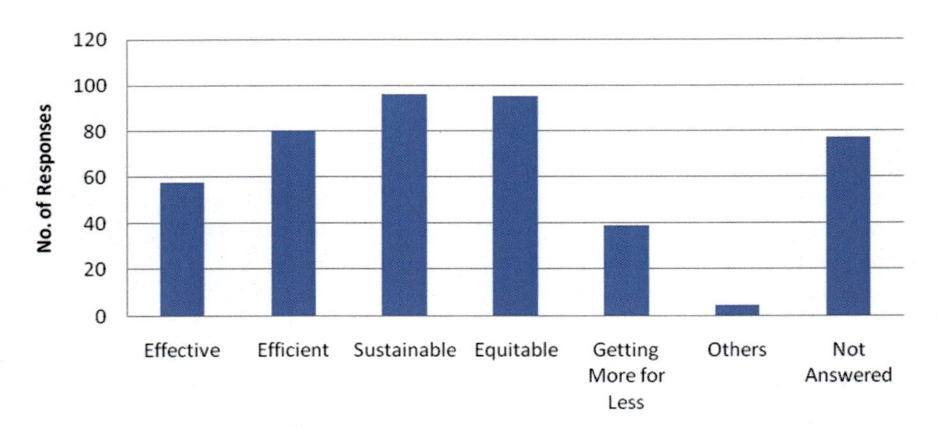

Fig. 15.2 Residents interpretation of smart city

and 15.8 show the age wise distribution of responses for smart city interpretation.

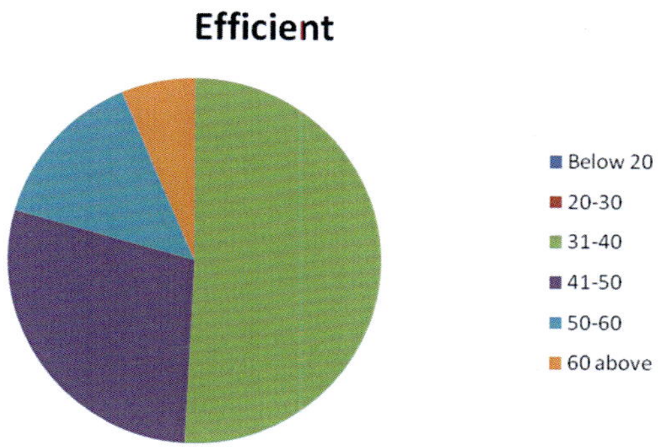

Fig. 15.3 Age wise distribution of responses for smart city interpretation

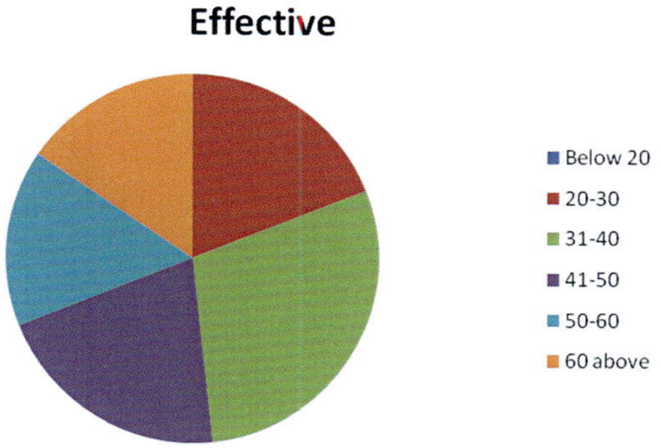

Fig. 15.4 Age wise distribution of responses for smart city interpretation

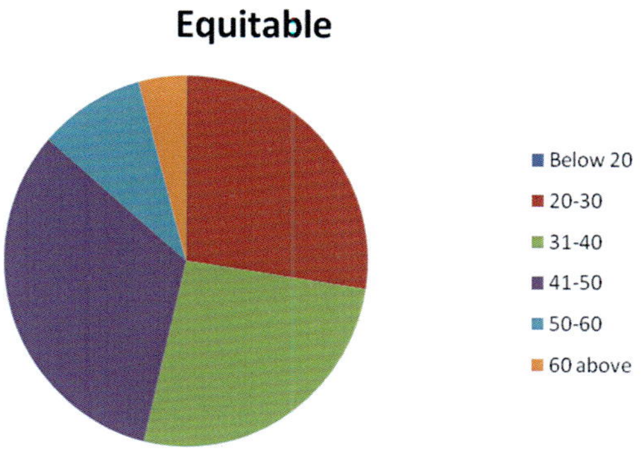

Fig. 15.5 Age wise distribution of responses for smart city interpretation

15.8 Discussion

The obtained result shows that the population of PMC is unevenly distributed in all the administrative wards and therefore the random sampling method cannot be

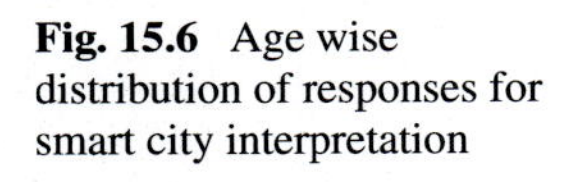

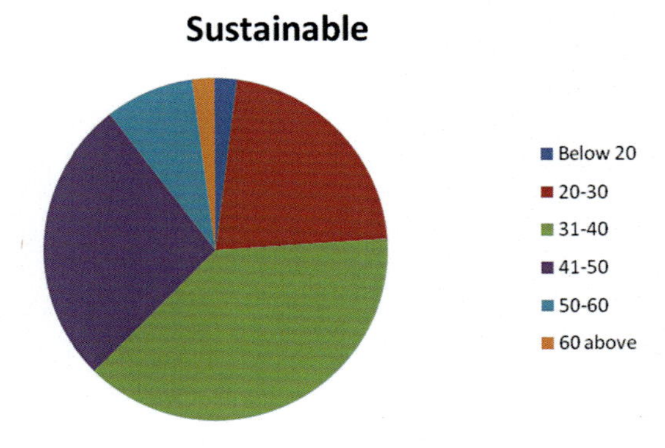

Fig. 15.6 Age wise distribution of responses for smart city interpretation

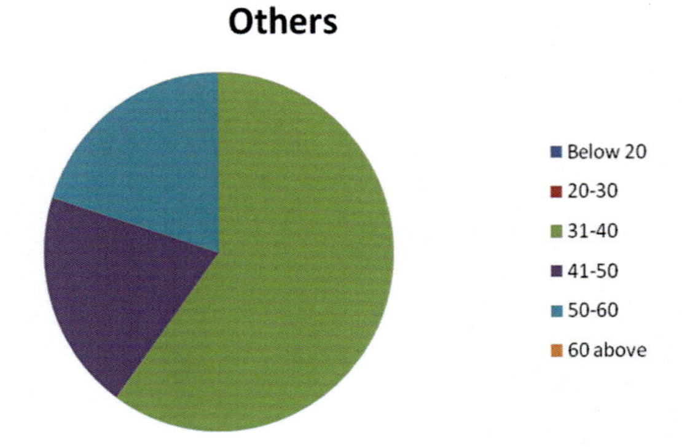

Fig. 15.7 Age wise distribution of responses for smart city interpretation

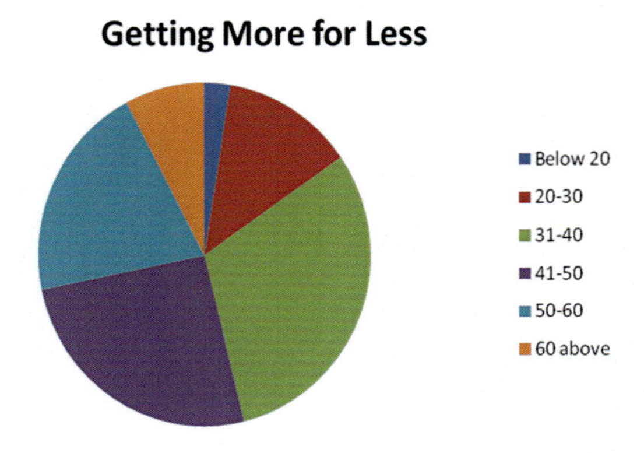

Fig. 15.8 Age wise distribution of responses for smart city interpretation

seen effective in studying the social category wise responses. It is therefore important that population structure has to be taken into consideration in implementing redevelopment planning strategy. As shown in Table 15.2, maximum unanswered questionnaires are seen in particular administrative wards; it is therefore needed to understand the literacy level and educational attainment of the population at each administrative ward. Also the structure of the ward has needs to be taken into consideration because as seen in Table 15.2 that Kasba–Vishrambagh ward has more unanswered questionnaires and the ward is mostly characterized by commercial activity so that the respondents were mostly business personnel as seen in Table 15.3 and may not have time to respond. It has been observed that the population in some administrative wards are much more involved in Private sector mostly the adjacent wards to Kasba–Vishrambagh and Bhavani Peth while the dependency level is also lower in Kasba–Vishrambagh ward. The maximum level of awareness about a smart city and its pillars is reported in Bhavani Peth and Bibwewadi, i.e., 100% while the minimum awareness level is observed in Ghole Road (53%) and Wanawadi (53%) Administrative Wards (Table 15.4). As compared to a smart city and its pillars and sustainability concept, it is observed that maximum awareness is about the smart city and its pillars than that of the concept of sustainability (Table 15.5). The interpretation of a smart city shows that mostly the respondents perceive sustainable and equitable environment and hence it is the responsibility of policy makers to plan the smart city strategy according to the interests of city's stakeholders. The age wise distribution of responses shows that 31- 40 age group have high no. of responses for all the five different expectations from smart city (Figs. 15.3, 15.4, 15.5, 15.6, 15.7 and 15.8).

15.9 Conclusion

Pune ranks first in the ease of living index 2018. Hence there is a strong possibility that the wave of population will welcome the change. On the one hand, Pune is considered as the most liveable city in contrary its ranking being a smart city shows negative results. Therefore, there is a need to find the gap between planned strategy of development and an actual scenario of development. This gap points toward the city's progress in its drive toward the achievement of the city's vision in terms of both administrator's vision and resident requirements. This gap will also denote the level to which the city would need to improve to become truly a smart city according to its main functions. Changing population at a faster pace creates various problems and pressure on available resources. To achieve smart city goals and for the better administration and planning, it is essential to correlate various urban services with the present population and future population. Besides this, each part of the city should be treated separately as to cater to the demands of local people. The population structure has to be considered for implementation of any development strategy. As in the case of present study area the total population of STs is considerably low. Hence it is important that smart city strategies should consider inclusive development. There is a need to make people aware to the Smart City Components, its major pillars, and take an initiative to increase more participation of citizens in the process of making a smart city.

Declarations

Funding: Indian Council of Social Science Research, New Delhi

Conflicts of Interest/Competing Interests: Not Applicable

Availability of Data and Material: Pune Municipal Corporation, Pune Smart City Development Corporation Limited, PMC Open Data Store

Code availability: Arc-map ESRI, MS-office, ZOTERO

Authors' Contributions: This research article is one of the outputs of ICSSR funded project which basically deals with sustainability aspects in Pune Municipal Corporation. The author contributes in the form of processes of rejuvenation in the study area with consideration of achieving sustainability parameters. Considering the constant change in the urban population the basic or essential services distribution and population distribution can help for better administration. In the present scenario of Covid-19 spread if population dynamics and health infrastructure management is done appropriately, then it can be a great assistance to administrators.

References

Baffoe G, Mutisya E (2015) Social sustainability: a review of indicators and empirical application. Environ Manag Sustain Dev 4(2):242. https://doi.org/10.5296/emsd.v4i2.8399

Bhattacharya S, Rathi S (2015) Reconceptualising smart cities: a reference framework for India. Center for Study of Science, Technology and Policy, September 2015. https://niti.gov.in/writereaddata/files/document_publication/CSTEP%20Report%20Smart%20Cities%20Framework.pdf

City Population Census (2001/2011) P.M.C., Pune

Colantonio A (2007) Social sustainability: an exploratory analysis of its definition, assessment methods, metrics and tools. In: Oxford Brookes University, 2007/01: EIBURS Working Paper Series, 37 July 2007

Doğu FU, Aras L (2019) Measuring social sustainability with the developed MCSA model: Güzelyurt case. Sustainability 11(9):2503. https://doi.org/10.3390/su11092503

Ease of Living Index (2018) Ministry of Housing and Urban Affairs (MoHUA), Government of India

Ho DCW, Yau Y, Law CK, Poon SW, Yip HK, Liusman E (2012) Social sustainability in urban renewal: an assessment of community aspirations. UrbaniIzziv 23(1):125–39. https://doi.org/10.5379/urbani-izziv-en-2012-23-01-005

Janubova B, Gress M (2018) Urbanization of poverty and the sustainable development of urban areas in Chile, p 14

Křižnik B (2018) Transformation of deprived urban areas and social sustainability: a comparative study of urban regeneration and urban redevelopment in Barcelona and Seoul. UrbaniIzziv 29(1):83–95. https://doi.org/10.5379/urbani-izziv-en-2018-29-01-003

Martinez-Fernandez C, Kubo N, Noya A, Weyman T (2012) Demographic change and local development: shrinkage, regeneration and social dynamics. OECD

Murgante B, Borruso G (2015) Smart cities in a smart world. In: Stamatina ThR, Pardalos PM (eds) Future city architecture for optimal living, vol 102, Springer optimization and its applications. Springer International Publishing, Cham, pp 13–35. https://doi.org/10.1007/978-3-319-15030-7_2.

Omann I, Spangenberg JH (2002) Assessing social sustainability. In: 7th biennial conference of the international society for ecological economics, vol 20. Sousse, Tunisia
Severson M, de Vos E (2018) A measurement framework: social sustainability in social and affordable housing. In: Conference Proceedings, vol 17

Chapter 16
Urban Wetlands: Opportunities and Challenges in Indian cities—A Case of Bhubaneswar City, Odisha

Prashna Priyadarsini and Ashis Chandra Pathy

Abstract Urban wetlands function differently than those in natural areas. In natural wetland, the water level is not changing rapidly until and unless natural events occur in urban context, and the water level of wetlands can fluctuate more rapidly due to anthropogenic activities which affect the ecosystem services of the wetland. The landscape of urban wetlands does not only have the role of carbon sink, water accumulation, cleaning and drainage but also it binds the nature with city dwellers. Yet, as built up spaces within urban areas have increased, these treasures have undergone a drastic decline. Analysis of published land use and land cover data from 22 cities by Wetlands International South Asia team, indicates that during 1970 to 2014, every one square kilometre increase in built-up area matched up with a loss of 25 ha wetlands. The main thrust of the study is to analyse the major causes behind wetland loss in the capital city and also to assess the wetland ecosystem services for the existing wetlands, its conservation and management.

Keyword WRAP (Wetland Rapid Assessment Procedure) · Wetland canopy · Wetland buffer · Wetland hydrology · Green infrastructure · Rainwater harvesting (RWH)

16.1 Introduction

Natural wetlands are in declined rapidly between 1970 and 2015, both inland and marine/coastal wetlands declined by 35%, three times the rate of forest loss. Contrast to this, human-made wetlands almost doubled in this period, now forming 12% of wetlands. The state of Odisha has 78,440 wetlands, of which 66% are Inland wetlands and 24% are Coastal wetlands. 6,90,904 ha of land was covered by total wetlands. Out of 66% inland wetlands, 34% are natural and 32% are manmade. Out of 24%

P. Priyadarsini
Dept. of Planning, College of Engineering and Technology, Bhubaneswar, India

A. C. Pathy (✉)
Dept. of Geography, Utkal University, Bhubaneswar, India

N. C. Jana et al. (eds.), *Livelihood Enhancement Through Agriculture, Tourism and Health*, Advances in Geographical and Environmental Sciences,
https://doi.org/10.1007/978-981-16-7310-8_16

Coastal Wetlands, 20% are Natural and 4% are manmade. 66,000 wetlands are of Small Categories because they have area of <2.25 ha. Bhubaneswar city is known as the "Temple City" and has more than 500 temples mostly confined to the old part of the city. The favourable physical environment created by surrounding rivers and streams along with undulating topography of the city has facilitated the existence of numerous water bodies which are a part of the excellent drainage network. It has been observed that there has been the progressive disappearance of wetlands from the city area, within the expanding municipal limits and from the neighbourhood also. This approach has been adversely impacting the drainage system of the city. Water logging and Urban Flooding, which was never heard of, have now become a regular feature during monsoon. Bhubaneswar is now known as one of the hottest city of India which is partly attributed due to the rapid loss of wetlands known for their ability to moderate local climate.

16.2 Study Area

The present area of study is the Bhubaneswar city bounded by Bhubaneswar Municipal Corporation (BMC) limits. This is in fact a part of a greater urban complex formed by Cuttack Municipal Corporation, Khurda Municipality, Jatni Municipality and Pipili NAC located at boundary corners with Bhubaneswar being at centre (Table 16.1). It is located between latitude 20°125′46″ to 20° 225′03″ N and longitude 85°45′24″ to 85°54′21″ E in the eastern part of Odisha (Fig. 16.1).

Table 16.1 Information regarding study area

District	Khorda
Demography	
Total Population 2001 (nos.)	6,47,302
Total Population 2011 (nos.)	8,40,834
Population Density (persons per km²)	2131
Number of Households (nos.)	2,01,873(Census 2011)
Avg. Household Size (nos.)3	4
Sex Ratio4	892
Slum Information	
Number of Slum settlements (nos.)	Authorized Slums: 116
	Unauthorized Slums: 320
Slum Population 2011 (nos.)	3,01,611
Slum Population as a percentage of total population (%)	35
Location, Climate & Topography	
Area (km²)	186
Climatic Zone	Tropical savanna climate
Soil Characteristics	Deltaic Alluvial
Ground Water Table (below ground level) (m)	7 m below ground level
Avg. max Temperature (°C)	37.2°C
Avg. min Temperature (°C)	15.6°C
Annual mean Rainfall (mm)	1436.1

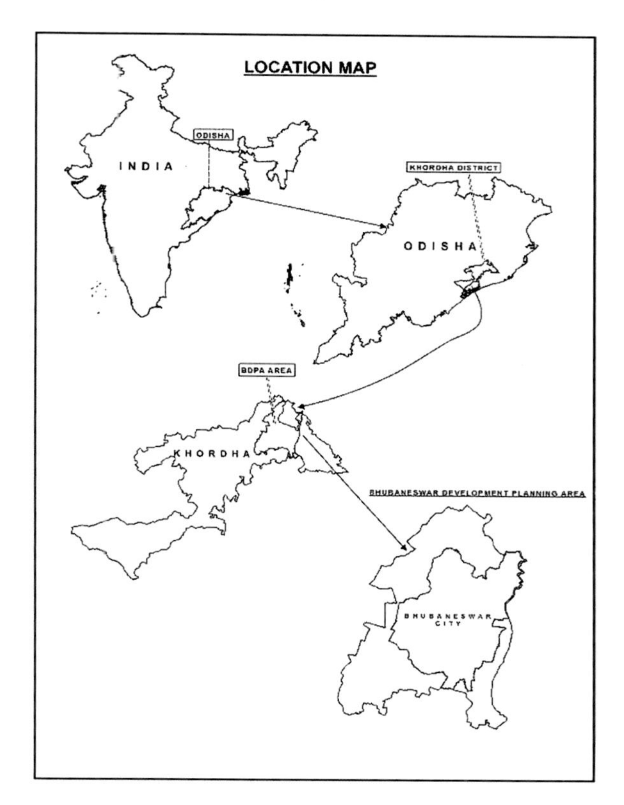

Fig. 16.1 Location of the study area

16.3 Objectives of the Study

Main objectives of this work are as follows:

1. To identify the existing Wetlands of the city.
2. To analyse wetland changes in the city.
3. To assess the wetland ecosystem services for the existing wetlands.
4. To find out the causes of wetland changes in the city and measures to conserve them.

16.4 Database and Methodology

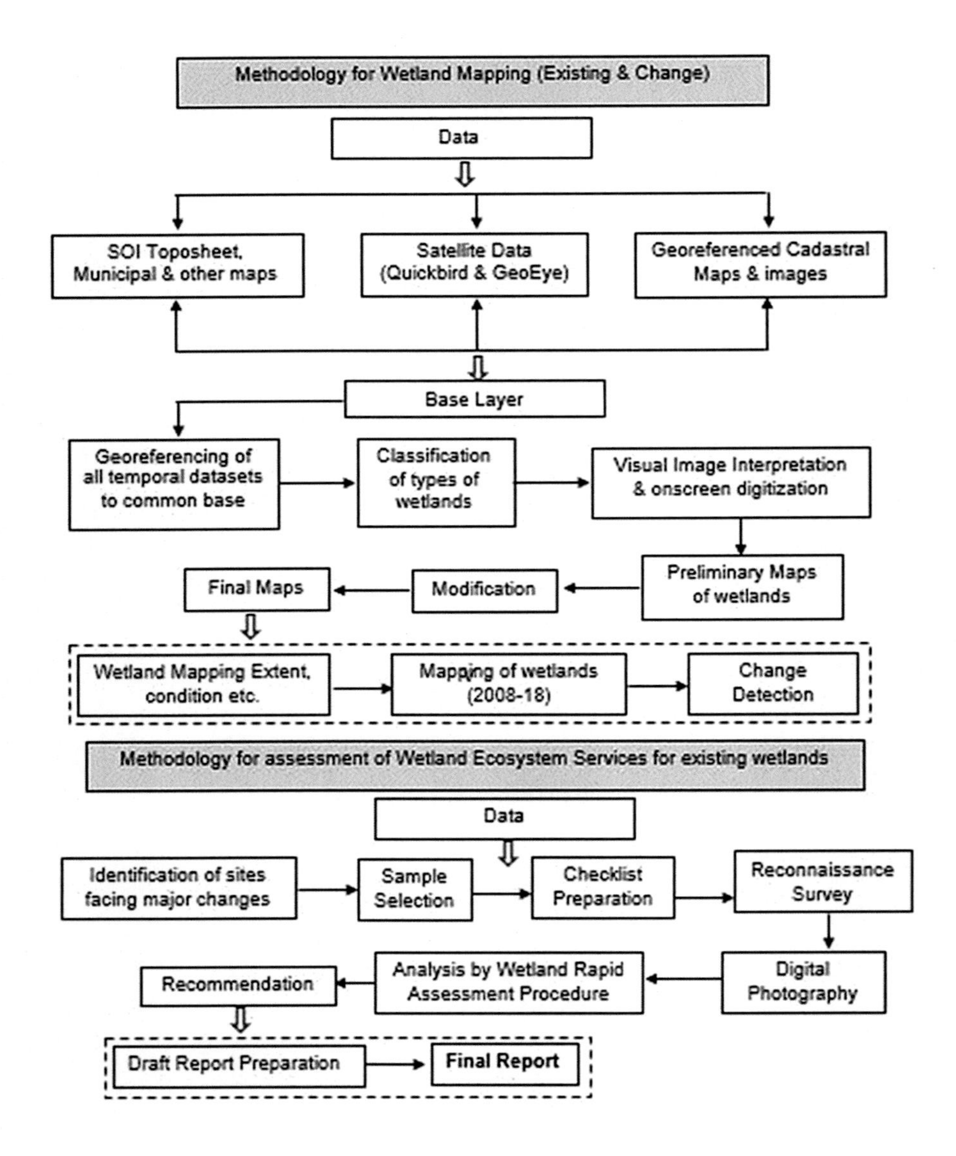

16.5 Identification of Sites

Through the help of satellite imageries, preliminary identification of existing wetlands got identified. Those wetlands include both natural and manmade wetlands. The same approach has been adopted for the identification of wetlands during the year 2008 and 2013 (Fig. 16.2), in order to analyse the changes that the city faced during the given decade.

The city is facing maximum degradation in both natural and manmade wetlands in NORTH & SOUTH-EAST ZONES. Hence, maximum number of samples from these two zones is taken into consideration for checking the degradation factors & ecological health of city wetlands. From the above selected two zones, only Man-made Wetlands were selected for study as it is possible to get water quality data regarding them from secondary sources. Man-made Wetlands include Tanks/Ponds (Perennial and Dry), waterlogged areas found in the City. An exception to this is that, city LAKES are also included because they have significant contributions to the city.

Around 127 waterbodies were found in the city limit, out of which here only public ponds are considered i.e. excluding private ponds due to unavailability of data. Ponds having legal issues are excluded. Such as

- Maa Duladei Pond (Mancheswar—Litigation issue)
- Chili Pokhari (Raja Rani—Litigation issue)
- 7th Battalion Pond (Samantarapur—Restricted Entry)

After deducting private ponds, there only left **67** ponds/tanks. **3** major lakes of the city also included in the study i.e. Ekamrakanan Lake, Vanivihar Lake and the BDA nicco park Lake.

YEAR	CATEGORY	NORTH ZONE	S-E ZONE	S-W ZONE	TOTAL NO.
2008	Natural Wetland	44	26	17	87
	Lakes	1	1	1	3
	Ponds	49	76	24	149
	Waterlogged areas	5	6	5	16
2013	Natural Wetland	30	13	12	55
	Lakes	1	1	1	3
	Ponds	40	73	22	135
	Waterlogged areas	8	13	11	32
2018	Natural Wetland	25	10	11	46
	Lakes	1	1	1	3
	Ponds	36	72	21	127
	Waterlogged areas	14	22	13	49

contributions to the city.

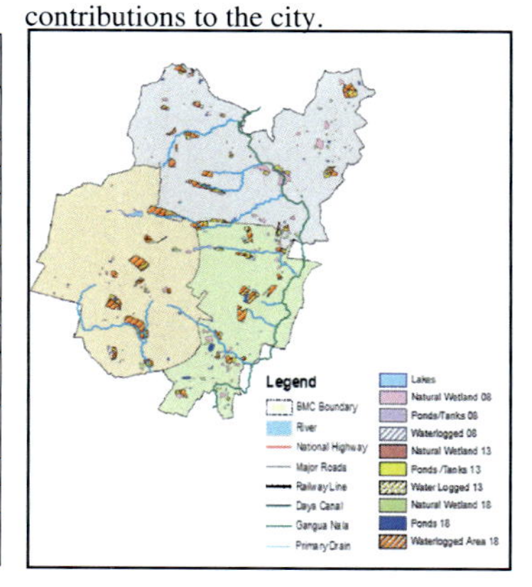

Fig. 16.2 Zone wise change in number of wetlands (2008–18)

Table 16.2 Selected ponds/tanks for study

Sl No	Name	Location	Area in (ha)	Perimeter in (m)
1	Bindusagar	Old town	8.84	1252
2	Kapileswar tank	Kapileswar	1.09	494
3	Fishery tank	Laxmi Sagar	7.43	1440
4	Damana Tungi pond	Damana	0.11	129.4
5	Maa Badei tank	Palasuni	0.2	181
6	Nilakantheswar pond	Baramunda	0.79	332
7	Nayapalli pond	Nayapalli	0.63	317
8	Matha pond	GGP colony	0.19	174
9	Gosagareswar tank	Gautam Nagar	0.18	171
10	Sundarpada tank	Sundarpada	5.49	1090
11	Jharpada pond	Jharpada	0.31	220
12	Gadakana pond	Gadakana	0.25	203.7
13	OUAT pond 1	Baramunda	0.33	226.4
14	OUAT pond 2	Baramunda	0.22	118
15	OUAT pond 3	Baramunda	0.24	197
16	Rangamatia pond	Garkana	1.3	488

Source BMC, Bhubaneswar

16.6 Sample Selection

Selection of ponds for sample survey was done on the basis of these following parameters (Table 16.2). Such as utility of citizens (domestic/rituals/heritage), technical availability (availability of water/source), necessity and availability of surrounding land, encroachments, population density. Hence, 16 tanks/ponds were selected as sample, which is approximately 25% of the total selected tanks. 17 waterlogged areas are identified which are considered as wetlands under manmade types.

16.7 About WRAP

"Habitat Assessment Variable" methodology is a series of discussions one for each WRAP assessment variable. Each variable is a rating index containing a set of calibration descriptions and corresponding score points; **3** is considered the best a system can function and **0** is for a system that is severely impacted and is exhibiting negligible attributes. An evaluator also has the option to score each variable in half (0.5) increments. This provides the flexibility to score a variable that is not accurately described or fitted by the calibration description. Half increments are utilized on the point scale from 0.5 through 2.5.

Table 16.3 Status of existing wetlands (2018)

Existing wetlands (2018)	
Types	Area covered (in ha)
Natural wetland	5.58
Lakes	9.81
Ponds	141.2
Waterlogged areas	1835.5

Each applicable variable is scored: the scores are totalled (åV) and then åV is divided by the total of the maximum score for that variable (åVmax). The final rating score for "Habitat Assessment Variables" will be expressed as a number between 0 and 1. Here Wildlife Utilization (WU), Wetland Canopy (WC), Wetland Ground Cover (WGC), Wetland Buffer (WB), Wetland Hydrology (WH), and Wetland Quality Input and Treatment (WQIT) are taken as variables for ecosystem service analysis through reconnaissance survey.

– The formulae of final rating score are as follows:

WRAP Score = sum of the scores for the rated variables (V)
sum of maximum possible scores for the rated variables (Vmax)
also expressed as ***åV/åVmax***.

16.7.1 Mapping of Existing Wetlands

The existing wetlands of the city delineated with the help Quickbird and GeoEye (Table 16.3). Then the information got imported to GIS layer. The major types of wetlands found in the city (Figs. 16.3, 16.4 and 16.5) are inland wetlands (Level I) i.e. natural wetlands include lakes, river, waterlogged areas and man-made wetlands include ponds/tanks, waterlogged area (Table 16.4).

16.7.2 Spatio—Temporal Changes of Wetlands Over a Decade

16.7.3 Factors Affecting Wetland Changes

- **Increase in the rate of Urbanization** (Figs. 16.6 and 16.7)
- **Increase in Population**

Population has played vital role for the cultural manifestation of landscape as per the demand of the people. The distribution and density of the population has been increased from 1991 to 2021 (Figs. 16.8 and 16.9).

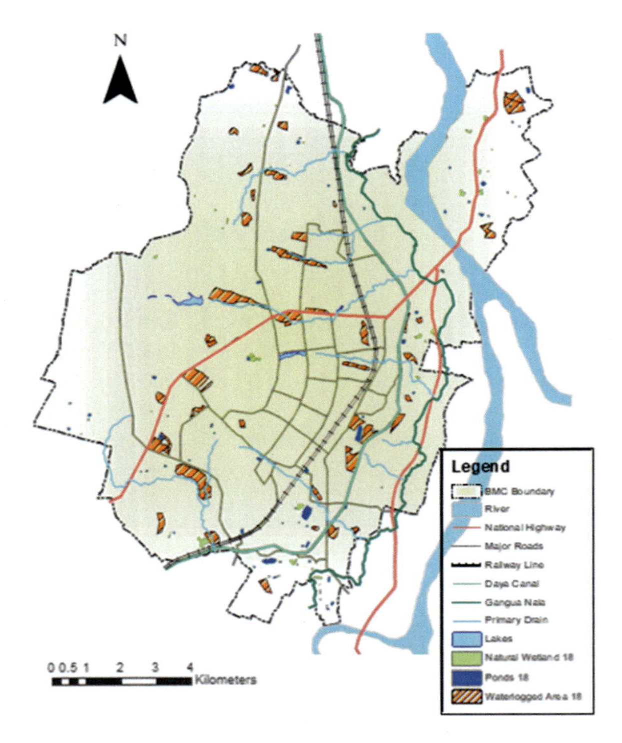

Fig. 16.3 Mapping of existing wetlands (2018)

- **Violation in the city**

  The city has witnessed violation in change of slum pockets from 1991 to 2011 which shows the slum population and growth rate has been increased over decades (Fig. 16.10).

- **Encroachment on wetlands**
  - The entire area of wetland connecting Ratha Road (Fig. 16.11), Mausima Temple Square and Lingaraj Temple (Fig. 16.12) was partly converted into housing colonies whereas another patch of wetland is being filled in an attempt to reclaim it. Earth and waste material is being dumped into the wetland so that a house may be built on it, and the waste is actually municipal garbage (Figs. 16.13, 16.14, 16.15 and 16.16).
  - Natural water body 'Nayapalli Haza' has shrunk to **3 acres** from **7 acres.** Due to construction activities on this 4 acres of land.
  - '**Fishery pond**' at Laxmisagar of **18.99 acres** has now shrunk to **12 acres** and '**Sundarpada Haja**' of **17.56 acres** shrunk to **11 acres**.

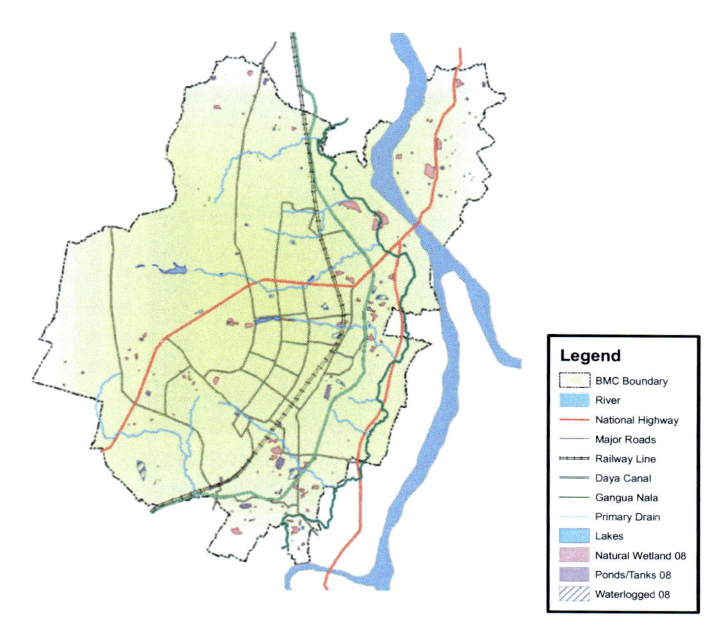

Fig. 16.4 Status of wetlands (2008)

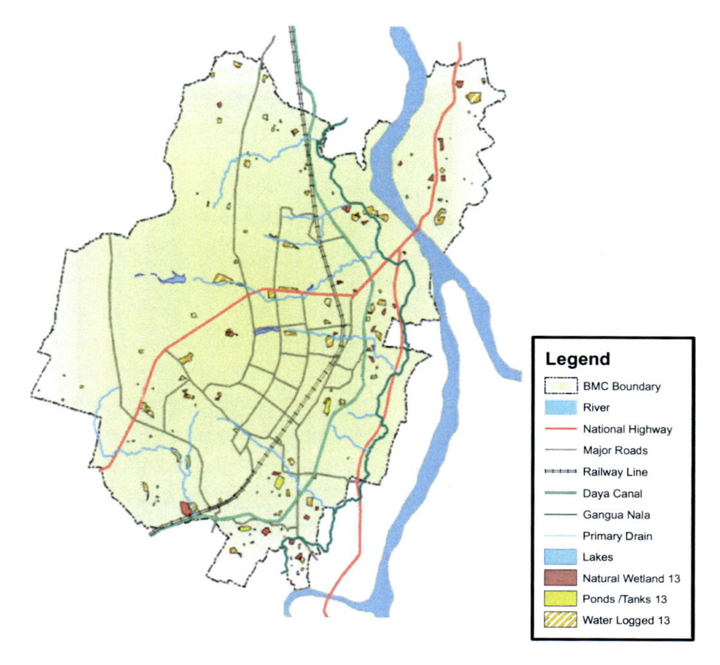

Fig. 16.5 Status of wetlands (2013)

Table 16.4 Total wetlands in the study area

Status of wetlands			
Year	Category	Area covered (in ha)	% of total area
2008	Natural wetland	137.7	1.02
	Lakes	9.81	0.07
	Ponds	163.3	1.21
	Waterlogged areas	492.75	**3.65**
	Total	**803.56**	**5.95**
2013	Natural wetland	86.8	0.64
	Lakes	9.81	0.07
	Ponds	147.15	1.09
	Waterlogged areas	972.06	7.2
	Total	**1215.82**	**9.04**
2018	Natural wetland	5.58	0.03
	Lakes	9.81	0.05
	Ponds	141.2	0.75
	Waterlogged areas	1835.5	**9.87**
	Total	**1992.41**	**10.7**

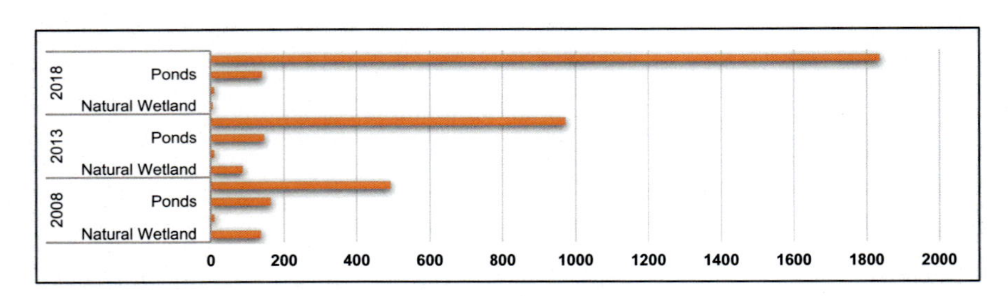

Fig. 16.6 Wetland changes (2008–18)

- **Change in Land Use**

  The violation in land use in the city has been increasing day by day Geetika and Anushree (2017). Here is the brief information about land use changes as per CDP, 2030 and the current use (Table 16.5). The deviation itself shows the level of violation. Maximum number of encroachments are occurring around wetlands and waterbodies because these are the soft targets for urban developments as the land prices are less.

- **Waterlogged areas**

  Most of the city is having a very low elevation. In addition to this, a prominent rainy season coupled with a lack of drainage system (only 45% coverage of drains), the city faces water logging issues at numerous places (Table 16.6). The areas are

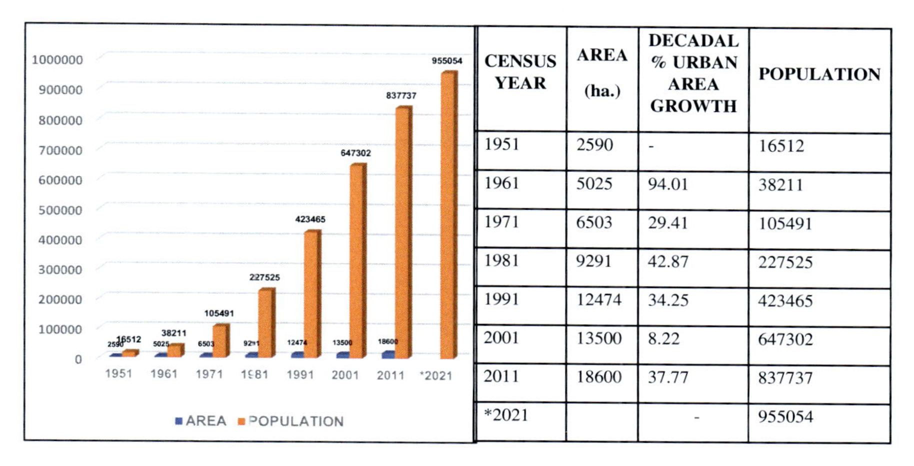

CENSUS YEAR	AREA (ha.)	DECADAL % URBAN AREA GROWTH	POPULATION
1951	2590	-	16512
1961	5025	94.01	38211
1971	6503	29.41	105491
1981	9291	42.87	227525
1991	12474	34.25	423465
2001	13500	8.22	647302
2011	18600	37.77	837737
*2021		-	955054

Fig. 16.7 Urban area and population change in the city (*Source* Various Census reports of GOI, 1951–2011)

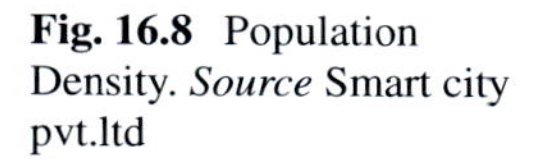
Fig. 16.8 Population Density. *Source* Smart city pvt.ltd

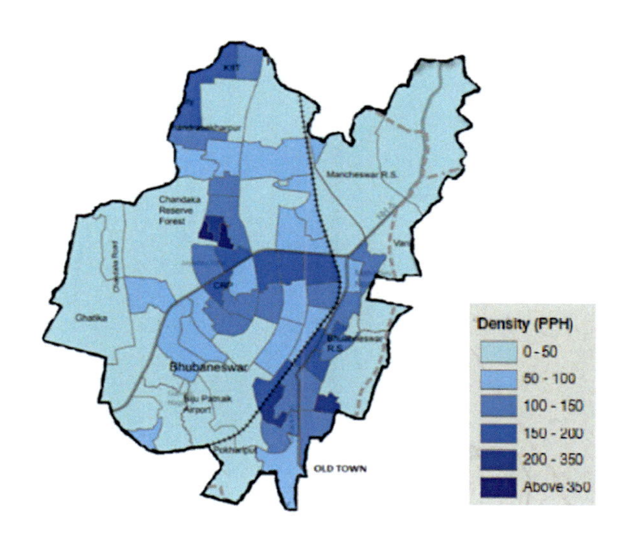

follows:

- Waterlogging prone areas where visits were made are Behera Sahi (Nayapalli), ISKCON, Ekamra Kanan, Nicco Park Nala at Acharya Vihar, Fire Station Chak.
- (Near Paika Nagar) and Satabdi Nagar Chhak.
- Palasuni, GGP Colony, Laxmisagar and Jharpada.

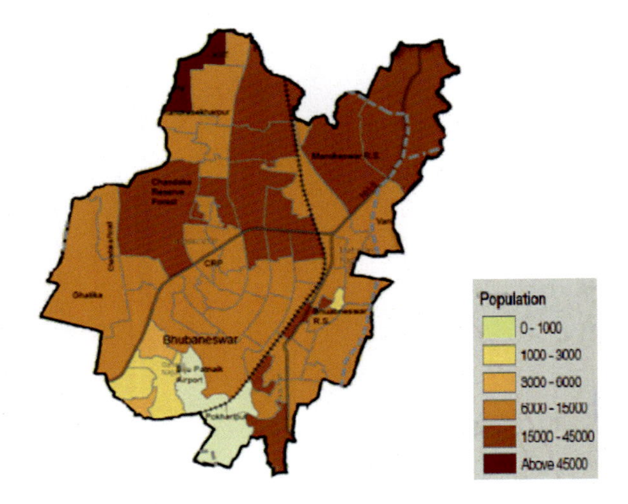

Fig. 16.9 Population Distribution. *Source* Smart city pvt.ltd

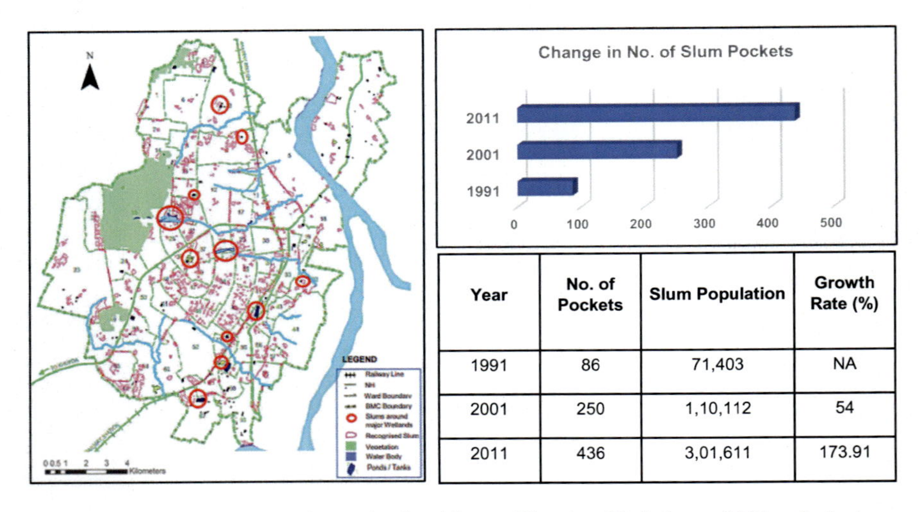

Year	No. of Pockets	Slum Population	Growth Rate (%)
1991	86	71,403	NA
2001	250	1,10,112	54
2011	436	3,01,611	173.91

Fig. 16.10 Major slums around waterbodies (*Source* Planning, Violation and Urban Inclusion, a case study for Bhubaneswar)

– The cause of water logging in most of the above areas was that the drains were fully covered and the inlets into the drains were either too small or choked by solid waste.

- **Encroachment on Drains**

Fig. 16.11 Natural wetland along Rath road (*Source* Physical Survey, March 2019)

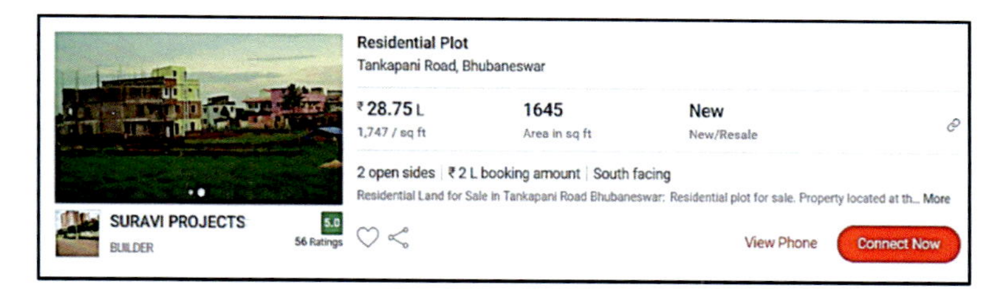

Fig. 16.12 An advertisement for sale of the wetland (*Source* Google photos)

Fig. 16.13 Encroachment along fishery pond (*Source* Physical Survey, March 2019)

The Comprehensive Drainage Plan prepared by the Bhubaneswar Development Authority indicates that the major drains of the city are **drain no. 01, 04, 08, 10** (Fig. 16.17). The widths of the master drains vary from 2 to 4 m at starting points and 4 to 13 m at outfall points. The average velocity of water flow in these drains vary from 2.5 to 3 m/sec. The depth of drain varies around 1.5 m at start and around 3 m at outfall points. The drain no. 01 discharges maximum water at outfall followed by drain 04 and 10 (Fig. 16.18).

Fig. 16.14 Encroachment along Nicco park lake. (*Source* Physical Survey, March 2019)

Fig. 16.15 OUAT pond. (*Source* Physical Survey, March 2019)

Fig. 16.16 Damana Tungi Matha Pond. (*Source* Physical Survey, March 2019)

Table 16.5 Land use violations in the city

WATERBODIES USE ZONE (W)				
AS PER CDP 2030			ACTITIES GOING ON	AS PER STUDY AREA
Uses / Activities Permitted	Uses/Activities Permissible on application to the Competent Authority	Uses/Activities Prohibited		Current use
Rivers, Canals	Fisheries	Use/activity not specifically related to water bodies use not permitted herein.		Fisheries, Sand Extraction, Solid waste dumping, discharge of waste water, Encroachment along them
Streams, Water springs	Boating, water theme parks, water sports, lagoons	All uses not specifically permitted in column (a) and (b)		NA
Ponds, Lakes	Water based resort with special bye laws	NA		Waste Dumping, Encroachment, discharging domestic waste water, infrastructures upon ponds.
Wetland, Aquaculture Ponds	Any other use/activity incidental to Water Bodies Use Zone is permitted	NA		Waste Dumping, Encroachment, discharging domestic waste water, infrastructures upon ponds
Reservoir	NA	NA		NA
Waterlogged areas / Marshy Area	NA	NA		Encroached, High densified residential area & others used zone

Table 16.6 Waterlogging points of the city

Area	No. of points	No. of times waterlogging in year
Key road intersections	Siva Nagar, Sriram nagar, Lane, Laxmi sagar, Sundarpada	4
Along roads (50 m length or more)	Nayapalli, Acharya vihar, Near Isckon Temple, Jgannath nagar, Bhagabat sandhan	2
Locality (afecting more than 50 HH)	Brahmeswarpatna, Gada mahavir road	<25

Source AMRUT Report

16.7.4 Assessment of Wetland Ecosystem services

Quality Level of all Parameters

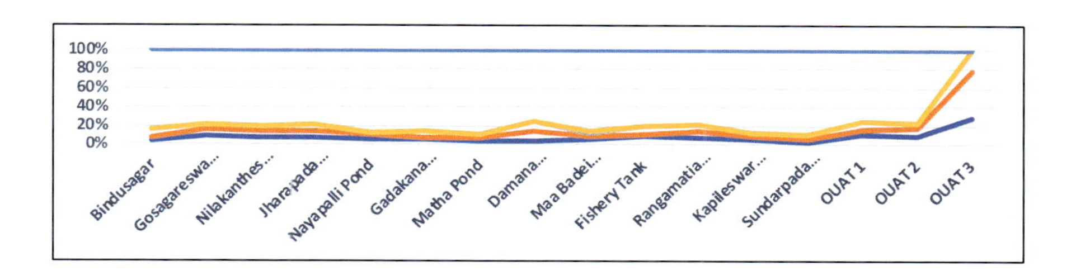

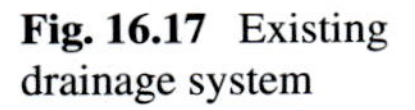

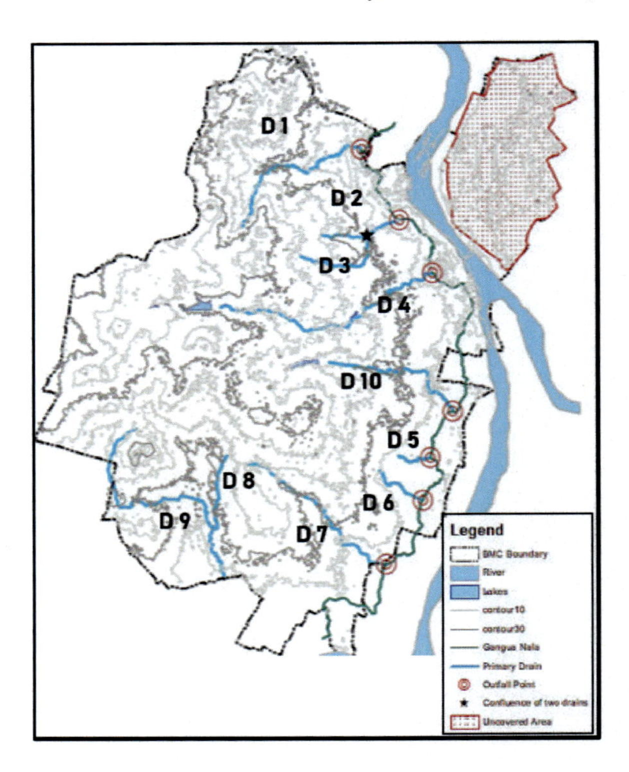

Fig. 16.17 Existing drainage system

1. Wetland loss (except waterlogging areas) is **1.47%** (2.3% in 2008 to 1.8% in 2018). Maximum Percentage of loss in Natural Wetlands i.e. **0.99%** over that decade.
2. Rapid increase in Water logging areas i.e. **4.75%**. Maximum Percentage of increase in Waterlogged wetlands due to rapid urbanisation, encroachment along the drains so also on the natural wetlands.
3. During the same decade the Municipality Boundary has also been increased **51 sq km i.e. 5100 ha**.
4. High pH value results in damage of aquatic lives so also increase in Plankton and Mosses and low pH value causes disappearance of fishes. High BOD means there is high influence of sewage i.e. presence of great amount of micro-organisms. Hard water is not a health risk indicator rather shows that water is not fit for bathing and washing purposes. Low DO results from presence of higher amount of algae and bacteria in the water which ultimately kills aquatic organisms. Too much nitrate leads to restriction of oxygen transport to the bloodstream and can cause overgrowth of aquatic plants and algae which results in decrease in animal and plant diversity (Fig. 16.19).

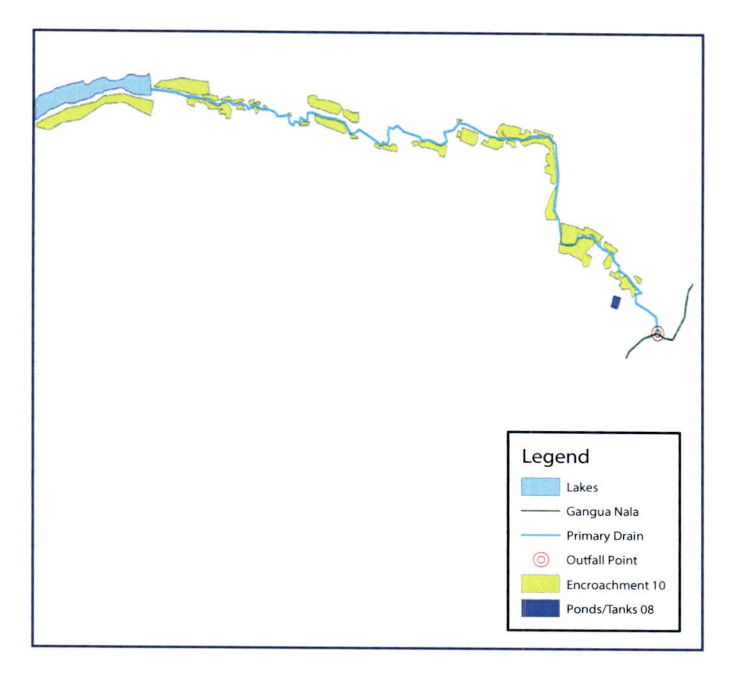

Fig. 16.18 Encroachment pattern along drain no 10

16.7.5 Decoding Collected Data

Calculating the data collected through reconnaissance survey for selected samples of wetlands (Tables 16.7, 16.8, 16.9 and 16.10). Here are the results.

WQIT Calculation done by considering surrounding **LAND USES (LU)** and onsite **PRE-TREATMENT (PT)** (Table 16.7).

WQIT = (% surrounding × LU1) + (%* surrounding × LU2) + … (n) = LU & (%* surrounding LU × PT1) + (%* surrounding LU × PT2) + … (n) = PT

Hence, WQIT = (LU total + PT total)/2(%* expressed in decimal)

LU = (0.75 × 1) + (0.5 × 1.5) + (0.20 × 1.5) = 1.8 & PT = (0.75 × 0) + (0.5 × 1) + (0.20 × 0) = 0.5

WQIT (BINDUSAGAR) = (1.8 + 0.5)/2 = 1.4

Hence, **WRAP Score** = sum of the scores for the rated variables (V)/sum of maximum possible scores for the rated variables (Vmax) i.e.

WRAP = V/Vmax and WRAP (Bindusagar) = 6.4/18 = **0.35**.

WQIT Calculation done by considering surrounding **LAND USES (LU) & onsite PRE-TREATMENT (PT)** (Table 16.8).

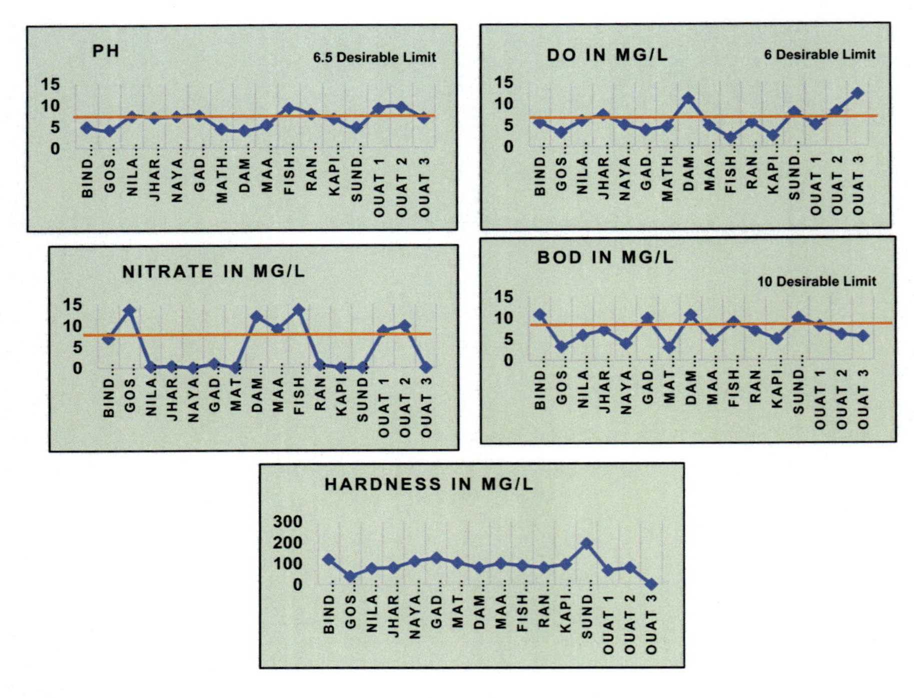

Fig. 16.19 Existing water quality analysis of the selected samples

Table 16.7 Calculation for Bindusagar

Bindusagar	WU	WC	WGC	WB	WH	WQIT
	1	1	1	1	1	1.4

Table 16.8 Calculation for Kapileswar

Kapileswar tank	WU	WC	WGC	WB	WH	WQIT
	1	2	1	2	2	2.3

Table 16.9 Calculation for Fishery tank

Fishery tank	WU	WC	WGC	WB	WH	WQIT
	0	0	0	0	0	0.9

WQIT = (% surrounding × LU1) + (%* surrounding × LU2) + … (n) = LU & (%* surrounding LU × PT1) + (%* surrounding LU × PT2) + …(n) = PT Hence, WQIT = (LU total + PT total)/2(%* expressed in decimal)

LU = (0.70 × 3) + (0.5 × 2) + (0.25 × 2) = 3.6 & PT = (0.70 × 3) + (0.5 × 0) + (0.25 × 0) = 2.1

Table 16.10 Calculation for Damana pond

Damana pond	WU	WC	WGC	WB	WH	WQIT
	0	0	0	0	0	0.9

WQIT (KAPILESWAR) = (3.6 + 2.1)/2 = 2.3

Hence, **WRAP Score** = sum of the scores for the rated variables (V)/sum of maximum possible scores for the rated variables (Vmax) i.e. WRAP = V/Vmax, So WRAP (Kapileswar) = 10.3/18 = **0.57**.

So, WQIT = (LU total + PT total)/2

LU = (0.75 × 1) + (0.20 × 1.5) + (0.5 × 1.5) = 1.8 & **PT** = 0 (no treatment)
WQIT (FISHERY TANK) = (1.8 + 0)/2 = 0.9
Hence WRAP SCORE = V/Vmax
WRAP (Fishery Tank) = 0.9/12 = **0.07**.
Like wise, WRAP SCORE of rest Ponds & Lakes has been carried out.
Nilakantheswar pond = 0.46
Nayapalli pond = 0.25
gosagareswara = 0.08
Sundarpada tank = 0.67
Damana pond = 0.05
Matha pond = 0.13
Ekamra lake = 0.48

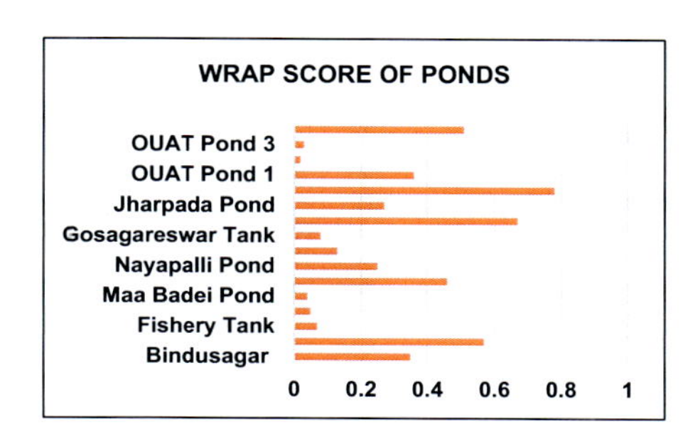

Jharpada pond = 0.27
Gadakana pond = 0.78
Rangamatia pond = 0.51
Ouat (1,2,3) = 0.36, 0.02, 0.03
Vanivihar lake = 0.05
Bda nicco park lake = 0.03

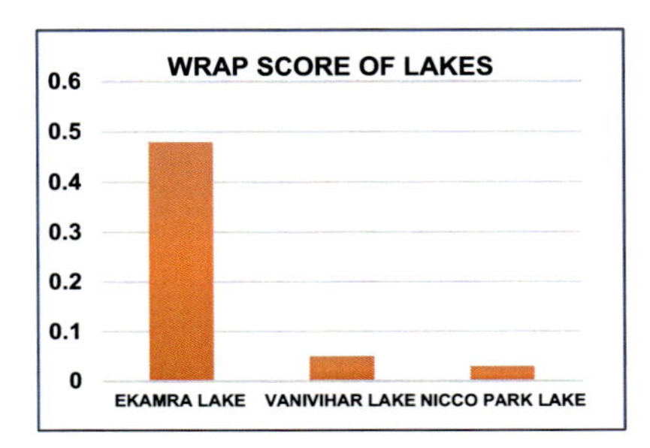

1. The wetland scores the least will need highest prioritization regarding its ecological health i.e. the **OUAT Pond 2**. The wetland scores the most must have best ecological health as compared to others i.e. the **Gadakana pond**. Here, Nicco park scores the lowest i.e. it needs higher prioritization.

16.8 Recommendations

16.8.1 City Level

(a) Removal of all types of encroachments of Wetlands, Lakes and Storm water Drains. Eviction should be done in phases with adequate time given to seek alternative economic opportunity and infrastructure. Registration of encroachers providing them basic infrastructure like Housing Colony, water and electricity. Rehabilitation packages needed to be provided.

EXISTING

(b) Maintaining Buffer around Wetlands including ponds, lakes and also along drains. Demarcating Drains, wetlands using pillars or fencing and protecting them from unwanted uses. It should be declared as **"NO DEVELOPMENT ZONE"** i.e. **50 m** from Primary Drains, **25 m** from Secondary Drains, **15 m** from Tertiary Drains, **75 m** around Lakes.

PRPOPOSED

Fig. 16.20 Major water sensitive techniques

(c) Preparing a Proper **Storm Water Drainage Plan** for the city, which includes increasing Drain Coverage (1036 km as per CDP), improving inlets into existing drains by unclogging inlet pipes, stopping the waste water entry into the Storm water drain, maintaining existing drains and increasing number of drains and adopting water sensitive techniques (Fig. 16.20).

(d) No disposal of wastes into the wetlands and only treated sewage into them and stopping entry of waste waters into waterbodies and storm water drain. Government of India (2013a, b) In the unsewered area i.e. **ward no 4** construction septic tanks with soak pits are needed and all the households should be connected to sewer network and Providing Dustbins around water bodies. Desilting drains, waterbodies in regular interval (at least before monsoon). Proposal should be developed to restore existing Natural Wetlands, Ponds and lakes (Fig. 16.21, 16.22, 16.23, 16.24, 16.25, 16.26 and 16.27).

(e) Restoration of existing natural wetlands, Ponds and lakes.

16.8.2 Community Level

(a) To Prepare Communities to handle their Local Wetlands.

(b) Adopting Local Wetlands

(c) Public Awareness Generation among target groups and adopting techniques in household level.

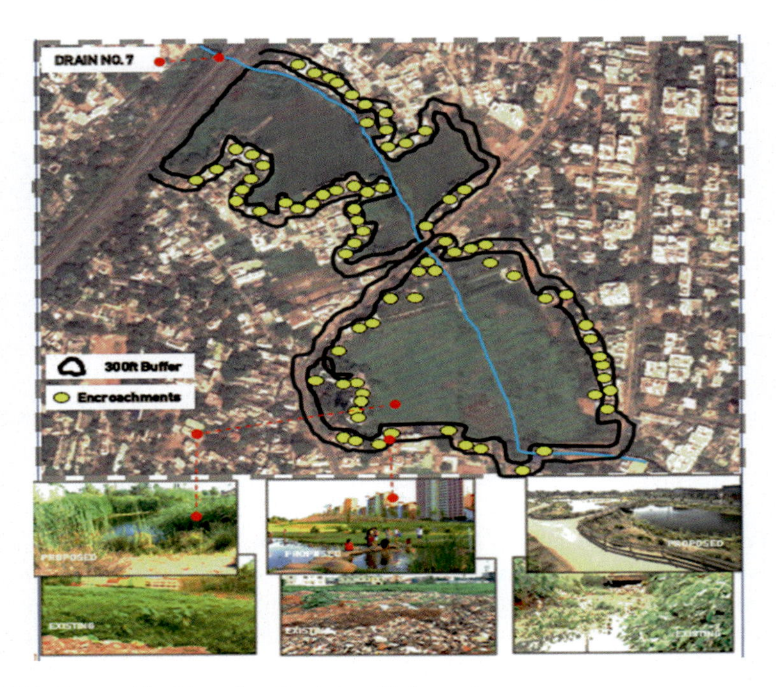

Fig. 16.21 Proposal for Natural wetland along Ratha road

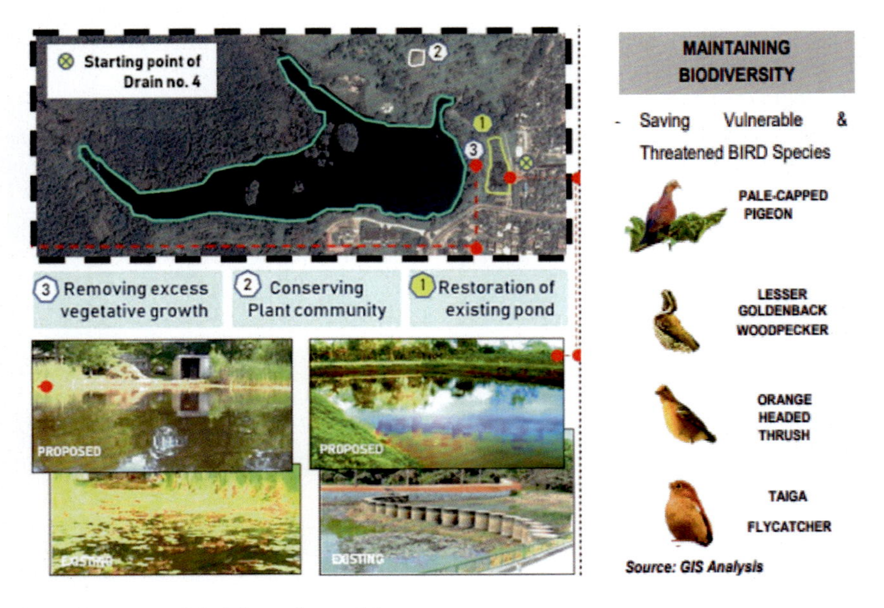

Fig. 16.22 Proposal for Ekamrakanan lake

Fig. 16.23 Proposal for Damana pond

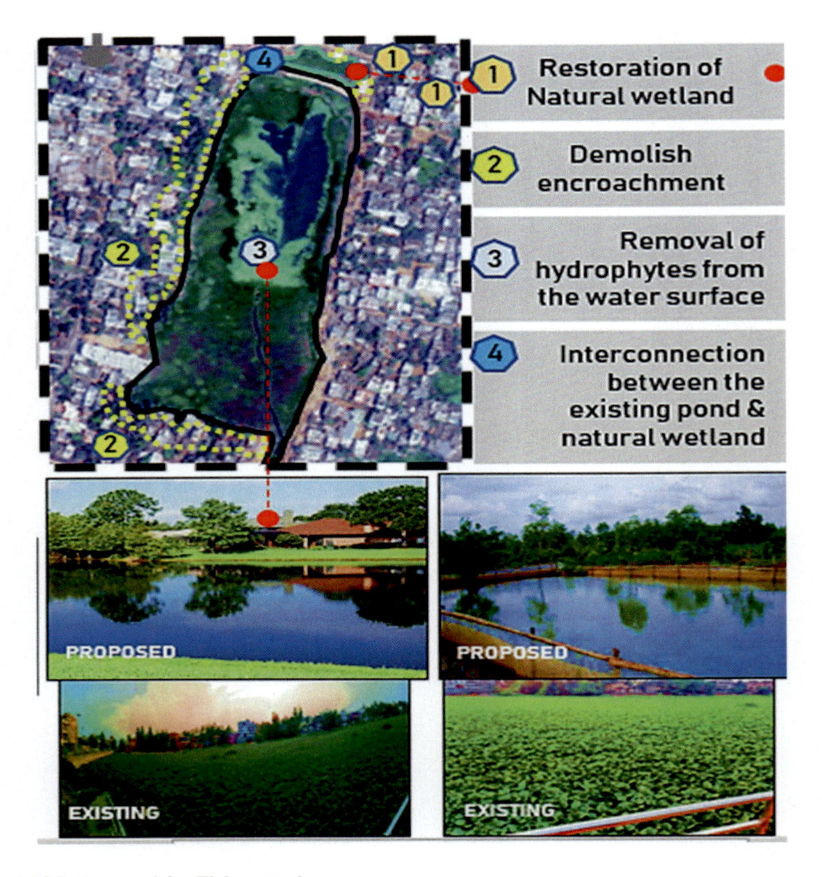

Fig. 16.24 Proposal for Fishery tank

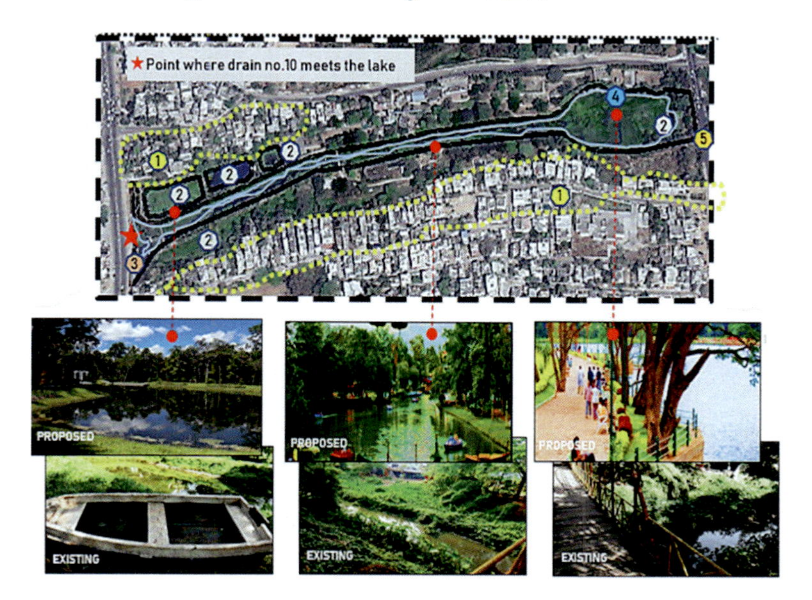

Fig. 16.25 Proposal for Vanivihar lake

Fig. 16.26 Proposal for Nicco park lake

Fig. 16.27 Proposal for OUAT pond

16.8.3 Policy Level

- Good Governance and Digitization of records.
- Violation of regulatory and prohibitory activities as per wetlands (conservation and management) rules, 2010; regulatory wetland framework, 2008: and Polluter pays Principle (National Environmental Policy 2006).
- Prevention of Pollution of lakes. (National Water Policy 2002) and Discharge of untreated sewage of lakes. Government of Odisha (2015)

16.9 Conclusion

Wetlands being diverse in nature, the conservation measures vary accordingly. Though a number of rules, summits have been organised just to save these precious natural resources in past few years, but the ever-growing urbanised societies are failing to cope with the degrading situation. It has been estimated that India will lose more than half of its present wetlands in 20 to 30 years. This above study is a tiny step towards conservation of wetlands in the capital city of Odisha. Because it is necessary to make a start towards this issue in the major cities and towns which are facing huge changes day by day. Here, an exploratory survey on 16 ponds and 3 natural lakes revealed that 98% of the present wetlands were found in very bad state. High pH, high BOD, low DO, hard water and too much nitrate are the characteristics of the present wetland water quality. So the work is based on three major levels such as (a) City level, (b) Community level and (c) Policy level, with an target to achieve sustainability of the measures that to be taken. The major concern is that to spread awareness among target groups i.e. the section of people who interact directly with the wetlands and the responsible Government should take up by incorporating good governance and rules and regulations in order to make some strict changes among the citizens regarding the natural resources. These basic and grassroot level changes can bring a huge amount of revolution regarding the wetlands whether it's a rural or urban area. We need to share the thought that **"Wetlands are not Wastelands rather these are Prized lands."**

References

Geetika A, Anushree D (2017) Planning, 'violations', and urban inclusion: a study of Bhubaneswar. Yuva & Indian Institute of human Settlement. Research Report
Government of India (2013a) Advisory on conservation & restoration of waterbodies in urban areas. CPHEEO. Ministry of Urban Development
Government of India (2013b) National wetland Atlas: wetlands of international importance under Ramsar convention. Ministry of Environment & Forests
Government of Odisha (2015) Detailed project report integrated sewerage system for Bhubaneswar city under JNNURM
National Environmental Policy (2006) Govt of India Report
National Water Policy (2002) Govt of India Report

Chapter 17
Analyzing Spatial Inequalities of Amenities in Jammu City Using Geo-Informatics

Rajender Singh, Kavleen Kaur, Sarfaraz Asgher, Davinder Singh, and Sandeep Singh

Abstract Since the partition of India and Pakistan, the City of Jammu has emerged as a most favorable and suitable destination to settle down, for the people who came from the other side of the border during 1947 and 1965 war between India and Pakistan and also for the people from inside the country thereafter. During this period the city of Jammu has shown unprecedented growth of population from 157,708 pers. in 1971 to 576,195 pers. in 2011. Keeping in view the flow of migrants and the widespread growth of population in Jammu City, current research is an attempt to analyze the availability and disparities in the spatial distribution of basic amenities using simple geospatial technique among the different wards of Jammu City. The outcome of the research shows that basic amenities among the different wards of Jammu City are not uniformly distributed. With a composite score of more than 43.47, about 26% of the total wards show high levels of overall development. There are only seven wards falling in the medium line. On the contrary, a highly drastic picture has been observed, as more than 63% of the total wards (i.e., 45 wards) show low to very low levels of overall development. Broadly speaking, old city wards are found to have adequate urban amenities when compared to peripheral wards. The findings of the current research are extremely useful to decision-makers for the comprehensive formulation of urban policy.

Keywords India · Pakistan · Jammu city · Unprecedented growth · Composite score · Overall development

17.1 Background

The City of Jammu has emerged as a most favorable and suitable destination to settle down, for the people who came from the other side of the border during 1947 and 1965

R. Singh (✉) · K. Kaur · S. Asgher · D. Singh
Department of Geography, University of Jammu, Jammu, India

S. Singh
Lecturer in Geography, School Education Department, Government of J&K, Jammu, India

N. C. Jana et al. (eds.), *Livelihood Enhancement Through Agriculture, Tourism and Health*, Advances in Geographical and Environmental Sciences,
https://doi.org/10.1007/978-981-16-7310-8_17

war between India and Pakistan. Apart from this it also supports huge population migrated within the erstwhile state of Jammu and Kashmir, and it is important to mention that it accommodates or is a home to more than 39,000 registered migrant families of Kashmiri Pandits in the year 1990 (Datta 2016). Jammu City being the state winter capital is acting as a magnetic force for the population of other districts of Jammu Division, as people of these districts wish to settle in Jammu City mainly for the following reasons: (a) Jammu is the winter capital of the State of Jammu and Kashmir as well as the economic hub for the whole Jammu Division, (b) Due to rapid growth of commercial and industrial activities it attract workforce from the surrounding districts, (c) Jammu is well advanced in Educational, Healthcare, and infrastructure facilities. All these factors are responsible for an unprecedented growth of Jammu City over the last four decades. During this period the city of Jammu has shown unprecedented growth of population from 157708 pers. in 1971, 549791 pers. in 2001 to 576195 pers. in 2011. The current study is an attempt to analyze the availability and disparities in the spatial distribution of basic amenities among the different wards of Jammu City, keeping in mind the flow of migrants and the widespread growth of population in the city.

17.2 Introduction

Being a key factor behind quality of life, amenities are defined as site- or region-specific goods and services that affect regional attractiveness for its workforce (Mulligan and Carruthers 2011). There are inequalities between spatial units just as there are between individuals. Basic services such as electricity, drinking water, sanitation, health care, and solid waste management are all important factors in determining urban quality of life (Bhagat 2010). In United States the first attempt to study urban amenities and quality of life was done after creating an elaborate index to rate the "goodness of life" in about some 310 American cities (Thorndike 1939). During the 1970s, Berry and Horton argued that amenities were playing an increasingly important role in the ongoing urban transformation of the American Sunbelt. Planners and administrators must monitor the chaotic growth pattern and changing land use along the urban–rural fringe, as well as within the densely populated urban core, in order to provide basic amenities and infrastructure for the complex and dynamic urban environment (Farooq and Ahmed 2008). Others have examined patterns of accessibility to certain services and the geographic relationship between service deprivation and area deprivation.

Urban inequality seems moderate for a majority of nations. However, there seems to be some evidence that urban inequality is greater in developing countries (Kim 2008). In India, urbanization is accelerating, with cities expected to reach 600 million people by 2030 and, according to the United Nations Population Fund, a large proportion of the population moving to urban areas by 2050 (Rana et al. 2018). Many empirical findings have shown that facilities are unequally distributed in our communities, resulting in a never-ending struggle for the vast majority of people to gain access to these infrastructures in order to improve their quality of life. The cities in developing countries have a wide range of spatial structures. In some areas, amenities are adequate, while in others, arrangements are inadequate. The results of the HRD Survey carried out by the Department of Economics of the University of Dares Salaam and the World Bank is very much true to many cities in Developing Nations. The Survey shows that the city runs the risk of splitting in two. On the one hand, the privileged population groups have access to quality infrastructure and services. On the other hand, the vast majority of low-income people are confronted with poor urban facilities and prohibitively expensive public transportation (Olvera et al. 2003). Living standards vary greatly depend on the availability and accessibility of basic infrastructure, both within and between regions and localities. The goal of urban planning is to ensure that all people have access to adequate and equitable services. Before we attempt to project and plan the future development of existing facilities in any region, we must first understand their nature and pattern of distribution.

Jammu District has about 12% of the population of Jammu and Kashmir, making it the most populous district in the state. Nearly half of the population lives in urban areas, while the other half lives in rural areas throughout the district of Jammu (Census of India 2011). Jammu City has spatially grown very fast in the last 3 to 4 decades predominately toward its southern and western direction. Most of this growth has been under unplanned residential use (Economic Reconstruction Agency Govt. of Jammu and Kashmir 2012). During the last decade, large influx of pilgrims to Mata Vaishno Devi and Shri Amarnath Yatra has been observed, thus increasing the demand for residential, commercial, recreational development, and transit facilities in the city (Master Plan Jammu 2032). This put a great pressure on the city's social amenities, resulting in an uneven distribution of basic services. This uneven distribution of basic amenities, affects the physical quality of life of the city dwellers. Therefore, it becomes very important to highlight the inequalities in the spatial distribution of basic amenities among the different wards of Jammu City which will be helpful to future planning for a balanced development.

17.3 Study Area

Jammu is the winter capital of Jammu and Kashmir and is commonly known as the "city of temples." It lies between 32°38'–32°48' North latitude and 74°47' - 74°50' East longitude. At an average elevation of 1030 feet above sea level, it is located on the banks of the river Tawi (Hussain 2006). The present extent of the Jammu City is 167.38 sq. km while the planning area is 287.92 sq. km (Jammu Master Plan 2021). It is administered by a Municipal Corporation. Jammu was a Municipal Committee during the 2001 Indian census, and was upgraded to a Municipal Corporation in September 2003. Jammu city is divided into 71 wards, 47 of which are in old Jammu City and 24 in new Jammu City (Fig. 17.1). Jammu, like the rest of North-Western India, has a humid subtropical climate (Koppen: Cwa), with extreme summer highs of 46 °C (115 °F) and temperatures occasionally falling below freezing point in the

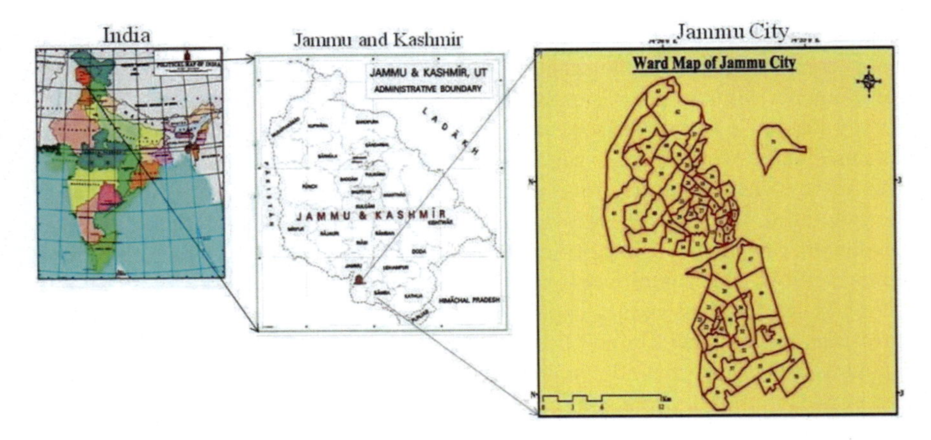

Fig. 17.1 Location map of study area

winter months. The hottest month is June, with average highs of 40.6 °C (105.1 °F), while the coldest month is January, with average lows of 7 °C (45 °F).

17.4 Objectives of the Study

Keeping in view the background of the study area, the present endeavor is focused to study the inequalities in the inter-ward distributional pattern of amenities in Jammu City, and to prioritize the wards for future development planning based on homogeneous levels of development.

17.5 Data Set

The following sources were used to collect and generate various types of data:

The Survey of India toposheets (1971) on the scale of 1:50000 are used to delineate the study area. Since the study relied more on secondary sources of data, the data pertaining to various variables have been gathered accordingly from various departments. Population data and its various attributes were obtained from Census Department; data related to educational institutions has been collected from Directorate of School Education and Directorate of Higher Education, Jammu. Whereas, the data related to fire service stations was obtained from the department of Fire and Emergency Services. Data related to different types of health institution was gathered from Directorate of Health Services Jammu. Jammu Municipal Corporation provided the data related to Solid waste collection points and data related to the locations of Government ration stores in Jammu City was obtained from Consumer Affairs and Public Distribution (CAPD) department.

17.6 Methodology

In order to achieve the objective of finding out disparities, in the distribution pattern of basic amenities among the different wards of Jammu City, the data so collected has been analyzed and represented through the following set of steps (Fig. 17.2).

First Step: **Base Map Preparation and data Segregation**; a base map was generated, georeferenced, and transferred to real world's coordinates to further show the

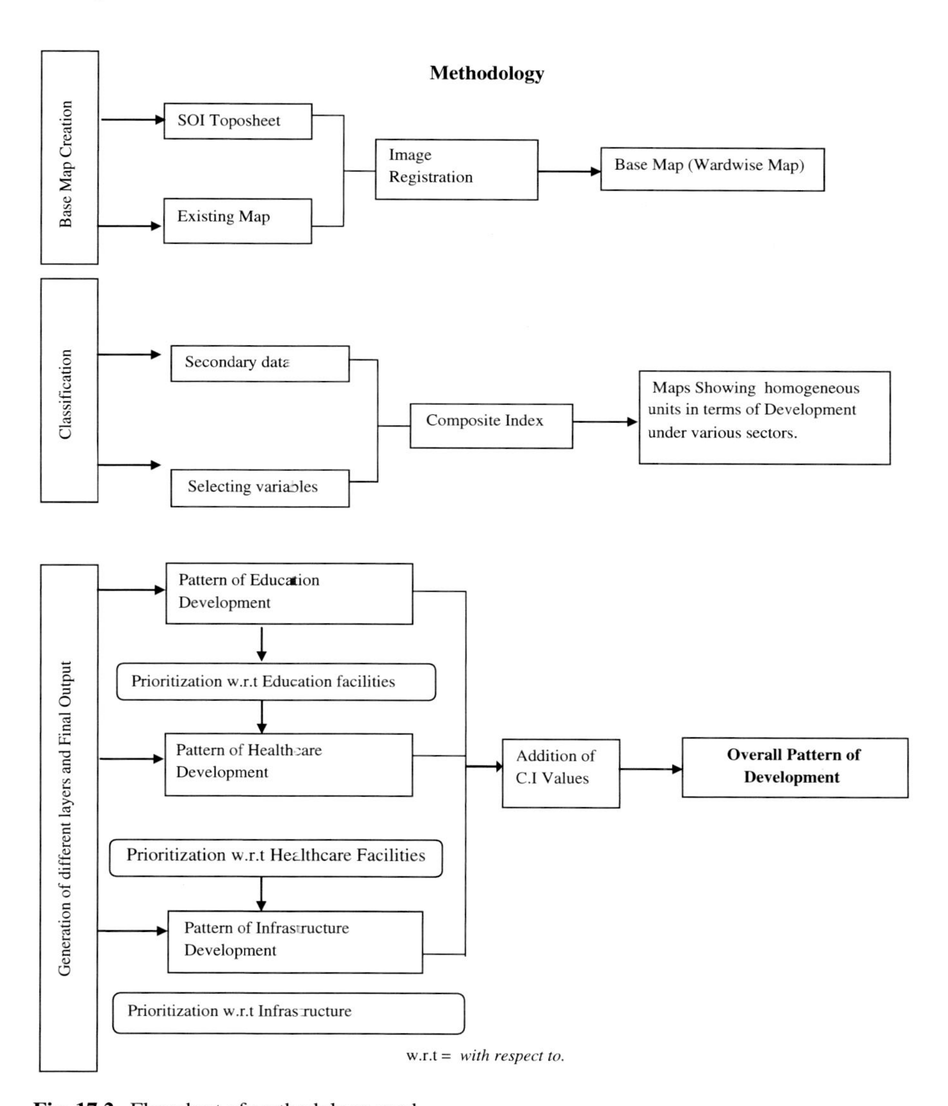

Fig. 17.2 Flowchart of methodology used

distributional pattern of amenities in Jammu City. Furthermore, the data has been divided into three main categories: education, infrastructure, and healthcare, and fourteen variables have been carefully chosen under each of these three headings, keeping in mind the regional personality of the study area and elements of regional development, to provide a concrete picture of spatial variation in the distribution of basic amenities.

Second Step: **Compositing of variables;** since the indicators of development which varies from one region to another in their occurrence are not equally important, different weights are, therefore, assigned to different indicators related to education, infrastructure, and healthcare facilities rooted in an understanding of the development dynamics of the region and based on informed judgment of the researchers. The values of all the indicators cannot be added directly as these are incomparable in terms of units and range of values and as such the values were made scale free by dividing the value of each variable by the respective mean of that column. Subsequently the scale-free values were added to arrive at the composite index for each ward. In the present study the index for all the three sets of indicators, i.e., education, infrastructure and healthcare has been worked out to see the disparities at each set. Finally, to determine the overall spatial pattern of development in the study area, a composite index of development is necessarily derived by combining all indicators related to three sectors of development in order to measure the overall levels of development, which is ultimately useful in area prioritization.

Third Step: Classification and Prioritization on the basis of Homogeneity; The wards have been divided into five categories based on their composite values, i.e., highly developed, developed, moderately developed, less developed, and very less developed by taking standard deviation as class interval from the mean. Wards have been prioritized based on the composite index value in each development sector. The wards with low composite score face shortage of services which needs to be a priority as compared to the wards with high composite index value possessing adequate facilities.

Fourth Step: Presentation of data; after the tabulation and processing of data using different statistical techniques the results so obtained was graphically represented in the GIS environment using Arc GIS software.

17.7 Data Analysis and Discussion

17.7.1 Spatial Patterns of Inequalities in Educational Development

Human resource is a dynamic resource with great potential for growth and development. However, it needs to be sharpened through the process of learning, knowledge, and skill, which in turn are enabled by the art of reading, writing, and thinking. Education reflects the socio-economic status and cultural setup of a class or community and its principal functions are that of fitting humanity to take control of its development. The best way to understand spatial pattern of educational development is to study the education level. Education and its aspects indicate the level of social development which sets the condition for economic development. Many scholars claim that education as an indicator serves the best as it not only shows the level of social awakening but also the level of human resource development which is inevitable for socio-economic development of any area or region.

Education rate (E1), a general index of cultural and technological advancement, serves as an inevitable indicator of social and economic development. New idea and skills cannot be effectively communicated to minds which are untrained to receive them and make use of them, whether it is family planning or improvement of sanitary standards or any program of social security, or any move which requires change of attitude and habits of life, it must make sense to the people. Thus, the analysis of education is a fairly reliable index of socio-cultural and economic advancement.

Indicator E2, number of educational institutions, has been adopted to examine how the educational facilities have been spatially distributed in different wards of Jammu City.

It is clear from Table 17.1 that about sixteen wards come in the category of very high levels of educational development. These wards include Golepuli Pore camp road, Chand Nagar, Kanji House, Deeli, Gujjar nagar, Patta paloura, New plot, Bohri, Channi biza, Digiana, Gangyal, Panjtirthi, and Jullakha mohalla, all of them standing at the first place. In the next band of high category, there is only one ward, namely, Paloura centre. There are six wards falling in the medium level of educational development, namely, Resham Garh colony, Lakshmi Nagar, Talab Tillo, Dogra hall, Channi Himmat, and Pacca Danga. In the lower categories, there are 48 wards among which 5 are in "low" and 43 in "very low" category, namely, Nanak Nagar, Chinore, Sanjay Nagar, Bagwati Nagar, Sanjay Nagar, Talab Khatikan, Sainik Colony, Rehari Colony, Partap Garh, Mast Garh, Paloura, Krishna Nagar, Greater Kailash, Narwal, Channi himmat house, Amphalla, Janipur centre, Janipur south, etc. (Fig. 17.3)

It is evident from the above discussion and Table 17.2 that only 24% of the total wards are better developed and about 8% of the wards are moderately developed in

Table 17.1 Indicators of education development

Code	Indicators	Weightage
E1	1.Percentage of literates to total population	
E2	2.Educational Facilities	
	2.(a) Primary school	1
	2.(b) Middle school	2
	2.(c) Secondary school	3
	2.(d) Higher secondary school	4
	2.(e) Government Colleges	5

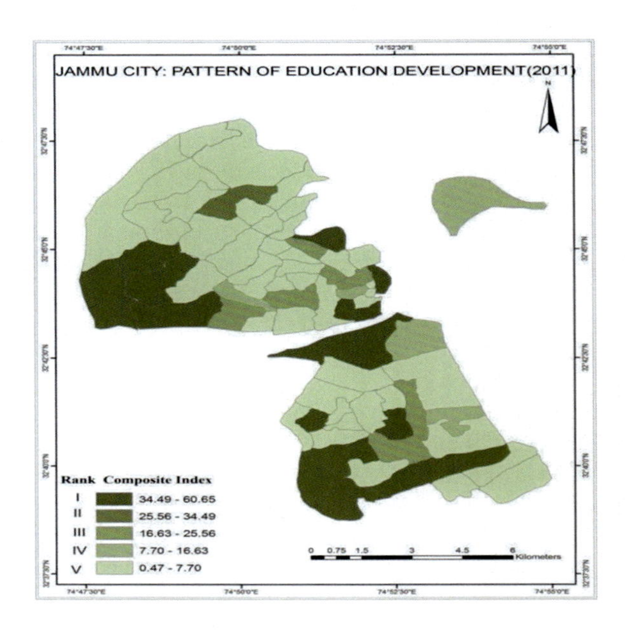

Fig. 17.3 Spatial distribution of educational factor score

Table 17.2 Pattern of educational development

Category	Value of composite index	Rank	Number of wards
Very high	> 34.49	I	16
High	25.56 – 34.49	II	1
Medium	16.63 – 25.55	III	6
Low	7.7 – 16.62	IV	5
Very low	> 7.7	V	43

terms of education. Whereas 68 % of wards have a very low level of educational development. The majority of these underdeveloped wards are found in the outskirts of Jammu City.

17.7.2 *Spatial Patterns of Inequalities in Infrastructural Development*

Infrastructure can be defined as activities that provide a society necessary services to conduct daily life and to engage in productive activities. Assessment of infrastructural facility of any area gives a good deal of information about the overall socio-economic development along with the dimension and degree of its expandability toward surrounding areas. Infrastructural development of different wards of the study area has been measured by using nine indicators, ration depot, LPG outlet, cooperative fair price, bank branch, post office, family welfare centre, anganwari centre, police station, and fire station. After combining the weights assigned to all nine indicators, they were finally projected as a single index. Keeping in view the overall scenario of Jammu City, the indicators like availability of ration depot and anganwari centre still has relevance to reflect the overall development scenario (Table 17.3).

A glance at Table 17.4 shows that, as far as infrastructural facilities are concerned, ward Mast garh, Talab khatikan, Krishna Nagar, Gandhi Nagar, Kanji house, Nai basti, Channi himmat, Rehari colony, Moholla Malhotria, Panjtirthi, and Jullakha Mohalla stand at first place. In the next band of High category, there are seven wards, namely, Paloura chinnore, Pacca danga, Gujjar nagar, Rehari colony , Gangyal ,Greater Kailash and Channi Biza. Moreover there are thirteen wards, namely, Amphalla, Fattu chowagan, Sainik colony 1, Sainik colony 2, Dogra hall, Bakshi nagar, Subash nagar, Resham garh, Bohripaloura centre, Durga nagar, and Lower Muthi which show moderate levels of infrastructure development. On contrary to this, about 47 wards show low levels of infrastructure development. The wards lie under this category are Digiana, Lakshmi nagar, Shastri nagar, New plot, Channi Himmat, Talab tillo, Chandnagar, Gangyal, Partap garh, Bahu west, Upper muthi, Patta paloura, Janipur, Shivnagar, Rajpura magotrian, Narwal, Bahu east, Chinore_2, Poonch house , Moholla ustad, Janipur centre, etc. (Fig. 17.4). In a nutshell it has

Table 17.3 Indicators of infrastructural development

Code	Indicators	Weightage
I1	Ration depot	**1**, if the said facility is available **0**, if the said facility is not available
I2	Police station	
I3	Fire station	
I4	Bank branch	
I5	LPG – outlet	
I6	Anganwari centre	
I7	Cooperative fair price	
I8	Family welfare centre	

Table 17.4 Composite index of infrastructural characteristics in Jammu City

Category	Composite index	Rank	Number of wards
High value	> 13.26	I	14
High	11.33 – 13.25	II	7
Medium	9.41 -11.32	III	13
low	7.49 – 9.40	IV	10
Very low	< 7.49	V	27

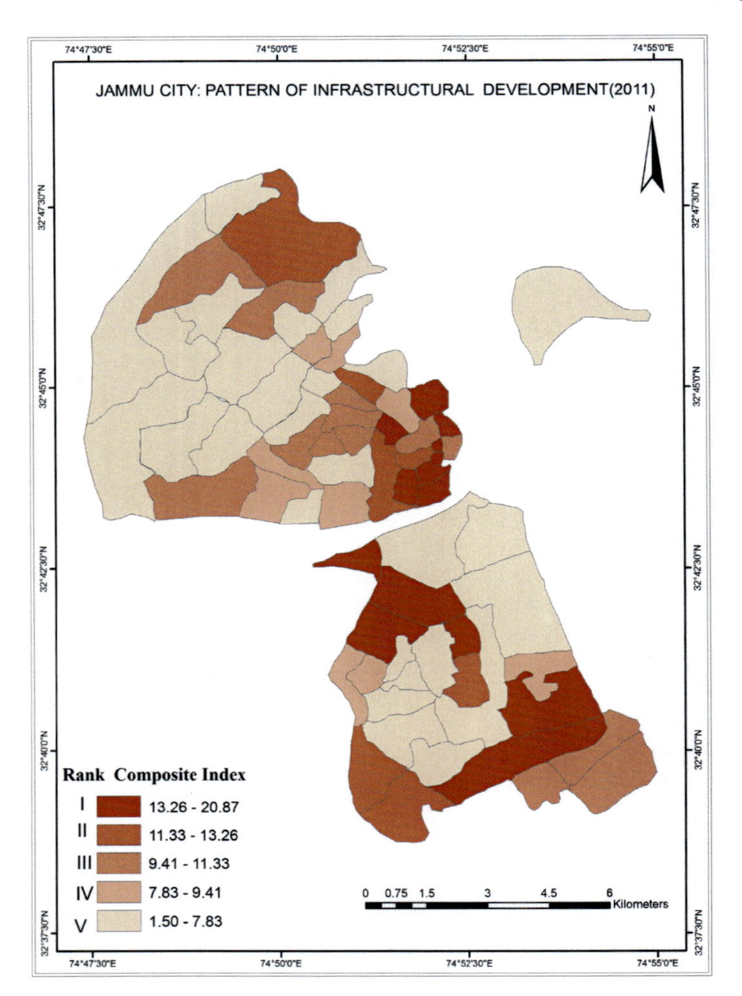

Fig. 17.4 Spatial Distribution of infrastructural facilities factor score

been noticed that about 52% of the total wards show low levels of development and are mostly located at the outskirt of the city and only few located at the city centre.

17.7.3 Spatial Patterns of Inequalities in the Development of Healthcare Facilities

Health can be defined as physical, mental, and social wellbeing, and as a resource for living a full life. Population health is an important component of human development and is critical for a country's economic growth and stability. In most of the studies dealing with development healthcare indicators has taken under Infrastructure category, but in the present study we tried to highlight the status of healthcare facility in more detail as such we decided to study healthcare indicators under independent category.

Health development of different wards of the study area has been measured by using following indicators: hospitals, primary health centre, dispensary, and subcentre. These indicators have been grouped together into a single index after assigning weightages to them (Table 17.5).

It is evident from Table 17.6 that as per the health facilities are concerned only 28% of the total wards (i.e., 20 wards) show high-level development, and these wards are Mohalla ustad, Partap garh, Pacca danga, Lakshmi nagar, Mohalla malhotrian, Channi himmat, Janipur, Bakshi nagar, Gurah bakshi, Rajpura, Talab tillo north, Talabtillo south, Janipur. Amphalla, Resham garh, and Gandhi nagar. There are about eleven wards which show moderate levels of healthcare development, and the wards falling in the medium line include eleven wards, namely, Gandhi Nagar, Gangyal, Dogra hall, Lower muthi, Krishna nagar, Panjitirthi, Rehari colony, Gole puli, Mast garh, New plot, and Chinore. However, more than 56% of the wards, namely, Digiana, Jullakha mohalla, Channi biza, Nanak nagar, Chand nagar, Nanak nagar, Channi himmat, Shiv nagar, Kanji house, Gujjar nagar, Talab khatika, Shastri nagar, Deeli, Upper Muthi, Toph sherkha, Subash nagar, Paloura centre, Poonch house, Nai basti, Patta paloura, Sainik colony, Bahu, Sanjay nagar, etc. show low levels of healthcare development (Fig. 17.5).

Table 17.5 Indicator of healthcare facilities

Code	Indicators	Weightage
H1	Subcentre	1
H2	Dispensary	2
H3	PHC	3
H4	Hospital	4

Table 17.6 Composite index of healthcare characteristics in Jammu City

Category	Composite index	Rank	Number of wards
High value	> 12.97	I	16
High	10.30 – 12.96	II	4
Medium	7.63 – 10.29	III	11
low	4.96 – 7.62	IV	15
Very low	< 4.96	V	25

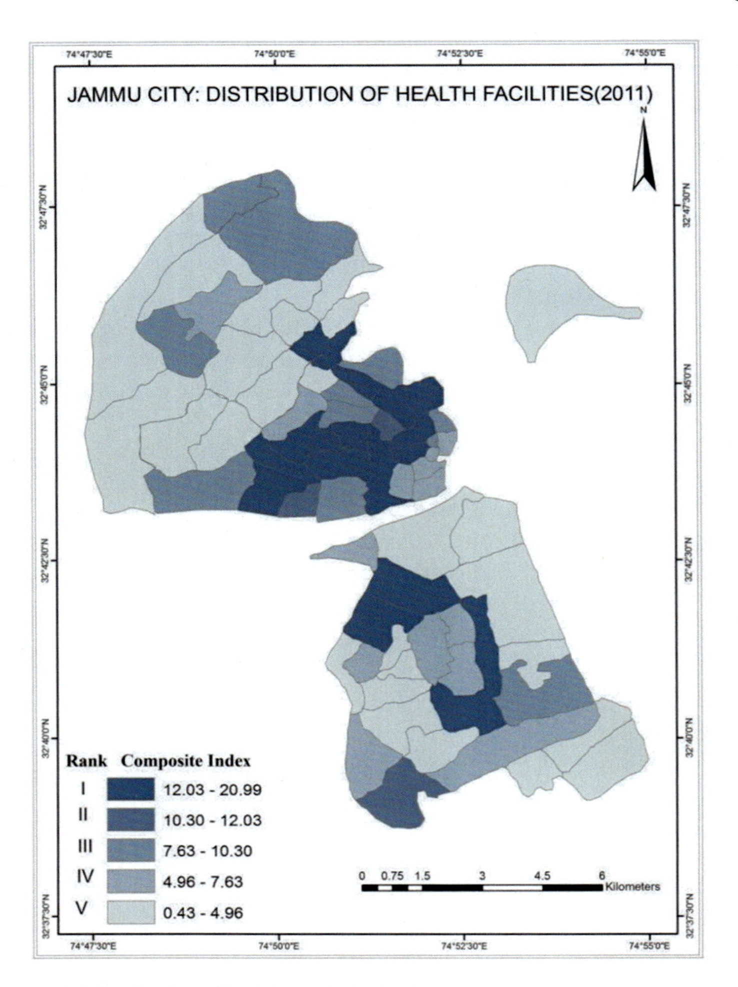

Fig. 17.5 Spatial distribution of healthcare facilities factor score

17.7.4 Pattern of Overall Development

The previous sections cover the levels of educational development, infrastructure development, and development of healthcare facilities individually. In order to see at a glance the overall spatial pattern of development in the study area, all the indicators are taken collectively.

In defining the overall development of the wards, different composite indexes of different indicators (health, education, and infrastructure) have been grouped into a single index. So as far the overall development is concerned, with composite scores of more than 43.47 (Table 17.7 and Fig. 17.6)) nearly about 26% of the total wards show high levels of overall development and the wards under these categories

Table 17.7 Composite index of overall development

Category	Composite index	Rank	Number of wards
Very high	> 53.26	I	14
High	43.47 – 53.26	II	5
Medium	33.68 – 43.46	III	7
Low	23.89 – 33.67	IV	20
Very low	< 23.89	V	25

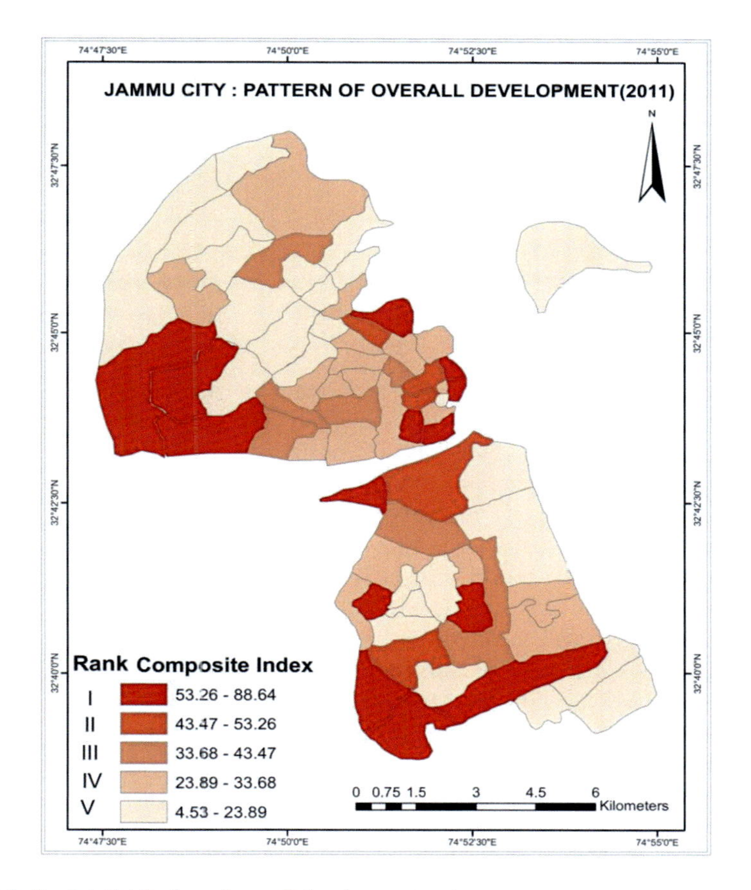

Fig. 17.6 Spatial distribution of overall development pattern

are Gole pulli, Kanji House, Deeli, Gujjar nagar, Chand nagar, Panjtirthi, Gangyal, Jullakha Mohalla, Channi Biza, New plot, Digiana, Bohri, Shastri Nagar, Digiana Jeevan Nagar, Bahu, Pakka danga, Mohalla Malhotrian, and Lakshmi nagar south. The ward falling in the medium line include seven wards, namely, Channi himmat, Paloura centre, Resham garh colony, Talab tillo North, Talab tillo South, Gandhi nagar north, and Dogra hall. On the other hand, a drastic picture has emerged, with over 63 % of the total wards (i.e., 45) showing low to very low levels of overall development. The wards that come under this category include Gandhi nagar south, Partap garh, Krishna nagar, Gurah bakshi nagar, Mast garh, Bakshi nagar, Rehari colony, Talab khatikan, Mohalla ustad, Channi Kamala, Chinorekern Roop Nagar, rehari Colony North, Rehari Colony South, Amphalla, Bhagwati nagar, Janipur North, Rajpura Mangotrian, Lower muthi, Nai Basti, Channi Himmath Housing Colony, Bahu, Fattu Chakkan, Subash nagar, Durga nagar, Sainik Colony Sec. A, B, C, Nanak Nagar, Greater Kailash, Janipur South, Shiv Nagar, Chinorekeran Bantalab, Upper muthi, Toph sherkhaina, Paloura top, Sainik Colony Sec. D, E, F, Baraiupper dharmal, gangyal, Narwal, Nanak Nagar, Poonch house, Sanjay Nagar, Nanak nagar, Paloura, Janipur central, Janipur west, Tawi Vihar housing colony, etc. The majority of Jammu's highly developed wards are located near the core of old Jammu City. In contrast, less developed wards are mostly found on the outskirts of the city.

17.7.5 Prioritization Based on the Levels of Development for Future Planning

In order to assess the levels of development the results so obtained is divided into five categories (i.e., very high, high, medium, low, and very low) based on the composite core values. However, entire Jammu City can be divided into three priority zones, i.e., first, second, and third on the basis of their composite score values of the different parameters taken under study. The simple principal followed for division is that prioritization is indirectly proportional to the composite score values. First Prioritization Zone includes all the wards which lie at IV and IV Categories (i.e., low and very low category) in terms of the levels of development. Similarly, wards of III and II Categories (i.e., High and Medium Categories) would be the part of Second Prioritization Zone whereas wards under very high category would be selected as Third Prioritization Zone for future development planning. This criterion of prioritization will be implemented for all the three parameters, i.e., education, infrastructure, and health taken under study apart from the overall development.

17.8 Conclusion

The outcome of the research shows that basic amenities among the different wards of Jammu City are not uniformly distributed. Around sixteen wards are classified as having very high levels of educational development. It is evident from the above discussion that only 24% of the total wards are better developed. On the other hand about 68% of wards are very less developed in terms of educational development.

The majority of these underdeveloped wards are found on the outskirts of Jammu City.

In terms of infrastructure development only 30% of the total wards show high to very high levels of development. On contrary to this, 52% of total wards (i.e., about 47 wards) lie in the category of low and "very low" category of infrastructure development. As per the health facilities are concerned only 28% of the total wards (i.e., 20 wards) show high-level development, However, more than 56% of the wards show low levels of healthcare development.

After combining all the indicators discussed above it has been noticed that about 27% of the total wards have high levels of overall development while 10% have a moderate level of development. However, a drastic picture has emerged, with more than 63 percent of total wards (i.e., 45) showing low to very low levels of overall development.

In the concluding line it must be stated that, in comparison to peripheral wards, old city wards have adequate urban amenities. The planning body for Jammu City should fill existing gaps in the provision of facilities and, in the future, keep up with the city's rapid growth in both time and space. The findings of the study are providing useful information to decision-makers in the formulation of comprehensive urban policies.

References

Adekunle, Aderamo, Aina (2011) Spatial inequalities in accessibility to social amenities in developing countries: a case from Nigeria. Australian J Basic Appl Sci 5(6):316–322p. ISSN 1991–8178

Anderson K, Pomfret R (2004) Spatial inequality and development in Central Asia. Research Paper No. 2004/36. Helsinki: United Nations University World Institute for Development Economics Research

Berry BL, Horton FE (1970) Graphic perspectives on urban systems Englewood Cliffs

Bhagat RB (2010) Access to basic amenities in urban areas by size class of cities and towns in India. Int Inst Popul Sci Mumbai 400088:2010

Datta Ankur (2016) Contributions to Indian Sociology 50(1):52–79. SAGE Publications Los Angeles/London/New Delhi/Singapore/Washington DC DOI: https://doi.org/10.1177/0069966715615024

Economic Reconstruction Agency Government of Jammu and Kashmir for the Asian Development Bank April 2012

Farooq S (2008, March) Ahmad Urban Sprawl Development around Aligarh city: a study Aided by Satellite Remote Sensing and GIS. J Indian Soc Remote Sens 36:77–88.P- 77

Henderson JV, Shalizi Z, Venables AJ (2001) Geography and development. J Econ Geog 1(1):81–105

Kim S (2008) Spatial inequality and economic development. The International Bank for Reconstruction and Development/The World Bank On behalf of the Commission on Growth and Development Commission on Growth and Development, World Bank [Working Paper No.16]

Mulligan GF, Carruthers JI (2011) Amenities, quality of life, and regional development. In: Investigating quality of urban life. Springer, pp 107–133

Nripendra P Rana, Sunil Luthra, Sachin Kumar Mangla3, Rubina Islam, Sian Roderick, Yogesh K. Dwivedi (2018) Barriers to the development of smart cities in Indian context. Information Systems Frontiers, pp 503–525

Olvera LD, Plat D, Pascal P (2003) Transportation conditions and access to services in a context of urban sprawl and deregulation. The case of Dar es Salaam, Transport Policy 10(4):287–298

Savita S (1988) Regional structure, processes and patterns of development (A Case Study of the Chota Nagpur Region), Concept Publishing Company, New Delhi

Thorndike EL (1939) Your City' Harcourt, Brace and Company, New York, (Book)

Vyas PR (1991) Social amenities and regional development. Rawat Publications, Jaipur

Chapter 18
Land Suitability Analysis for Settlement Concentration in Fringe Area of Siliguri Town, West Bengal (India)—A GIS-Based Multi-Criteria Decision-Making Approach

Sanu Dolui and Sumana Sarkar

Abstract One of the crucial questions among the urban planners is to determine suitable locations for future urban expansion, especially in areas adjacent to large cityscapes. Plain land, fertile soil, an opportunity for a livelihood, and a decent transportation system have always encouraged human habitation; however, adverse physical environments and inadequate livelihood opportunities have always constrained urban expansion. The present study area, Siliguri town, stretched over Darjeeling and Jalpaiguri districts which is the third largest urban agglomeration in the Indian state of West Bengal and continueously growing at a rapid pace. After considering various socio-economic, environmental and physical factors, the final suitability map of settlement concentration was prepared using remote sensing technique, AHP and FAHP method. Among the selected factors, seven are found as favourable factors, viz. elevation, slope, distance from the river and road, distance from main settlement patches, changes in an existing built-up area, night-time light images; and three are discouraging factors, viz. dense forest cover, river flood-prone area, distance from tea garden and protected landscape. The generated thematic maps of these criteria were standardized and given weights according to their importance to each other using a pairwise comparison matrix applying AHP and FAHP method. The final suitable map was classified into four suitable zones; in the highly suitable zone, 92.66 percent of pixels are matched with both the weightage methods. This study revealed that FAHP was marginally more useful than AHP in detecting future urban expansion. This research may be useful for optimizing land use planning and help urban planners in the decision-making process.

Keywords Urban expansion · Siliguri town · AHP · FAHP · Land suitability map · Multi-criteria decision-making

---

S. Dolui (✉) · S. Sarkar
Department of Geography, The University of Burdwan, Burdwan, WB, India

S. Sarkar
e-mail: ssarkar@geo.buruniv.ac.in

N. C. Jana et al. (eds.), *Livelihood Enhancement Through Agriculture, Tourism and Health*, Advances in Geographical and Environmental Sciences,
https://doi.org/10.1007/978-981-16-7310-8_18

18.1 Introduction

Human settlement refers to clusters of houses in a place where people are working and living together and transforming the surrounding landscape at a rapid pace (Dutta 2011). The population is unevenly disseminated among various regions in our earth; some places are sparsely populated due to physical and social-economic obstacles, while others are densely populated due to locational advantages. Even within a city, population is extremely concentrated in some areas while low-density households can be seen outside the city proper. Among the numerous factors that determine population distribution in a region, physical environment may be the most crucial component among all the geographic parameters. Apart from physiographic and climatic condition, a number of socio-economic and demographic factors have an impact on population dispersion. The ever-increasing population growth, especially in urban areas, is posing numerous challenges to the implementation of sustainable urban development. Rapid population growth, industrialization and economic development are stimulating urban expansion which engulfing the city's adjacent rural areas.

Siliguri town experiences phenomenal growth evolving from a tiny village to one of the largest urban agglomerations in the country and has become an important city for the whole of northeast India. The strategic location of Siliguri town had a significant impact in the growth of this town. Siliguri is bounded by Darjeeling District on the North, Bihar and Bangladesh on the South, Jalpaiguri on the East and Nepal on the West. Siliguri serves as a commercial center and an important node of communication, as it is well connected with the states, viz. Sikkim, Assam, Arunachal Pradesh. This city is located in very close vicinity with international borders of Nepal, Bhutan, Bangladesh, due to its accessibility to northeast India and its strategic importance, this city is known as the Gateway to Northeast India (Ghosh 2017; World Public Library).

Land suitability analysis is basically a mapping technique that determines the most suitable location for the activities to be planned based on certain predetermined parameters, which is beneficial in planning and managing land resources (FAO 1976). Geographical Information System (GIS) and Multi-Criteria Decision-Making (MCDM) are one of the most popular techniques used to analyze the suitability of a sites for urban development. A decent amount of research has been conducted over the last two decades, to determine the suitability of a particular site with MCDMs techniques, was found to be effective, flexible and widely accepted (Ferretti and Pomarico 2012). The integration of GIS software and multi-criteria decision analysis is an efficient way for evaluating land suitability that greatly reduces time and effort (Aburas et al. 2017). Due to rigorous environmental regulations and government policies, exploring new location for urban development becoming a challenging task for the researcher. The resulting outcome from site suitability analysis produces the most suitable areas after filters out less desirable sites. Researchers adopted various weights methods; among those the most popular one is analytical hierarchy process (AHP) proposed by Saaty (1980) based on a pairwise comparison matrix.

As a MCDM method AHP can compute qualitative and quantitative data. While applying the AHP model, the main challenging task was to the right judgment of relative weight while comparing factors to each other based on their importance. AHP technique proved its robustness in numerous studies and transparently reflected the spatial decision-making process (Ferretti and Pomarico 2012).

Though AHP is one of the popular weightage methods, it has been criticized for its inability to adequately handle fuzziness, ambiguity and vagueness in expert decision-making (Deng 1999). Since decision-makers usually feel more comfortable to give judgments in intervals rather than expressing in a single crisp numeric value. The conventional AHP approach may not be able to completely represent a style of human thought through a single judgmental value (Kabir and Hasin 2013). As the results of AHP solely depend on the expert's observation, sometimes fuzziness, uncertainty and biasedness appeared in decision-making process. To address the shortcomings of conventional AHP, the fuzzy modification of AHP (FAHP) was introduced and successfully employed in many land use suitability studies. Previous literature suggested that Fuzzy AHP methods have been successfully implemented in several research to determine optimal sites such as to identify suitable areas for cultivation of fodder crops like Wheat (Mohammadrezaei et al. 2013; Abbasi et al. 2019), Barley (Hamzeh et al. 2014), Rice (Maddah et al. 2017) and also for plantation of cash crops like Rubber (Van et al. 1996), Citrus (Mokarram et al. 2018), Eucalyptus (Rujee Rodcha et al. 2019), Coffee (Nurfadila et al. 2020) and also various socio-economic decision-making issues, such as the success of an E-Commerce site (Kong and Liu 2005), Landfill siting (Changa 2007), Hospital site selection and health care system (Vahidnia et al. 2009; Ameryoun et al. 2014). FAHP is also used in the exploration of different issues related to an urban area; in China it evaluates population quality (Zhou et al. 2014), solar farms exploitation (Noorollahi 2016), sustainable transport development decision (Moslem et al. 2019), location selection of shopping malls (Erdin et al. 2019), suitable sites for urban development (Ullah et al. 2014; Santosh et al. 2018) and selection of fire stations (Wang 2019). Several researchers have found that Fuzzy AHP is also suitable for solving tourism-related issues, such as the location of international tourist hotels (Chou et al. 2008), potentiality of island tourism (Noraida et al. 2012). A comparative analysis of AHP and FAHP method for analyzing site suitability mapping of settlement concentration has been presented in this paper (Table 18.1). The variables and weights used in the AHP and FAHP methods were similar, but the calculation procedure differed. This research aims to combine GIS techniques with analytical hierarchy process (AHP) and fuzzy analytical hierarchy process (FAHP) to generate a final land suitability map of settlement concentration in both methods based on selected criteria.

18.2 A Brief Review of the Study Area

Siliguri town and its surrounding rural landscape extended from 88°15′E to 88°40′E and 26°36′N to 26°52′N approximately and shared its parts with Darjeeling and

Table 18.1 Comparison of weights in AHP and FAHP based on Saaty's AHP method (1980) and Chang's Fuzzy AHP (1992)

The intensity of importance in AHP[a]	Numerical rating	Reciprocal	Triangular fuzzy AHP	Reciprocal judgment of FAHP	Explanation
1	Equal importance	1	(1, 1, 1)	1, 1, 1	Two activities contribute equally to the objective
3	Moderate importance	1/3	(2, 3, 4)	$\frac{1}{4}, \frac{1}{3}, \frac{1}{2}$	Experience and judgment slightly more favoured one activity over another
5	Strong importance	1/5	(4, 5, 6)	$\frac{1}{6}, \frac{1}{5}, \frac{1}{4}$	Judgment strongly favour one activity than another one
7	Very strong or demonstrated	1/7	(6, 7, 8)	$\frac{1}{8}, \frac{1}{7}, \frac{1}{6}$	One factors strongly dominant or favoured over another factor
9	Extreme importance	1/9	(9, 9, 9)	$\frac{1}{9}, \frac{1}{9}, \frac{1}{9}$	The evidence of one factor that extremely dominated and the other one negligible

[a]Intensities of 2, 4, 6 and 8 are the value of intermediate intensities

Jalpaiguri districts in the Indian state of West Bengal (www.siliguri.gov.in). Physiographically, this area is situated in the *Terai* region (moist land) at the base of Himalaya, covered with swampy grassland and dense tropical deciduous forest (Nbtourism.tripod.com). River Mahananda flows through the middle of towns and bifurcates the study area into two equal parts (www.wbtourismgov.in). Summer months are warmer with an average temperature of 35 °C and winter months are basically cool with an average temperature of 10 °C, sometimes temperature even drops below 5 °C. The overall average annual temperature remains 23.7 °C throughout the year. This region received abundant rainfall during the monsoon period (IMD report, https://mausam.imd.gov.in/). Siliguri has witnessed an unprecedented increase in the rate of population growth during the last three decades for its geo-strategic location. Siliguri is the second-largest city in Northeast India after Guwahati, and it is known as the "Gateway to Northeast India" because of its easy access to all of the states in northeast India (Fig. 18.1). The total population of the Siliguri urban agglomeration is 10,57,438 people (http://www.censusindia.gov/2011). The city has transformed

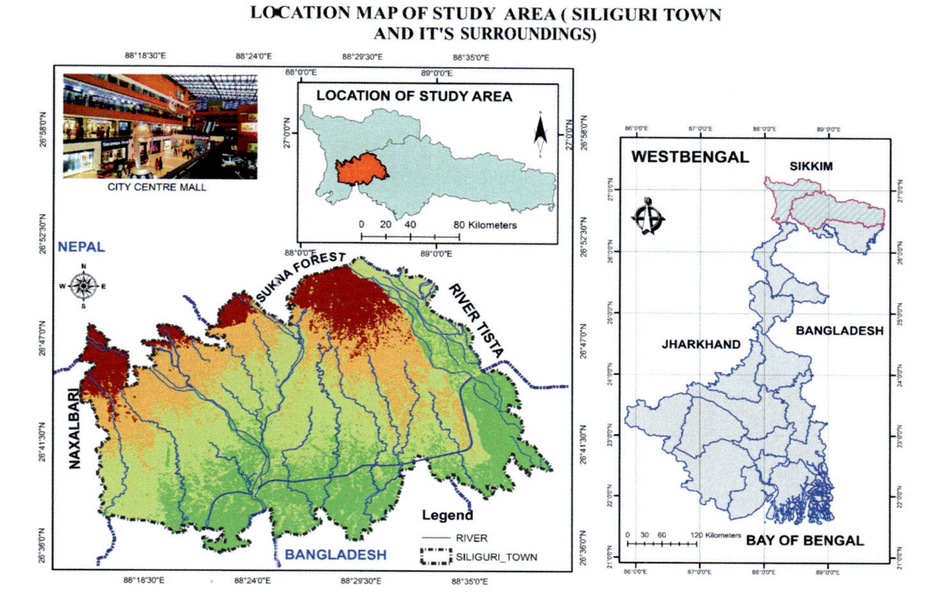

Fig. 18.1 Location map of study area

from a sleepy small town to a booming commercial centre in northeast India with rapid growth (www.siligurismc.in).

18.3 Materials and Methods

At first base map was prepared using NATMO map with the help of QGIS (3.10.0 version) software and verified by a survey of India toposheet downloaded from Nakshe portal (https://soinakshe.uk.gov.in/). Aster DEM data of 90 m resolution have been downloaded and then subset operation was performed to demarcate the area of interest (AOI). After the aster image has been geometrically and atmospherically adjusted, the slope and elevation maps are generated from the input data. For calculation of EBBI index, Landsat-TM5 (1990, 2000, 2010) and Landsat8 image (OLI sensors) of 2018 downloaded from the USGS site and ENVI 5.4 are used for satellite image correction and rectification. Night-time satellite image Luojia 1–01 was also downloaded. Data sources are given in Table 18.2. Road map, River map, Protected area and Tea garden map and Flood-prone area map were produced using open street map data and google earth images with Quantum GIS 3.10.2 being used for digitization. In this study, site suitability analysis was performed to identify suitable areas for urban development based on positive factors (exaggerate) and negative factors (constraints) for settlement concentration. Different thematic maps for each

Table 18.2 Selected variables for site suitability analysis of settlement concentration in Siliguri town

Factors	Driving forces	Techniques used	Sources
1	Elevation	Surface elevation map generated (DSM)	Aster Dem 90 m (2011) https://glovis.usgs.gov/
2	Slope	Slope map generated	
3	Road	Road density map (Kernel Density tool)	Open street map https://www.openstreetmap.org
4	Changes in the existing built-up area	EBBI (Enhanced Built-Up and Bareness Index)	Landsat image (1990, 2000, 2010, 2018) https://earthexplorer.usgs.gov/
5	Distance from settlement patches	Digitization of settlement patches	From Google earth image https://earth.google.com/
6	Distance from river	Digitization of river courses	
7	Night-time light images	Data downloaded from Hubei Data and Application Center	Luojia1 01 (2019, September) (http://59.175.109.173:8888/,)
8	Distance from tea garden and protected area	Digitization of protected area and Tea garden	From google earth image https://earth.google.com/
9	Distance from reserved forest	Digitization of reserved area	
10	Distance from flood prone river	Digitization of flood affected area of Teesta river	

criterion were prepared in domain of ArcGIS software. After preparing the suitability maps, weights are given by comparing each factor according to expert advice using AHP and FAHP criteria weights and maps were combined using ArcGIS tools (Fig. 18.2). For the final suitability model, weighted overlay analysis was performed using criteria weights found in AHP and FAHP methods (Fig. 18.3).

18.3.1 Identification of Influencing Factors

The most useful criteria for site suitability analysis for urban development are described below, along with their relative importance.

Elevation: Dwellers' less prefer hilly terrain and rugged topography for building settlements due to high construction cost. The steep slope in mountain areas restricts the availability of land for agriculture, development of transport, industries and other economic activities which may be the most discourageable factors for human habitation. Plainland and fertile agricultural soil, on the other hand, are the ideal areas for human settlement (Small and Cohen 1998). The elevation map was created to

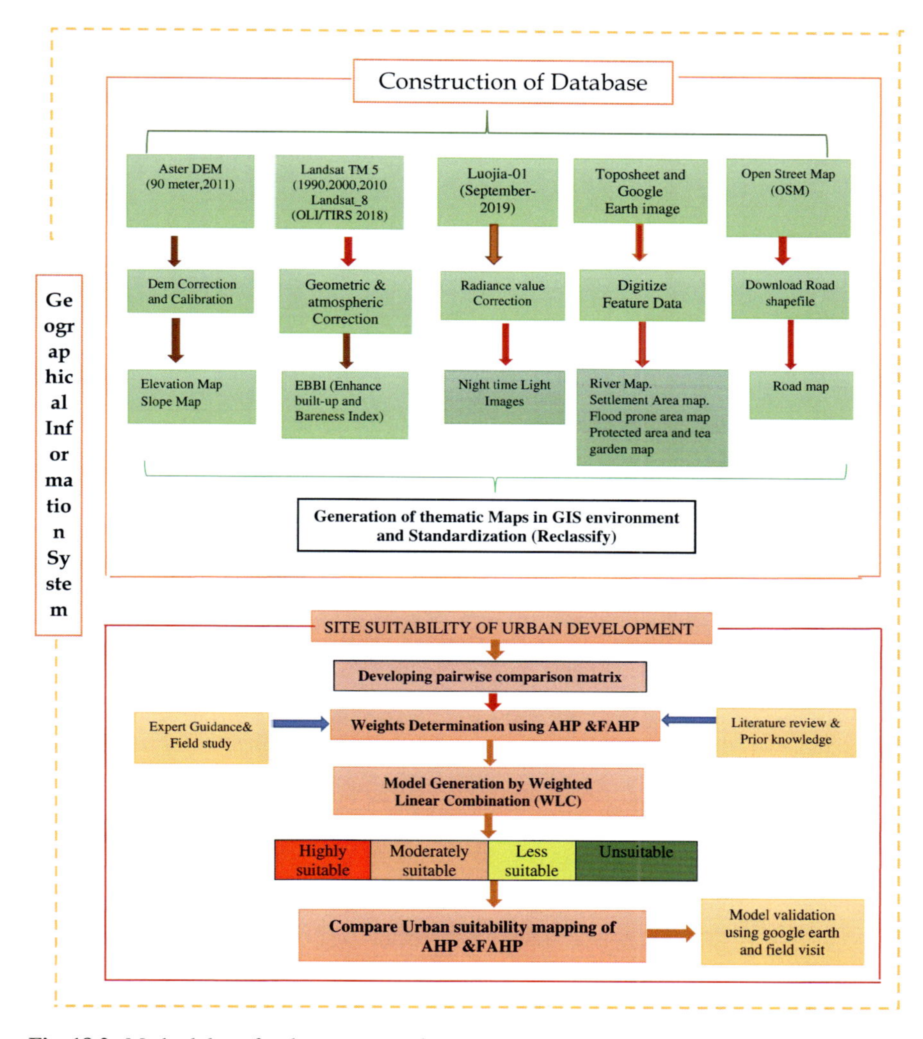

Fig. 18.2 Methodology for the present study

better understand the physiography of the current area, and it was reclassified into four categories based on their suitability for habitation. The classified elevation zones range from (1) 15–85 m, (2) 85–185 m, (3) 155–225 m, (4) 225–285 m (Fig. 18.4).

Slope: Slope is an important criterion for locating suitable sites for urban growth. People used to build their houses on a gentle slope because steep slopes imposed major challenges for the construction of houses, and there is always a high risk of natural disasters such as floods, landslides and earthquakes. Slope less than 10° were considered as most suitable for settlement concentration (Rawat 2010). Steep slopes offer limited opportunities to grow crops and restrict livelihood opportunities.

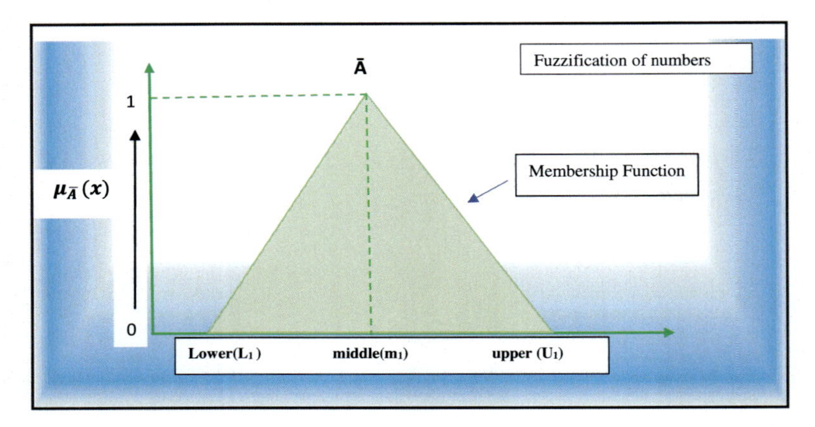

Fig. 18.3 Triangular fuzzy number representation

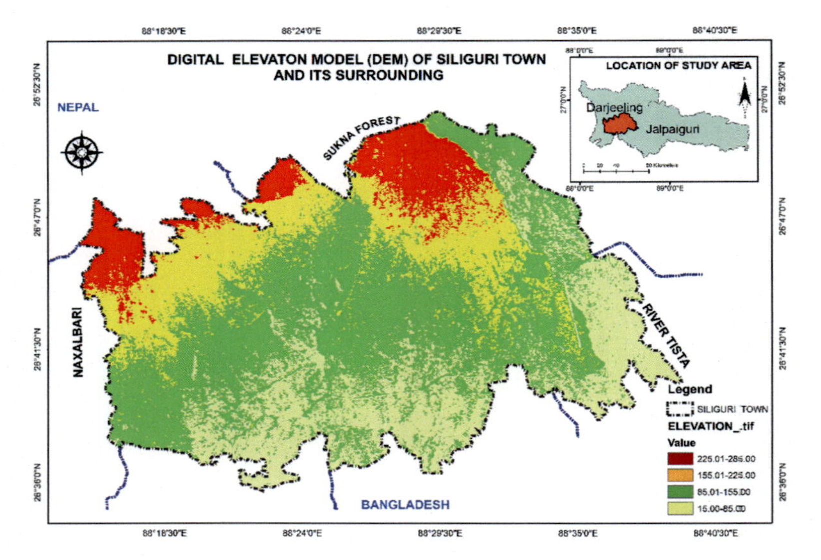

Fig. 18.4 Elevation map

Land with steeper slopes (>16 degrees) and with dense forest cover was found to be unsuitable for human habitation in this study area (Tao et al. 2017). There are very few settlements found in such areas. The slope suitability map was prepared and maximum weights was given to gentle slope areas. Based on suitability, the slope map is classified into four categorical classes such as (1) 0°-8°, (2) 8°–16°, (3) 16°–24° and (4) 24°-62° (Fig. 18.5).

Distance from the river: Ancient human civilization evolved along riversides. Sources of water may refer to as a beneficial factor when human beings choose where to live (Fang et al. 2019). Most of the urban agglomeration clustered around river for

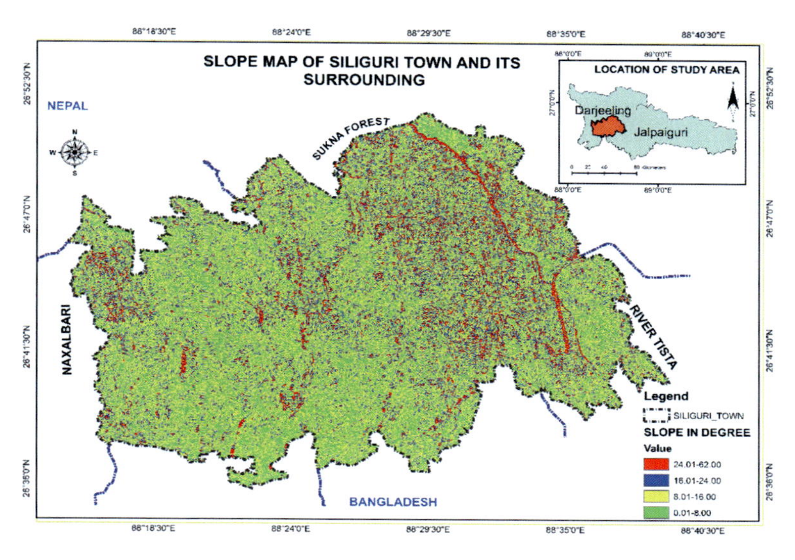

Fig. 18.5 Slope map

variety of reason , river accessibility beneficial for navigation, agriculture, business and industrial purposes (Ceola et al. 2015; Kummu et al. 2011). The relationship between human settlement and river depends upon accessibility (physical location) and utilization of river water. Mahananda, Teesta, Panchai rivers are flowing through the study area, and apart from the flood-affected area of Teesta river, settlements are concentrated along riversides. A river map was prepared, and distances were calculated from river channel. As shown in Fig. 18.6, the river map was classified into four categorical classes such as (1) 0–500 m, (2) 500–1000 m, (3) 1000–1500 m, and (4) 1500–2000 m.

Change in built-up area: Cities in India are expanding at an anxious pace, especially in large urban centres and as a result, the built-up areas also increased. As discussed earlier, Siliguri town expanding outwards at a rapid pace in the last three decades in an unplanned way. The Enhanced Built-Up and Bareness Index (EBBI) was used to quantify urban expansion for the last 28 years (1990–2018). This index was chosen because of its unique ability to differentiate between built-up and bare ground using a single equation. The EBBI index using following wavelengths and band combination: 0.83 m (NIR infrared), 1.65 m (Short wave infrared), and 11.45 m (Thermal infrared) are used to map built-up area and Bare land surfaces (Herold et al. 2004). On the basis of reflection and absorption properties in built-up and bare land surfaces these wavelengths were selected. The EBBI index value showed Positive value for built-up and barren land, negative value for vegetation and water bodies (As-syakur Abd and Rahman et al. 2012). The value of EBBI index deviates from −1 to +1. The EBBI index is calculated here using Landsat image data with following formula:

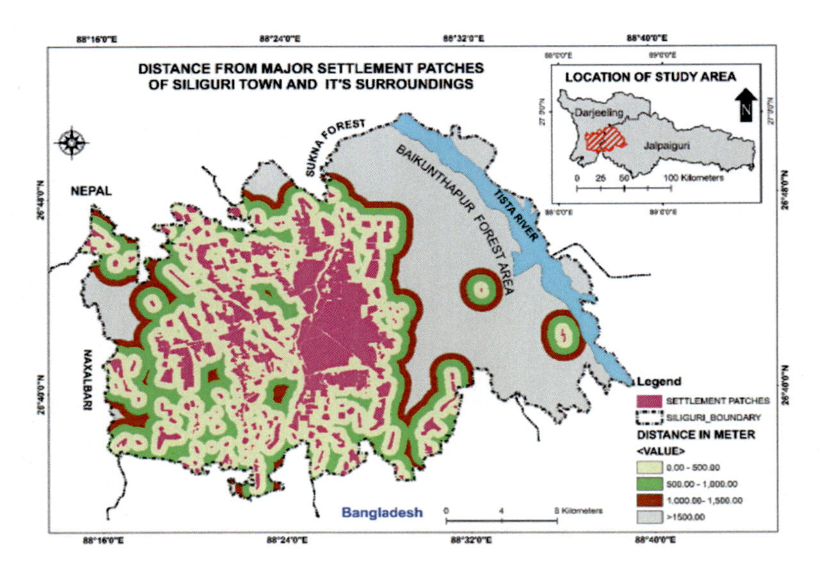

Fig. 18.6 Distance from settlement patches

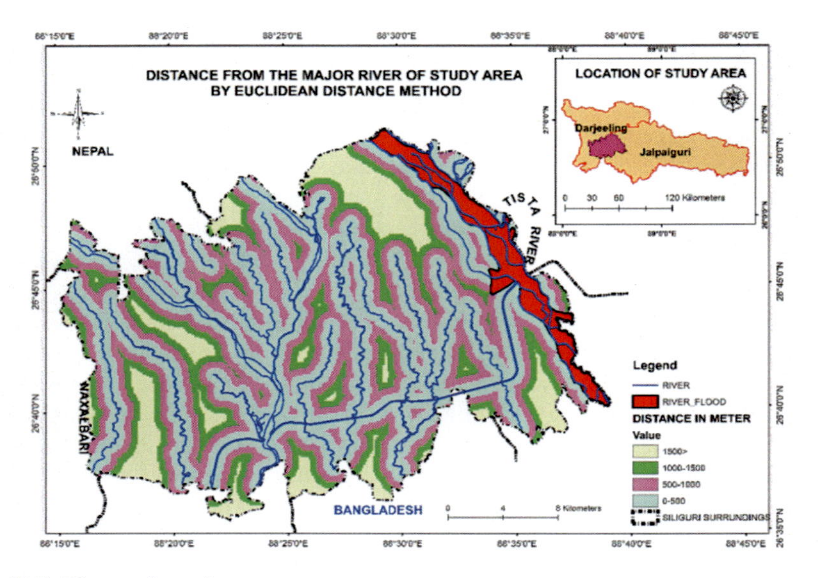

Fig. 18.7 Distance from river

$$EBBI = \frac{Band5 - Band4}{10\sqrt{Band5 + Band6}} \tag{18.1}$$

Changes in the built-up area was calculated and mapped between (1990 and 2018) by change detection method. Finally, changes in built-up areas (EBBI index) map was categorized into four classes (Fig. 18.7) based on index value.

Distance from settlement patches: Settlement patches are defined as spatial arrangements of continuously built-up houses in an area. Urban area is witnessing a large continuous clustering of settlements, and at a distance from the main urban area, there are several small-sized patches, usually 1 to 2 km radius (Paolo and Martin 2009). People like to live together in close proximity. As a result of the urbanization process, city areas are naturally expanding outside of these settlement patches and even within a village, the settlement growing outside these patches. The distance from main patches of the settlement were calculated and given maximum weights to those areas which have nearer to main settlement patches. Distance from settlement patches map is further divided into four classes as shown in Fig. 18.8 such as (1) 0–500 m, (2) 500–1000 m, (3) 1000–1500 m, and (4) >1500 m.

Road density: An well functioning transport network is always beneficial for socio-economic development. An efficient road network always shortens the travel time and costs of traveling which is beneficial for economic development of surrounding areas. In many cities it can be observed that ribbon types of urban development along the routes of road networks radiate from a main urban centre. The density of roads is highest at city-centre; the density of roads decreases as one moves away from the city. Road density is calculated using kernel density tools. High concentration of settlements associated with dense road networks and low density

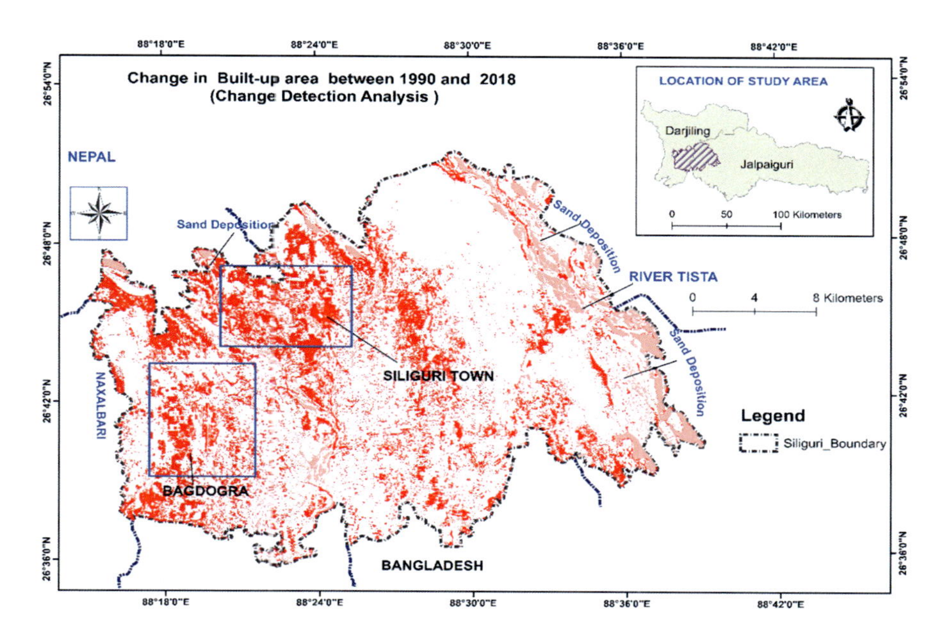

Fig. 18.8 Change in built-up area between 1990 and 2018 by enhance built-up and bareness index

of road observed outside city proper due to dense forest cover and other physical hindrances. As shown in Fig. 18.9, road density map is further categorized into four classes according to density, such as (1) 0–5 sq.km, (2) 5–10 sq.km, (3) 10–15 sq.km and (4) 15–20 sq.km.

Night-time light images: (Distance from source of lights) Stable night light images provide us information about the human settlement and landscape modification. Night light images are the most accurate way to efficiently mapping the urban area (Elvidge et al. 2001). It can be used to estimate urban growth and monitor the spatial extent of the peri-urban development of the large cities (Elvidge et al. 1997, 2001; Small and Elvidge 2013). At night, the brighter portion indicates the urban core area and the darker portion indicates extreme rural areas. In this study Luojia 1–01 satellite image with a spatial resolution of 130 m and a swath of 250 km was used. It can capture high-precision night-time light imagery with a dynamic range of up to 14 bits at night. As shown in Fig. 18.10, radiance value of Luojio1-01 satellite are classified into four categories (1) 0–2000, (2) 2000–4000, (3) 4000–10,000, and (4) 10,000–15,450.

Distance from the flood-prone areas: Riverine floods are the most common and widespread of natural hazards. The potentiality of flood casualties and damages has been increased for those who live close to flood-affected areas. River Teesta which runs through the present study area is a flood-affected river surrounded by dense forest, so the combination of dense forest and river flood contribute to low population density in the immediate surroundings (Pal. and Biswas 2016). Density of the population is less in areas adjacent to Gajoldoba wetland and river Teesta as water overflows during the monsoon period. A map is prepared based on distance

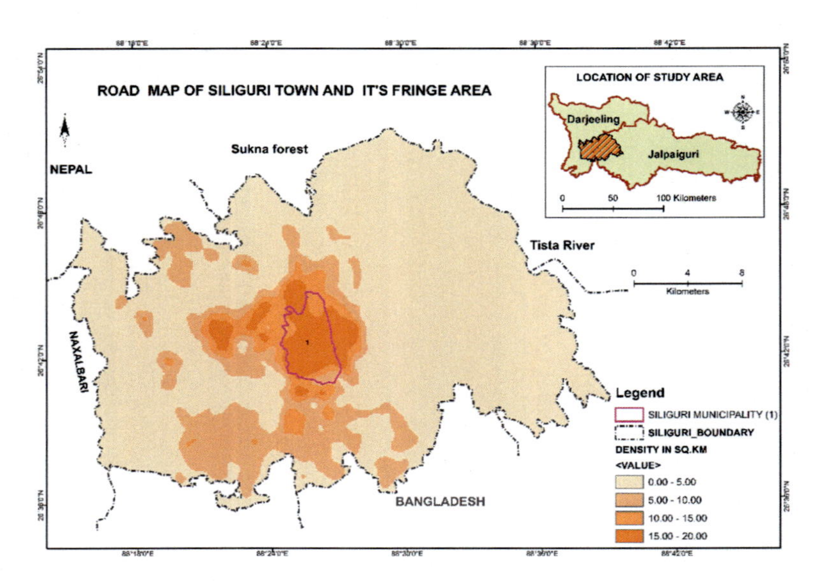

Fig. 18.9 Road density map

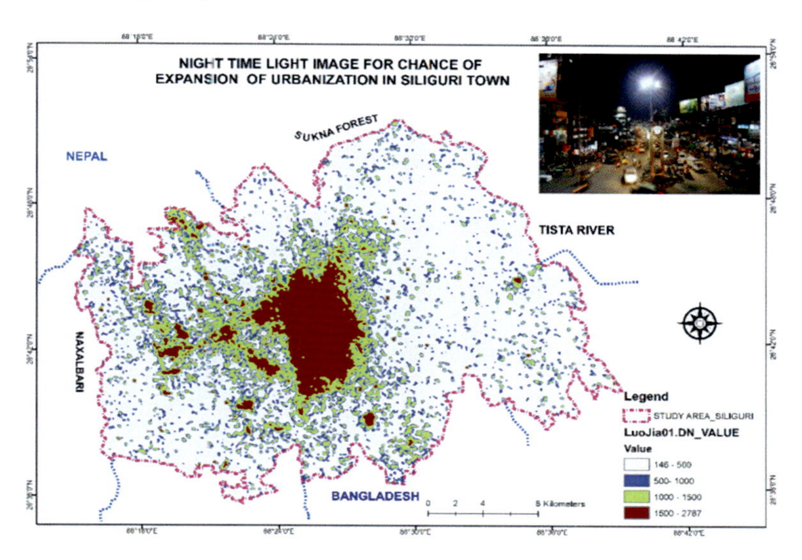

Fig. 18.10 Luojia1-01 Night-time light data

from the flood-affected area and maximum weights are given to those situated at a distance from the flood-prone areas (Banik 2012). As shown in Fig. 18.11, distance from flood-prone area is classified into four categories (1) 0–1000 m, (2) 1000–8000 m, (3) 8000–16,000 m, (4) 16,000–32,000 m.

Distance from tea garden and protected area: Tea gardens play a vital role in the economy of this area, as this region requires a steady supply of both male and female labourers for a variety of jobs. Plucking of tea leaves is a specialized job performed mostly by women (Jannat 2017). Multi-ethnic tribal groups of peoples migrated from Nepal, Bangladesh, Assam, Bihar for working as a labourers in tea garden in the hilly tracts of Darjeeling district as well as in the plain areas of Siliguri and Jalpaiguri. Since tea gardens are the backbone of the economy and mainly protected so there is a lesser chances of conversion in the near future. Map of the tea garden was prepared and lower weights were assigned to those regions close to tea garden.

The Siliguri corridor is basically a narrow strip of land which connects North-eastern parts of India with rest of the Indian territory, also known as the chicken's neck. Siliguri town is located very close in proximity to four neighboring countries: Nepal, Bangladesh, Bhutan and China (Gokhale 1998). As a sensitive area, this area is heavily patrolled by the Indian Army, the Border Security Force and the West Bengal Police. Due to its geo-political and strategic importance, lots of protected areas such as different army offices, BSF training centre, police training school and Indian air force bases at Bagdogra are located and very few chances of future conversion in those areas. As shown in Fig. 18.12, distance from tea garden and protected area is classified into four categories (1) 0–1500 m, (2) 1500–2500 m, (3) 2500–5000 m, (4) 5000–16,000 m.

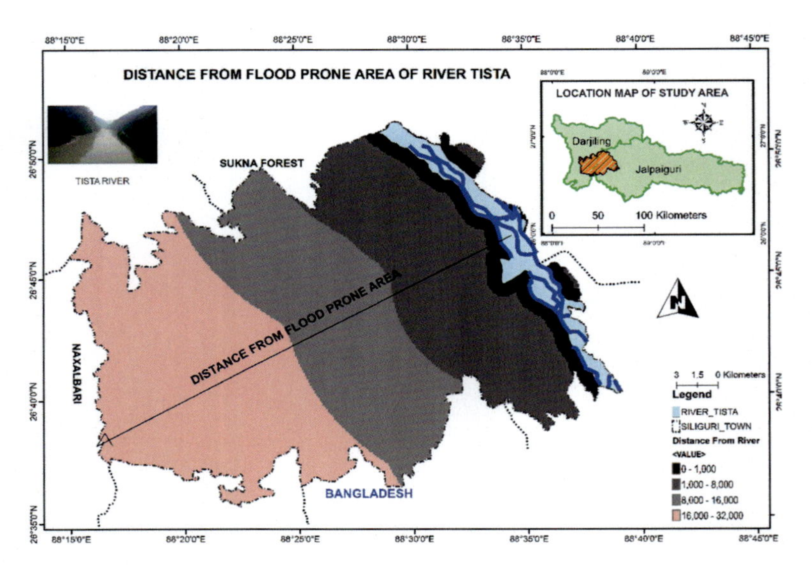

Fig. 18.11 Distance from flood prone area

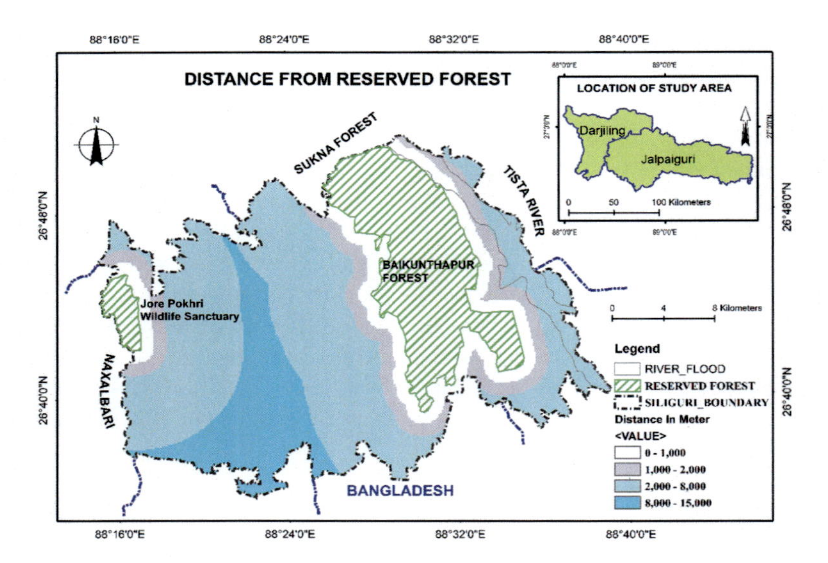

Fig. 18.12 Distance from protected area and tea garden

Distance from reserved forest: The terms reserved forest or protected forest reveals that there is some degree of protection for the wild flora and fauna and there is some restriction to build houses and other activities, so there are lesser chances of conversion in the near future. Dalkajhar forest, Sukna forest range, Baikunthapur forest and Mahananda Wildlife Sanctuary are some forest ranges and protected areas in this region that are largely protected and surrounded by tribal villages, so their

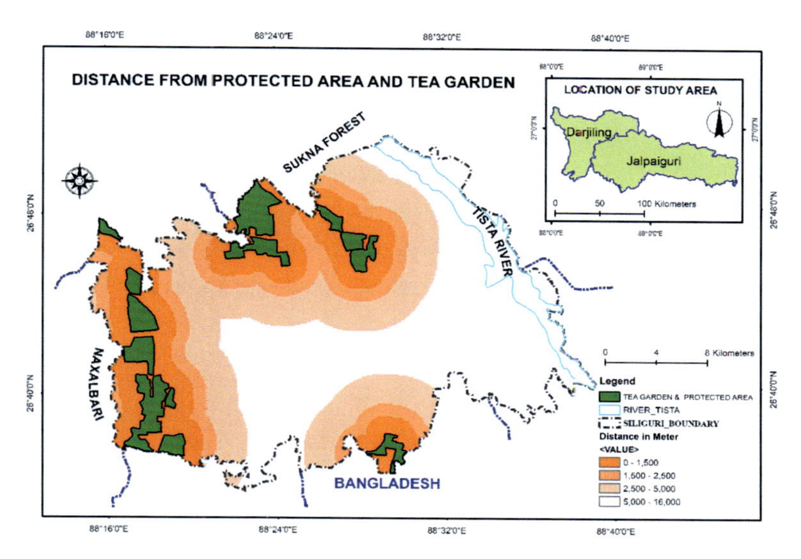

Fig. 18.13 Distance from reserved forest

chances of conversion to settlement are less. Initially, these forest areas are mapped, and distances are measured, with the maximum weightage given at a distance of 4 km from the forest. As shown in Fig. 18.13 distance from reserved forest is classified into four categories: (1) 0–1000 m, (2) 1000–2000 m, (3) 2000–8000 m, (4) 8000–15,000 m.

18.3.2 Determination of Weights

As discussed in the introduction section for site suitability mapping, there are several methods for determining weights for the criteria. In this study, weights were calculated using the analytical hierarchical process and fuzzy analytical hierarchical process weightage methods. Each of these weightage methods has its own advantages and disadvantages as discussed earlier; comparing both methods are expected to yield a more prominent outcome. The factors for this study were chosen based on previous literature that emphasized the determination of suitable sites for urban development. Furthermore to determining criteria weights local experts who are working in development authority, town planning departments, and local municipal authorities are interviewed.

18.3.2.1 Construction of Pairwise Comparison Matrix

A pairwise comparison matrix was developed with the help of the criteria and sub-criteria. After constructing pairwise comparison matrix, each criteria and sub-criteria were normalized. The AHP and FAHP weightage methods were used to determine relative weights.

18.3.2.2 Standardization of Weights Using AHP and FAHP

To perform weighted overlay analysis, all variables must be standardized in the same units as these are in different units. Standardization makes the measurement units uniform, and the scores lose their dimension along with their measurement unit (Effat and Hassan 2013). All the vector layers were converted to raster further reclassified according to suitability after performing weighted overlay analysis which finally gave the suitability map. The reclassify tool in ArcGIS Spatial Analyst is used to standardize the values of all criteria for comparison. Weights were assigned to the selected factors based on their significance in settlement concentration, as determined by expert opinions and the authors' judgement using the AHP and FAHP process.

18.3.3 Analytical Hierarchical Process (AHP)

Saaty introduced the Analytical Hierarchical Process (AHP) in 1980 as an efficient decision-making tool. The Pairwise Comparison Method (PCM) had been used to derive criteria weights or priority vectors. Since all the criteria have different degrees of importance while compared to each other, a matrix was constructed, where each criterion is compared with the other criteria based on relative importance, on a scale from 1 to 9. Weights are assigned with a help of expert guidance and intense field survey based (Satty 1990). Since the accuracy of the judgement can be statistically checked, this approach becomes more reliable. The calculation steps of AHP are given below.

Step 1: Compare the factors: The experts' perspectives were used to build a pairwise matrix. Calculation of pairwise comparison matrix is expressed in following equation:

$$\begin{bmatrix} C_{11} & C_{12} & C_{13} \\ C_{21} & C_{22} & C_{23} \\ C_{31} & C_{32} & C_{33} \end{bmatrix} \tag{18.2}$$

where C_{11} denotes respective value of ith row (first row) and jth column(first column) in this comparison matrix.

Step 2: Complete the matrix: The columns values were summed up individually in the matrix, the following equation expressed the column sum of each pairwise matrix:

$$C_{ij} = \sum_{i=1}^{n} C_{ij} \ldots\ldots\ldots\ldots \tag{18.3}$$

Step 3: Matrix normalization: after that, the following equations can be used to express the normalization for each column value:

$$X_{ij} = \frac{C_{ij}}{\sum_{i=1}^{n} C_{ij}} = \begin{bmatrix} X_{11} & X_{12} & X_{13} \\ X_{21} & X_{22} & X_{23} \\ X_{31} & X_{32} & X_{33} \end{bmatrix} \tag{18.4}$$

Step 4: Weight determination: After normalization, the row sum in the normalization matrix was divided by the total number of respective criteria. This process of computing resulted weights of priority vector are expressed as follows:

$$W_{ij} = \frac{\sum_{j=1}^{n} X_{ij}}{n} = \begin{bmatrix} W_{11} \\ W_{12} \\ W_{13} \end{bmatrix} \tag{18.5}$$

Step 5: Calculate the consistency ratio (CR.) Priority weights of each criteria can be used only after determining the consistency ratio (CR) value, and the credibility of judgmental value can only be evaluated using the CR value. As suggested by Saaty, the comparison matrix was considered to be consistent when the CR value is less than 0.10 (10%). When the calculated CR value appeared more than 0.10, the pairwise matrices were considered as adequately consistent, and the resulted weights of the criterion found unacceptable for further calculation implying that the pairwise matrix is needed to be re-examined for the comparisons (Saaty 1980).

The initial consistency vector was obtained by multiplying the pairwise matrix by weights vectors:

$$CV = \begin{bmatrix} C_{11} & C_{12} & C_{13} \\ C_{21} & C_{22} & C_{23} \\ C_{31} & C_{32} & C_{33} \end{bmatrix} * \begin{bmatrix} W_{11} \\ W_{12} \\ W_{13} \end{bmatrix} = \begin{bmatrix} C_{11}W_{11} + C_{12}W_{21} + C_{13}W_{31} \\ C_{21}W_{11} + C_{22}W_{21} + C_{23}W_{32} \\ C_{31}W_{11} + C_{32}W_{21} + C_{33}W_{33} \end{bmatrix} \tag{18.6}$$

Lambda(λ) max: The principal eigenvector ($\lambda_{\max}$) was derived by taking the average value of all consistency vector. The principal eigenvalue ($\lambda_{\max}$) was calculated using the following equation (Kanga et al. 2017):

$$\lambda_{max} = \sum_{i}^{n} CV_{ij} \tag{18.7}$$

Table 18.3 Value of the random index RI

n	1	2	3	4	5	6	7	8	9	10	11	12	13	14	15
RI	0.00	0.00	0.52	0.89	1.11	1.25	1.35	1.40	1.45	1.49	1.51	1.54	1.56	1.57	1.58

n = order of the matrix. *Source* Saaty (2008)

Consistency Index (CI) was computed to measure how far a matrix has deviated from consistency. Han and Tsay (1998) explained that the value of λ_{max} was required in calculating the consistency ratio. This was obtained by calculating matrix product of the pairwise comparison matrix of weighted vectors and then adding all elements of the resulted vector. After that a Consistency Index (CI) was calculated as

$$CI = \frac{\lambda_{max} - n}{n - 1} \tag{18.8}$$

where n was the number of criteria and λmax is the biggest eigenvalue (Han and Tsay 1998; Malczewski 1999).

Random Index (RI) was the consistency index of a pairwise comparison matrix generated randomly. Random index solely depends on the number of elements that were compared. RI index value for consistency index is shown in Table 18.3.

Consistency Ratio (C. R): Consistency of relative weights assigned for each criteria is finally judged through CR value (Kanga et al. 2017; Kayet et al. 2018). The final consistency ratio was calculated by comparing the CI with the Random Index (Malczewski 1999).

$$CR = \frac{CI}{CR} \tag{18.9}$$

The consistency ratio was designed in such a way it reflects a reasonable level of consistency in the pairwise comparisons. If the consistency ratio (CR) is 0.10 or less, the AHP analysis can be continued. If the consistency value is greater than 0.10, the judgmental value must be revised to counter the cause of the inconsistency.

18.3.4 Fuzzy Analytic Hierarchy Process

Zadeh (1965) was the first to introduce fuzzy set theory to express the ambiguity of human thought in decision-making in order to improve precision. As previously mentioned, the fuzzy AHP technique is an innovative weightage technique that evolved from the traditional AHP. Generally, decision-makers usually feel confident enough to make interval judgments instead making judgement in single numeric value. There were several methods for calculating priority vectors in FAHP, few of them are mentioned below: Van Laarhoven and Pedrycz (1983) designed the first

Table 18.4 Linguistic variables used in triangular fuzzy numbers (TFNs) for pairwise comparison

Saaty's scale of relative importance	Linguistic variables	Triangular fuzzy numbers	Reciprocal triangular fuzzy numbers
$C_{ij} = 9$	Absolutely important (AI)	(9, 9, 9)	(1/9, 1/9, 1/9)
$C_{ij} = 7$	Strongly important (SI)	(6, 7, 8)	(1/8, 1/7, 1/6)
$C_{ij} = 5$	Fairly important (FI)	(4, 5, 6)	(1/6, 1/5, 1/4)
$C_{ij} = 3$	Weakly important (WI)	(2, 3, 4)	(1/4, 1/3, 1/2)
$C_{ij} = 1$	Equally important (EI)	(1, 1, 1)	(1, 1, 1)
$C_{ij} = 2, C_{ij} = 4, C_{ij} = 6, C_{ij} = 8$	Intermediate value between two adjacent scale	(1, 2, 3) (3, 4, 5), (5, 6, 7), (7, 8, 9),	(1/3, 1/2, 1/1), (1/5, 1/4, 1/3), (1/7, 1/6, 1/5), (1/9, 1/8, 1/7)

fuzzy extension of the AHP method with triangular fuzzy numbers, geometric mean method (Buckley 1985), Fuzzy Extent analysis and synthetic decision (Chang 1992), Entropy-Based Fuzzy AHP (Chang 1996), fuzzy least square Method (Xu 2000), the direct fuzzification of the Method of (Csutora and Buckley 2001), Mikhailov's fuzzy preference programming (Mikhailov 2003) and two-stage logarithmic programming (Wang et al. 2005) Fuzzy Delphi Method (Hsu 2010).

Different researcher in their respective studies introduced different Fuzzy AHP methods. Buckley's (1985) geometric mean method is used to calculate criterion weights in this present research as it was easy to implement and is reliable. The following steps are necessary for the implementation of Buckley's Method.

Step 1: Considering expert opinion and linguistic variable: after discussions with experts on various suitable factors of urban expansion weights for each criterion are allocated. Instead of using mathematic equation, linguistic expressions were used such as "Very good", "good", "bad", and "very bad" during the implementation of the FAHP method, and it is easier and better way of reflecting human thinking style. Decision-makers compares or give the relative weight to each criteria or alternatives according to a linguistic variable as per Table 18.4.

Step 2: Pairwise comparison matrices were used to compare all parameters in the hierarchical structure. Linguistic expressions corresponding to pairwise comparison matrices are allocated by considering which was more important among the two criteria. Calculations of the pairwise triangular fuzzy comparison matrix are given below.

$$\breve{A}^{k} = a_{ij}^{k} = \begin{bmatrix} \tilde{a}_{11}^{k} & \tilde{a}_{12}^{k} & \cdots & \tilde{a}_{1n}^{k} \\ \tilde{a}_{21}^{k} & \tilde{a}_{22}^{k} & \cdots & \tilde{a}_{2n}^{k} \\ \cdots & \cdots & \cdots & \cdots \\ \tilde{a}_{m1}^{k} & \tilde{a}_{m2}^{k} & \cdots & \tilde{a}_{mn}^{k} \end{bmatrix} \quad (18.10)$$

where $\tilde{A}$ denotes pairwise comparison matrix, where $\overset{\sim k}{\mathbf{a}}_{12}$ indicated first decision-maker preference of first criteria over second criteria. $\mathbf{a_{ij}} = (\mathbf{l_{ij}}, \mathbf{m_{ij}}, \mathbf{u_{ij}})$ and calculation of reciprocal fuzzy number $\overset{\sim -1}{A}\ (\boldsymbol{l}, \boldsymbol{m}, \boldsymbol{u})^{-1} = \left(\frac{1}{u_{ji}}, \frac{1}{m_{ji}} \frac{1}{l_{ji}}\right)$ for i, j = 1 ….., n and i ≠ j criterion.

$$a_{ij} = \begin{cases} i > j & (1, 1, 3), (1, 3, 5), (3, 5, 7), (5, 7, 9), (7, 9, 9) \\ i = j & (1, 1, 1) \\ i < j & (1/3, 1, 1), (1/5, 1/3, 1), (1/7, 1/5, 1/3), (1/9, 1/7, 1/5), (1/9, 1/9, 1/7) \end{cases}$$

Wang et al. (2005).

where i>j indicates that criterion i is more important than criterion j, i = j denotes that both criteria are equally important, and i<j denotes that criterion j is more important than criterion i.

Step 3: Where there was more than one expert involved in decision-making process, the preferences of each expert $\overset{\sim k}{\mathbf{a}}_{\mathbf{ij}}$ were averaged $\tilde{\mathbf{a}}_{\mathbf{ij}}$ was calculated as per the given equation:

$$\tilde{\mathbf{a}}_{\mathbf{ij}} = \sum_{\mathbf{k}=1}^{\mathbf{K}} \mathbf{a}_{\mathbf{ij}}^{\mathbf{k}} \tag{18.11}$$

Step 4: Step 3: A revised pairwise comparison matrix based on averaged preferences is shown below.

$$\tilde{\mathbf{A}} = \begin{bmatrix} \tilde{\mathbf{a}}_{11} & \tilde{\mathbf{a}}_{12} & \cdots & \tilde{\mathbf{a}}_{1n} \\ \tilde{\mathbf{a}}_{21} & \tilde{\mathbf{a}}_{22} & \cdots & \tilde{\mathbf{a}}_{2n} \\ \cdots & \cdots & \cdots & \cdots \\ \tilde{\mathbf{a}}_{m1} & \tilde{\mathbf{a}}_{m2} & \cdots & \tilde{\mathbf{a}}_{mn} \end{bmatrix} \tag{18.12}$$

Step 5: Geometric mean: As proposed by Buckley (1985) determination of fuzzy geometric mean of each criterion could be estimated using the following formula:

$$\tilde{\mathbf{Z}} = \left[\prod_{\mathbf{j}=1}^{\mathbf{n}} \tilde{\mathbf{a}}_{\mathbf{ij}}\right]^{1/\mathbf{n}} \quad \ldots\ldots \mathbf{i} = 1, 2 \ldots\ldots \mathbf{m} \tag{18.13}$$

Step 6: Determination of fuzzy weights: The fuzzy weights ($\tilde{W}_i$) of each criterion were obtained by finding vector summation of each $\tilde{Z}_i$ and that findings are the (-1) power of summation vector, finally multiplying each $\tilde{Z}_I$ with reverse vectors to determine the fuzzy weight of a criterion.

$$\tilde{w}_i = \tilde{Z}_i \otimes \left(\tilde{Z}_1 \oplus \tilde{Z}_2 \oplus \ldots \tilde{Z}_n\right)^{-1} = (1_i,\ m_i,\ u_i) \quad (18.14)$$

Step 7: Centre of area (COA): Fuzzy weights composed of fuzzy triangular numbers (FTN), so it is needed to be transformed into crisp value using by Centre of Area (COA) method, Chou and Chang (2008) proposed this de-fuzzification techniques.

$$\mathbf{M_i} = \frac{(\mathbf{lw_i} + \mathbf{mw_i} + \mathbf{uw_i})}{3} \quad (18.15)$$

Step 8: Fuzzy normalization: $\boldsymbol{M_i}$ weights should be normalized, as they are crisp and a non-fuzzy number. It can be done using the following equation:

$$\mathbf{NW_i} = \frac{\mathbf{M_i}}{\sum_{\mathbf{i}=1}^{\mathbf{n}} \mathbf{M_i}} \quad (18.16)$$

Following these seven steps one can estimate normalized weights of both criteria and the alternatives. On the basis of these results, the criterion with highest weights was found to be most important.

Weighted linear combination for suitability mapping of settlement concentration by AHP and FAHP tchniques: All selected criteria layers (elevation, slope, road, change in existing built-up area, settlement patches, river, night-time light images, tea garden and protected area, reserved forest and flood prone area) were integrated using ArcGIS raster calculator to delineate site suitability of urban growth (Tables 18.5, 18.6, 18.7, 18.8, and 18.9). Weights were assigned according to the importance of the criteria of each layer for suitability mapping of AHP and FAHP method (Table 18.10). In the AHP method, the C.R matrix of the ten parameters was 0.0633. Therefore, pairwise comparison matrix used here in AHP method seemed to have enough internal consistency to be considered it acceptable. Since raster data format are less complicated rather than vector data format, all criteria layers were converted to raster format for further analysis (Chang 2006). Thereafter all criteria layers were multiplied by their respective weights to prepare final land suitability map using ArcGIS weighted overlay tools (Voogd 1983). Finally, summation of all criteria map and weights generate land suitability map. Equation for weightage overlay analysis is given below (Eastman et al. 1995):

$$S_i = \sum_{i=1}^{n} (w_i \times x_i) \quad (18.17)$$

$$S = \sum (\text{Wi} * \text{X}i) \text{ or } \sum (\text{criteria map1} * \text{weight}) + (\text{criteria map2} * \text{weight}) + (\text{criteria map3} * \text{weight})$$

Table 18.5 Pair-wise comparison matrix for AHP

Factors	Elevation (F1)	Slope (F2)	Road (F3)	Change in existing built-up area (F4)	Settlement patches (F5)	River (F6)	Night-time light images (F7)	Tea garden and protected area (F8)	Reserved forest (F9)	Flood prone river (F10)
Elevation (F1)	1	3	0.5	0.5	0.5	2	0.5	6	7	5
Slope (F2)	0.333	1	0.25	0.5	0.333	0.5	0.333	2	3	4
Road (F3)	2	4	1	2	2	2	0.5	4	4	5
Change in existing built-up area (F4)	2	2	0.5	1	0.333	0.5	0.5	4	4	6
Settlement patches (F5)	2	3	0.5	3	1	0.5	2	4	5	6
River (F6)	0.5	2	0.5	2	2	1	2	4	5	6
Night-time light images (F7)	2	3	2	2	0.5	0.5	1	4	4	5
Tea garden and protected area (F8)	0.167	0.5	0.25	0.25	0.25	0.25	0.25	1	2	2

(continued)

Table 18.5 (continued)

Factors	Elevation (F1)	Slope (F2)	Road (F3)	Change in existing built-up area (F4)	Settlement patches (F5)	River (F6)	Night-time light images (F7)	Tea garden and protected area (F8)	Reserved forest (F9)	Flood prone river (F10)
Reserved forest (F9)	0.143	0.333	0.25	0.25	0.2	0.2	0.25	0.5	1	2
Flood prone river (F10)	0.2	0.25	0.2	0.167	0.167	0.167	0.2	0.5	0.5	1
SUM	**10.343**	**19.083**	**5.95**	**11.667**	**7.283**	**7.617**	**7.533**	**30**	**35.5**	**42**

Table 18.6 Normalized weight matrix for AHP

Factors	F1	F2	F3	F4	F5	F6	F7	F8	F9	F10	Row Sum	Consistency measure (λ)	Criteria Weights	Rank
Elevation (F1)	0.097	0.157	0.084	0.043	0.069	0.263	0.066	0.2	0.197	0.119	1.295	10.7521	0.1295	5
Slope (F2)	0.032	0.052	0.042	0.043	0.046	0.066	0.044	0.067	0.085	0.095	0.572	10.5901	0.0572	7
Road (F3)	0.193	0.21	0.168	0.171	0.275	0.263	0.066	0.133	0.113	0.119	1.711	11.1335	0.1711	1
Change in existing built-up area (F4)	0.193	0.105	0.084	0.086	0.046	0.066	0.066	0.133	0.113	0.143	1.035	10.9698	0.1035	6
Settlement patches (F5)	0.193	0.157	0.084	0.257	0.137	0.066	0.265	0.133	0.141	0.143	1.577	11.1276	0.1577	2
River (F6)	0.048	0.105	0.084	0.171	0.275	0.131	0.265	0.133	0.141	0.143	1.497	10.9072	0.1497	3
Night-time light images (F7)	0.193	0.157	0.336	0.171	0.069	0.066	0.133	0.133	0.113	0.119	1.49	10.9537	0.149	4
Tea garden and protected area (F8)	0.016	0.026	0.042	0.021	0.034	0.033	0.033	0.033	0.056	0.048	0.343	10.5802	0.0343	8
Reserved forest (F9)	0.014	0.017	0.042	0.021	0.027	0.026	0.033	0.017	0.028	0.048	0.274	10.608	0.0274	9
Flood prone river (F10)	0.019	0.013	0.034	0.014	0.023	0.022	0.027	0.017	0.014	0.024	0.206	10.8715	0.0206	10
Sum	**1**	**1**	**1**	**1**	**1**	**1**	**1**	**1**	**1**	**1**	**10**	**10.849**	**1**	

Value of lambda max (λ) = 10.8494, Consistency index (CI) = 0.0943, Consistency ratio (CR) = 0.0633

Table 18.7 Pairwise comparison matrix with relative fuzzy (Part- 1)

Factors	F1			F2			F3			F4			F5		
	L	M	U	L	M	U	L	M	U	L	M	U	L	M	U
Elevation (F1)	1.000	1.000	1.000	2.000	3.000	4.000	0.333	0.500	1.000	0.333	0.500	1.000	0.333	0.500	1.000
Slope (F2)	0.250	0.333	0.500	1.000	1.000	1.000	0.200	0.250	0.333	0.333	0.500	1.000	0.250	0.333	0.500
Road (F3)	1.000	2.000	3.000	3.000	4.000	5.000	1.000	1.000	1.000	1.000	2.000	3.000	1.000	2.000	3.000
Change in existing built-up area (F4)	1.000	2.000	3.000	1.000	2.000	3.000	0.333	0.500	1.000	1.000	1.000	1.000	0.250	0.333	0.500
Settlement patches (F5)	1.000	2.000	3.000	2.000	3.000	4.000	0.333	0.500	1.000	2.000	3.000	4.000	1.000	1.000	1.000
River (F6)	0.333	0.500	1.000	1.000	2.000	3.000	0.333	0.500	1.000	1.000	2.000	3.000	1.000	2.000	3.000
Night-time light images (F7)	1.000	2.000	3.000	2.000	3.000	4.000	1.000	2.000	3.000	1.000	2.000	3.000	0.333	0.500	1.000
Tea garden and protected area (F8)	0.143	0.167	0.200	0.333	0.500	1.000	0.200	0.250	0.333	0.200	0.250	0.333	0.200	0.250	0.333
Reserved forest (F9)	0.125	0.143	0.167	0.250	0.333	0.500	0.200	0.250	0.333	0.200	0.250	0.333	0.167	0.200	0.250
Flood prone river (F10)	1.000	0.200	3.000	0.200	0.250	0.333	0.167	0.200	0.250	0.143	0.167	0.200	0.143	0.167	0.200
Sum	6.851	10.343	17.867	12.783	19.083	25.833	4.100	5.950	9.250	7.210	11.667	16.867	4.676	7.283	10.783
Geometric mean	108181			18.4711			6.0881			112364			7.1613		

Table 18.8 Pairwise comparison matrix with relative fuzzy (Part- 2)

Factors	F6			F7			F8			F9			F10		
	L	M	U	L	M	U	L	M	U	L	M	U	L	M	U
Elevation (F1)	1.000	2.000	3.000	0.333	0.500	1.000	5.000	6.000	7.000	6.000	7.000	8.000	4.000	5.000	6.000
Slope (F2)	0.333	0.500	1.000	0.250	0.333	0.500	1.000	2.000	3.000	2.000	3.000	4.000	3.000	4.000	5.000
Road (F3)	1.000	2.000	3.000	0.333	0.500	1.000	3.000	4.000	5.000	3.000	4.000	5.000	4.000	5.000	6.000
Change in existing built-up area (F4)	0.333	0.500	1.000	0.333	0.500	1.000	3.000	4.000	5.000	3.000	4.000	5.000	5.000	6.000	7.000
Settlement patches (F5)	0.333	0.500	1.000	1.000	2.000	3.000	3.000	4.000	5.000	4.000	5.000	6.000	5.000	6.000	7.000
River (F6)	1.000	1.000	1.000	1.000	2.000	3.000	3.000	4.000	5.000	4.000	5.000	6.000	5.000	6.000	7.000
Night-time light images (F7)	0.333	0.500	1.000	1.000	1.000	1.000	3.000	4.000	5.000	3.000	4.000	5.000	4.000	5.000	6.000
Tea garden and protected area (F8)	0.200	0.250	0.333	0.200	0.250	0.333	1.000	1.000	1.000	1.000	2.000	3.000	1.000	2.000	3.000

(continued)

Table 18.8 (continued)

Factors	F6			F7			F8			F9			F10		
	L	M	U	L	M	U	L	M	U	L	M	U	L	M	U
Reserved forest (F9)	0.167	0.200	0.250	0.200	0.250	0.333	0.333	0.500	1.000	1.000	1.000	1.000	1.000	2.000	3.000
Flood prone river (F10)	0.143	0.167	0.200	0.167	0.200	0.200	0.333	0.500	1.000	0.333	0.500	1.000	1.000	1.000	1.000
Sum	4.843	7.617	11.783	4.817	7.533	11.367	22.667	30.000	38.000	27.333	35.500	44.000	33.000	42.000	51.000
Geometric mean	7.5749			7.4437			29.5641			34.9509			41.3470		

Table 18.9 Average weights and normalized weights, percentage and rank of each criteria in FAHP

Factors	Fuzzy geometric mean ($\tilde{Z}$)			Fuzzy weight ($\tilde{W}_i$)			Average weights	Normalized weight criterion (N_i)	Percentages	Rank
	L	M	U	L	M	U				
Elevation	1.1147	1.5475	2.2938	0.0620	0.1228	0.2637	0.1495	0.1263	12.63%	**5**
Slope	0.5394	0.7490	1.0960	0.0300	0.0594	0.1260	0.0718	0.0607	6.07%	**7**
Road	1.4310	2.1919	3.0085	0.0796	0.1740	0.3458	0.1998	0.1688	16.88%	**1**
Change in existing built-up area	0.9162	1.3195	1.9482	0.0510	0.1047	0.2239	0.1265	0.1069	10.69%	**6**
Settlement patches	1.3887	2.0107	2.8058	0.0772	0.1596	0.3225	0.1864	0.1575	15.75%	**2**
Distance from river	1.2089	1.8541	2.6489	0.0672	0.1471	0.3045	0.1730	0.1461	14.61%	**3**
Night-time light images	1.2311	1.8541	2.6360	0.0685	0.1471	0.3030	0.1729	0.1461	14.61%	**4**
Tea garden and protected area	0.3298	0.4480	0.6123	0.0183	0.0356	0.0704	0.0414	0.0350	3.50%	**8**
Reserved forest	0.2732	0.3527	0.4745	0.0152	0.0280	0.0545	0.0326	0.0275	2.75%	**9**
Flood prone river	0.2664	0.2732	0.4573	0.0148	0.0217	0.0526	0.0297	0.0251	2.51%	**10**
Total	**8.6994**	**12.6005**	**17.9813**				**1.1836**	**1.0000**	1.00	
Reciprocal fuzzy number	**0.0556**	**0.0794**	**0.1150**							

Table 18.10 Computation of criteria, sub-criteria weights in AHP and FAHP methods

Main criteria	Sub-criteria	Area (SQ.KM)	Area in percentage	Suitability Scale	Sub-criteria weights	AHP weights	FAHP weights
Elevation in meter	15–85	167.27	24.07	4	0.50	0.1295	0.1263
	85–155	309.93	44.60	3	0.28		
	155–225	146.79	21.12	2	0.13		
	225–285	70.96	10.21	1	0.08		
Slope in degree	0–8.00	191.96	28.24	4	0.40	0.0572	0.0607
	8.00–16.00	332.44	48.92	3	0.35		
	16.00 24.00	121.33	17.85	2	0.16		
	24.00–62.00	33.89	4.99	1	0.09		
Distance from settlement patches	0–500 (meter)	51.96	11.47	4	0.53	0.1711	0.1688
	500–1000	95.97	21.19	3	0.25		
	1000–1500	50.37	11.12	2	0.13		
	1500>	254.57	56.21	1	0.09		
Distance from river sources	0–500	345.24	49.69	4	0.39	0.1035	0.1069
	500–1000	180.42	25.97	3	0.28		
	1000–1500	95.67	13.77	2	0.20		
	1500–2000	73.41	10.57	1	0.14		
Road density	0–0.50 (SQ.KM)	607.19	87.34	1	0.06	0.1577	0.1575
	0.50–1.00	44.45	6.39	2	0.13		
	1.00–1.50	26.59	3.82	3	0.18		
	1.50–2.50	16.93	2.43	4	0.64		
Change in built-up area of the city	−0.74–(−0.34) water	32.11	4.80	1	0.18	0.1497	0.1461

(continued)

Table 18.10 (continued)

Main criteria	Sub-criteria	Area (SQ.KM)	Area in percentage	Suitability Scale	Sub-criteria weights	AHP weights	FAHP weights
	−0.33–(−0.05) vegetation	349.79	52.29	1	0.18		
	0.04–0.28 (built-up area)	153.05	22.88	4	0.41		
	0.28–0.65 (sand)	134.00	20.03	2	0.24		
Night-time light Images	0–2000	460.60	66.17	1	0.12	0.1490	0.1461
	2000–4000	67.36	9.68	2	0.10		
	4000–10,000	89.54	12.86	3	0.16		
	10,000–15,450	78.54	11.28	4	0.62		
Distance from reserved forest	0–1000 (meter)	220.97	16.42	1	0.10	0.0343	0.0350
	1000–2000	69.12	13.68	2	0.12		
	2000–8000	311.44	11.76	4	0.60		
	8000–15,000	93.18	58.17	3	0.17		
Distance from flood prone river (Tista)	0–1000	75.23	10.83	1	0.07	0.0274	0.0275
	1000–8000	177.45	25.54	2	0.14		
	8000–16,000	189.26	27.24	4	0.40		
	16,000–32,000	252.71	36.37	4	0.40		
Distance from tea garden and protected area	0–1500	178.63	25.71	1	0.05	0.0206	0.0251
	1500–2500	73.00	10.51	2	0.11		
	25,000–5000	164.55	23.69	4	0.48		
	5000–16,000	278.40	40.08	3	0.36		

$$+ \text{ (criteria map4} * \text{weight) } \ldots\ldots \text{ (criteria mapN} * \text{weight)} \quad (18.18)$$

Where $\mathbf{S_i}$ denotes final suitability map, where $\mathbf{w_i}$ represents criteria map and $\mathbf{x_i}$ represent criteria weights. For present investigation suitability map is divided into four suitability classes based on their suitability score.

Validation: The criteria were weighted and scored in terms of their importance for settlement concentration in this study. The overall consistency ratio (CR) of AHP method was 0.0633 which was below the 0.10 ratio, suggesting that the judgement had a reasonable level of consistency. High suitable area of both the methods AHP and FAHP are intersected and it is found that 92.66% of the pixel are matched. The validation method is used here to determine about what percentage of current settlement patches fall under suitability map of highly suitable and moderately suitable zone. The final suitability map produced from the AHP and FAHP model was verified to ensure accuracy of model by actual settlement map of study area. Existing settlement areas were extracted from the land use map of Siliguri town. Using the "geometry intersection" tool in Arc GIS software, highly suitable and moderately suitable class was intersected with the layer of existing map of settlement area. Results suggest that AHP >94.63% and FAHP >96.89% percentage of areas fell within the actual settlement areas modelled "highly suitable" and "model suitable", it suggests the validity of model (Fig. 18.18).

18.4 Results and Discussion

Uncontrolled urban expansion always poses a great threat not only to sustainable city development but also creates different environmental problems. The unexpected growth of cities engulfs the agricultural land, forest and wetland area and also influence local climate. As a result, selecting an appropriate site for urban expansion becaming a challenging task for the urban planner. In this study a suitability map has been prepared by AHP and FAHP weightage methods after considering ten factors of settlement concentration (Figs. 18.14, 18.15). As per objectives, same weights were given to both AHP and FAHP method to make a logical comparison. Though AHP (0.0633) method ultimately gives a better consistency ratio (CR) value than the CR value of FAHP (0.080) but as discussed earlier people are more confident to give interval judgment. Higher consistency in statistics does not always ensure better accuracy for site suitability mapping. It appeared from model validation that the accuracy of FAHP method was 96.89%, while accuracy of AHP method was 94.63%, though both the method performed well in matching of pixel over 92%. Interestingly factors weights in both are so similar that all the criteria in both methods have same ranking according to their importance. Here some outcomes from this study are discussed: Road density or Road connectivity (1st rank) was found as a most important factor for the concentration of settlement found in this region; density of settlement is highly correlated with the density of roads. Distance from settlement

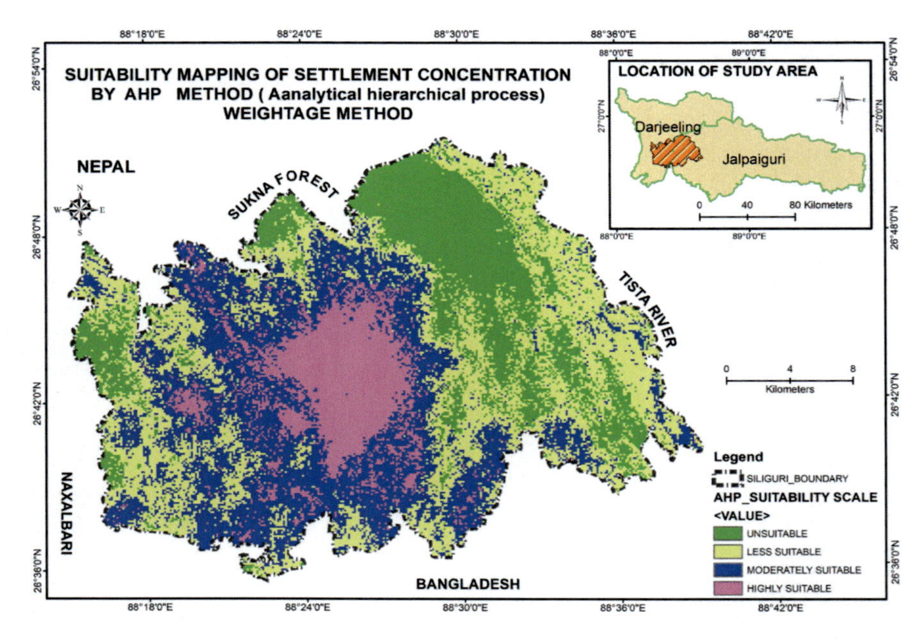

Fig. 18.14 Suitability mapping by AHP weightage method

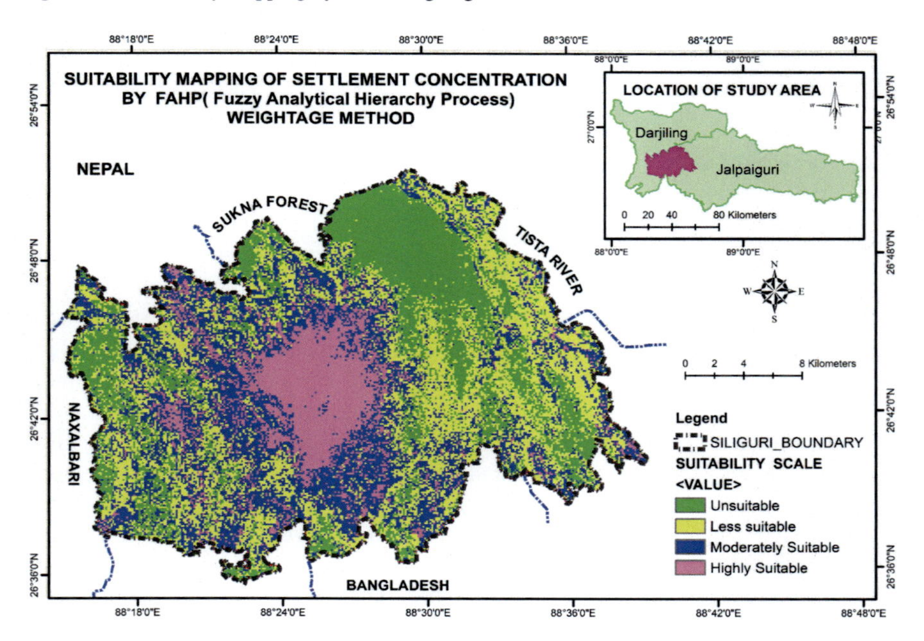

Fig. 18.15 Suitability mapping by FAHP weightage method

patches ranked 2nd as in most of the cases settlement is expanding adjacent to the existing patches of settlements. Proximity to the rivers ranked 3rd because these are the only source of water in this region and night-time light sources ranked 4th as it is found suitable to detect very accurately the expansion of urban spaces. While comparing resulted weights of AHP and Fuzzy AHP methods, it was found that there was no significant difference in resulted weights (Figs. 18.16 and 18.17; Tables 18.10

Fig. 18.16 Share of area under different suitability scale in AHP method

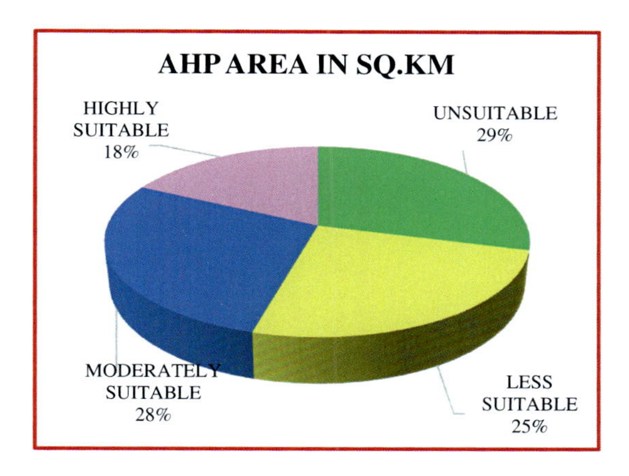

Fig. 18.17 Share of area under different suitability scale in FAHP method

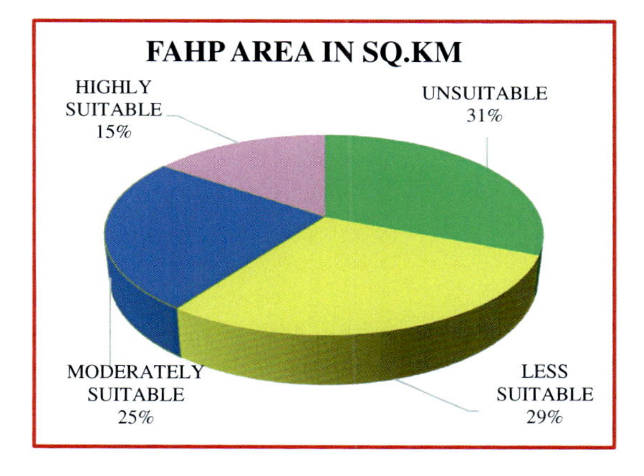

Table 18.11 Suitable area under AHP method

FAHP	Area in sq.km
Unsuitable	212.69
Less suitable	197.93
Moderately suitable	170.10
Highly suitable	104.78
Total	685.51

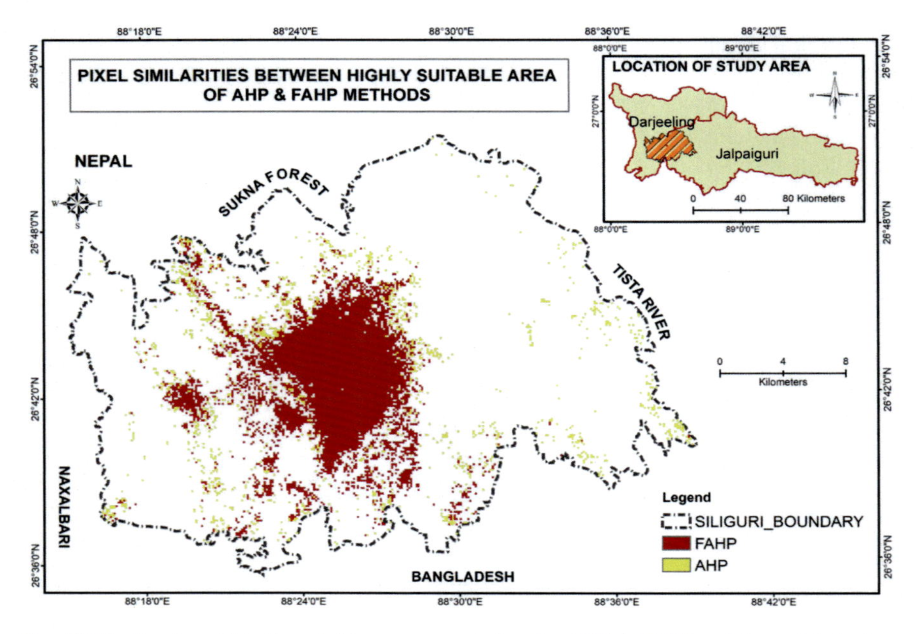

Fig. 18.18 Overlay of highly suitable area generated under AHP & FAHP weightage Methods

Table 18.12 Suitable area under FAHP method

FAHP	Area in sq.km
Unsuitable	198.92
Less suitable	172.42
Moderately suitable	192.23
Highly suitable	121.92
Total	685.50

and 18.11). Pearson correlation coefficient value of 0.85 justified that there exists a positive correlation between AHP and the FAHP scores (Fig. 18.18 and Table 18.12).

18.5 Conclusion

Urban sprawl and rapid growth of urban population always present a menace to the environment and human beings, disrupting environmental equilibrium. The present study demonstrates the efficiency of Remote Sensing and GIS technology and, Multi-criteria decision-making techniques for producing site suitability map of settlement concentration in Siliguri town and surrounding rural landscapes. Though a number of research have already been conducted to investigate the application of the fuzzy analytical hierarchy process in decision-making and planning purposes ranging from physical to environmental science, unfortunately, no conclusive evidence was found in the previous literature to show how far Fuzzy AHP was better than AHP. As humans are confident in making judgment in interval rather than the crisps numerical value, most of the researchers agreed with the fact that FAHP provides better resultant outcomes than the classical AHP. This study concludes that, after comparing both weightage methods, FAHP with the superior capability of addressing inconsistency and fuzziness in geospatial datasets, and also comparing with the current map of settlement patches achieved better accuracy in the modeling process. It can be said that FAHP method is slightly more accurate for site suitability mapping of settlement concentration. In this present study, an attempt has been made to capture the difference in results obtained by using AHP and FAHP models, respectively. From this study, if factors were chosen carefully and if weights were assigned with a deeper understanding of study areas, the resultant outcomes would be equally good irrespective of whether fuzzy mathematics is embedded with AHP or not. Policymakers should pay more attention to the selected criteria and sub-criteria; it would be beneficial achieving a balance between economic development and environmental concern. However, this result was not validated with AUC curve or RMSE method in this research. Therefore, further improvements such as validation mechanisms and other intelligent methods are still open for future investigation.

References

Abbasi NA, Ali MNHA, Abbasi B, Soomro SA, Nangraj NAK, Sahto JGM, Morio SA (2019) Assessment of agricultural land suitability using fuzzy set method. Pak J Agric Res 32(2):252–259. https://doi.org/10.17582/journal.pjar/2019/32.2.252.259

Aburas MM, Abdullah SHO, Ramli MF, Asha'ari ZH (2017) Land suitability analysis of urban growth in Seremban Malaysia, using GIS based analytical hierarchy process. Proc Eng 198:1128–1136. https://doi.org/10.1016/j.proeng.2017.07.155

Ali NH, Sabri IA, Ismail NM (2012) Rating and Ranking criteria for selected islands using fuzzy analytic hierarchy process (FAHP)

Ameryoun A, Zaboli R, Haghoost AA, Mirzae T, Tofighi S, Shamsi MA (2014) Applying Fuzzy Analytic Hierarchy Process (F.A.H.P.) in healthcare system. Int J Med Res Rev 2(6):610–617. https://doi.org/10.17511/ijmrr.2014.i06.25

As-syakur AR, Adnyana IW, Arthana IW, Nuarsa IW (2012) Enhanced Built-Up and Bareness Index (E.B.B.I.) for Mapping Built-Up and Bare Land in an Urban Area. Remote Sens 4:2957–2970. https://doi.org/10.3390/rs4102957

Banik S (2012) Socio economic impact of Gajoldoba Teesta multiple river valley project. The Journal of Bengal Geographer, pp. 46–52, October, 2012.www.dassonopen.com

Buckley JJ (1985) Fuzzy hierarchical analysis. Fuzzy Sets Syst 17(3):233–247. https://doi.org/10.1016/0165-0114(85)90090-9

Boender CGE, de Graan JG, Lootsma FA (1989) Multi-criteria decision analysis with fuzzy pairwise comparisons. Fuzzy Sets Syst 29(2):133–143. https://doi.org/10.1016/0165-0114(89)90187-5

Chang D (1992) Extent analysis and synthetic decision, optimization techniques and applications. World Scientific, Singapore 1:352–355

Chang D (1996) Applications of the extent analysis method on fuzzy A.H.P. Eur J Operat Res 95:649–655. https://doi.org/10.1016/0377-2217(95)00300-2

Chou S-Y, Chang Y-H (2008) A decision support system for supplier selection based on a strategy-aligned fuzzy SMART approach. Expert Syst Appl 34:2241–2253. https://doi.org/10.1016/j.eswa.2007.03.001

Csutora R, Buckley JJ (2001) Fuzzy hierarchical analysis: the Lambda-Max method. Fuzzy Sets Syst 120:181–195. https://doi.org/10.1016/S0165-0114(99)00155-4

Ceola S, Laio F, Montanari A (2015) Human-impacted waters: new perspectives from global high-resolution monitoring. Water Resour Res 51(9):7064–7079. https://doi.org/10.1002/2015WR017482

Cinzano P, Falchi F, Elvidge CD (2001) The first World Atlas of the artificial night sky brightness. Monthly Not Royal Astronom Soc 328(3, December):689–707. https://doi.org/10.1046/j.1365-8711.2001.04882.x

Chang DY (1992) Extent analysis and synthetic decision. Optimization Techniques and Applications, World Scientific, Singapore 1, 352–355, 1992. 352–355

Chang Ni-Bin, Parvathinathan G, Breeden Jeff B (2008) Combining G.I.S. with fuzzy multicriteria decision-making for landfill siting in a fast-growing urban region. J Environ Manag 87:1153. https://doi.org/10.1016/j.jenvman.2007.01.011

Ceren E, Akbaş H (2019) A comparative analysis of fuzzy TOPSIS and geographic information systems (GIS) for the location selection of shopping malls: a case study from Turkey. Sustainability. https://doi.org/10.3390/su11143837

Chougale Santosh, Krishnaiah Chikkamadaiah, Deshbhandari Praveen (2018) Site suitability analysis for urban development using GIS based multicriteria evaluation technique: a case study in Chikodi Taluk, Belagavi District, Karnataka, India. IOP Conference Series: Earth and Environmental Science. DOI: https://doi.org/10.1088/1755-1315/169/1/012017

Chou TY, Hsu CL, Chen MC (2008) A fuzzy multicriteria decision model for international tourist hotels location selection. Int J Hosp Manag 27:293–301. https://doi.org/10.1016/j.ijhm.2007.07.029

Climate data. https://mausam.imd.gov.in/. Retrieved from 21 march 2019

Cohen EJ, Small C (1998) Hypsographic demography: the distribution of human population by altitude November 24. Appl Phys Sci 95:14009–14014, November. https://doi.org/10.1073/pnas.95.24.14009

Deng Hepu (1999) Multi-criteria analysis with fuzzy pairwise Comparison. Int J Approx Reas IJAR 21:726–731, vol. 2. https://doi.org/10.1016/S0888-613X(99)00025-0

Dutta B, Giunchiglia F, Maltese V (2011) A facet-based methodology for geo-spatial modeling. In: Claramunt C, Levashkin S, Bertolotto M (eds) GeoSpatial semantics. GeoS 2011. Lecture Notes in Computer Science, vol 6631. Springer, Berlin, Heidelberg. https://doi.org/10.1007/978-3-642-20630-6_9

Effat, Hala, Hassan Ossman (2013) Designing and evaluation of three alternatives highway routes using the Analytical Hierarchy Process and the least-cost path analysis, application in Sinai Peninsula, Egypt. The Egyptian Journal of Remote Sensing and Space Science, 16, pp 141–151. http://dx.doi.org/https://doi.org/10.1016/j.ejrs.2013.08.001

Erensal YC, Oncan T, Demircan ML (2006) Determining key capabilities in technology management using fuzzy analytic hierarchy process: a case study of Turkey. Inf Sci 176:2755–2770. https://doi.org/10.1016/j.ins.2005.11.004

Elvidge CD, Baugh KE, Kihn EA, Kroehl HW, Davis ER (1997) Mappingcity lights with nighttime data from the D.M.S.P. Operational Linescan System, Photogrammetric Engineering and Remote Sensing 63:727–734. URL: http://worldcat.org/issn/00991112

Fang Yu, Jawitz James (2019) The evolution of human population distance to water in the USA from 1790 to 2010. Nat Commun. https://doi.org/10.1038/s41467-019-08366-z

Flora and Fauna of north Bengal. Nbtourism.tripod.com. Retrieved 21 March 2019.

Ferretti V, Pomarico S (2012) An integrated approach for studying the land suitability for ecological corridors through spatial multicriteria evaluations. Environ Dev Sustain 15:859–885. https://doi.org/10.1007/s10668-012-9400-6

Food and Agriculture Organization (1976) A framework for land evaluation. FAO soils bulletin 52, FAO, Rome, 79p. [Outlines the basic principles of the FAO approach to land evaluation and land use planning]

Gao N, Li F, Zeng H, van Bilsen D, De Jong M (2019) Can more accurate night-time remote sensing data simulate a more detailed population distribution? Sustainability (Switzerland) 11(16). https://doi.org/10.3390/su11164488

Gokhale Nitin A (1998) Chicken's Neck, All choked up. Outlook. Retrieved 27 February 2011

Ghosh Atig (2017) Fluid futures: migrant labour and trafficked lives in Millennial Siliguri. Tata Institute of Social Sciences, Patna Centre Takshila Campus, p 1, 26. https://tiss.edu/

Grimm, N. B., Faeth, S. H., Golubiewski, N. E., Redman, C. L., Wu, J., Bai, X., & Briggs, J. M. (2008). Global change and the ecology of cities.Science, 319(5864), 756–760. Vol. Doi: https://doi.org/10.1126/science.1150195.

Herold Martin, Gardner Margaret, Roberts Dar (2003) Spectral resolution requirements for mapping urban areas. Geoscience and Remote Sensing, IEEE Transactions on. 41:1907–1919. Doi: https://doi.org/10.1109/TGRS.2003.815238

Herold Martin, Gamba Paolo (2009) Global mapping of human settlement: experiences, datasets, and prospects. Taylor & Francis series in remote sensing applications C.R.C. Press. https://doi.org/10.1201/9781420083408

Hamzeh S, Mokarram M, Alavipanah SK (2014) Combination of Fuzzy and A.H.P. methods to assess land suitability for barley: Case Study of semi-arid lands in the southwest of Iran. Desert 19(2):173–181. Doi: https://doi.org/10.22059/jdesert.2014.52346

Han WJ, Tsay WD (1998) Formulation of quality strategy using analytic hierarchy process. In Proceedings of the Twenty Seven Annual Meeting of the Western Decision Science Institute, University of Northern Colorado, Greeley, CO, pp 580–583

Hsu Y, Lee C, Kreng V (2010) The application of fuzzy delphi method and fuzzy AHP in lubricant regenerative technology selection. Expert Syst Appl 37:419–425. https://doi.org/10.1016/j.eswa.2009.05.068

https://siliguritimes.com/gajoldoba-police-outpost-submerged-in-flood-water/

India's Forest Conservation Legislation: Acts, Rules, Guidelines, from the Official website of: Government of India, Ministry of Environment & Forests. date of accessed. 12/02/2020

Kabir Golam, Hasin M (2013) Multi-criteria inventory classification through integration of fuzzy analytic hierarchy process and artificial neural network. Int J Ind Syst Eng (I.J.I.S.E.) 14:74–103. Article in Press. DOI: https://doi.org/10.1504/IJISE.2013.052922

Kamas W, Naji H, Hasan A (2017) Evaluation of urban planning projects criteria using fuzzy AHP technique. J Eng 23(5):12–26. https://joe.uobaghdad.edu.iq/index.php/main/article/view/38

Kayet N, Chakrabarty A, Pathak K, Sahoo S, Dutta T, Hatai BK (2018) Comparative analysis of multi-criteria probabilistic FR and AHP models for forest fire risk (FFR) mapping in Melghat tiger reserve (MTR) forest. J for Res 31:1–15. https://doi.org/10.1007/s11676-018-0826-z

Kanga S, Tripathi G, Singh SK (2017) Forest fire hazards vulnerability and risk assessment in Bhajji forest range of Himachal Pradesh (India): a geospatial approach. J Remote Sens GIS 8:1–16

Kong F, Hongyan L (2005) Applying fuzzy analytic hierarchy process to evaluate successfactors of e-commerce. Int J Infor Syst Sci 13(4):406–412. http://www.math.ualberta.ca/ijiss/

Kummu M, de Moel H, Ward PJ, Varis O, Perc M (2011) How close do we live to water? A global analysis of population distance to freshwater bodies. PLoS ONE 6(6):e20578. https://doi.org/10.1371/journal.pone.0020578

Maddahi Z, Jalalian A, Kheirkhah Zarkesh MM, Honarjo N (2017) Land suitability analysis for rice cultivation using a GIS-based fuzzy multi-criteria decision-making approach: central part of Amol District. Iran Soil Water Res 12:29–38. https://doi.org/10.17221/1/2016-SWR

Mohammadrezaei N, Pazira Ebrahim, Sokoti R, Ahmadi Abbas (2013) Comparing the performance of fuzzy AHP and parametric method to evaluate of land suitability of wheat production in the southern plain of Urmia. Int J Agron Plant Product 4(12):3438–3443. http://www.ijappjournal.com

Mokarram M, Mirsoleimani A (2018) Using Fuzzy-AHP and order weight average (OWA) methods for land suitability determination for citrus cultivation in ArcGIS (Case study: Fars province Iran). Physica A. https://doi.org/10.1016/j.physa.2018.05.062
Moslem S, Ghorbanzadeh O, Blaschke T, Duleba S (2019) Analysing stakeholder consensus for a sustainable transport development decision by the Fuzzy AHP and interval AHP. Sustainability 11(12):3271. https://doi.org/10.3390/su11123271
Malczewski J (1999) G.I.S. and Multicriteria Desision Analysis, John Wiley and Sons Inc, New York, NY
Mikhailov L (2003) Deriving priorities from fuzzy pair wise comparison judgements. Fuzzy Sets Syst 134(3):365–385. https://doi.org/10.1016/S0165-0114(02)00383-4
Noorollahi E, Fadai D, Akbarpour Shirazi M, Ghodsipour S (2016) Land suitability analysis for solar farms exploitation using GIS and fuzzy analytic hierarchy process (FAHP)—A Case Study of Iran. Energies, 9(8):643. https://doi.org/10.3390/en9080643
Nusrat Jannat (2017) Investigation of settlement pattern and dwelling system of the tea workers. Community in Chittagong Region, Bangladesh International Journal of Engineering and Innovative Technology (I.J.E.I.T.) 6(10). doi:https://doi.org/10.17605/osf.io/s3dmr
Nurfadila JS, Baja S, Neswativ R, Rukmana D (2020) Evaluation of land suitability for coffee plants based on fuzzy logic in Enrekang district. Conference: I.O.P. Conference Series: Earth and Environmental Science. DOI: https://doi.org/10.1088/1755-1315/486/1/012069
Pal R, Biswas SS, Mondal B, Pramanik MK (2016) Landslides and floods in the Tista Basin (Darjeeling and Jalpaiguri Districts): Historical Evidence, Causes and Consequences
Rawat et al. (2010) Terrain characterization for land suitability analysis of the Igo River Basin, Eastern Himalaya, Arunachal Pradesh, India, Asian Journal of Geo informatics, Vol. 10, No. 4. http://203.159.29.7/index.php/journal/article/view/10/9
Rodcha R, Tripathi NK, Shrestha RP (2019) Comparison of cash crop suitability assessment using parametric, A.H.P., and F.A.H.P. Methods Land, M.D.P.I. Open Access J 8(5):1–22, May. DOI: https://doi.org/10.3390/land8050079
Siliguri-The gateway of Northeast India. www.siliguri.gov.in. Retrieved 22 March 2019
Saaty TL (1980) The analytic hierarchy process: planning, priority setting, resource allocation. McGraw-Hill, New York, NY, p 437
Saaty TL (1977) A scaling method for priorities in hierarchical structures. J Math Psychol 15:234–281. https://doi.org/10.1016/0022-2496(77)90033-5
Saaty TL (1990) How to make a decision: the analytic hierarchy process. Eur J Oper Res 48(1):9–26. https://doi.org/10.1016/0377-2217(90)90057-I
Saaty TL (2008) Making and validating complex decisions with the AHP / ANP 2. J Syst Sci Syst Eng 14:1–36
Siliguri corridor 'vulnerable', warns security expert. D.N.A. 22 July 2007. Accessed 30 May 2008
Siliguri (2020) World Public Library. http://www.worldlibrary.org/articles/siliguri. (last accessed February 12, 2020)
Siliguri-about location. www.wbtourismgov.in. Retrieved 8 June 2019
Siliguri-description. www.siliguri.gov.in. Retrieved 8 June 2019
"Siliguri metropolitan area"www.siliguri.gov.in. Retrieved 5 June 2019.
Siliguri- the gateway to the northeast India. www.siligurismc.in. Retrieved 8 June 2019
Small C, Pozzi F, Elvidge C (2005) Spatial analysis of global urban extent from DMSP-OLS night lights. Remote Sens Environ 96:277–291. https://doi.org/10.1016/j.rse.2005.02.002
Small C, Elvidge C (2013) Night on Earth: mapping decadal changes of anthropogenic night light in Asia. Int J Appl Earth Obs Geoinf 22:40–52. https://doi.org/10.1016/j.jag.2012.02.009
Tao J, Chen H, Xiao D (2017) Influences of the natural environment on traditional settlement patterns: a case study of Hakka traditional settlements in eastern Guangdong province. J Asian Archit Build Eng 16(1):9–14. https://doi.org/10.3130/jaabe.16.9
Ullah Kazi, Mansourian Ali (2014) Evaluation of land suitability for urban land-use planning: case study Dhaka City. Transactions in G.I.S. 20(1). https://doi.org/10.1111/tgis.12137

Urban Agglomerations/Cities having population 1 lakh and above (PDF). Provisional Population Totals, Census of India 2011. censusindia.gov. Retrieved 21 October 2011

Van R, Tang H, Groenemans R, and Sinthurahat S (1996) Application of Fuzzy logic to land suitability for rubber production in Peninsular Thailand. Geoderma 70:1–19. https://doi.org/10.1016/0016-7061(95)00061-5

Vahidnia MH, Alesheikh A, Alimohammadi A (2009) Hospital site selection using fuzzy AHP and its derivatives. J Environ Manage 90:3048–3056. https://doi.org/10.1016/j.jenvman.2009.04.010

Van Laarhoven PJM, Pedrycz W (1983) A fuzzy extension of Saaty's priority theory. Fuzzy Sets Syst 11:229–241. https://doi.org/10.1016/S0165-0114(83)80082-7

Voogd H (1983) Multicriteria evaluation for urban and regional planning. Pion, London

Wang W (2019) Site selection of fire station in cities based on geographic information system and fuzzy analytic hierarchy process

Wang YM, Yang JB, Xu DL (2005) A two-stage logarithmic goal programming method for generating weights from interval comparison matrices. Fuzzy Sets Syst 152:475–498. https://doi.org/10.1016/j.fss.2004.10.020

Xu R (2000) Fuzzy least square priority method in the analytic hierarchy process. Fuzzy Sets Syst 112(3):395–404. https://doi.org/10.1016/S0165-0114(97)00376-X

Zadeh LA (1965) Fuzzy sets. Inf Control 8:338–353. https://doi.org/10.1016/S0019-9958(65)90241-X

Zhou D, He Y, Bai C, JichaoL, V (2014) Research on evaluation for population quality in china based on FAHP. J Appl Sci 14:296–300. Doi: https://doi.org/10.3923/jas.2014.296.300

Chapter 19
GIS-Based Healthcare Accessibility Analysis—A Case Study of Selected Municipalities of Hyderabad

Srikanth Katta and B. Srinagesh

Abstract The spatial access and dynamics of a changing population in urban areas with changing healthcare needs require frequent and logical methods to evaluate and assist in primary healthcare access and planning. Spatial or geographical access is an important aspect in the planning process. Healthcare accessibility analysis based on GIS is a logical method which can be applied to test the degree to which equitable access is obtained. In reality, a person will always go to their closest facility; GIS analysis is, however, based on this assumption of this rational choice. Inputs to the analysis are supplied in the form of healthcare facilities and demand estimates in the form of people who are actually seeking the healthcare service. Hyderabad healthcare system is a dual system made up of private and public healthcare facilities. Private healthcare system is expensive and only affordable to rich class. In the present study, GIS analysis is applied to determine three distinct demand scenarios based on a combination of three variables: (a) Household income groups, (b) Age criteria, (c) Chronic diseases, and Healthcare emergencies. GIS is used to determine catchment or buffer areas for each healthcare facility, allocating demand to its closest healthcare facility limiting access based on facility capacity and accessibility through a road network. The catchment or buffer area analysis results from each of the three demand scenarios are compared with actual situations in the form of nearest facilities and mapped origins of number of users at each facility. The major objective of the study is to show the use of GIS to quantify and improve the access to healthcare resources in terms of availability (supply of services which meets the population needs) and Accessibility (physical access along with travel time and cost) in Circle No.9 of GHMC, Hyderabad.

Keywords GIS · Accessibility · Hospital locations · Demand areas · Travel time

S. Katta (✉) · B. Srinagesh (✉)
Department of Geography, University College of Science, Osmania University, Hyderabad, India

N. C. Jana et al. (eds.), *Livelihood Enhancement Through Agriculture, Tourism and Health*, Advances in Geographical and Environmental Sciences,
https://doi.org/10.1007/978-981-16-7310-8_19

19.1 Introduction

Healthcare is very important for community well-being. Variations in the accessibility, obtainability and affordability of the people to healthcare facilities, however, make for differences or disparities in healthcare in different population groups. The differences or difficulties in geographic access to healthcare facilities can be attributed to location of facilities, transportation services and population distribution in a specified area. The accessibility to healthcare is an important aspect of human health, and it is a complicated process and mainly depends on the nature of population which they need to service. One of the most important features affecting the health status of the population is the distance between the populated places and the location of healthcare facilities. Travel time to reach the health care facilities is also an important factor and the time distance most often is related to the route distance, though there may be exceptions to the case. Also in developing countries like India the route distance may not be a direct derivative of the actual terrestrial distance as traffic snags and bottlenecks may be the direct straight line relation between the two variables. Travel time especially in the case of health emergencies also determines the survival rates of people and hence the efficacy of the services.

The present study focuses on the concept of location of right type of healthcare facilities within the optimum travel time. This again mainly depends upon the spatial arrangement of the population, healthcare facilities and road connectivity. The study and analysis of this spatial structure forms the basis for the measurement of various problems and healthcare incongruence due to shortage of required healthcare facilities which are basically needed for the local people.

Generally pregnant women, cardiac patients and children face a lot of difficulties in getting proper healthcare within designated time intervals due to poor accessibility to healthcare services. There is a need to project the status of healthcare services in a region to reveal the actual needs of the healthcare services by the people in that particular region. GIS can easily facilitate this. The components like hospital locations, population and road network provide the estimated travel time between a hospital location and a patient. Here the real-time travel estimates and network analyst tool in GIS are used to get the estimated travel time between source and destination, which are specific to the population groups residing in defined locality boundaries.

The optimal travel time in the service areas for the various specialties in healthcare are defined by the service zones generated using ARCGIS. These service zones are mainly considered within a 1 km radius from each specialty and are divided based on 5, 10, 15 and 20 min travel times within the specified radius in the 3 circles of Greater Hyderabad Municipal Corporation area under consideration. The datasets used here are Greater Hyderabad Municipal Corporation Circle and Ward boundaries, Hospital Locations data (Only Specialties), Google Imagery and Road Network. Here the travel time-based services will not be common to all the specialties, because every specialty in healthcare has an optimal time for treatment for lowering the mortality rates and for the safety of the patient's health. The following specialties are taken up for the study in Circle No. 9:

1. Dentists
2. Dermatologists
3. General Physicians
4. Gynecologists
5. Pediatricians and
6. Cardiologists.

19.2 Study Area

Hyderabad, the capital city of Telangana state, is spread over an area of 260 km^2. The city lies at 17.36°N Latitude and 78.47°E Longitude. The city is located in the Deccan Plateau and has an average height of 536 m above sea level. The city falls under Seismic Zone II and therefore seismically has the lowest chance of earthquakes. Musi River shows through the centre of city. The elevation of the city falls gradually from west to east. The highest elevation point in the city is in Banjara Hills area, which is around 672 m (Fig. 19.1).

Originally the city was founded on the bank of Musi River and has spread extensively on both banks of the river from the time of its origin. The historic old city which has Charminar and Mecca Masjid lies on the southern bank of the river. As

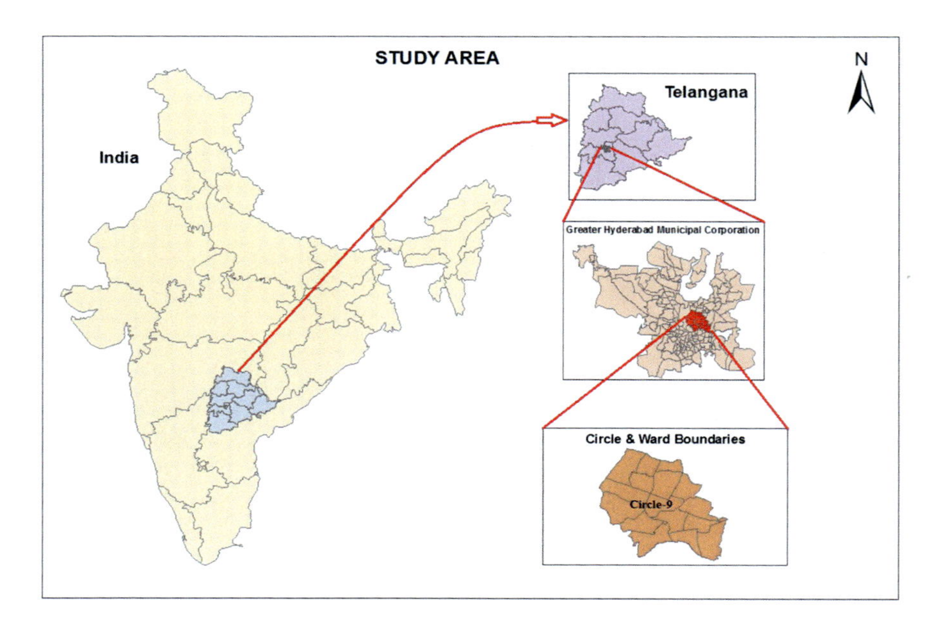

Fig. 19.1 Study area

Table 19.1 Shows the details of circle and total number of wards

Circle No.	Total no of wards	Name of the ward
9	16	Himayatnagar
		Barkatpura
		Kachiguda
		Golnaka
		Amberpet
		Bagh Amberpet
		Vidhyanagar
		Nallakunta
		Bagh Lingampally
		Adikmet
		Ramnagar
		Musheerabad
		Bholakpur
		Gandhinagar
		Kavadiguda
		Domalguda

the time progressed many Government buildings and landmarks came into existence and as a result the city center shifted to the Northern side of the river.

Secunderabad, another city, lies adjacent to Hyderabad and therefore Hyderabad and Secunderabad are called as "twin cities" as they are closely linked with each other. The two cities are separated by a manmade lake Hussain Sagar. Osman Sagar and Himayat Sagar were the major sources of drinking water for the city of Hyderabad. Musi River was called as Muchukunda River in olden days. The oldest bridge called the Puranapul is across this river. It was built by Qutub Shahi of Golconda.

In circle-9, Himayathnagar and Kachiguda are the commercial areas and the remaining areas are mostly residential (Table 19.1).

19.3 Objectives of the Study

1. To locate and connect the specialty hospitals, general physicians, nursing homes, in the study area with the help of GIS and to analyze the availability and non-availability of medical services in relation to the population and demand areas.
2. To find the number of existing specialty services in the study area and to estimate the travel time of the patients to reach the healthcare facilities from their point of locations and to identify the areas which are in short of important specialty services.

3. To analyze the existing ambulance services in the study area and suggest better ways for effective performance of ambulance services and to find the patients in emergency situations.
4. To convey the information to the public and healthcare professionals on the use of healthcare network in an adequate and cost-effective way.

19.4 Hypothesis

1. GIS helps in locating hospitals, general physician, pharmacies and community care center with ease and also helps in providing prospective locations for each one of them.
2. Hospitals can network and pool up their resources.
3. GIS helps by improving the efficiency of Emergency Services.
4. GIS helps to disseminate information about healthcare network among the people and healthcare professionals for effective functioning.

19.5 Methodology

The data sources like administrative boundaries, hospital locations and emergency services are used in the study. Figure 19.2 represents the methodology chart.

Hospital Location Capture

Hospital location data for the study area is collected with the help of Android **mobile** application **MapIt**. This is an open-source application which can be easily downloaded from the Internet. After installing the application a quick launch icon appears on the mobile screen from which we can work on the application and start the data collection process.

Administrative Boundaries

Initially the GHMC ward map and circles map are collected from the Municipality and georeferenced and digitized. For roads, the database is created from high-resolution google imagery. Google imagery is downloaded from google earth pro in 179 tiles and then georeferenced using the corresponding coordinates.

Hospital Data

The hospital location information has been collected using the "MapIt" Android mobile application. After collecting the location of hospital on the mobile the point data is exported in to Kml format and then to shape file format. The attribute data like hospital name, address, available specialties, and contact numbers have been collected in the data sheet separately.

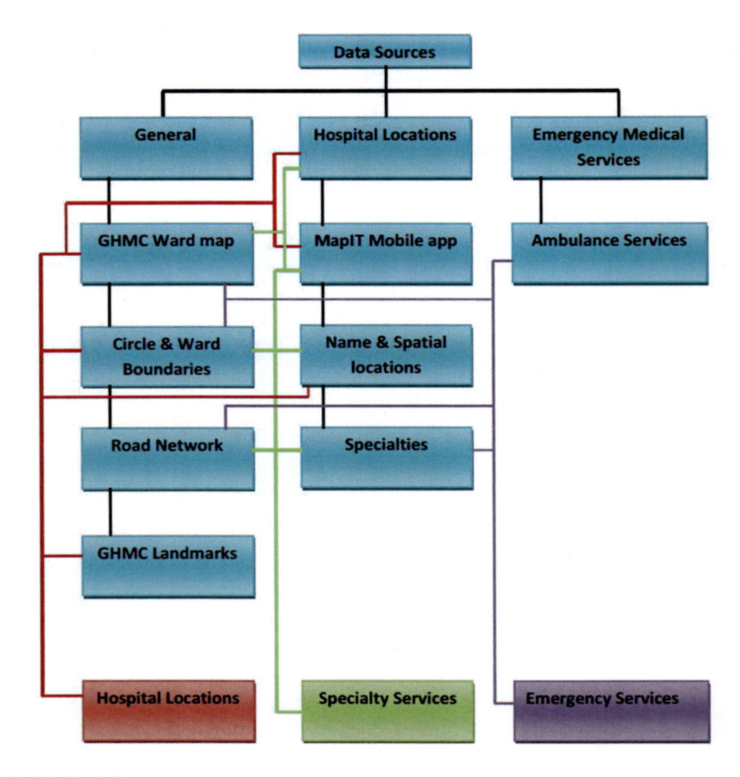

Fig. 19.2 Methodology

Specialty and Doctor

The locational details of hospitals along with the data of available specialties and Doctors are used in the study of specialty-wise service in an area.

Emergency Medical Services

Ambulance services are major in Emergency Medical services:

- GHMC ward map, minor and major road network is used for the study.
- Current locational details of Ambulance services and the area covered by them in shortest time are studied and new standby locations have also been proposed for better and faster services. GHMC ward map and road network is used for the study.

Table 19.2 Circle-9 population details

S. No.	Ward name	Population	Area (km^2)	Population density
1	KAVADIGUDA	28,887	1.45	19,922
2	HIMAYATNAGAR	40,354	1.77	22,798
3	DOMALGUDA	32,690	1.37	23,861
4	VIDYANAGAR	27,719	1.16	23,895
5	BARKATPURA	30,388	1.17	25,972
6	KACHIGUDA	37,320	1.39	26,848
7	RAMNAGAR	34,899	1.27	27,479
8	AMBERPET	41,720	1.42	29,380
9	BAGH AMBERPET	24,826	0.82	30,275
10	BAGHLINGAMPALLY	36,526	1.14	32,040
11	NALLAKUNTA	32,531	0.98	33,194
12	GOLNAKA	26,762	0.8	33,452
13	GANDHINAGAR	26,593	0.76	34,990
14	ADIKMET	34,664	0.94	36,876
15	BHOLAKPUR	31,697	0.68	46,613
16	MUSHEERABAD	28,426	0.51	55,737

19.6 Circle 9 Hospital Locations

This circle includes 16 wards like Kavadiguda, Himayatnagar, Domalguda, Vidyanagar, Barkatpura, Kachiguda, Ramnagar, Amberpet, Bagh amberpet, Baghlingampally, Nallakunta, Golnaka, Gandhinagar, Adikmet, Bholakpur and Musheerabad (Table 19.2).

19.7 Circle-9

Circle-9 is located in the central zone of Greater Municipal Corporation of Hyderabad with 16 wards of Musheerabad, Bholakpur, Adikmet, Gandhinagar, Golnaka, Nallakunta, Baghlingampally, BaghAmberpet, Amberpet, Ramnagar, Kachiguda, Barkatpura, Vidyanagar, Domalguda, Himayatnagar and Kavadiguda. These wards have been arranged in descending order of population density. All the sixteen wards have well connected road network with good transportation facilities (Fig. 19.3). The details of specialty services like Dentists, Dermatologists, General Physicians, Gynecologists, Ophthalmologists, Pediatricians and Cardiologists in circle-9 are analyzed below.

Heat maps are generated for all the specialty services with 1 km radius from each point location. The location of hospitals is represented in terms of concentration of

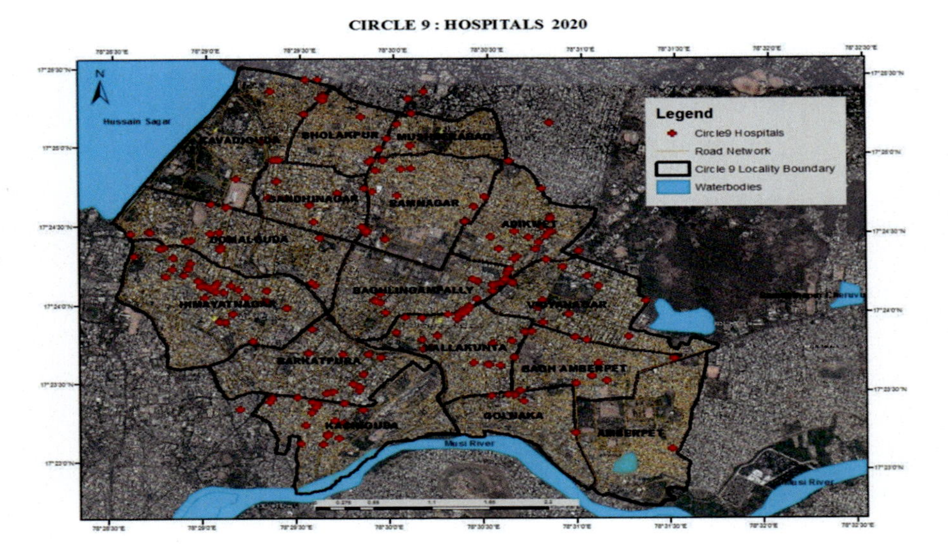

Fig. 19.3 Hospital locations

hospitals, and public access to these gynecology services are shown in four categories of access zones like:

1. Very high access zone: Represented as white color zones on the map.
2. High access zone: Represented as brown color zones on the map.
3. Moderate access zone: Represented as green color zones on the map.
4. Low access zone: Represented as light green color zones on the map.
5. Service Limits of proposed care centers: Represented as pink color on the map.

19.7.1 Gynecology Services

There are as many as 27 well-known gynecology services in circle-9, of which 5 hospitals are located in Amberpet, 1 in Baghlingampally, 4 in Bharkatpura, 1 in Domalguda, 5 in Gandhinagar, 5 in Himayatnagar, 1 in Kachiguda, 1 in Nallakunta and 4 in Vidyanagar area. Figure 19.4 represents the gynecology services and access zones in terms of nearest gynecology facilities and estimated travel times for the people to access the gynecology services.

1. **Very high access zone**: This zone represents the gynecology facilities within 5 min travel time. The density or incidence of gynecology services is comparatively more in this region. This zone forms the core area of the region with important commercial and residential areas. It finds its positioning in parts of Barkatpura, Kachiguda, Himayatnagar and Domalguda. These areas are very important commercial, educational and residential areas.

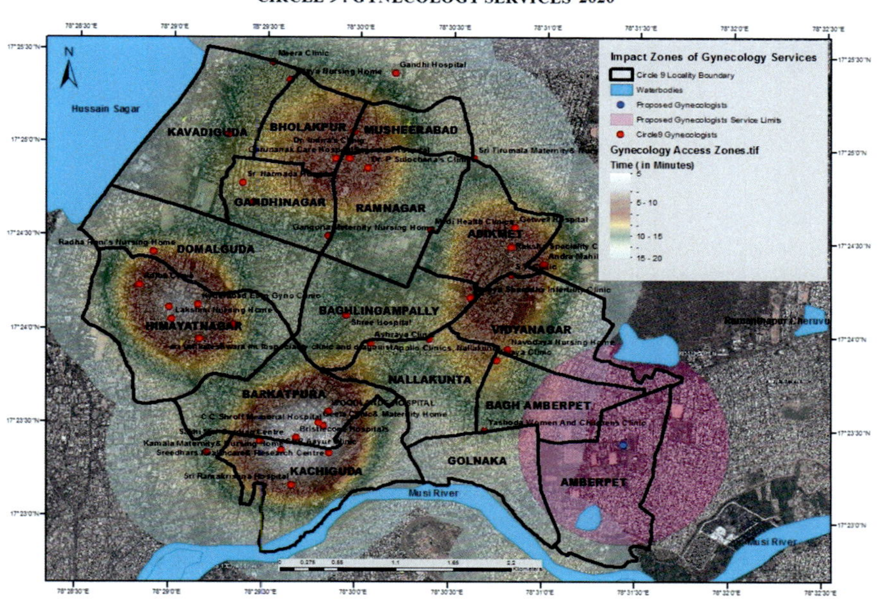

Fig. 19.4 Gynecology services

2. **High access zone**: It represents the gynecology facilities within 5–10 min travel time. This zone adjoins the white colored zone with comparatively lower density of gynecology services when compared to the white colored zone. Parts of Bholakpur, Musheerabad, Gandhinagar, Ramnagar, Adikmet, Vidyanagar and Nallakunta fall under this zone.
3. **Moderate access zone**: It represents the gynecology facilities within 10–15 min travel time. This zone abuts the brown colored zone with minimum gynecology services but comes under the influence of the brown colored zone. This zone is partially remote in location when compared to the white and brown colored zones. People in this zone depend mostly on personal transport for accessing the gynecology facilities.
4. **Low access zone**: It represents the gynecology facilities within 15–20 travel time. This zone is around the green colored zone with very less gynecology services, and this zone comes under the buffer area of 1 km radius of the gynecology services located in the core area of the region. People in this zone completely depend on personal transport for accessing the gynecology services. The remaining wards come under the influence of service zones of adjacent hospitals in the circle.

Some south eastern parts of circle-9 are not under the service zones of any of the existing gynecology services, so one new gynecology service is proposed in the south-eastern region with 1 km radius from the point of location so that no area in the circle is left un-provided for.

19.7.2 General Physicians

There are 41 well-known general physician services in circle-9, of which 6 hospitals are located in Amberpet, 3 in Baghlingampally, 3 in Bharkatpura, 4 in Domalguda, 6 in Gandhinagar, 6 in Himayatnagar, 6 in Kachiguda, 3 in Musheerabad, 2 in Nallakunta, 1 in Ramnagar and 1 in Vidyanagar area.

Figure 19.5 represents the general physician services and access zones in terms of nearest general physician facilities and estimated travel times for the people to access the general physician services. *Heat maps* that have been generated for the general physician services with 1 km radius from each point location help in analyzing the scenario with regard to the general physician services. The location of hospitals is represented in terms of density of hospitals, and public access to these general physician services is shown in four categories of access zones like:

1. **Very high access zone**: It represents the general physician facilities within 5 min travel time. The occurrence of general physician services is comparatively more in this region. This zone forms the core area of the region with important

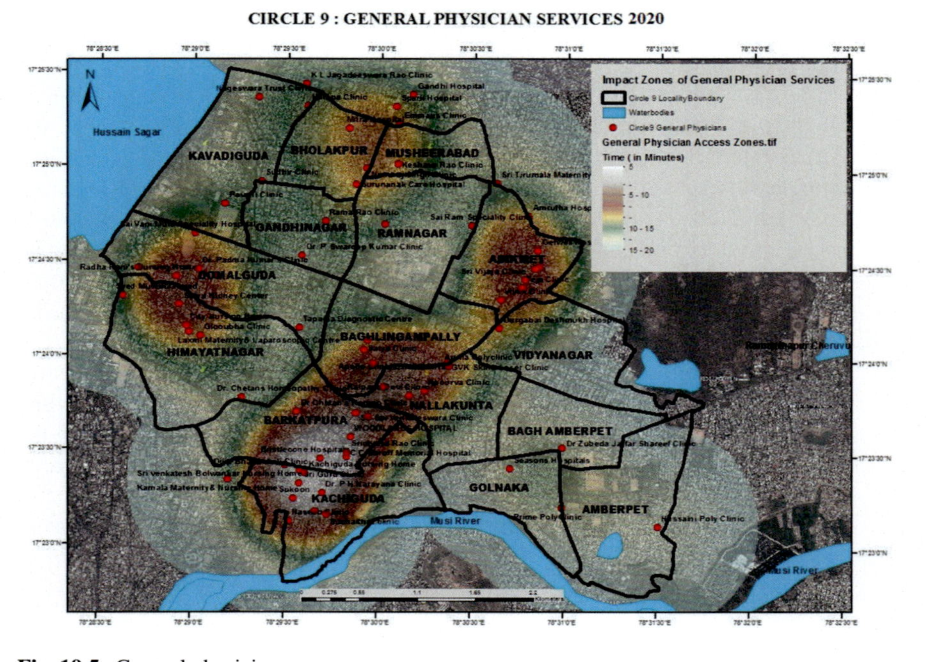

Fig. 19.5 General physicians

commercial, educational and residential activities. These zones are found in parts of Barkatpura, Kachiguda, Baghlingampally and Nallakunta.

2. **High access zone**: It represents the general physician facilities within 5–10 min travel time. This zone is adjoining the white colored zone with comparatively lower density of general physician services when compared to the white colored zone. Parts of Bholakpur, Musheerabad, Adikmet, Vidyanagar, Himayatnagar and Domalguda fall under this zone.
3. **Moderate access zone**: It represents the general physician facilities within 10–15 min travel time. This zone is around the brown colored zone with relatively less number of general physician services. This zone is partially out-of-the-way when compared to the white and brown zones. People in this zone depend mostly on personal transport for accessing the general physician facilities.
4. **Low access zone**: It represents the general physician facilities within 15–20 min travel time. This zone is bound within the green colored zone with no general physician service, and this zone comes under the buffer area of 1 km radius of the general physician services located in the core area of the region. People in this zone completely depend on personal transport for accessing the general physician services. The remaining wards come under the influence of service zones of adjacent hospitals in the circle.

19.7.3 Ear Nose Throat Hospitals

There are 10 well-known Ear Nose Throat services in circle-9, of which 1 hospital is located in Bharkatpura, 3 in Himayatnagar, 1 in Kachiguda, 2 in Musheerabad, 1 in Nallakunta and 2 in Vidyanagar area.

Figure 19.6 represents the ENT hospitals and access zones in terms of nearest ENT facilities and estimated travel time for the people to access the ENT services. *Heat maps* are created for the ENT services with 1 km radius from each point location. The location of hospitals is represented in terms of density of hospitals, and public access to these ENT services is shown in four categories of access zones like:

1. **Very high access zone**: It represents the ENT facilities within 5 min travel time. The density or presence of ENT services is comparatively more in this region. This zone forms the core area of the region with commercial, educational and residential areas of importance. These zones are found in parts of Bholakpur, Musheerabad, Ramnagar, Gandhinagar, Himayatnagar, Domalguda, Baghlingampally and Barkatpura.
2. **High access zone**: It represents the ENT facilities within 5–10 min travel time. This zone surrounds the white colored zone with comparatively lower density of ENT services when compared to white colored zone. Parts of Nallakunta and Domalguda fall under this zone.
3. **Moderate access zone**: This zone represents the ENT facilities within 10–15 min travel time. This zone girdles the brown colored zone with comparatively less number of ENT services. This zone is partially interior when compared to

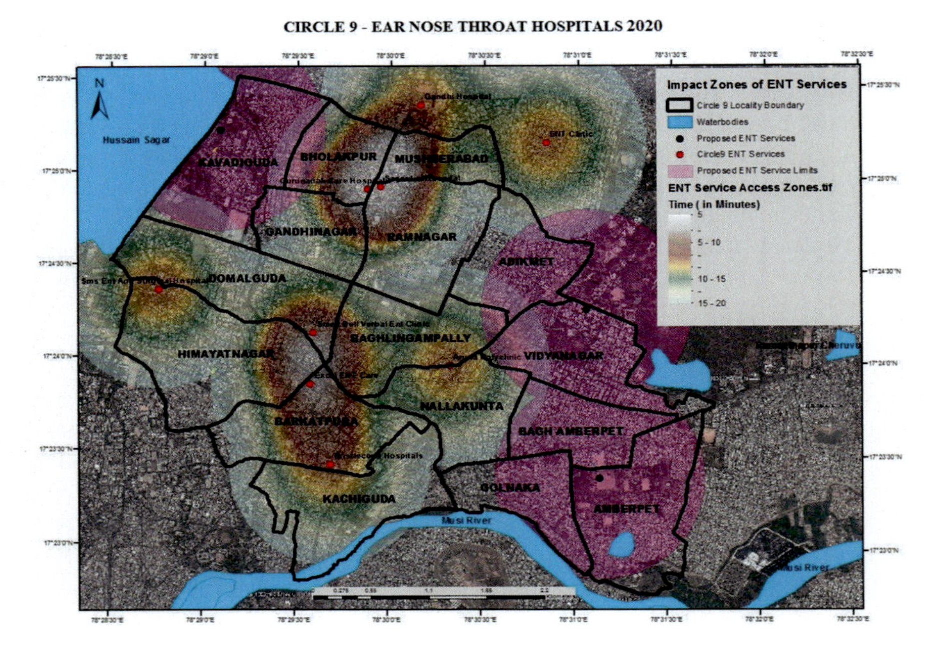

Fig. 19.6 ENT hospitals

the white and brown zones. People in this zone depend mostly on personal transport for accessing the ENT facilities.

4. **Low access zone**: This zone represents the ENT facilities within 15–20 min travel time. This zone is around the green colored zone with no ENT services, and this zone comes under the buffer area of 1 km radius of the ENT services located in the core area of the region. People in this zone wholly depend on private transport for accessing the ENT services. The remaining wards come under the influence of service zones of adjacent hospitals in the circle.

19.7.4 Dermatology Services

There are 15 well-known Dermatology services in circle-9, of which 1 hospital is located in Bharkatpura, 8 in Himayatnagar, 1 in Kachiguda, 1 in Amberpet, 1 in Baghlingampally and 2 in Ramnagar.

Figure 19.7 represents the dermatology services and access zones in terms of nearest dermatology facilities and estimated travel times for the people to access the dermatology services. Heat maps are generated for the dermatology services with 1 km radius from each point location. The location of hospitals is represented in terms of density of hospitals, and public access to these dermatology services are shown in four categories of access zones like:

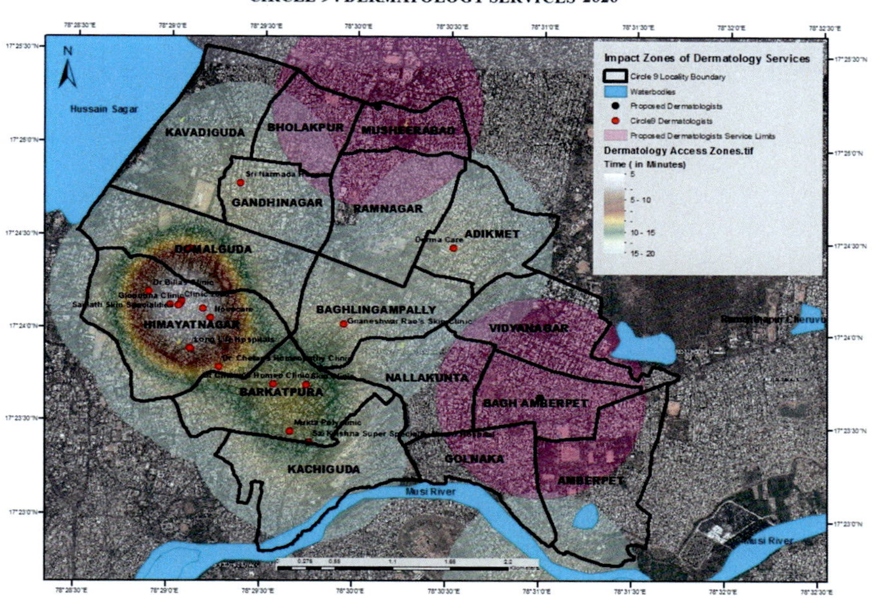

Fig. 19.7 Dermatology service

1. **Very high access zone**: Represents the dermatology facilities within 5 min travel time. The density or presence of dermatology services is comparatively more in this region. This zone forms the core area of the region with commercial, educational and residential activities of relative significance. These zones are found in parts of Himayatnagar and Domalguda.
2. **High access zone**: It represents the dermatology facilities within 5–10 min travel time. This zone is around the white colored zone with comparatively lower density of ENT services when compared to white colored zone. Parts of Barkatpura fall under this zone.
3. **Moderate access zone**: It represents the dermatology facilities within 10–15 min travel time. This zone is around the brown colored zone with likely less number of dermatology services. This zone is partially inaccessible when compared to the white and brown zones. People in this zone depend mostly on own transport for accessing the dermatology facilities.
4. **Low access zone**: It represents the dermatology facilities within 15–20 min travel time. This zone is around the green colored zone with no dermatology services, and this zone comes under the buffer area of 1 km radius of the dermatology services located in the core area of the region. People in this zone completely depend on private transport for accessing the dermatology services. The remaining wards come under the influence of service zones of adjacent hospitals in the circle.

19.7.5 Dental Services

There are as many as 73 well-known Dental services in circle-9, in which 16 hospitals are present in Amberpet, 1 in Barkatpura, 5 in Domalguda, 4 in Gandhinagar, 17 in Himayatnagar, 5 in Kachiguda, 4 in Musheerabad, 10 in Nallakunta, 3 in Ramnagar and 8 in Vidyanagar.

Figure 19.8 indicates the dental services and access zones in terms of nearest dental facilities and estimated travel time for the people to access the dental services. *Heat maps* are generated for the dental services with 1 km radius from each point location. The location of hospitals is represented in terms of concentration of hospitals and public access to these dental services is shown in four categories of access zones like:

1. **Very high access zone**: This zone represents the dental facilities within 5 min travel time. The density or presence of dental services is comparatively more in this region. This zone forms the core area of the region with vital commercial, educational and residential activities. This zone is found in parts of Nallakunta, Baghlingampally, Vidyanagar, Himayatnagar and Domalguda.
2. **High access zone**: It represents the dental facilities within 5–10 min travel time. This zone adjoins the white colored zone with rather lower density of dental

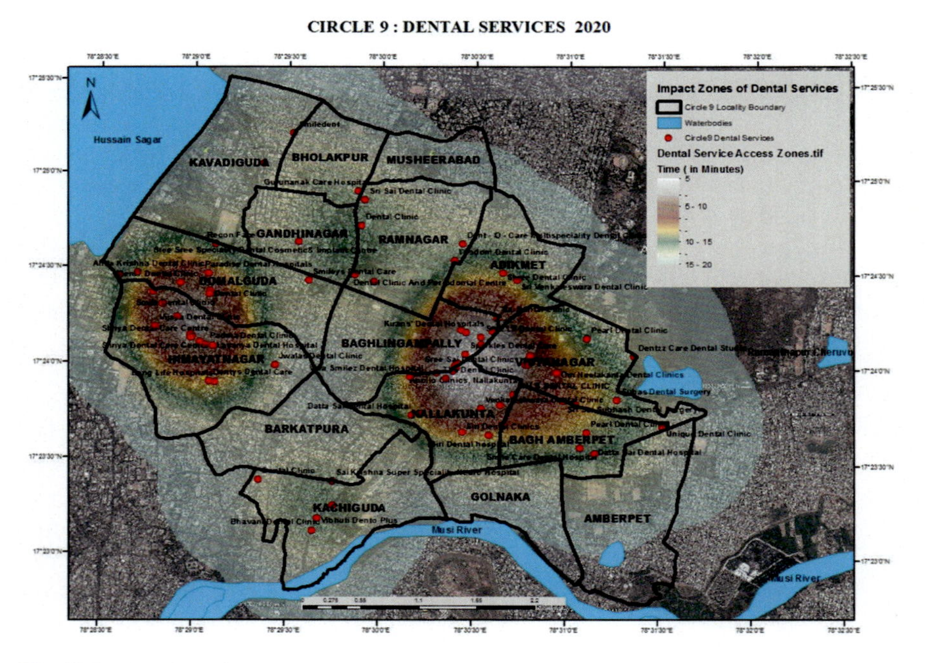

Fig. 19.8 Dental services

services when compared to the white colored zone. Parts of Kachiguda are included in this zone.

3. **Moderate access zone**: It represents the dental facilities within 10–15 min travel time. This zone abuts the brown colored zone with hardly one or two dental services. This zone is partially inland when compared to the white and brown colored zones. People in this zone depend mostly on individual transport for accessing the dental facilities.
4. **Low access zone**: It represents the dental facilities within the 15–20 min travel time. This zone is around the green colored zone with no dental services, and this zone comes under the buffer area of 1 km radius of the dental services located in the core area of the region. People in this zone completely depend on personal transport for accessing the dental services. The remaining wards come under the influence of service zones of adjacent hospitals in the circle.

19.7.6 Cardiology Services

Cardiac cases require quick attention as the time lag can at times prove to be fatal. There are as many as 7 well-known cardiology services in circle-9, of which 2 hospitals are in Barkatpura, 1 in Domalguda, 1 in Himayatnagar, 2 in Musheerabad and 1 in Vidyanagar. Figure 19.9 represents the cardiac services and access zones in terms of nearest cardiac facilities and estimated travel times for the people to access the cardiac services. Heat maps are generated for the cardiac services with 1 km radius from each point location. The location of hospitals is represented in terms of compactness of hospitals, and public access to these cardiac services is shown in four categories of access zones like:

1. **Very high access zone**: It represents the cardiac facilities within 5 min travel time. The density or presence of cardiac services is comparatively more in this region. This zone forms the core area of the region with important commercial and educational establishments and residential areas. This zone is found in parts of Gandhinagar, Bholakpur, Musheerabad, Ramnagar, Barkatpura, Kachiguda, Adikmet and Vidyanagar.
2. **High access zone**: It represents the cardiac facilities within 5–10 min travel time. This zone is around the white colored zone with comparatively lower density of cardiac services when compared to the white colored zone. Parts of Domalguda and Himayatnagar fall under this zone.
3. **Moderate access zone**: It represents the cardiac facilities within 10–15 min travel time. This zone is around the brown colored zone with less number of cardiac services. This zone is partially interior when compared to the white and brown zones. People in this zone mostly depend on personal transport for accessing the cardiac facilities.
4. **Low access zone**: Represents the cardiac facilities within 15–20 min travel time. This zone is about the green colored zone with very less cardiac services, and this zone comes under the buffer area of 1 km radius of the cardiac services

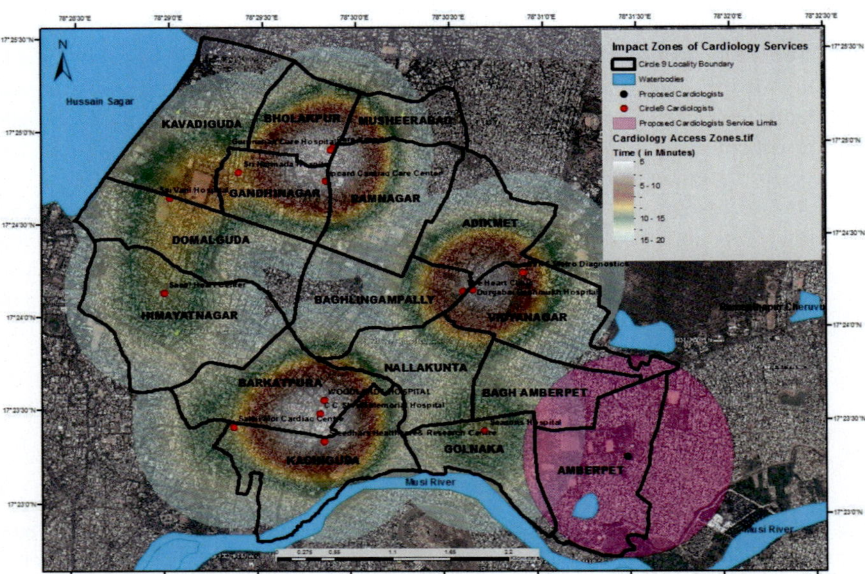

Fig. 19.9 Cardiology services

located in the core area of the region. People in this zone completely depend on private transport for accessing the cardiac services. The remaining wards come under the influence of service zones of adjacent hospitals in the circle.

A new cardiac care center is proposed in the areas of Amberpet and Baghamberpet area which can serve 1 km radius from the point of its location.

19.7.7 Pediatric Services

There are as many as 9 well-known pediatric services in circle-9, of which 2 hospitals are in Amberpet, 2 in Barkatpura, 1 in Domalguda, 2 in Gandhinagar, 1 in Musheerabad and 1 in Vidyanagar.

Figure 19.10 represents the pediatric services and access zones in terms of the nearest pediatric facilities and estimated travel time for the people to access the pediatric services. *Heat maps* are generated for the pediatric services with 1 km radius from each point location. The location of hospitals is represented in terms of density of hospitals, and public access to these pediatric services is shown in four categories of access zones like:

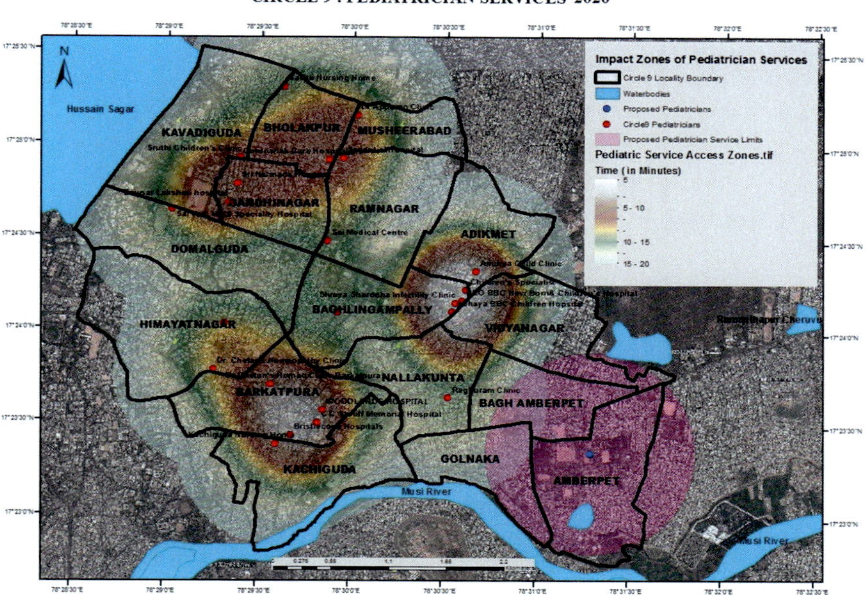

Fig. 19.10 Pediatrician services

1. **Very high access zone**: It represents the pediatric facilities within 5 min travel time. The density or presence of pediatric services is comparatively more in this region. This zone forms the core area of the region with important commercial, educational and residential activities. This zone is found in parts of Barkatpura, Kachiguda, Adikmet, Vidyanagar and Baghlingampally.
2. **High access zone**: It represents the pediatric facilities within 5–10 min travel time. This zone is around the white colored zone with comparatively less number of pediatric services when compared to the white colored zone. Parts of Bholakpur, Gandhinagar, Kavadiguda, Musheerabad and Ramnagar fall under this zone.
3. **Moderate access zone**: This zone represents the pediatric facilities within 10–15 min travel time. This zone is around the brown colored zone with less number of pediatric services. This zone is partially interior when compared to the white and brown colored zones. People in this zone depend mostly on personal transport for accessing the pediatric facilities.
4. **Low access zone**: It represents the pediatric facilities within 15–20 min travel time. This zone surrounds the green colored zone with very less number of pediatric services, and this zone comes under the buffer area of 1 km radius of the pediatric services located in the core area of the region. People in this zone completely depend on own transport for accessing the pediatric services. The

remaining wards come under the influence of service zones of adjacent hospitals in the circle.

Two new pediatric care centers are proposed in the areas of Amberpet and Baghamberpet which can serve 1 km radius from the point of its location.

19.8 Conclusion

The study identifies the localities with less number of healthcare facilities and identifies the demand areas. The wards which are having highest population densities must have more number of healthcare facilities compared to other wards; according to the study Uppal and Bholakpur in circle 2, Musheerabad and Gandhinagar in circle 9, Borabanda and Rahmathnagar in circle 10 have less number of healthcare facilities when compared to their population densities. The hospitals must be developed such that they can be a part of network in their concerned ward boundaries and spatial access will be easy for the people. This will prevent the people and patients from travelling longer distances for cheaper and effective medical care. Many of the existing hospitals are out of the network, which means they are not having proper medical reimbursement schemes or not under private ownership which offers medical care with cheaper prices. So, all the hospitals in a locality with major specialities must be brought under a group of hospitals. This system also solves the traffic congestion problems during the transport or shifting of patients during the emergencies. Some areas in the study area are facing shortage of public transport, so means of public transport must be provided for easy and faster access to healthcare facilities. The paying capacity of the people in 50% of the study area is less, so along with the healthcare schemes, the state government should start one government healthcare centre for each ward in the study area for quality and cheaper medical services.

Chapter 20
Dynamics of Disease Diffusion: A Critical Analysis of Dengue Outbreak in Kolkata and Adjacent Areas

Teesta Dey

Abstract While medical advancement has conquered most infectious diseases in the twenty-first century, the majority of the tropical and sub-tropical developing countries are still fighting against various vector-borne diseases as an inevitable consequence of the climate change. The predominance of vector-borne diseases in Kolkata has become a continuous threat to human health. Rapid increase of disease incidence, proliferation, and fatality rates among dengue patients creates multi-dimensional effects on the socio-political scenario of urban daily lives. In the last 10 years, the city has experienced a gradual transformation of Dengue Ecology among the city dwellers. Over time, the virus has changed its temporal Disease–Population Dynamics including its transmissibility, rate of replication, infectivity, and virulence. Major Dengue outbreaks in this city exhibit the ineffective surveillance, improper urban environmental planning, vulnerable living status, insufficient political strategies, poor perception quality, and ignorance. High rate of Dengue induced death poses an important question regarding the effectivity of health planning in a metro city like Kolkata. In addition to that, discrepancies in dengue-related reports and data have become associated with disease ecology and the health politics of Kolkata. In this context, an attempt has been made to focus on the Spatio-temporal transformation of the Dengue Virus, to analyze the severity pattern among the patients, to generate dengue-prone area mapping based on spatial autocorrelation method, to correlate the spatiality factor of Dengue pockets with local socio-economic conditions, and finally to study the changing Dengue Dynamics in this city from a geographical perspective.

Keywords Disease ecology · Infectivity · Virulence · Dengue diffusion · Health politics

---

T. Dey (✉)
Kidderpore College, Kolkata, India

N. C. Jana et al. (eds.), *Livelihood Enhancement Through Agriculture, Tourism and Health*, Advances in Geographical and Environmental Sciences,
https://doi.org/10.1007/978-981-16-7310-8_20

20.1 Introduction

The direction of human civilization has taken a turn after the outbreak of the Covid-19 pandemic. The decade has been witnessing a possible transformation of entire livelihood patterns all over the world. Disease and survival have become an inseparable aspect of human existence in twenty-first century. The variation in Covid-19 virulence pattern in different developed and developing countries of the world has led to expansion of the study of *Disease Ecology* and *Spatial Epidemiology* within the realm of Medical Geography. While the developed nations are now shivering due to the advent of Novel Corona Virus, the third world countries of equatorial and tropical humid regions are struggling against this virus while battling other various vector borne diseases simultaneously.

India is now playing an important role in the emerging global geo-strategic issues related to Covid-19. The comparatively low contamination rate among the common people from Covid-19, usage of Bacille Calmette-Guérin (BCG) vaccine, supply of Hydroxychloroquine, rigorous medical research to combat the pandemic, etc. all are nothing but the result of earlier experiences that India always has had due to the prevalence of various infectious and vector borne diseases like Malaria, Dengue, Chikungunya, Kalazar, Japanese Encephalitis, etc. While Malaria has been a problem in India for centuries, Dengue fever has mostly affected people since last six decades. But 1990 onwards, India has been experiencing tremendous growth of Dengue fevers, Dengue Hemorrhagic Fever(DHF), Dengue Shock Syndrome (DSS), etc. with severe casualties. Dengue has become an endemic and public health concern in India where every year some particular geographical locations have to experience this disease and its terribly negative effects. Even in the twenty-first century, every year on an average 200 to 250 people die just due to a bite of mosquito in India. More than one lakh people every year in India are being affected by this fever (Government of India 2018). Dengue has now become a part of existence of many people in this country. West Bengal in the last four years has consistently exhibited the highest occurrence of dengue cases (Government of West Bengal 2016a) and now has become the most affected state in India. Before the Covid-19 Pandemic, Dengue was the most alarming vector borne disease both in India and in West Bengal. Kolkata, in fact, is the only mega city in India with a very high rate of dengue cases in the last seven years (Government of India 2018). The presence of dengue fever during almost nine months, from April to December, in Kolkata, accelerates the morbidity and fatality rates significantly. In addition to that Covid-19 pandemic has aggravated the entire scenario and common people across the city will be exposed to remarkable health hazards in near future if proper controlling programs are not implemented. Common people and state government both are now combating against Covid-19 pandemic and Dengue which pose a great threat to the physical and mental health of the citizens.

In this context, this research article will throw some lights on the proliferation, growth, extension, and severity of Dengue cases, on developing dengue concentration mapping for controlling the disease and finally on the critical assessment of the role of politics in Dengue controlling mechanism in Kolkata as a part of West Bengal.

20.2 Seasonality and Society: Scene Behind Dengue Proliferation in India

Proliferation of any kind of vector borne disease depends mostly on locational and climatic attributes of any area. The tropical regions are endowed with high level of temperature and humidity that render the growth of the vectors and accelerate the rate of spread of the diseases. Birth, growth, and effectivity of any vector are controlled mostly by average rainfall, temperature, and humidity levels (Bisht et al. 2019). The graphs of average temperature and rainfall pattern of Kolkata in the last 120 years exhibit a gradual rising trend (Government of West Bengal 2016a) that leads to the creation of favorable conditions for rapid and consistent growth, evolution, proliferation, and mutation of different vectors, especially the *Aedes Aegypti* and *Aedes Albopictus* female mosquitoes causing Dengue in the city. Recent research reveals that these species are continuously mutating and causing serious health hazards to common people. These Dengue viruses (DENV) belong to family *Flaviviridae*, and there are four serotypes of the virus referred to as DENV-1, DENV-2, DENV-3, and DENV-4 (Bandyopadhyay et al. 2013). Over the years Kolkata has been experiencing all the varieties of serotypes which create complexities in combating with the virus both by the patients and the doctors. Seasonality is another important factor that acts significantly on the effectivity of the vectors. High levels of average temperature, humidity, and rainfall between June and September in Kolkata exert direct influence on dengue proliferation (Fig. 20.1). This is mostly the monsoon season in Kolkata with average monthly rainfall of 300 mm and temperature of 28 °C (Sharma et al. 2014).

Consequently, highest dengue case incidence and fatality rates are observed during the Monsoon and post Monsoon periods (Chandra 2015). Noticeably in the last few years, a notable spatial and temporal diffusion of dengue has been observed where

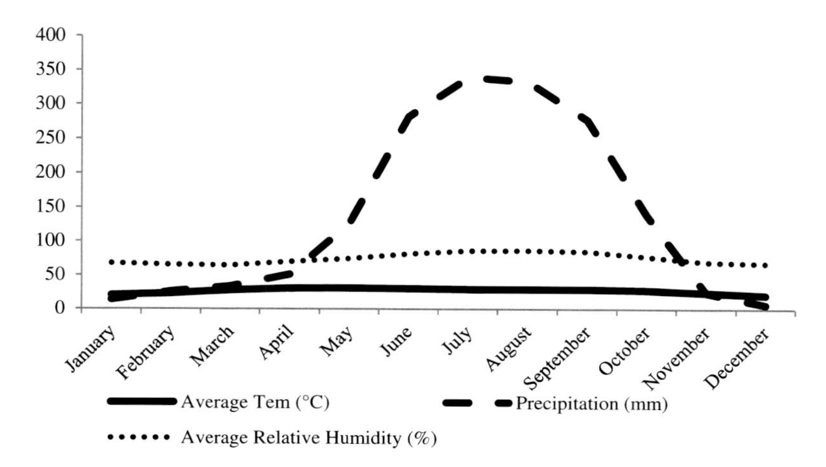

Fig. 20.1 Average monthly weather phenomena in Kolkata (1901 to 2019)

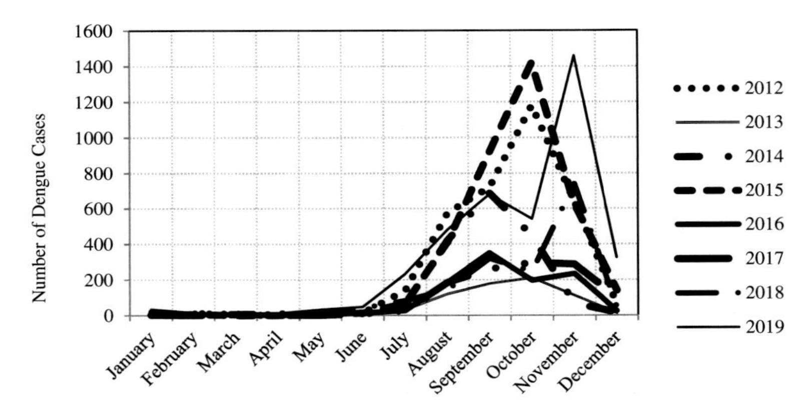

Fig. 20.2 Month-wise changing pattern of dengue cases in Kolkata (2012–2019)

the seasonal effectivity pattern of these vectors has been extended up to the winter season (Fig. 20.2). The Dengue outbreak survey also reveals it is gradual spreading from Kolkata to the adjacent areas very rapidly every year. Now in Kolkata and its adjacent localities, dengue incidence and fatality cases have been extended even up to December. Surprisingly low temperature in winter season has loosened its impact on controlling the breeding and effectivity of the vectors leading to incidence of the disease even in the first three months of the year. Dengue has actually now becomes a year-round disease in Kolkata.

Apart from these climatic aspects, at present disease ecology also depends on the social environment. Poverty, high rate of rural to urban immigration, proliferation of slums, congested residential locality, poor living conditions, crowding, poor sanitation, accumulation of wastes, clogged sewerage system, dominance of built environment, truncated neighborhood environment, low education level, lack of awareness, water logging condition, ineffective development policies, and lack of urban management accelerate dengue dissemination in this city.

20.3 Temporality in Spatial Disease Diffusion

Dengue became one of the virulent vector borne diseases with severe causalities in India since 2005 (Chatterjee et al. 2013). The rising growth curves of dengue cases and deaths exhibit its gradual dominance in the sphere of vector borne diseases among common people. Back in 2005, maximum dengue cases were observed in West Bengal, but later on, Punjab and Maharashtra consistently exhibited dominance in dengue cases and fatality rates (Gupta et al. 2012). But noticeably since 2016, West Bengal solely became the most dengue affected state in India with highest number of cases. Maharashtra on the contrary represents highest dengue fatality cases in most

Table 20.1 Spatio-temporal variations in dengue cases in India

Year	States with highest case incidences	States with highest fatalities
2010	Punjab	Haryana
2011	Punjab	Punjab, Odisha
2012	Tamil Nadu	Tamil Nadu
2013	Kerala	Maharashtra
2014	Maharashtra	Maharashtra
2015	Punjab	Kerala
2016	West Bengal	West Bengal
2017	West Bengal	Maharashtra
2018	West Bengal	Maharashtra
2018	West Bengal	West Bengal

of the years. In 2019, West Bengal recorded highest dengue cases till date (Table 20.1).

The identity of Bengal as a Dengue–prone state reveals the failure of dengue awareness programs among the common people and also the failure of the authority. To remove this tag, West Bengal government started rigorous awareness and cleaning programs every year from March through many task forces. But considering the extent and magnitude of dengue proliferation in West Bengal, such actions seem to be insufficient. Noticeably in West Bengal, maximum cases are observed only in two districts viz. North 24 Parganas and Kolkata. These two districts reflect more than 50 percent of the total cases in West Bengal. A steady and sharp growth of dengue cases from 2010 to 2019 clearly explains the severity levels of this disease in these two districts (Fig. 20.3). Darjeeling in North Bengal and Howrah and Nadia in South Bengal are also prone to high dengue cases. The spatio-temporal representation of dengue concentration in West Bengal reveals a clean spatial concentration pattern in eastern part of Gangetic Bengal. The amalgamation of favorable locations, climatic, geographical, and anthropological conditions are mostly responsible for such severity of Dengue cases in West Bengal.

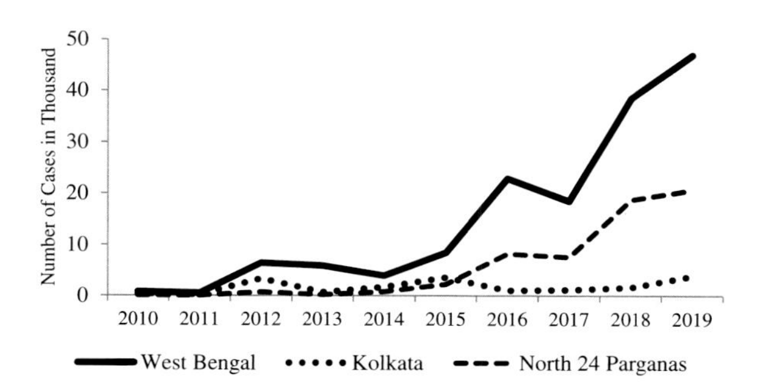

Fig. 20.3 Trend of dengue cases in the most affected districts of West Bengal (2012–2019)

20.4 Dengue Virulence in Kolkata

The Identity of West Bengal as the most dengue affected state since 2016 reveals an intense crisis of proper execution of dengue eradication programs. In West Bengal Dengue Case Incidence Rate varies significantly among the districts. Mostly the southern part of Bengal is prone to this disease. In North Bengal, Darjeeling shows inconsistent vulnerability. The spatio-temporal disease mapping shows a clear pattern of Dengue case concentration in only three districts of West Bengal viz. Kolkata, North 24 Parganas and Howrah. These three adjacent districts consist of more than three fourth of the state's total dengue cases. Such data reveals a strong positive relation among population density, rate of urbanization, concentration of slums, and degree of dengue proliferation (Chandra 2018). Since 2012, a consistently high rate of dengue occurrence is observed in Kolkata and in North 24 Parganas. Since 2016, the Dengue incidence rate has also increased in the Howrah district. Additionally Nadia, East and West Medinipur and Murshidabad districts are also prone to this disease. But the case incidence and fatality rates are highest in North 24 Parganas in the last five years. The last ten years' data reveals consistent growth of dengue cases in Kolkata and North 24 Parganas districts. There are certain pockets in North 24 Parganas where every year consistently high rate of dengue cases and deaths are found. These are *Habra-Ashoknagar, Deganga, Bongaon, Bidhannagar, Rajarhat-New Town, South Dum Dum Municipality,* etc. (Fig. 20.4). In 2019, 17 areas experienced severe dengue cases within North 24 Parganas. Noticeably the urban spaces are more susceptible to dengue vulnerability than the rural localities.

Despite the intense work of Dengue eradication action teams, in 2019, West Bengal, especially Kolkata and North 24 Parganas districts experienced highest ever recorded dengue cases. North 24 Parganas had more than 20,000 dengue patients whereas Kolkata had nearly about 6000 patients affected by dengue. In 2019 almost 70 people died from dengue. In 2019, the disease started spreading since the month

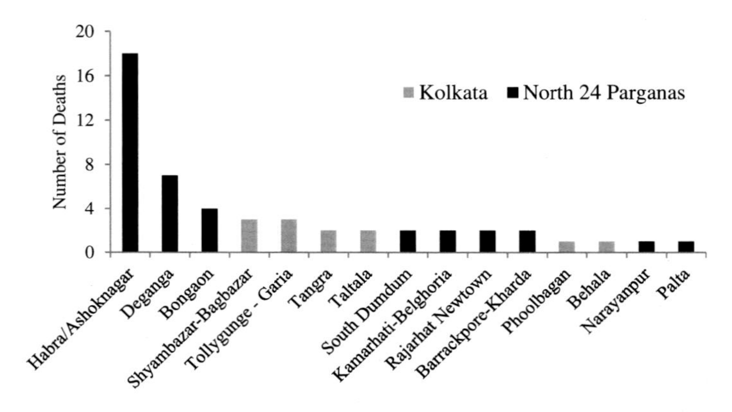

Fig. 20.4 Locality-wise variations in dengue induced deaths in two most dengue prone districts of West Bengal (2019)

of June and reached its peak twice, first in August and then again in November. Such dual peak formation represents gradual genetic mutation and expansion of effectivity of the vectors. Surprisingly both cases and deaths reached the apex in November, which is a winter month in Kolkata. Approximately 39 People died in North 24 Parganas followed by 12 persons in Kolkata and 4 persons in Howrah in 2019. If the occurrence of dengue cases and deaths are fitted in a space–time scale, a clear picture of space–time specific concentration of dengue virulence can be observed. The spatio-temporal assessment of dengue induced death reveals that in Habra-Ashoknagar deaths happen from 28th July to 6th August, in Deganga during second week of August and Third week of November, in Kolkata during third week of September and early December. Such spatial–temporal clustering of dengue induced death reveals that area specific disease analysis should be done on a regular basis. Specific locality based dengue management plan is of utmost requirement at the moment; otherwise it can cause several deaths in future. In Kolkata the dengue case incidence and fatality cases of 2018 and 2019 reveals that south-central and south-eastern part of the city are more vulnerable (Fig. 20.5). Such locality based analysis can help in implementing management schemes for dengue control.

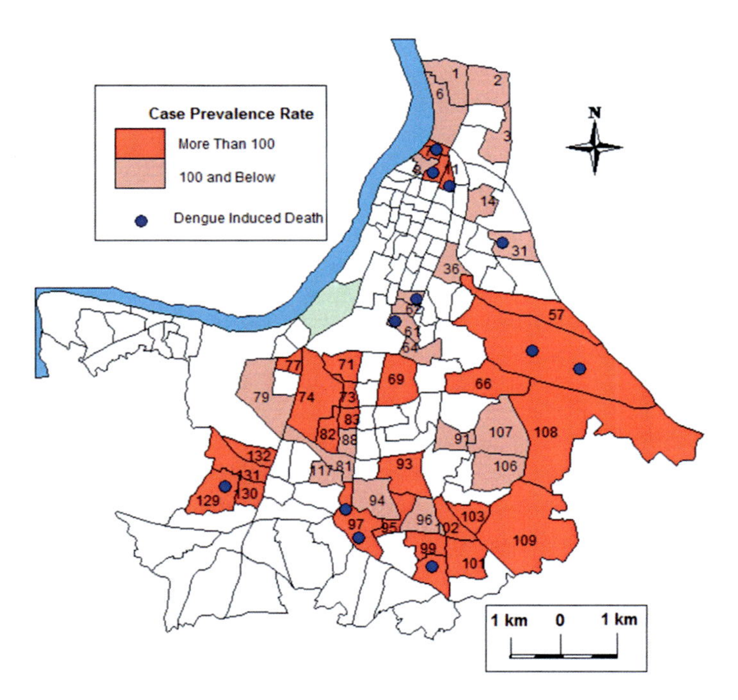

Fig. 20.5 Dengue concentration pattern in Kolkata municipal corporation (2019)

20.5 Dengue Dissemination Mapping in Salt Lake City Based on Spatial Autocorrelation Method

Space–time concentration of dengue death reveals that investigation and identification of the geographical clustering of dengue cases and instant controlling of the disease is of utmost requirement to control the number of fatalities. Although such cluster analysis is nearly always ineffective in identifying the causes of disease, it often has to be used to address public concern about the disease virulence (Olsen et al. 1996). For that purpose an intense spatial analysis and dengue mapping is required. Such mapping can help in the identification of the most vulnerable zones and simultaneously strict plans can be implemented. In case of disease mapping, the most effective spatial method is spatial autocorrelation technique (Delmelle et al. 2015). This method helps in understanding the degree to which one dengue patient is similar to other nearby patients of same disease. For that Moran's I Index has been used which assumes spatial autocorrelation based on Geographer Waldo R Tobler's first law of Geography—"Everything is related to everything else but near things are more related than distant things". Spatial autocorrelation measures how close diseased persons are in comparison with other diseased persons (Jarup 2004).

To follow the mentioned principle, an attempt has been made to find out the dengue clusters in Salt Lake City, a satellite town under Bidhannagar Municipal Corporation that is adjacent to the mother city of Kolkata. Based on the 1857 blood test reports of dengue positive cases in 2019 from 38 diagnostic centers of Salt Lake City, spatial auto-correlation method has been applied to identify the Dengue Hotspot zones where maximum spatial similarity of patients has been found. The blood samples have been tested by NS1-Antigen (ELISA) and Dengue IgM antibodies (MAC ELISA) methods. All confirmed sero-positive dengue patients residing within Salt lake have been recorded in relation to their residential address. Five percent of the total cases have also been reconfirmed by visiting their houses. Block-wise dengue positive cases have been plotted on the geo-referenced map of Salt Lake for analyzing the spatial clustering of the epidemiological and demographic phenomena. For the assessment of the spatial correlation between the Dengue cases, Moran's I method has been used. To evaluate the strength of spatial autocorrelation between the positive dengue cases, Moran's Global Spatial Autocorrelation Index (Moran's I) has been used. Moran's I is based on the cross product of the deviations from the mean and is calculated for n observation on a variable x at i,j location. While positive spatial autocorrelation refers to similar values near one another, negative spatial autocorrelations refers dissimilar values occurring near one-another. As Moran's I is used as a global measure to detect spatial clustering in a broad area, Local Moran's I is used for smaller spatial units. This analogous local measure is known as LISA (Local Indicator of Spatial Association) for measuring the local spatial autocorrelation indices where the variable values are high and the area is geographically homogeneous (Ganguly et al. 2018). LISA Matrices allow measurement of neighborhood relations for each region. This technique will help to delineate all the micro areas with similar and dissimilar concentration pattern of dengue patients. Moran's I can be classified as

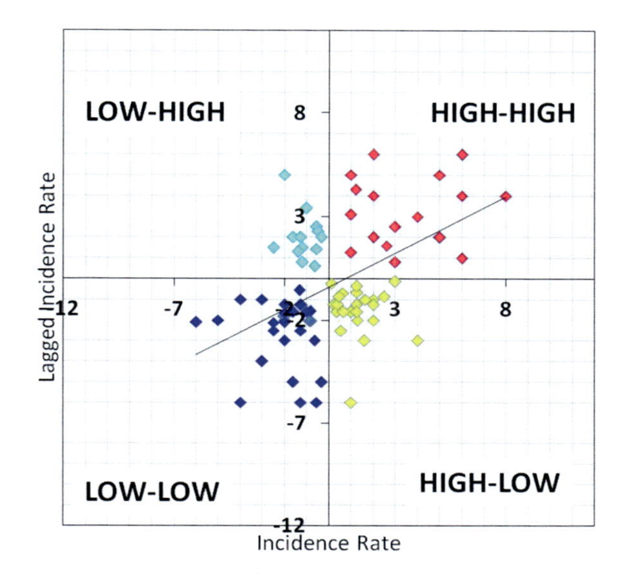

Fig. 20.6 Scatter plots of Moran's I

positive, negative, and no auto-correlation. Positive spatial auto-correlation occurs when similar values cluster together on a map (Martinez-Beneito 2013). This can be of two types, viz. similarities of areas with high dengue cases and similarities of areas with low dengue cases. On the contrary, when dissimilar values cluster together i.e. mixing of high and low dengue cases, it reveals negative spatial auto-correlations (Fig. 20.6).

Based on the scattered plots of Local Moran's I values, four quadrants can be identified with specific dengue concentration pattern. These are as follows.

20.5.1 High–High Concentration Zone

This is the clustering of blocks with high dengue incidence rates with other similar blocks with high dengue incidence cases. The spatial similarity with high rate of dengue cases represents the Dengue Hotspot zones. The high-high or Dengue Hotspot zone has been identified in CA, DA, EA, FA, FB, and FC blocks in western part adjacent to Duttabad Slum area and AG, AH, AJ, AK, AL, BJ, CJ, BK, BL, CL, CN, CP, CM, and BP blocks adjacent to *Keshtopur Canal*. Owing to congestion, unhygienic environment, prevalence of water logging conditions, etc. proper breeding grounds of the vectors can be observed in the slums and areas adjacent to the canal. These areas are of maximum importance for analysis to eradicate the problems. Otherwise within a few weeks this area will face high casualty rates.

20.5.2 Low–Low Concentration Zone

This is the clustering of blocks with low dengue cases with other blocks of low dengue incidence rates. These areas are identified as Dengue Cold Spot zones due to relatively lesser possibility of dengue virulence. The objective of the state authority is to convert all the dengue prone zones into such Cold Spot zones with least cases of dengue fever. Few scattered blocks are under low–low or cold spot zones where regular surveillance is not required. The central part of the residential segment including BB, BC, BD, BE, CB, CC, CD, CE, DB, DC, DD, BF, CF, DF, ED, FD, FE, GC, GD, HA, HB, IA, IB, EM, EN, GM, and GN fall under this category (Fig. 20.7).

Fig. 20.7 Dengue cluster map based on local Moran's I of Salt Lake City (2019)

20.5.3 High–Low Concentration Zone

The negative Local Moran's I value encompasses all those areas which have high potentials to be converted into dengue affected areas very rapidly. Clustering of blocks with high dengue incidence rates and blocks with low incidence rates cover 28 blocks in a scattered way. These blocks are AB, BA, AC, AE, AF, BE, BH, CE, CG, DE, DG, CK, DK, DL, EB, GA, GB, BN, DP, EP, FF, GE, IC, JB, JD, KB, KC, LA, and LB. These areas are identified as Spatial Outliers.

20.5.4 Low–High Concentration Zone

There are approximately 11 Blocks demonstrating low variables and dissimilarities of dengue cases where mixing of low and high number of dengue patients can be observed. These are also known as spatial Outliers. It includes AA, AD, AE, EC, BG, DJ, EE, HC, DN, GP, JA, JC, and HC. Maximum areas in Salt Lake City fall under spatial outlier zones where regular monitoring of the dengue cases is highly required. Any changes in the degree of auto-correlation will immediately need further dengue controlling steps.

20.5.5 Zero Significant Zone

Here blocks with zero or nil spatial auto-correlation are found. Such ideal blocks can rarely be found in reality, and hence only some blocks can be identified where the Moran's value lies near to the neutrality. These blocks have higher chances to be declared either Hotspot or Cold spot zones depending on their variable values.

Spatial autocorrelation mapping techniques are now used significantly for spatial epidemiological analysis (Rican and Salem 2010). Dengue eradication programs need such type of micro-level spatial concentration analysis through which instant plans can be implemented to reduce the dengue effectivity.

20.6 Dengue, Death, and Politics

While dengue severity with high morbidity and fatality cases bother the state significantly, it becomes a curse for the common people when such a disease becomes politicized. Since last five years dengue has become the most critical challenge for the state officials. Frequent Dengue induced deaths caused severe agitation and became a major issue for the opposition parties for quite a long period. This situation became aggravated when massive discrepancies were observed in the government data-base

Table 20.2 Discrepancies in dengue statistics in Kolkata city

Year	Govt. registered dengue cases	Actual number of dengue cases	Govt. registered dengue induced death cases	Actual number of dengue induced death cases
2015	3610	6000	1	13
2016	1063	1686	4	12
2017	1172	2374	8	13
2018	1640	3240	6	20
2018	3862	6424	3	12

regarding the number of dengue cases and dengue induced deaths. In Kolkata, as an example, significant variations in dengue cases can be seen in the last five years. While the government has declared near about 4000 cases only in Kolkata in 2019, many newspapers and health centers reveal that the actual number of cases is near about 6500. On an average 1000 to 1500 dengue cases are not registered in the government database every year (Table 20.2). But the situation exacerbates when huge difference is found in the fatality cases. Although the death certificates reveal dengue as the cause of death, government still delays in identifying the reasons and publishing other data. Such discrepancies are also associated with lack of immediate hospitalization of many dengue patients, poor availability of platelets, lack of proper medication, and care in the hospitals, more dependency on private hospitals, deprivation of the poor, and finally misinterpretation of actual reason of death by using medical jargons to hide the dengue severity levels. Doctors have used such terms as "febrile illness bleeding disorder", "multiple organ failure", "high fever with thrombocytopenia (low platelet count)", or simply "unknown fever" in death certificates instead of directly mentioning dengue, allegedly under pressure from the state government (Bhattacharya, Nov 11, 2017). Such complaints have also been voiced by practicing doctors themselves and one of them, Dr. Arunachal Dutta Chowdhury, posted at Barasat District Hospital, was even suspended by the government for claiming on Facebook that hospitals were short-staffed, patient care was poor and that actual numbers were being suppressed (Banerjie Nov 12, 2017). As a result, there is a great deal of confusion regarding the actual number of fatalities caused by dengue disease in West Bengal and such confusion stems from contradictory statements coming from the highest level of the administration. For example, after claiming on 25 October 2017 that 34 people had died from Dengue in West Bengal, Chief Minister Mamata Banerjee herself revised the figure to 13 by claiming that 27 other deaths were caused by "malaria, dengue, or swine flu" and that no final count could be confirmed before verifying blood sample as the deaths had occurred in private hospitals (Talukdar, Nov 14, 2017). Such numbers were abysmally low compared to ground reports documenting hundreds of deaths. One report from Sougata Mukhopadhyay even spoke of 150 deaths in Deganga block of North 24 Parganas (27 October 2017). Quite naturally, there has been a great deal of uncertainty among the public regarding the true extent of the dengue disease and its virulence, leading to as many as eight petitions filed in the Calcutta High Court, one of which even contained copies of death certificates that mentioned dengue as the cause of death even though names of those victims were missing from the official list compiled by the state (Banerjie, Nov 25, 2017). Such politicization of dengue and death cannot be a viable option to eradicate this vector borne disease properly from the state. The state officials are now much more alarmed regarding the possible severity of dengue cases in coming years and various area management schemes have already been applied to control this disease. Common people should be made properly aware of the virulence of the dengue by providing them with the actual data despite the national political pressure. Awareness among the common people is the primary management procedure to combat Dengue.

20.7 Resisting Dengue: Means and Methods

The role of State Government in controlling Dengue virulence has been quite prominent for the last few years. The local urban governmental authorities are now actively participating in Dengue combating Programmes. Over time, with changing disease ecological pattern, the authorities have been taking new innovative steps to control the disease. World Health Organization (WHO) provides various Vector Management procedures on a regular basis which are meticulously followed by the authorities (WHO 2009). Despite such implementation of standard updated dengue control measures, the disease seems to be continuing its virulence in West Bengal. This renders obvious the presence of some lacunae on the controlling or management procedures. One needs to combine all the possible management procedures together to combat the disease. The variety of methods for resisting this disease can be grouped under certain categories for clear analysis and for easy future implementation purposes. These are as follows.

20.7.1 Multi-level Effective Organizational Functions

A well-structured organization proves to be very effective in managing dengue crisis. Regular monitoring of the vector control activities and formation of Rapid Action Teams (RAT) in different Boroughs and Municipalities has been done by the state government. These teams act rampantly to detect the loopholes and reduce the dengue virulence. There must be collaboration among the various departments like health, drainage, water supply, etc. for rapid controlling of dengue in the vulnerable localities. Intense micro level field based and home to home survey for the detection of breeding grounds of mosquitoes by the health workers is the need of the hour. Regular spraying of Larvicidal oil and gases in all possible inaccessible areas is very important for the common people. Integration of various chemical and non-chemical vector control methods, usage of larvivorous Guppy Fish in tanks and other waterbodies to reduce

Aedes Aegypti mosquito, spraying of Bacillus Thuringiensis Israelnsis (BTI) bacteria to destroy Aedes Aegypti mosquitoes, door to door pyretheum spraying, releasing Wolbachia-carrying mosquitoes to the environment to reduce the breeding capacity of the Aedes Aegypti mosquito, etc. are some of the techniques used against the virus. Administrative workforce should improve the physical conditions of the slums, must ensure regular garbage clearance and provide good sanitation facilities for the slum dwellers to reduce the probabilities of mosquito breeding. Regular visits and investigation of the major social structures and organizations like schools, colleges, hospitals, workplaces, markets, etc. to prevent the breeding of the mosquito species is highly required. Special government cell must be developed to render help to everyone who are susceptible to the dengue. To prevent mosquito breeding at the construction sites or any other localities special decrees can be issued against the land owner if required. Finally more trained human resources must be used in the dengue eradication programs by the state authorities.

20.7.2 Efficient Medical Units

Improved medical facilities and proper coordination among the health centers is very important for treating the dengue patients well and to reduce the dengue induced mortality rates. Involvement and strength of Doctors, nurses, and government health workers must be increased at a significant rate, at least during the peak season. The dengue patients must be admitted to the hospitals without any discrepancies. Availability of beds and medical services in different government hospitals for the dengue patients must be recorded and updated in any app or website so that patients can be admitted easily in hospitals without facing any trouble. All the government and private hospitals must provide optimal medical resources to the patients. Facilities of free dengue detection, formation of special health camps, free platelet count facilities are all now provided by the state authority. Information of blood tests for any vector borne diseases must also be collected from the private pathology labs also to prepare the total severity chart of the Dengue and other VBD per year. It will help in projecting the situation in future. Importance has also now been given on alternate Ayurveda and Homeopathy treatments to cure the Dengue patients. Implementation of New Real-time Polymerase Chain Reaction (RT-PCR) technology to confirm dengue virus in a blood sample within few hours is of great importance now. Various medical research units are working rigorously to find any vaccine against dengue.

20.7.3 Proficient Entological Research

Entomological surveillance of dengue vectors have been a key aid in understanding the population persistence and dynamics inclusive of relative abundance at a spatiotemporal scale (Banerjee et al. 2013). Entological survey and research have been conducted in different localities to find out the disease pattern and to detect genetic variety of various vectors and their effects on human bodies. Regular recording and preparation of database on potential mosquito breeding areas, type and rate of occurrence of Vector Borne Diseases, etc. helps significantly in future entological research to combat the disease. Various quantitative techniques have been used to categorize areas into different vulnerable zones like ***House Index*** (Percentage of houses positive for Aedes Aegypti breeding containers), **Breteau Index** (number of Aedes Aegypti breeding containers per 100 houses), **Container Index** (Percentage of water containers positive for *Aedes Aegypti* larvae), **Pupal Index** (number of containers with *Aedes aegypti* pupae per 100 houses), etc. (Sharma et al. 2014). These techniques are now being used significantly in spatial epidemiological analysis. Regular application of these techniques can also identify the quality of ongoing anti-mosquito drive taken in a locality. Mapping of vulnerable areas (ward) of each and every year can help in detecting the locational pattern of dengue virulence and various controlling steps can be taken accordingly.

20.7.4 Social Awareness and Guidance

Dengue can be eradicated from any locality mostly by inculcating awareness among the residents. Social mobilization and empowerment of the communities can reduce the vulnerability levels of Dengue fever. Increasing awareness by regular supplement of dengue related local data and conditions can generate responsibility among the people to use mosquito nets and to clean the stagnant water and garbage in their surroundings. Initiatives are now being taken by Schools to increase awareness among the children. Mass awareness campaigns through social media, TV programs, FM radio services, Documentary films, newspapers, leaflets, and by other methods can also reduce the possibility of breeding of the mosquitoes (Biswas et al. 2014). Every year the campaigns should start from the month of March and must continue up to October.

20.7.5 Prominent Infrastructural and Technological Application

New technologies are emerging to combat with dengue in different ways. Special Mobile App has been developed for detecting and spreading information of potential breeding grounds of Aedes Aegypti of different localities. People can now inform about the dengue prone localities directly to the government by using various apps and websites. For intense micro level survey Drone technology is now used for the detection of potential mosquito breeding grounds in inaccessible areas. Drones are also used to spray larvicidal oil and gases in all possible inaccessible areas.

The WHO has recommended the Integrated Vector Management Schemes (IVM) which seeks to improve the efficiency, cost effectiveness, ecological soundness, and sustainability of disease vector controls. IVM requires the establishment of principles, decision-making criteria, and procedures of vector controls together with timeframes and targets (WHO 2012a).

20.8 Conclusion

In 2012 dengue ranked as the most important mosquito borne viral disease in the world (WHO 2012b). Dengue is now a major public health concern for a majority of the residents of West Bengal. Dissemination of Dengue has had huge impact on our physical and mental health, personal spheres of lives, socio-economic conditions, and of course on our very existence. While Covid-19 has opened various new research fields in medical geography, Dengue dissemination has already become an important research aspect of spatial epidemiology in various tropical countries (Mutheneni et al. 2017). Climate change, proliferation of vector borne diseases, Dengue dissemination, disease severity, vulnerability of common people, death and politics about the disease—everything has become relevant themes of disease studies in the Indian context. Further Dengue related research and new strategies must include a holistic approach covering all possible physical and anthropogenic spheres of our environment. Effective medical policies including capacity building, resource allocation and mobilization, integrated surveillance, improved health services, outbreak preparedness, community awareness, facilitating new planning, etc. must be implemented in West Bengal to reverse the trend of dengue severity.

References

Bandyopadhyay B et al. (2013) A comprehensive study on the 2012 dengue fever outbreak in Kolkata, India. Hindawi Publishing Corporation, International Scholarly Research Notices Virology

Banerjie M (2017) Bengal government doctor suspended for facebook comments on dengue. NDTV.com, Nov 12 2017. https://www.ndtv.com/india-news/bengal-government-doctor-suspended-for-facebook-comments-on-dengue-1774273, last accessed 2020/6/19

Banerjie M (2017) No clarity on number of dengue deaths in Bengal, Say Petitioners. NDTV.com. Nov 25 2017. https://www.ndtv.com/india-news/no-clarity-on-number-of-dengue-deaths-in-bengal-say-petitioners-1779738, last accessed 2020/6/19

Banerjee S et al (2013) Pupal productivity of dengue vectors in Kolkata, India: implications for vector management. Indian J Med Res 137:549–559

Bhattacharya A (2017) Behind West Bengal's dengue disaster is a lax state government and, possibly, a smarter virus. Scroll.in. Nov 11, 2017. https://scroll.in/pulse/857395/behind-west-bengals-dengue-disaster-is-a-lax-state-government-and-possibly-a-smarter-virus. last accessed 2020/6/19

Bisht B et al. (2019) Influence of environmental factors on dengue fever in Delhi. Int J Mosq Res 6(2):11–18

Biswas D et al. (2014) Three unique initiatives of KMC for Dengue prevention in Kolkata city, India, 2014. J Harm Res Med Health Sci 1(1):83–86
Chandra R (2018) A brief scenario on the emergence and occurrence of dengue fever in the Slum Dwelling Areas of Kolkata, West Bengal. Int J Sci Res 7(2):208–212
Chatterjee S et al. (2013) An overview of dengue infection during 2000–2010 in Kolkata, India. Art Dengue Bull 37:77–86
Delmelle E et al. (2015) Space-time visualization of dengue fever outbreaks. In: Kanaroglou P et al (eds) Spatial analysis in health geography, pp 85–99. Ashgate Publishing Company, USA
Ganguly KS et al. (2018) Spatial clustering of dengue fever: a baseline study in the city of Kolkata. Int J Health Med Sci 4(10):170–187
Government of India (2018) Central Bureau of Health Intelligence: National Health Profile 2018. Ministry of Health and Family Welfare, India
Government of West Bengal (2016a) State Bureau of Health Intelligence: Health on the March, 2015–16. Directorate of Health Services, West Bengal
Government of West Bengal: (2016b) West Bengal State Action Plan on Climate Change. Government of West Bengal, Kolkata
Gupta N et al (2012) Dengue in India. Indian J Med Res 136:373–390
Jarup L (2004) Health and environment information system for exposure and disease mapping, and risk assessment. Nat Inst Environ Health Sci 112(9):995–997
Martinez-Beneito MA (2013) A general modeling framework for multivariate disease mapping. Biomet Trust, Oxford J 100(3):539–553
Mukhopadhyay S (2017) Bengal's dengue crisis: has the Mamata Banerjee Government Buried Its Head In Sand?. News18.com. October 27, 2017, https://www.news18.com/news/india/bengals-dengue-crisis-has-the-mamata-banerjee-government-buried-its-head-in-sand-1559531.html. last accessed 2020/6/19
Mutheneni SR et al. (2017) Dengue burden in India: recent trends and importance of climatic parameters. Emerg Mic Infect 6:1–10
Olsen SF et al. (1996) Vector cluster analysis and disease mapping: why, when and how? A step by step guide. BMJ: Brit Med J 313(7061):863–866
Rican S, Salem G (2010) Mapping disease. In: Brown T et al. (eds.) A companion to health and medical geography, pp. 96–110. Wiley-Blackwell, United Kingdom
Sharma SN et al. (2014) Vector borne diseases in Kolkata Municipal Corporation (KMC): achievements and challenges. J Commun Dis 46(2):68–76
Singh PK, Dhiman RC (2012) Climate change and human health: Indian Context. J Vector Borne Dis 49(June):55–60
Talukdar S (2017) Dengue outbreak in Bengal: Mamata Banerjee's focus is on controlling the spread of news, not virus. Firstpost.com. Nov 14 2017. https://www.firstpost.com/politics/dengue-outbreak-in-bengal-mamata-banerjees-focus-is-on-controlling-the-spread-of-news-not-virus-4209371.html, last accessed 2020/6/19
World Health Organization: Dengue Guidelines for Diagnosis, Treatment, Prevention and Control. World Health Organization and Special Programme for Research and Training in Tropical Diseases, France (2009)
World Health Organization (2012a) Global Strategy for Dengue Prevention and Control 2012–2020
World Health Organization (2012b) Handbook for Integrated Vector Management, WHO, 12–14

Chapter 21
Tuberculosis Patients in Malda District of West Bengal, Eastern India: Exploring the Ground Reality

Tapan Pramanick, Deb Kumar Maity, and Narayan Chandra Jana

Abstract Tuberculosis is the leading casue of death worldwide. It is caused by the bacteria Mycobacterium tuberculosis and is usually disseminated through air. This paper strives to analyse the socio-demographic condition of tuberculosis patients and their behavioural patterns based on the empirical field-based study conducted in 2019 in the Malda district of West Bengal, India. The study was conducted among 167 samples (89 male and 78 female), selected through a simple random sampling technique. It was found that 54.49% (91) patients were Category-I, 16.76% (28) Category-II, 25.75% Category-IV (Multi-Drug Restant-MDR), and 2.99% in the category-V (Extensive Drug Resistant-XDR). XDR patients were very critical in conditions. Inspite of the severity of the disease, more than 74% of patients were not using the mask. Furthermore, socio-demographic characteristics of the patients reveal that 34.73% of patients were of the age group 21–30 years while 7.18% from >60 years, 78% of the patients were illiterate and only 2.99% were graduates, more than 30% of patients were from the Scheduled Caste and the family income of more than 60.47% patients'was below <5000. Besides, 58.08% of patients were daily labour and 23.55% were *bidi* workers. The study reveals that the disease is widely prevalent among the educationally and socially backward population having lower awareness levels. Moreover, the majority of tuberculosis patients were malnourished, drug addicts, and smokers, which pose a serious concern for health planners. The outcome of this study will help the health planners and medical practitioners in identifying the major causes of the disease and guide them in mobilizing proper strategies for combating its spread.

Keywords Tuberculosis · Ground reality · Patients · Multi drug-resistant (MDR) · Extensive drug-resistant (XDR) · Behavioral aspects

T. Pramanick (✉) · N. C. Jana
Department of Geography, The University of Burdwan, Bardhaman, West Bengal, India

D. K. Maity
Department of Geography, West Bengal State University, Barasat, India

N. C. Jana et al. (eds.), *Livelihood Enhancement Through Agriculture, Tourism and Health*, Advances in Geographical and Environmental Sciences,
https://doi.org/10.1007/978-981-16-7310-8_21

21.1 Introduction

Mycobacterium tuberculosis was first identified on 24 March 1882 by Robert Koch (Al-Sharrah 2003). Tuberculosis is one of the deadliest infectious bacterial disease spread by mycobacterium tuberculosis. This bacteria initially attacks in the lungs and also affects other parts of the body such as joints, glands, intestine, skin, and meninges (Sharma et al. 2016; Griffiths et al. 2010). Some of the infections do not have symptoms in which case it is known as latent tuberculosis (Morrison et al. 2008). About 10% of latent infections turn into an active disease which left untreated. Chronic cough and cough with blood-containing sputum lasting more than two weeks, night sweats, fever, and weight loss is the main symptoms of tuberculosis. The other symptoms of tuberculosis are chest pain, tiredness, shortness of breath, and loss of appetite (Park 2005). TB is closely linked to both overcrowding and malnutrition, making it one of the major diseases of poverty (Kapoor et al. 2016). Drug adictors, who are associated with vulnerable people gatherings such as homeless shelters and prisons, poor and medically underpriviledge, ethnic minorities, childrens, and olderly people who came into contact with affected people and the medical persons who are taking care of the patients are more vulnerable portion of the society (Bass et al. 1994). Cigarate and bidi smoker people are also the border line people for this disease. Lung disease, diabetes mellitus, and alcholism can also increase the chances of tuberculosis disease. Ittransmits through air, persons to persons via droplet of throat and lungs with the active cases. One active patient may infect 10–15 peoples (Feng et al. 2002).

The first successful cases were found against tuberculosis in 1906 by Albert Calmette and Camille Guerin (Dubos and Dubos 1987). Mycobacterium tuberculosis is curable with the course of antibiotics medicine. The meaning of Directly Observed Treatment with Short Course Chemotherapy (DOTS) is to take medicine of tuberculosis in the presence of health staff. DOTS is the successful treatment system of tuberculosis patients. The primary aim of the Revised National Tuberculosis Control Program (RNTCP) was a cure at least 85% of every registered new sputum positive patients. Despite these global measures, the transmission of TB and mortality rate is significantly high in various countries and India is one of them. Due to its dangerous impact on the health of humans and widespread negative societal and economic influence, the World Health Organization stated TB as a global health emergency in 1993 (Zumla et al. 2013). In terms of death count, TB is the ninth-ranked disease globally and foremost reason from single transmittable agent remaining above HIV/AIDS. Approximately 1.3 million deaths of an HIV-negative person along with an extra 374,000 deceased among HIV-positive person were estimated in 2016 (Bloom and Murray 1992).

Several studies have been carried out to investigate the ground condition of TB patients, i.e. Rohit et al. discussed the socio-demographic profile of the TB patients registered at DTC Rewa of Central India (Dubos and Dubos 1987). Spatial–temporal distribution and influencing factors of pulmonary tuberculosisin China, during 2008–2015, was deliberated by Mao et al. (2019). Kapoor et al. (2016) elaborated about the pattern of socio-economic and health aspects among TB patients and controls. Globally, spatial nature of TB has been studied in different research work i.e. by Pramanick and Jana (2019), Touray et al. (2010), and Hassarangsee et al. (2015).

The present study was conducted in the all TU of Malda district, West Bengal, India. The objectives of the study are analysis of the socio-demographic profile and socio-economic status of the tuberculosis patients and their behavioural pattern.

21.2 Study Area

Malda District is located in the middle position of West Bengal, India. District headquarter is Englishbazar is also known as Malda. Latitudinal extension vary from 24°40′20″N to 25°32′08″N and longitudinal extension from 87°45′50″E to 88°28′10″E. This district covers 3733.66 km^2. According to the Census of India, 2011, the total population was 398,845 (Census of India 2011).

This district comprises two subdivisions Chanchal and MaldaSadar. This consists of thirteen police stations, fifteen community development blocks, two municipality, 146-Gram Panchayat, and 1798 mouzas. At present this district is divided into twenty Tuberculosis Unit (TU) namely Englishbazar Urban (EBU), Englishbazar Rural (EBR), Kaliachak-I, Sujapur, Baishnabnagar, Golapganj, Old Malda Urban (OMU), Old Malda Rural (OMR), Bangitola, Manikchak, Ratua-I, Ratua-II, Chanchal-I, Chanchal-II, Harishchandrapur-I, Harishchandrapur-II, Gazole RH, Hatimari, Habibpur, and Bamongola according to Revised National Tuberculosis Control Programme (RNTCP) guideline (Fig. 21.1).

21.3 Methodology

The present study was conducted among the all tuberculosis unit (TU) of Malda district during 2019 to identify the nature of spread among the people. RNTCP divides all India into TU's according to the number of population per TU. Based on these guidelines, this study area is divided into twenty tuberculosis unit and each TU comprises more than two lakh population. This study is conducted on some selected TUs of Malda district based on a field-based survey of demographic, socioeconomic as well as other characteristics of the surveyed population (Fig. 21.2).

21.3.1 Sample Survey-Based Field Study

Simple random sampling technique is used for this study. The field survey was conducted on randomly select 167 registered TB patients who were under treatment procedure during field survey time. Out of these 167 patients, 89 were male and 79

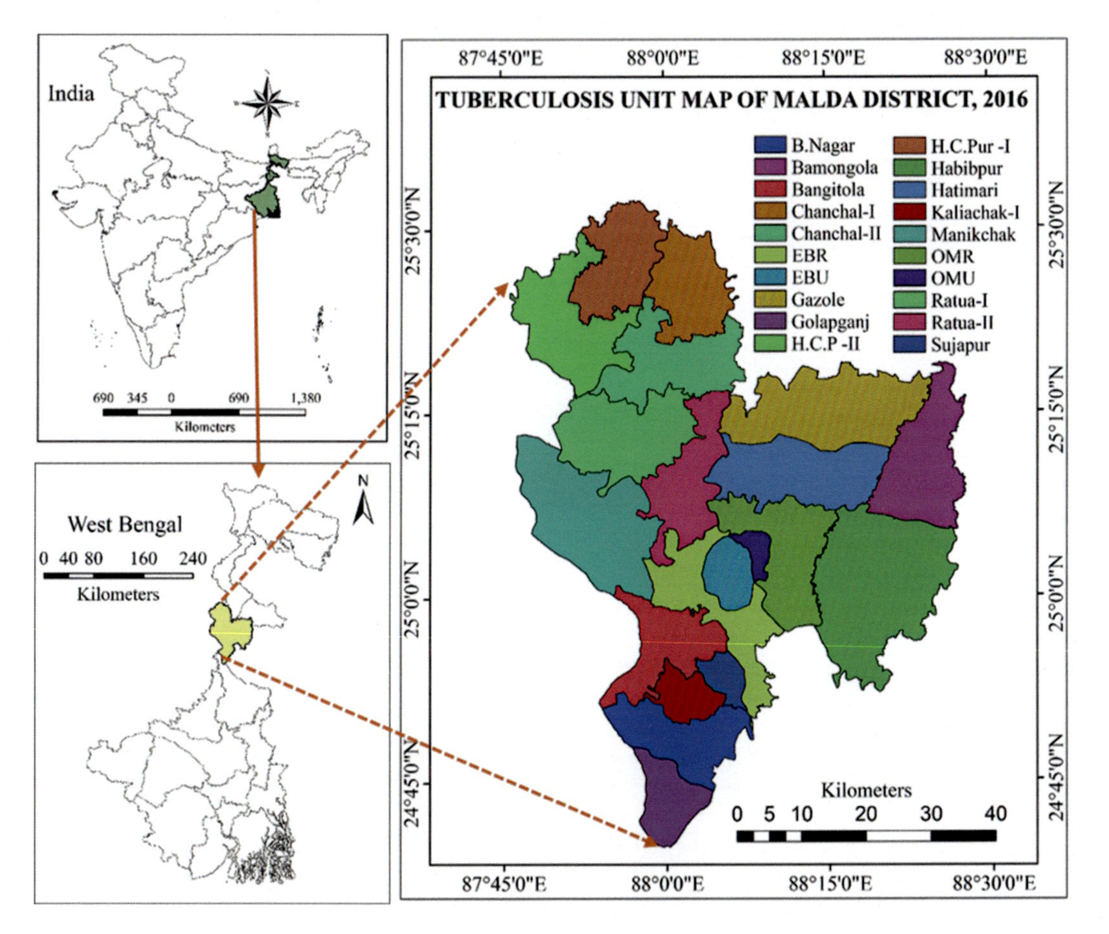

Fig. 21.1 Study area

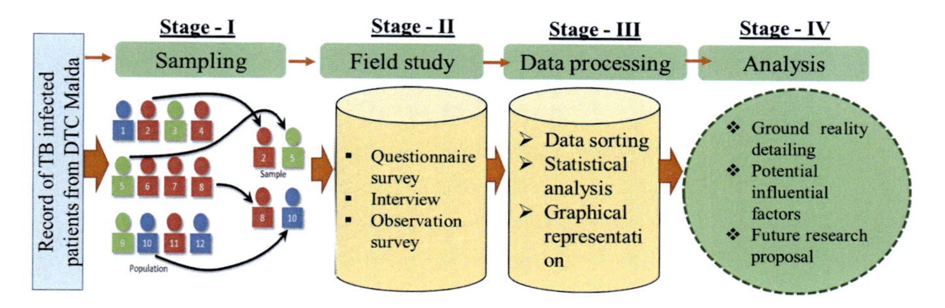

Fig. 21.2 Methodological workflow of the present study

were female. Total sampling size includes four categories of patients, i.e. 91 category-I patients, 28 category-II patients, 43 category-IV (MDR) patients, and 5 Category-V (XDR) patients. The data were collected using a structured questionnaire and visiting all the patients home for verbal conversation and understanding their surrounding environmental condition.

21.3.2 Data Entry and Analysis

Collected field survey-based data are used for identifying different characteristics of patients and their surrounding environments. Demographic and socioeconomic data are analyzed using some basic statistical techniques. But from some previous studies, it is found that several TB patients are directly related to the age group of the patients. To identify this dependency of several TB patients and on the age group of the patients, a Chi-Square test was conducted using statistical software packages. Graphical representations and mappings are prepared using different statistical software, i.e. SPSS v17, MS Excel 2016, and ArcGIS v10.3.1. Mainly this study is dependent on basic statistical techniques which helps to understand the nature of TB patients and their Socio-Demographic characteristics.

21.4 Results

21.4.1 Socio-economic Standing of TB Infected

The distribution of the sample population (patient) according to their age and sex has been represented in Table 21.1. The total study population was 167 patients of which 89 patients were male and 78 patients were female. The age-based analysis found the most share of TB infected in the age group of 21–30 years (58 individual) which is 34.73% of the total sample. Similarly, this was followed by the age group

Table 21.1 Age and sex-wise distribution of patients

Sl. No.	Age in years	Male	Female	Total
1	<10	7	6	13
2	11–20	12	11	23
3	21–30	31	27	58
4	31–40	13	11	24
5	41–50	9	9	18
6	51–60	10	9	19
7	>60	7	5	12

Chi square = 0.22549, d.f. = 6, p value = 0.9997805

31–40 years, 11–20 years, and 41–50 years. Out of the sample population, the age group >60 years and <10 years was found to be least infected. This result may depict social mobility as one of the influential factors to trigger the transmission rate of TB. It is entirely rational that these two group of peoples have less mobility in society or potential transmitting areas due to age factor. After the old age group, the pediatric patients were second fewer 13 (7.78%). Age-sex wise distribution was not dependent, and age-sex wise distribution was not statistically significant (p value > 0.05).

In the case of educational status, 46.70% (78) patients were unschooled and rest have attained a different level of education. 8.98% have completed primary education, 10.77% have upper primary schooling, secondary and higher secondary schooling completed by 10.77% and 12.57% respectively and graduate degree and above qualification have been achieved by only 2.99% infected. From the perspective of social category, among the surveyed population, 37.12% were Schedule Tribe (ST), 14.97% were Other Backward Classes (OBC), 17.36% were General (Gen), and 30.54% were scheduled caste (SC). Most of the infected were from the under-privileged category (60.47%). The analysis has shown a hierarchy of a decreasing number of infected with increasing earning amounts. Study regarding the size of the family does not seem to have a significant role in this context as 47.3% of respondents have family members <4, 28.14% responders live with 5–7 members in a family and 24.54% of the surveyed individuals family comprised with more than 8 members. Consideration of occupational dimension has found an attention-grabbing outcome. Five occupational categories have been recorded from the investigation of collected field data viz. daily wage labour, farmer, businessman, govt. employees, and *bidi* workers. The diseased magnitude was found quiet high among the daily wage worker (58.08% of the total sample) followed by *bidi* worker (23.35%), businessman (10.17%), government employees (5.99%), and farmer (2.39%) (Fig. 21.3a–e).

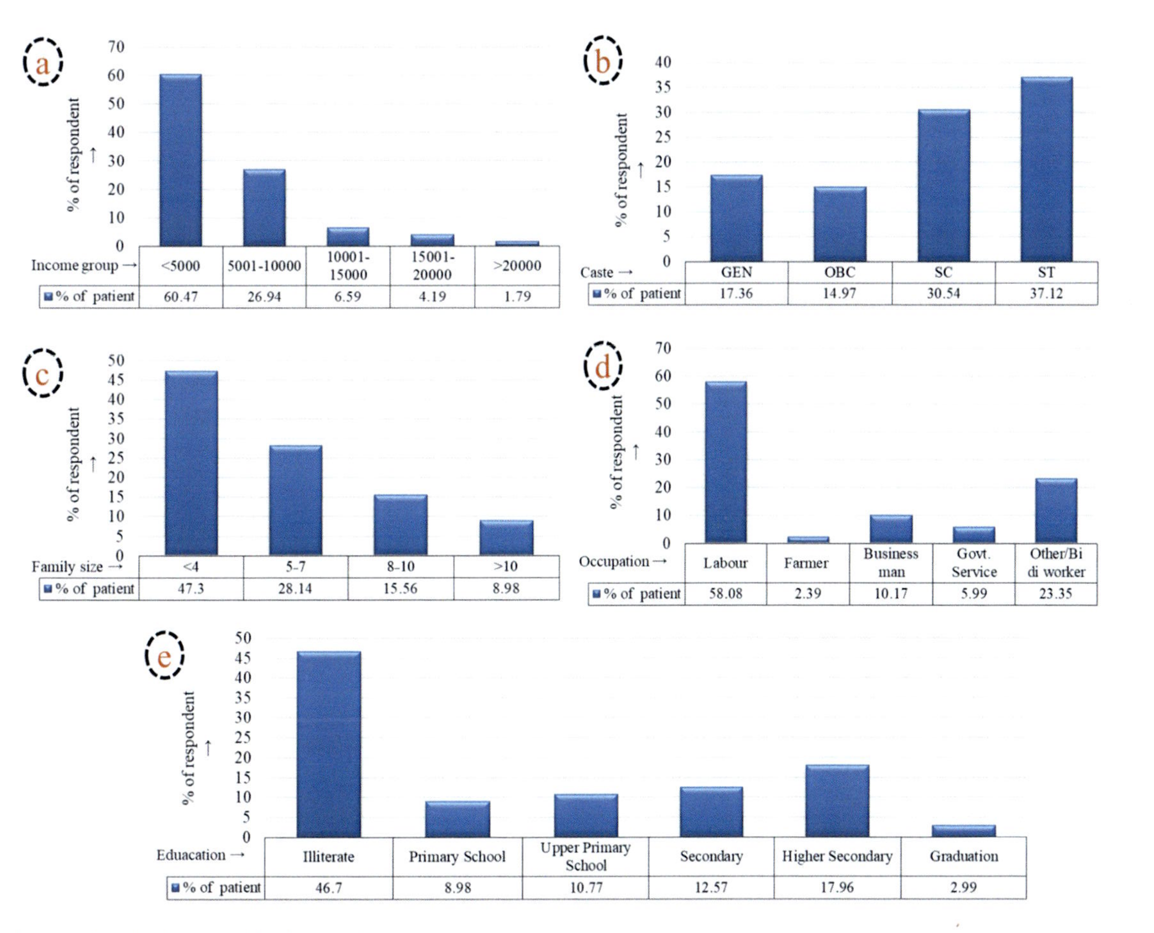

Fig. 21.3 Distribution of socio-demographic characteristics

21.4.2 *Behavioral Particulars of TB Infected*

The behavioral perspectives considered in this study were the treatment urgency of the patients. The behavioral measures of the patients approximate suffering duration, the gender of the patients, safeguard measure status, i.e., use of mask and quality of the mask, spitting condition, nearby infected patient, present treatment facility type, and distance to the primary health center (in km). The severity category of the disease among the respondents is detailed in Fig. 21.4. The study found that 54.49% (91) patients were under category-I, 16.76% (28) patients belong to Category-II, 25.75% were under Category-IV (MDR), and 2.99% were in the category-V (XDR). XDR patients were very critical in conditions. Tuberculosis is an infectious disease suffering from six months to two and a half years. Out of a total of 167 infected, 37 (22.15%) patients are suffered from three months who were newly affected. A total of 47 (28.14%) patients were suffering from four to six months, 20 (11.97%)

Behavioral aspect	Behavioral categry	No. of patient	% of patient
Severity category	Cat I	91	54.49
	Cat II	28	16.76
	Cat IV (MDR)	43	25.75
	Cat V (XDR)	5	2.99
Suffering from (months)	<3	37	22.15
	4-6	47	28.14
	7-12	20	11.97
	>12	63	37.72
Male-female ratio of patients	Male	89	53.29
	Female	78	46.7
Use of mask	N-95	8	4.79
	Normal mask	32	19.16
	Not used	127	76.04
Spiting Condition	Specific Place	42	25.14
	Anywhere	125	74.85
Nearby TB patients	Yes	118	70.65
	No	49	29.34
Treatment facility	Private	65	38.92
	Government	102	61.07
Distance to PHC (in km)	<4	67	40.11
	5-8	55	32.93
	>8	45	26.94

Fig. 21.4 Behavioral aspects of tuberculosis patients

patients affected since 7–12 months, and 63 (37.72%) patients were suffered more than one year. The male–female ratio in percentage among the surveyed infected was 54.49/45.50. Consciousness and responsiveness of the TB infected is one of the prime issues to control the transmission of TB bacteria as it could be beneficiated for both the infected and surrounding inhabitants also. However, the level of awareness in the present investigation has found to be very substandard. Regarding the use of the mask, only 23.95% were conscious of having proper use it while 76.04% of people were not using the mask. The concern of spitting in a specific place is essential to barrier the spread of TB; however, only 25.14% of patients found to thrown spit in the specific place or used container to through cough and rest have no specific point. The presence of nearby TB patients is very much unsafe for the peoples and the present analysis significantly revealed that 70.65% of the interviewed had infected persons within the family or adjacent areas. At present, government treatment is availing by 61.07% of respondents whereas 38.92% of patients relying on the private sector treatment. The distance of the health center from the house of the 40% patients is within four-kilometers, while 33% of patients have to avail service within five to eight kilometer and 27% of patients have to go more than eight kilometers to avail it (Fig. 21.4).

21.5 Discussion

The present inquiry was about the statistics and status of the tuberculosis patients of Malda district, West Bengal, India. The total study area is presently divided into twenty tuberculosis unit (TU) to provide the necessary treatment and spread cognizance regarding this disease. Each TU has more than two lakh population. The District Tuberculosis Center (DTC) is situated in the English bazar town, which is the district administrative headquarter to monitor tuberculosis ailment. The statistics were convened during April 2019, in which a total of 167 TB affected individuals were interrogated. Assembled information discloses that among them, 91 infected have the urgency of category-I, 28 people are under Category-II, 45 individuals are under category-IV or MDR, and the rest of 5 were under category-V (XDR). From the perspective of gender, out of total infected, 53.29% were male and 46.70% were female patients. One consistent output of the analysis was found that economically productive age groups (age between 21 and 50 years) are more infected in this district (Table 21.1) which may be deliberated to their higher social mobility and intense social interactions without precautions. Similar information was come out in the age-sex based study of TB infection by Rohit et al. who defined that the 21–30 age group was the most common in both male and female categories in this context. Furthermore, within this age group of infected, male patients were 63.15% and female patients were 36.84% (Table 21.1).

Besides the gender perspective, literacy view point was also analyzed which reconnoitered that majority of the affected peoples were illiterate (46.7%). In contrast, primary and upper education was accomplished by 8.98 and 10.77%, the secondary and higher secondary were attained by 12.57% and 17.96% respectively, and only 2% of persons have accomplished the graduation degree. This result magnets the necessity of education, awareness to live, lead a healthy life and to obey the safeguard measures for escaping from different diseases, specifically from TB in the present context. However, instantaneous provisions in this regard could be relevant such as public awareness programs, meant the need of cleanness, etc. In various studies of TB disease, illiteracy has been attributed as a major concern of diffusion of TB in different regions. The overall literacy rate of Maldais is 61.73%, and male and female literacy rate of this district are 66.24% and 56.96% respectively (Census of India 2011) which indicate low literacy rate influence for spreading the disease (Census of India 2011).

Although caste factor may not be a relevant concern for dispersal of TB, somehow, the caste-based literacy information from the Census of India (2011) and socioeconomic data can be relatable with this analysis. Among the surveyed individuals, 30.54% respondent belongs to Schedule Caste (SC), 37.12% belongs to Schedule Tribe (ST), 14.97% belongs to Other Backward Classes (OBC), and 17.36% persons were from General (Gen) category. The target sample was randomly chosen from infected patients in the study area. The results reflect two dimensions of clarification, i.e., the diffusion rate of the TB patients higher in the SC/ST community compared to other social categories or the patients from SC/ST community are more reliant on free treatment facilities from government which may reflect their socioeconomic standing. The economic illustration shows that 60.47% of the surveyed patients belong to a very less income group (<5000 rupees/month) and 26.96% have the monthly income between 5001 and 10,000 rupees. This outcome strongly recommends their underprivileged nutritional and health status as a hindrance to fighting against TB. The economic burden of the service availing TB patients from the government is a well-established fact in several kinds of literature which are also imitated by the occupational inquiry of the respondent (Oxlade and Murray 2012). Among the sample population, a large share of migrant labour and *bidi worker* may institute an association between TB infection propensity and professional aspect.

The disease is curable with appropriate treatment and medication. The patient of Cat-I has the opportunity to cure within a duration of six to eight months. Therefore, this study revealed the fact that a large count of the surveyed patients is under the Cat-I, which may suggestive for the betterment of precaution measures and own behavioural activities. Cat-IV (MDR) patient count was the second highest signifying deficiency of proper measures among the patients under Cat-I and Cat-II which is an issue of public health sensitivity in the present study area. The statistical inquiry regarding the level of awareness was not satisfactory for this district (Table 21.1). This disease spread through the air rapidly and a TB patient can infect ten to twelve people. In this study, 70.65% of the enquired individuals have TB infected persons nearby or within the family which may confermost of the tuberculosis patients were affected by their home or neighboring. So from the field observation and analysis, it can be identified that this disease has a domain area and this area is very active and contagious. Besides this, most of the patients have no consciousness about the need fora good quality mask (N 95). Spitting behavior was random for the most infected. These results of alertness found the need of further exercise of cognizance development practice among the patients. Sub-health center is the smallest unit of the health care system. Sub-center and primary health centers provide the treatment to tuberculosis patients through DOTs. In the urban areas of Malda district like English Bazar, Old Malda, they get the facility of the treatment within a short distance and short time. But in rural areas, the people face the problems to get the treatment quickly. Most of the primary health center does not have enough doctor to meet the need timely. This problem is found explicitly in the Harishchandrapur, some parts of Kaliachak, Bedrabad, and Chanchal also.

21.6 Conclusion

The present study identified the real socioeconomic conditions of the tuberculosis patients during 2019 of Malda District, Eastern India. The most of tuberculosis patients found working for age group 21–50 years. More than half of the patients were illiterate, low-income groups, daily labour. Another significant factor is that 23.35% of the patents were engaged a *bidi* workers (Plate 21.1a–c). Most of the patients cannot use masks and spit anywhere. ST and SC populations were more affected than the General Caste population. The main problems were the number of MDR and XDR TB patients increasing day by day. It may be told that this disease was found in the illiterate people, socially backward people, economically backward, underdeveloped area and unconscious people, malnutrition, alcohol abdicated and active and passive smokers also. This study may help the health planners, doctors, RNTCP for fulfilling the 85% patient to cure and researcher for further study.

Plate 21.1 **a** Female working people are making bidi at the afternoon after taking lunch while their children are playing near them without wearing any mask at Sujapur, Malda. **b** Homemakers are making bidi at their home with the minimum wages at Baishnabnagar, Malda. **c** Both male and females are making bidi in the afternoon, and the male persons are cutting the leaves and women are preparing bidi at Bedrabad, Malda

Acknowledgements The authors would like to appreciate to the Ministry of Health and Family Welfare, DTC, Malda and TB affected patients for continuous help, data sharing, permission, and support to carry the study.

References

Al-Sharrah YA (2003) The Arab tradition of medical education and its relationship with the European tradition. Prospects 33(4):413–425

Bass Jr, JB, Farer LS, Hopewell PC, O'Brien R, Jacobs RF, Ruben F, Thornton G et al (1994) Treatment of tuberculosis and tuberculosis infection in adults and children. American thoracic society and the centers for disease control and prevention. Am J Respir Crit Care Med 149(5):1359–1374. https://doi.org/10.1164/ajrccm.149.5.8173779

Bloom BR, Murray CJ (1992) Tuberculosis: commentary on a reemergent killer. Science 257(5073):1055–1064. https://doi.org/10.1126/science.257.5073.1055

Census of India (2011) https://www.census2011.co.in/census/district/6-maldah.html
Dubos RJ, Dubos J (1987) The white plague: tuberculosis, man, and society. Rutgers University Press
Feng Z, Iannelli M, Milner FA (2002) A two-strain tuberculosis model with age of infection. SIAM J Appl Math 1634–1656. https://doi.org/10.1137/S003613990038205X
Griffiths G, Nyström B, Sable SB, Khuller GK (2010) Nanobead-based interventions for the treatment and prevention of tuberculosis. Nat Rev Microbiol 8(11):827–834. https://doi.org/10.1038/nrmicro2437
Hassarangsee S, Tripathi NK, Souris M (2015) Spatial pattern detection of tuberculosis: a case study of Si Sa Ket Province, Thailand. Int J Environ Res Public Health 12(12):16005–16018. https://doi.org/10.3390/ijerph121215040
Kapoor AK, Deepani V, Dhall M, Kapoor S (2016) Pattern of socioeconomic and health aspects among TB patients and controls. Indian J Tuberc 63(4):230–235. https://doi.org/10.1016/j.ijtb.2016.09.011
Mao Q, Zeng C, Zheng D, Yang Y (2019) Analysis on spatial-temporal distribution characteristics of smear positive pulmonary tuberculosis in China, 2004–2015. Int J Infect Dis 80:S36–S44. https://doi.org/10.1016/j.ijid.2019.02.038
Morrison J, Pai M, Hopewell PC (2008) Tuberculosis and latent tuberculosis infection in close contacts of people with pulmonary tuberculosis in low-income and middle-income countries: a systematic review and meta-analysis. Lancet Infect Dis 8(6):359–368. https://doi.org/10.1016/S1473-3099(08)70071-9
Oxlade O, Murray M (2012) Tuberculosis and poverty: why are the poor at greater risk in India? PloS One 7(11). https://doi.org/10.1371/journal.pone.0047533
Park K (2005) Park's textbook of preventive and social medicine. Preventive Medicine in Obstet, Paediatrics and Geriatrics
Pramanick T, Jana NC (2019) Spatio-temporal analysis of tuberculosis in Malda District. West Bengal 8(2):3022–3038
Sharma A, Bloss E, Heilig CM, Click ES (2016) Tuberculosis caused by *Mycobacterium africanum*, United States, 2004–2013. Emerg Infect Dis 22(3):396. https://doi.org/10.3201/eid2203.151505
Touray K, Adetifa IM, Jallow A, Rigby J, Jeffries D, Cheung YB, Hill PC (2010) Spatial analysis of tuberculosis in an urban west African setting: is there evidence of clustering? Trop Med Int Health 15(6):664–672. https://doi.org/10.1111/j.1365-3156.2010.02533.x
Zumla A, George A, Sharma V, Herbert N (2013) WHO's 2013 global report on tuberculosis: successes, threats, and opportunities. Lancet 382(9907):1765–1767. https://doi.org/10.1016/S0140-6736(13)62078-4

Chapter 22
An Assessment Study on Hierarchical Integrity of Road Connectivity and Nodal Accessibility of Maternal Health Care Service Centres in Itahar Block, Uttar Dinajpur District, West Bengal

Madhurima Sarkar, Tamal Basu Roy, and Ranjan Roy

Abstract Access and reachability to healthcare centre is an important issue for lucrative delivery of health services to its recipient. The basic essence of emergency service facility is to assess the degree of reachability through its connectedness and accessibility. Better connectivity and accessibility would provide nodal services with greater extent. The first and foremost objective of this study is to recognize the spatial location of different health care service centres as a nodal service point with its linkage perspective in Itahar block, Uttar Dinajpur District. The study has given its emphasis on to recognize the relative location of maternal health care service centres, its network alignment, connectivity, and accessibility. The study put its effort to highlight the fact that only maternal health care services equipped with better quality is not enough to give its optimum until and unless the better accessibility is achieved through the said services to its recipient. The entire study involves in acquiring and analysis of the spatial data such as discrete location of the maternal healthcare centres, its weathered road connectivity, degree of reachability and to recognize the spatial extent of its services for each and individual healthcare service centre. In this regard the road network connectivity to each health care centre has been taken into consideration. The entire analysis has been carried out through the geospatial analysis techniques. A matrix algebra technique, different algorithm regarding network analysis, has been carried out to assess and evaluate the connectivity and accessibility of maternal healthcare service centres in Itahar block.

Keyword Maternal · Health · Network · Connectivity · Accessibility

M. Sarkar · T. B. Roy (✉)
Raiganj University, Raiganj, India

R. Roy
North Bengal University, Siliguri, India

N. C. Jana et al. (eds.), *Livelihood Enhancement Through Agriculture, Tourism and Health*, Advances in Geographical and Environmental Sciences,
https://doi.org/10.1007/978-981-16-7310-8_22

22.1 Introduction

Maternal healthcare service centre is an important and emergency service facility which helps to care about and improve the health of maternal community and their child (Navaneetham and Dharmalingam 2002). The desirable treatment received from the said centres is inevitable to the maternal community during antenatal and postnatal period (Johri et al. 2011; Ali et al. 2016). Procreation period of maternal community should be supervised with stipulated protocol. Any kind of negligence of taking the proper healthcare services may be fatal to the maternal community and its foetus (Shapiro et al. 2018). In this regard the access to maternal healthcare centre plays crucial role in maintaining the health and hygiene of maternal community (Mahajan and Sharma 2014). The equity of the services has a direct impact on the quality of life of individual. Usually, the burden of disease related to the maternal health during the pre- and post-partum period should be properly monitored and managed. Accessibility is concerned with the degree of reachability to a specific point. Indeed, it is determined by the connectedness of the said point of services (Mahajan and Sharma 2014; Cascetta et al. 2013). Accessibility to the healthcare centre is the ability of population to obtain specified set of health care services and accessibility measure the size of the population affected by the transport (Halden et al. 2000). Spatial accessibility to the maternal health centre is very important task for health planner because without proper access to the health care centre; it is not possible to deliver the better healthcare services to its people (Alake 2014). The matter of health of the mother is an emergency aspect, as most of the health aspects is matched with the institutional delivery system and its antenatal and postnatal complexities (Fekadu et al. 2018). The dissemination of healthcare services and its demand depend on the distance and time taken between the location of the individual and the health care centres (Wellay et al. 2018). In this regard the road connectivity and its accessibility to each centre play the key role for determining the overall spatial accessibility to the recipient (Neutens 2015). However, the proximity of individual is the key concern of the maternal healthcare service issue. In this regard each and every individual health care centre should be better linked with its neighbourhood. Health care facilities served from the different corner in any region can be divided into three broad categories that is sub centre, primary health care centre, and hospitals. Sub centre and primary healthcare services provide basic healthcare services, and hospital provides services for specialist health treatment as it is better equipped than the former. Health is one of the inevitable ingredient parameters in developmental aspects of a region. However, it has been evolved with better services with the passage of time. Nonetheless, the access of maternal health services in rural areas of developing countries remains in poor condition because of low availability of human resources (Iyengar et al. 2009). In India, the accessibility to healthcare facilities is extremely limited in many rural parts and backward regions (Saikia et al. 2014). The geographical location and the distribution of health service centre, road network, status of connectivity and accessibility are the key component in the utilization of maternal healthcare services (Otu 2018). Accessibility to the healthcare is concerned with ability of the people to obtain certain set of healthcare services (Levesque et al. 2013). Thaddeus and Maine (1994) postulated three delay models to establish the implication of importance of utilization of maternal healthcare services (Thaddeus and Maine 1994). Maternal mortality and morbidity remain a major problem in many developing countries like India (Ronsmans and Collin 2008). Although there has been improvement in maternal mortality rate since this issue was adopted by Union Nation of Millennium Development Goals (MDG) in the year 2000, this is an important issue that needs to be monitored (Chatterjee and Paily 2011). The well-being of mother is an important foundation of prospective nation. The optimum distribution of health centres might be effective to deliver the better healthcare services to the mothers. The regular visit to the centre may enhance the awareness of the mother about their do's and don'ts. Nowadays the proximity the service and its poly nuclear and hierarchical character are getting popularised in rural area. The traveling distance and time between location of maternal health care centre and location of individual is playing the key role for determining the better delivery of emergency services (Ghosh and Mistri 2016). Maternal health has been becoming a global concern, because the lives of millions of women in reproductive age can be saved through maternal health care services (Kifle et al. 2017). It encloses the health care dimensions of family planning, preconception, prenatal, and postnatal care in order to ensure a positive and make good experience and to reduce maternal mortality and morbidity. Access to maternal healthcare service centre and their connectivity measures the capacity of health system to reach the population without excluding part of it and timely use of personal health services to achieve the best health outcomes (Ghosh 2015). Ensuring a high degree of access to maternal healthcare centre improves mother's overall health status, extends life expectancy, reduces health inequalities, and lessens Maternal Mortality Rate (MMR) and gives the better health prospect for forthcoming generation.

22.2 Objectives of the Study

The broad objective of the study is to assess the role of road network and spatial accessibility of maternal health care service centre in utilization of maternal healthcare services. In order to fulfil the above broad objective, the study has followed some important essences as follows.

- To show the existing locational and distributional aspects of maternal health care centres in the study area.
- To show the spatial connectivity and accessibility of maternal health care centre within the study area.
- To determine the served area for each maternal health care centre and its functional gaps.

22.3 Database and Methodology

22.3.1 Study Setting and Data Sources

The entire work has been carried out under the ambit of geospatial technology, applied GIS methods to analyse spatial accessibility to maternal health care centre. In this regard the geo relational data model has been prepared. The discrete location of each maternal health care centre is positioned spatially, and its geometric relationship with weathered road network is conceptualised into topological map. The study is based on information accrued from Google Earth, Block map, reports from CMOH (Chief Medical Officer of Health) office of Uttar Dinajpur district, population census (2011), data from Itahar Rural Hospital, and primary healthcare centres. With the help of GPS, location of health centre has been identified.

22.3.2 Study Tool

Several datasets in non-digital format were collected and transformed into digital format and processed through ArcGIS and QGIS software. The ArcGIS program (version 10.4) was used to prepare the map of accessibility and proximity to the maternal health facility. QGIS program (version 3.4.4) was used to construct road network, spatial distribution of maternal health centre, and various connectivity measurements.

22.4 Methods

The road network map of Itahar block was created with the help of Google Earth which describes the pattern of road network, road length, etc. and enables to understand the spread of the road network over the study area. The location of health centre has been identified with the help of GPS and overlaid with the road network in QGIS platform to examine the relationship between the location of maternal health centres and road network. The threshold pressure of each maternal health care centre has been computed through ratio measurement at GP level. In order to accomplish the study about the degree of connectivity, the major important connectivity indices devised by Hanson and Kansky has been applied in the study which has its own applicable meaning in the domain of transport geography. Cyclometric number (μ), Alpha Index (α), Beta Index (β), Gamma Index (γ), and Eta Index (η) are the most fundamental properties of a network system. Cyclometric number defines number of circuits in the network, a high value of cyclometric number indicates highly connected network. Alpha Index is one of the significant measures of connectivity of a network which is adjusted from cyclometric number. It is the ratio between observed numbers of circuit to the maximum number of circuits in the network system. The alpha index value ranges between 0 and 1, and this measure may be written in percentage, thus it ranges from 0 to 100. The value of 100% will be considered as completely connected network. Beta Index is expressed by the relationship between numbers of link over the number of nodes. $\beta < 1$ indicates simple or tree type network structure, $\beta = 1$ indicates a connected network with one circuit and $\beta > 1$ denotes a complex network with more than one circuit. Gamma Index measures connectivity that varies from a set of nodes having no linkage to the one in which each node has an edge linked with other nodes in the graph. It also ranges from 0 to 1, or it may be written in the form of percentage. Eta Index is the ratio between total network distance and number of arcs, and it is very useful in examining shape and utility of a transport network system. In Eta index, block wise road distance is measured in kilometres and divided it by the observed number of edges. All the values of these measures have been transformed into standardized score. Composite Connectivity Score (CCS) has been calculated subsequently by adding respective Z score values of cyclometric number, alpha, beta, gamma, and eta index. Detour Index (DI) is measured with the help of actual route distance and straight-line distance, and the resultant ratio is converted into percentage. In this study detour value is calculated in respect of distances from all sub health centre and primary health centre to rural hospital to know the efficiency of the transport network or how well it overcomes the distance or the friction of distance. Increasing the value of detour indicates reducing accessibility from the rural hospital. Additionally, the low detour value signifies high accessibility and lesser degree of surface friction and vice versa. Here assuming the index in 100, 130, and 160%, and based on three isolines the entire resultant values are interpolated as <100 is high accessible zone, 100–130 and 130–160 is moderate accessible zone, and >160 is low accessible zone. Shimbel Index calculates minimum number of paths necessary to connect one node with all the nodes in a specific network system. It is not necessary to measure total number of paths between two health centres or between sub centre and household, but rather what are the shortest paths between them is necessary. Shimbel index is calculated with the different sub health care service centres and primary health care centres to rural hospital (hierarchically apex body in the study area in respect of maternal health care services) which is located at major junction point of the study area. Shimbel index evaluates accessibility from rural hospital to all the sub health centre and primary health care centre. The entire methods adopted in the study has been briefly devised as (Rodrigue et al. 2019)

$$\mu = (\mathrm{e} - \mathrm{v} + \mathrm{p}), \tag{22.1}$$

where

e is the total number of edges.
V is the total number of vertices.
P is the number of non-connected subgraphs.

$$\alpha = (\mathrm{e} - \mathrm{v} + \mathrm{p})/2\mathrm{v} - 5, \tag{22.2}$$

where
v is the number of edges.

$$\beta = E/V \tag{22.3}$$

where
E is the total number of edges.
V is the total number of vertices in the network.

$$\gamma = e/3(v - 2), \tag{22.4}$$

where
e is the number of edges.
v is the number of vertices in the network.

$$\eta = M/E, \tag{22.5}$$

where
M is the total network length in kms.
E is the observed number of edges.

$$DI = D(S)/D(T), \tag{22.6}$$

where
D(S) is the straight distance.
D(T) is the actual distance.

$$A_i = \sum_{j=1}^{N} d_{ij}, \tag{22.7}$$

where
A_i is the Shimbel Index.
d_{ij} is the shortest path between i node to j node.

To identify the closeness of the maternal health centre multi buffer ring method has been applied. For this application ArcGIS spatial analyst tool has been used to calculate distances around the health centres.

22.5 Study Area

The study is carried out in Itahar community development block of Raiganj subdivision under the Uttar Dinajpur district in the Indian state of West Bengal. Itahar CD Block is bounded by Kaliyaganj and Raiganj CD Blocks on the north, Dakshin Dinajpur district, Harirampur CD Block on the east, Malda district on the south and Bihar on the west. This block consists of 12 Gram Panchayat (GP). Total population of Itahar CB block is 303,678 of which 155,777 are male and 147,901 are female (census 2011). The literacy rate of this block is 58.95%. The proportion of Muslim community is higher in this region that is 51.98%, followed by Hindu (47.43%) and Christian (0.43%). 42 sub health centres, 3 primary health centres, and one rural hospital are present in Itahar block. Total ASHA karmi of this block is 254. 39 sub centres have its own building and 3 sub centres are rented and all sub centres have electricity connection.

22.6 Results and Discussion

22.6.1 Spatial Distribution of the Maternal Health Care Centres and Prevailing Road Network

One of the major aims of this study is to find out the spatial distribution pattern of healthcare centres within the study area. Availability of health care centre is imperative factor to manage the utilisation of maternal health care services. In reality the distribution of healthcare service centre is indeed skewed in nature. As in most of the cases the pregnant women belonging to scanty resource region, many times suffered from complications related to pregnancy and delivery due to less and timely access to transport system to reach the health care centre. There are only one sub centre present in Chhayghara GP, 5 sub centres in Durlavpur GP, 3 Sub centres in Durgapur GP, 2 sub centres in Gulandar-I GP, 3 sub centres in Gulandar-II GP, 4 sub centres in Itahar GP, 5 sub centres in Joyhat GP, 3 sub centres in Kapasia GP, 4 sub centres in Marnai GP, 4 sub centres in Patirajpur GP, 4 sub centres in Surun-I GP, and 2 sub centres in Surun-II GP (Fig. 22.1). The total 263 Asha worker engaged in these sub centres. A block rural hospital and three primary health care centres are present in the study area. Total road length of Itahar block is 732.42 kms. It is observed that, in case of road length per sq. km in Marnai GP (115.62 kms) region is in better position and lowest road network expansion has been observed Chhayghara GP (26.12 kms) region (Fig. 22.2). There are total 39 Sub Centre (SC), 3 Primary Health Centre (PHC), and one Rural Hospital (RH) present in Itahar block. But the delivery facility is only available in PHC situated in Marnai Gram Panchyat (GP) and RH situated in Itahar GP. Road network and its accessibility to the health care centre significantly affect maternal health care services within the study area.

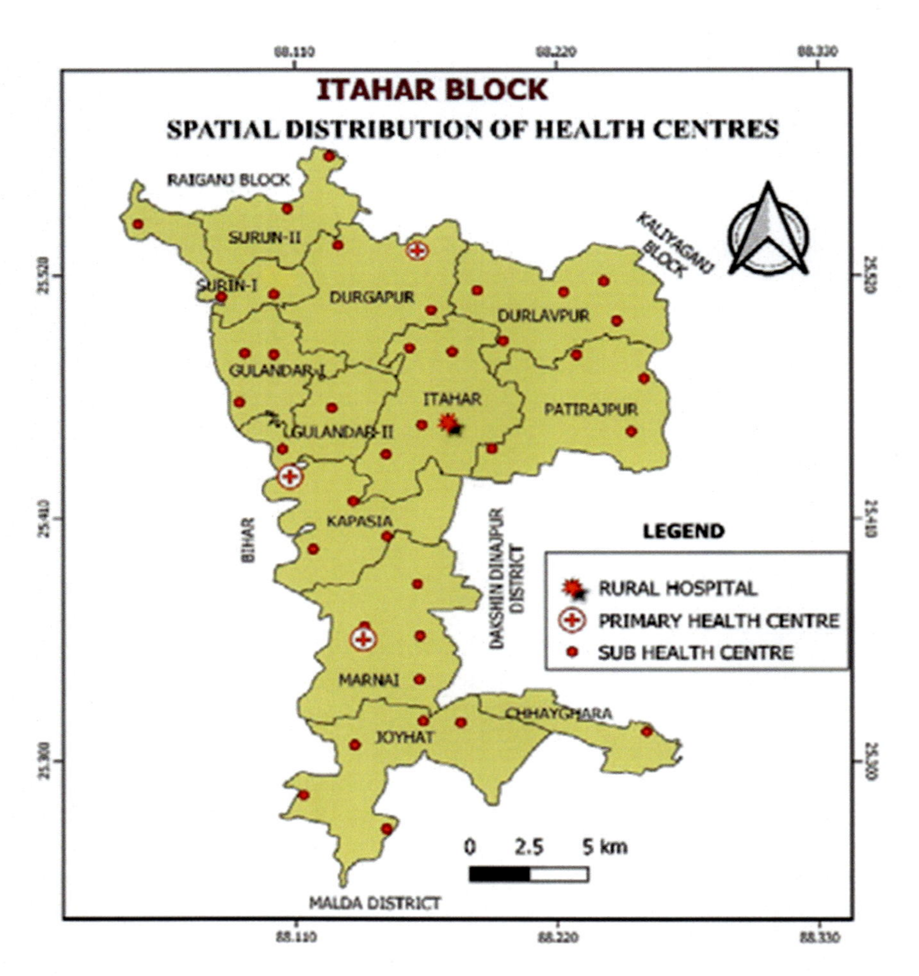

Fig. 22.1 Spatial distribution of health centres

22.6.2 Distribution of Health Centre and Population

The population distribution and availability of health centre is one of the major indices for measurement of health potentiality in a particular area. It is also essential to study the structure of the health care centre because a sub centre provides the basic or primary health care services to the maternal community at the grass-root level. As per population norms a sub centre is set up for every 5000 people in plain areas and for every 3000 population in hilly or desert or tribal areas, primary health care centre established to serve 20,000–30,000 people, and community health centre or block primary health centre set up to serve 80,000–120,000 population (Directorate General of Health Services 2006). Ratio between GP wise population and healthcare

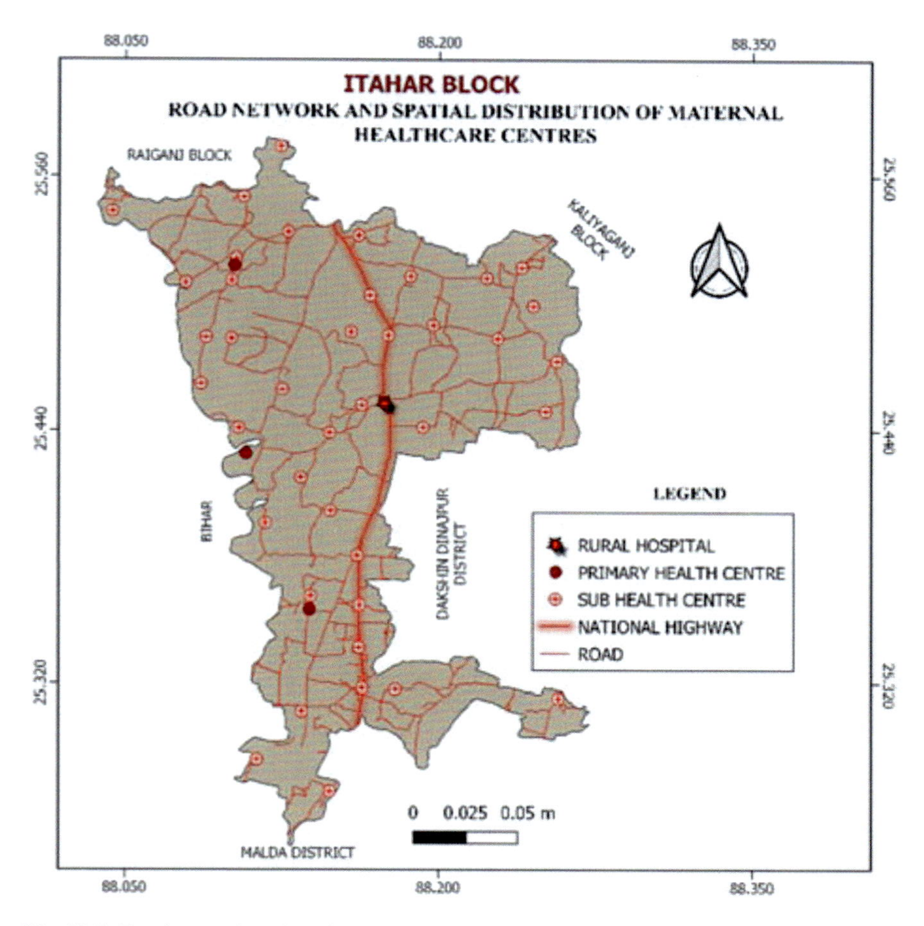

Fig. 22.2 Road network and spatial distribution of maternal healthcare centres

centre has been calculated which gives an overview about the population pressure over the health facility (Table 22.1). Health centre and population ratio is higher in Durgapur GP and Chayghara GP (Fig. 22.3). Total population of Durgapur GP is 32279, and number of health care centre is 3, that means each health centre of this GP provide services to 10,759 people. Total population of Joyhat GP is 30118 and 5 health care centres present in this GP, that means each health centre provide their services to 6023 people. So, the population pressure on health centre is highest in Durgapur GP, followed by Chhayghara GP, Gulandar-I GP, and Surun-II GP. Least population pressure on each health centre is recorded in Joyhat GP, followed by Durlavpur GP, Patirajpur GP, Surun-I GP, and Itahar GP.

Table 22.1 Health centre and population ratio

Name of the GP	Number of health centre	Population (2019)	Health centre and population ratio
1. Joyhat	5	30,118	1:6023.6
2. Chhayghara	1	9631	1:9631
3. Marnai	4	34,110	1:8527.5
4. Kapasia	4	32,659	1:8164.75
5. Gulandar-I	2	19,000	1:9500
6. Itahar	6	40,975	1:6829.16
7. Patirajpur	4	26,916	1:6729
8. Durlavpur	5	30,578	1:6115.6
9. Durgapur	3	32,279	1:10,759.66
10. Surun-II	2	19,000	1:9500
11. Surun-I	4	27,155	1:6788.75
12. Gulandar-II	3	20,835	1:6945

Source Itahar Rural Hospital, 2019

22.6.3 Measurement of Connectivity and Accessibility

The connectivity and the degree of accessibility to the maternal healthcare centre are not uniform for each and individual nodal centre for a particular region. The connectivity of a network may be defined as the degree of completeness of links between nodes. The role of sub centre or the primary health centre is to provide the combination of services such as registration of pregnancy, antenatal care, postnatal care, treatment of tiny or minor disease, prevention of malnutrition, counselling on diet, rest, hygiene and contraception which are very crucial and basic requirement for maternal community. Besides these services these health centre provides essential drugs like iron tablets, folic acid supplements for 12 weeks, treatment of anaemia, general examination such as blood pressure, height, weight, and tetanus toxoid injection which are rudimentary for every pregnant woman. These health centre also provide services related to pre-birth preparedness and complication readiness, breast feeding etc. As because of providing these essential services, connectivity and accessibility to these health centres is an important issue for the maternal community. For comprehensive analysis of road transportation network in administrative community block like Itahar, there is a need for analysis of some connectivity indices. Cyclometric number has been measured to discover the maximum number of independent cycles in a graph. Cyclometric number highest in Marnai GP (10) indicates highly connected network, and lowest value has been found in Gulandar-I GP (1) indicates poorly connected network. Alpha index has been measured to identify the complexity of the transport network system of this study region. It is a measure of connectivity which evaluates the number of cycles in a graph in comparison to maximum number of cycles in a network. In Itahar block the value of Alpha index varies from 4.4 to

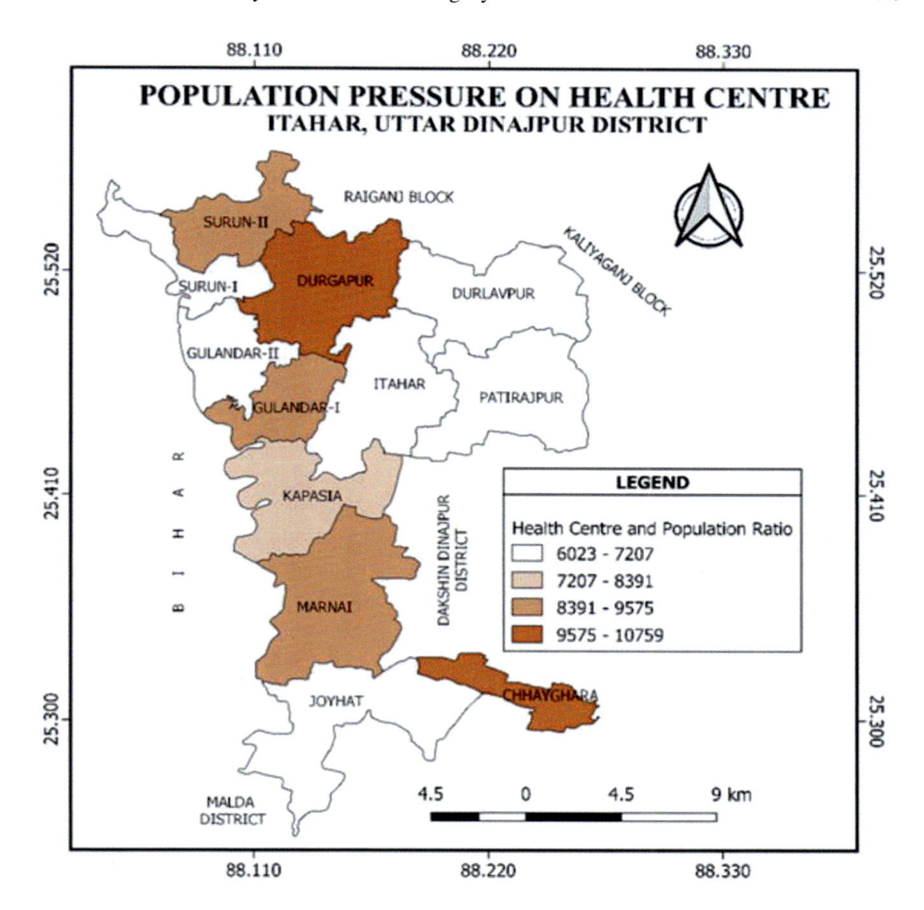

Fig. 22.3 Population pressure on health centre

13.4%. Highest alpha index value found in Itahar GP usually indicates the greater degree of connectivity than rest of the parts. Lowest alpha index found in Durlavpur GP means the degree of connectivity is much lower in this region. Beta index also reveals complexity and simplicity of a network. The value of Beta index in Itahar block ranges from 0.894 to 1.166. Beta value is highest in Itahar GP, which indicates a complicated network structure having more than one circuit. There are two GPs, viz. Surun-I and Gulandar-II, having beta value less than 1, that indicates very simple and lesser integrated networks. The Gamma index exhibits the relationship in between the number of observed links and the number of possible links of a network. Highest gamma index value found in Surun-II (42.5%), followed by Gulandar-I, Kapasia, Itahar. Lowest gamma value is found in Gulandar-II (33.3%). Eta index is designed to capture the structural relationship between the transport network as a whole and its routes as individual component of that network. Eta index has a great significance in examining the utility of a particular network. Gulandar-I has the highest eta value that indicates sparsely populated area and lack of connections with the en route nodes. An opening of new node within Gulandar-I transport network may reduce the value of eta index. The discussion indicates that none of the above measures can properly evaluate the network connectivity. Therefore, composite connectivity score (CCS) has been computed by adding all respective measures (Table 22.2). Based on CCS, Itahar block has been divided into five categories. Surun-I and Gulandar-I fall in very low connectivity zone, Joyhat, Chhayghara, Durlavpur, Patirajpur falls in low connectivity region. Medium connectivity region includes Gulandar-II GP. High connectivity zone comprises Surun-II, Durgapur and Marnai GP and very higher degree of connectivity is observed in Itahar and Kapasia GP (Fig. 22.4).

Various kinds of measures have been used by geographers to measure accessibility. One of the most popular measurements is the distance measured along the route from a centre point. Hence, a region adjacent to a centre or route is well connected and more accessible. With increasing the distance from the centre, accessibility become decreasing. Direct distance between centre (Rural Hospital) and other location (Sub Centre and Primary Health Care centre) can be determined along a straight line but most of the roads are curve line or irregular bends owing to intervened by unavoidable objects or condition like, agricultural fields, settlement, river, lake or pond, govt. buildings, etc. So, actual road distance and straight-line distance between two locations is not same and straight-line distance always underestimates the actual length of a route. High Detour Index (DI) value indicates indirect and sinuous connection between rural hospital and sub centre, low DI value denotes more often straight or direct road connectivity of sub centre to the rural hospital. The analysis of DI value

Table 22.2 Composite connectivity score of Itahar C.D. Block

Composite connectivity score						
Name of the GP	Alpha index	Cyclometric number	Beta index	Gamma index	Eta index	CCS
Joyhat	−0.83355	0.51599	−0.30201	−0.93511	−0.33797	−1.89265
Chhayghara	−0.19149	−0.5767	−0.19354	−0.06389	−1.76993	−2.79555
Marnai	0.11264	1.97291	1.03581	0.3543	−0.70982	2.76584
Kapasia	1.32918	0.15176	0.83092	0.94673	0.63437	3.89296
Gulandar-I	−0.52942	−1.30516	−0.7359	1.08612	2.35966	0.8753
Itahar	1.76849	1.60868	1.26481	0.91188	0.09854	5.6524
Patirajpur	−1.1039	−0.5767	−0.01276	−0.30783	0.71752	−1.28367
Durlavpur	−1.27286	−0.5767	−0.07302	−0.44723	−0.49734	−2.86715
Durgapur	0.65333	0.51599	0.39341	0.11035	0.34336	2.01644
Surun-I	−0.29287	−0.5767	−1.26621	−1.24875	−0.16475	−3.54928
Surun-II	1.09263	−0.21247	1.07197	1.39976	−0.59665	2.75524
Gulandar-II	−0.73218	−0.94093	−2.01347	−1.80633	−0.07699	−5.5699

Computed by Authors

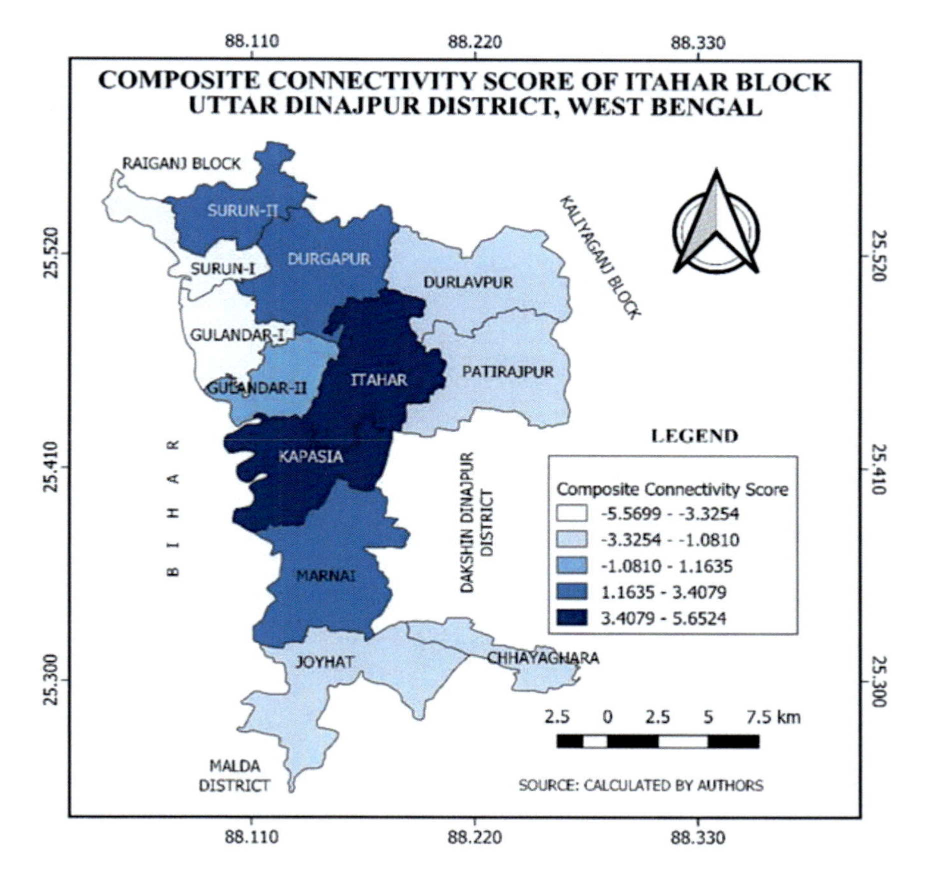

Fig. 22.4 Composite connectivity score

reveals that there is a lac hog straight line connection between Rural Hospital and other health facilities (Fig. 22.5). Highest value of 262% is recorded in Lalganj sub centre followed by Gulandar (222), Gopalpur (175), Indran (175), Damdoila sub centre (172), and Surun primary health centre (162). On the other hand, low detour value of Bangar (102), Sonapur (102), Baidara (103), Kukrakunda (106) sub centre indicates direct road connection. All these sub centres are located along the periphery of the block and lacking direct connectivity. Shortest path can be changed according to traffic flow on each segment, but Shimbel index calculates minimum number of paths necessary to connect a node to rest of the nodes in a network system. Shimbel index also helps to examine accessibility which can be derived from shortest path matrix shown in Fig. 22.6. From the Shimbel index it is observed that Mirjatpur sub centre, Kamlapur sub centre, Paikpara sub centre, Sripur, and Ghera sub centre having lower Shimbel value which indicates these are most accessible health centre from rural hospital. Higher Shimbel value found in Koachpara sub centre (21) in

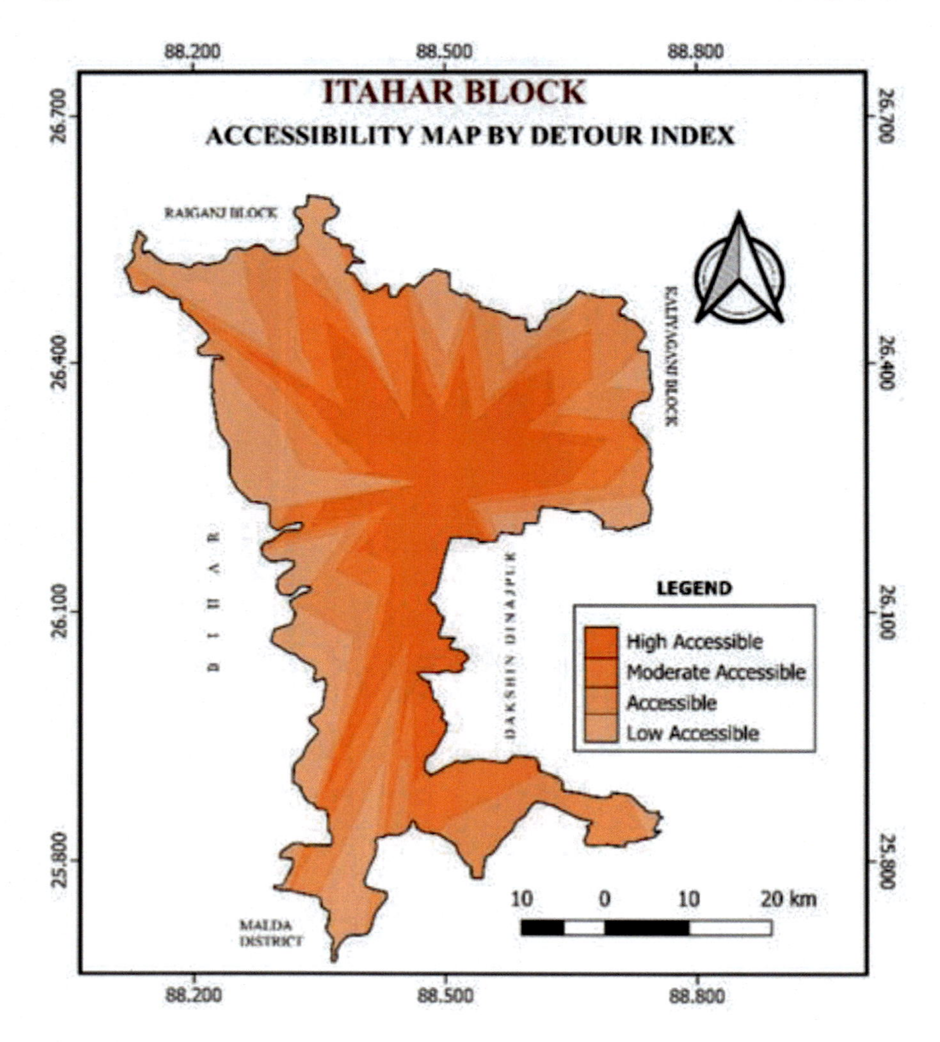

Fig. 22.5 Accessibility map by Detour Index

Chhayghora GP and Joyhat sub centre (23) and Kasba sub centre (22) of Joyhat GP, which are the outlying areas with sparse accessibility. This index is categorised into five zones varying from very high accessible zone to very low accessible zone. Most of the parts of Itahar GP and some parts of Durlavpur and Patirajpur GP fall in very high accessible zone. Itahar GP is the main junction point of the study area that is why accessibility is high in this area. Moderate accessibility is found in some parts of Marnai, Kapasia, Gulandar-II, Durgapur, Durlavpur, and Patirajpur GP. Very low accessibility zone is found in Joyhat, Chhayghara, Surun-I and Surun-II GP which are far away from the rural hospital.

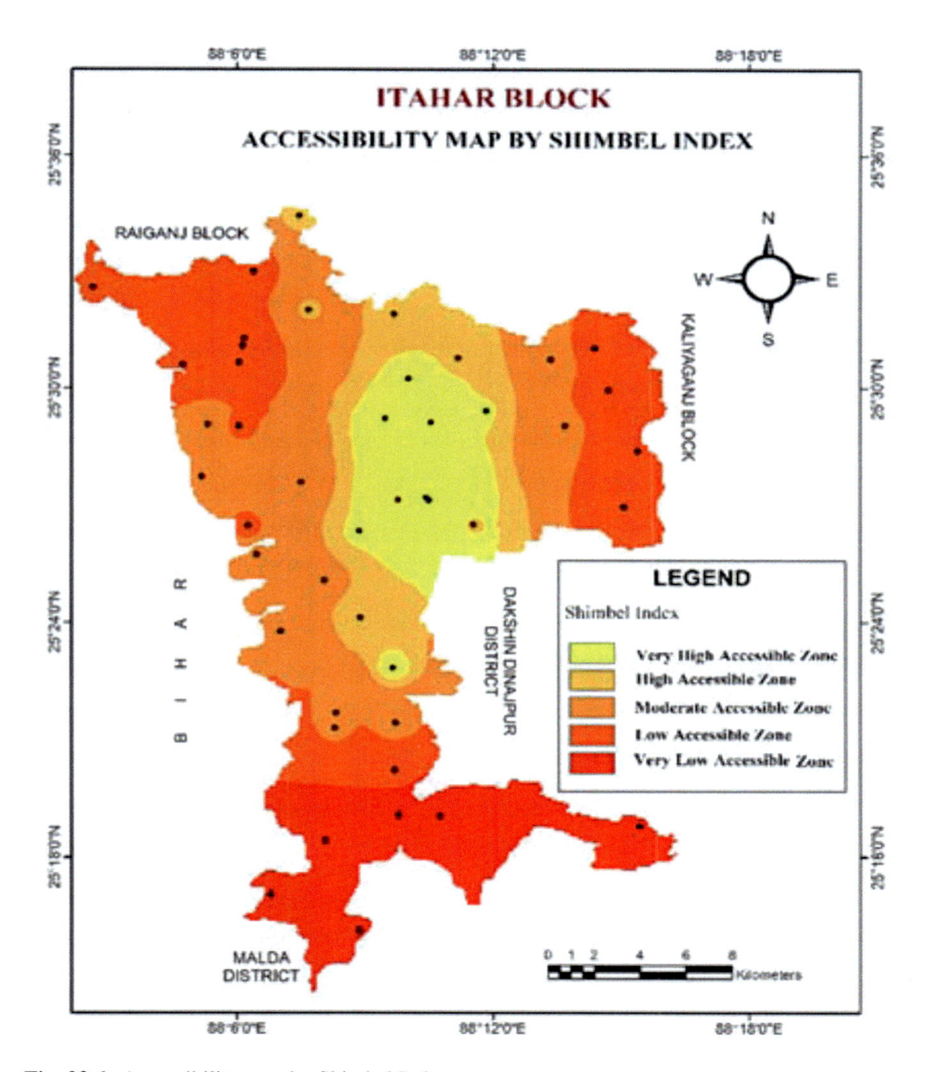

Fig. 22.6 Accessibility map by Shimbel Index

22.6.4 Proximity to Health Facility

Physical accessibility to maternal health facility has been influenced by some factors, such as long travel distance, health behaviour, and population determinants. The study aims to evaluate geographic proximity to the maternal health centre. One way of defining accessibility to maternal health centres is by knowing how far the mother live from their nearest health centre. In case of proximity, distance between provider and recipient is the main tool for measuring health centre's accessibility and identify barriers to timely interventions and treatments. The use of health care services is very much related to service accessibility and quality of service. Still in many parts of the world, health care accessibility is limited to urban areas from its peripheral part and hence the unfortunate consequences happen in the form of maternal morbidity and mortality. It is quite difficult to draw out a definite boundary of influence zone of health facilities. For this application, distance has been calculated around the health centres to identify the proximity to health centres. The service area has been contoured in below following map which (Fig. 22.7) shows that there are 3 rings covering the health care centres, each ring having 1 km width which shows the degree of influence of each and individual health care centre. Obviously, the proximity to health care centres its benefit is better achieved by the mother who resides within the tolerable distant limit that she could reach easily to the health care centre and receive the services from these health facilities. It is clear from Fig. 22.7 that several parts of this block remain left from the benefit of proximity of health care service centre. The yellow portion indicates outward area, where the health care centres are too far from this zone. Some parts of Joyhat, Marnai, Chhayghara, Kapasia, Durgapur, and Surun-II falls in this category. Health planner can use this model for the emergence of new health facility so that all women can be able to access the health facility.

22.7 Findings and Conclusion

One of the main goals of the health system is to achieve 100% coverage of maternal health care services. It is possible to achieve 100% maternal health care coverage by spatially planning the location of each health facility based on population demand for the maternal health services and measure distance between the location of health care centres and households. The benefit of spatial accessibility planning is that an impartial geographic distribution of health care service centre which can minimize oversupply and at the same time increases the integrity of access to health services for medically unserved areas. Major findings of this study are that well connected traffic axial route is comparably inadequate in this region and these are distributed with lesser density, resulting in weaker capability to deals with whole network. The accessibility along the National Highway embellishes much better province in the whole network system. This study has identified the major problems of the concerned health services related to connectivity and accessibility. Detour Index (DI) and Shimbel Index (SI) have become the prime factor in the analysis of accessibility. It was found that there are some areas where sub health centres and primary health centres are poorly connected with rural hospital. Some areas are situated far away from the maternal health facility in accordance with deplorable road condition. This can be considered as the reason behind the lower utilization of antenatal and post-natal care services. Sub centre provides only basic facilities, if the pregnant women suffering from complications, she must need advanced facilities which are available in the rural hospital (Singh et al. 2019). In these cases, patient is referred to higher order health facility. That is why connectivity and accessibility between sub centre

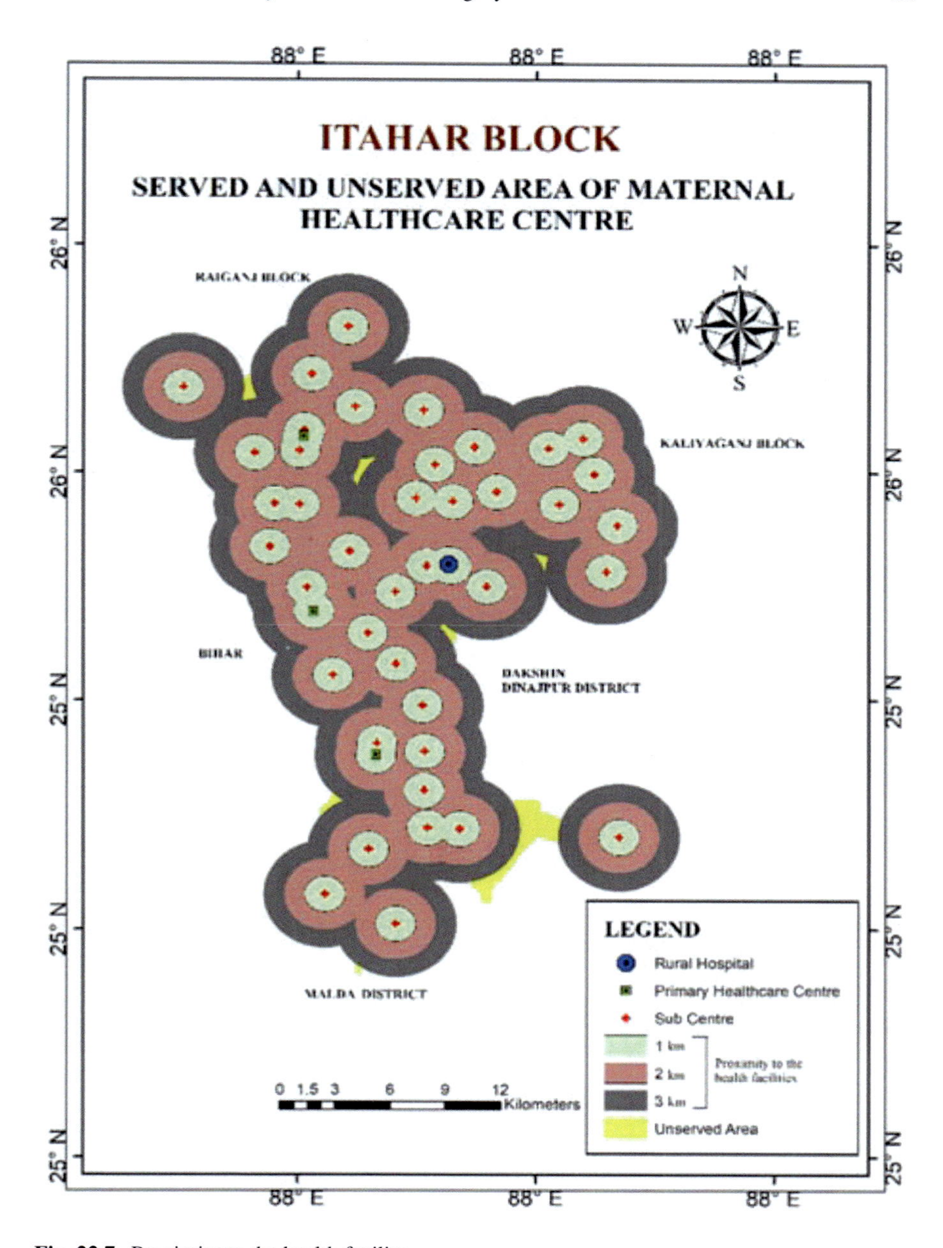

Fig. 22.7 Proximity to the health facility

and rural hospital is very essential. In the study area rural hospital is in the foremost position in respect of delivering services and provides much more opportunity to the maternal community. For this reason, higher order health facility attracts women living in the periphery area. Besides the distance factor, accessibility is affected by availability of transport, transport cost, travel time, road condition, etc. (Varela et al. 2019). But the allocation and distribution of health care service centre should be guided by standard norms, population strength, and demand. Finally, the study has given its stressed to focus onto foster the promotion of transport connectivity in the form of construction new roads and expansion of road network coverage, especially to those areas where maternal health centres are located. It will ensure the strengthening of the capability and effectiveness of health care services provided by maternal health facility of this region.

References

Alake MA (2014) Spatial accessibility to public maternal health care facilities in Ibadan Nigeria. Int J Soc Sci 26(1):13–28

Ali SA, Dero AA, Ali SA, Ali GB (2016) Factors affecting the utilization of antenatal care among pregnant women in Moba Lga of Ekiti State, Nigeria. Int J Tradit Complement Med 2(2):41–5. https://doi.org/10.35841/neonatal-medicine.2.41-45

Cascetta E, Cartenì A, Montanino M (2013) A new measure of accessibility based on perceived opportunities. Procedia Soc Behav Sci [Internet] 87(October):117–132. https://doi.org/10.1016/j.sbspro.2013.10.598

Chatterjee A, Paily VP (2011) Achieving millennium development goals 4 and 5 in India. BJOG an Int J Obstet Gynaecol. 118(SUPPL. 2):47–59. https://doi.org/10.1111/j.1471-0528.2011.03112.x

Directorate General of Health Services (2006) Indian Public Health Standards for sub-centers: guidelines, 2006;(March). http://mohfw.nic.in/NRHM/Documents/IPHS_for_SUBCENTRES.pdf

Fekadu GA, Kassa GM, Berhe AK, Muche AA, Katiso NA (2018) The effect of antenatal care on use of institutional delivery service and postnatal care in Ethiopia: a systematic review and meta-analysis. BMC Health Serv Res 18(1):1–11. https://doi.org/10.1186/s12913-018-3370-9

Ghosh A (2015) Impact of morphometric attributes and road networks in maternal health care services of Birbhum District, West Bengal Biswaranjan Mistri 6959(219):219–232. ISSN:2349-6959

Ghosh A, Mistri B (2016) Impact of distance in the provision of maternal health care services and its accountability in Murarai-ii block, Birbhum district. Sp Cult India 4(1):81–99. https://doi.org/10.20896/saci.v4i1.182

Halden D, McGuigan D, Nisbet A, McKinnon A (2000) Accessibility: review of measuring techniques and their application, p 107

Iyengar SD, Iyengar K, Gupta V (2009) Maternal health: a case study of Rajasthan. J Heal Popul Nutr 27(2):271–292

Johri M, Morales RE, Boivin JF, Samayoa BE, Hoch JS, Grazioso CF et al (2011) Increased risk of miscarriage among women experiencing physical or sexual intimate partner violence during pregnancy in Guatemala City, Guatemala: cross-sectional study. BMC Pregnancy Childbirth 11. https://doi.org/10.1111/1471-0528.13898

Kifle D, Azale T, Gelaw YA, Melsew YA (2017) Maternal health care service seeking behaviors and associated factors among women in rural Haramaya District, Eastern Ethiopia: a triangulated community-based cross-sectional study. Reprod Health [Internet] 14(1):1–11. https://doi.org/10.1186/s12978-016-0270-5

Levesque J-F, Harris M, Russell G (2013) Patient-centred access to health care. Int J Equity Health 12(18):1–9. https://doi.org/10.1186/1475-9276-12-18

Mahajan H, Sharma B (2014) Utilization of maternal and child health care services by primigravida females in urban and rural areas of India. ISRN Prev Med 2014:1–10. https://doi.org/10.1155/2014/123918

Navaneetham K, Dharmalingam A (2002) Utilization of maternal health care services in Southern India. Soc Sci Med 55(10):1849–1869

Neutens T (2015) Accessibility, equity and health care: Review and research directions for transport geographers. J Transp Geogr [Internet] 43:14–27. https://doi.org/10.1016/j.jtrangeo.2014.12.006

Otu E (2018) Geographical access to healthcare services in Nigeria—a review. J Integr Humanism 10(1):17–26. https://doi.org/10.5281/zenodo.3250011

Rodrigue J-P, Comtois C, Slack B (2019) Transportation and the spatial structure. Geogr Transp Syst: 49–94

Ronsmans C, Collin S (2008) Nutrition and health in developing countries. Nutr Health Dev Ctries. https://doi.org/10.1007/978-1-59745-464-3_2

Saikia D, Kalyani A, Das K (2014) Access to public health-care in the rural Northeast India. NEHU J [Internet] XII(2):77–100. http://dspace.nehu.ac.in/bitstream/123456789/12934/1/Journal_Jul_Dec14_Art5.pdf

Shapiro GD, Sheppard AJ, Bushnik T, Kramer MS, Mashford-Pringle A, Kaufman JS et al (2018) Adverse birth outcomes and infant mortality according to registered First Nations status and First Nations community residence across Canada. Can J Public Heal 109(5–6):692–699. https://doi.org/10.17269/s41997-018-0134-6

Singh S, Doyle P, Campbell OMR, Murthy GVS (2019) Management and referral for high-risk conditions and complications during the antenatal period: knowledge, practice and attitude survey of providers in rural public healthcare in two states of India. Reprod Health 16(1):1–14. https://doi.org/10.1186/s12978-019-0765-y

Thaddeus S, Maine D (1994) Too to walk: maternal mortality in. Soc Sci Med 38(8):1091–1110. https://doi.org/10.1016/0277-9536(94)90226-7

Varela C, Young S, Mkandawire N, Groen RS, Banza L, Viste A (2019) Transportation barriers to access health care for surgical conditions in Malawi a cross sectional nationwide household survey. BMC Public Health 19(1):1–8. https://doi.org/10.1186/s12889-019-6577-8

Wellay T, Gebreslassie M, Mesele M, Gebretinsae H, Ayele B, Tewelde A et al (2018) Demand for health care service and associated factors among patients in the community of Tsegedie District, Northern Ethiopia. BMC Health Serv Res 18(1):697. https://doi.org/10.1186/s12913-018-3490-2

Chapter 23
Strategies for Sustainable Tribal Development in Purulia District, West Bengal, India: A Socio-ecological Perspective

Somnath Mukherjee

Abstract Social ecology is simply an approach to understand the human world through some interconnected elements like ecology, social structure, culture, economy and polity. The social ecology can interpret the societal condition and many plans and policies for sustainable development of both man and environment can also be made by the study of social ecology. Social ecology discerns that the outrage of the development and growth have destroyed the ecological settings of the tribes. The tribes have been uprooted from their pristine land and compelled to live with mixed culture. The said actions affect both the tribal economy and the tribal identity. The Kharia Sabar of the eastern part of the Chotanagpur plateau is the most affected tribal community who lost their traditional habitat and was compelled to live with the stigma of criminality followed by extreme poverty and social exclusion. The present study explores some strategies and suggestions for the sustainable tribal development by understanding the life and livelihood of the tribe under socio-ecological lines.

Keywords Social ecology · Primitive accumulation · Cultural assimilation · Tribal identity · Criminal tribe · Self-reliance

23.1 Introduction

Social ecology is the scientific explanation of the interrelationship among ecological infrastructure and the four social models of the human society which are social structure, culture, economy and polity. Mukherjee (1942) initially articulated the scopes and contents of the concept of social ecology. Guha (2008) stressed on the diversification of the concept to understand different strata of human society under ecological lines. The present study takes this opportunity and helps to understand the indigenous tribal people of the Purulia District, West Bengal, India with the elements of the Social ecology. The district is also situated in the extended part of the eastern Chotanagpur plateau. The study focuses on the three important tribal communities

S. Mukherjee (✉)
Bankura Christian College, Bankura, West Bengal 722101, India

N. C. Jana et al. (eds.), *Livelihood Enhancement Through Agriculture, Tourism and Health*, Advances in Geographical and Environmental Sciences,
https://doi.org/10.1007/978-981-16-7310-8_23

and these are Santhal, Kharia Sabar and Birhor. The objective of this research is to provide some strategies and suggestions towards tribal development on the basis of the present situation of the tribal communities. It is observed that over the period of times the tribes have lost their pristine habitat due to the ecological degradation and the cruel oppression of the effects of globalization. They have been uprooted from their traditional ecological settings and forced to live with the mainstream societies. This leads to the world of deprivation, severe economic backwardness, high illiteracy rate, ill health and the identity crisis of the tribes. The Kharia Sabars become the most affected due to this. They are the most hated and humiliated tribe of this region with the lowest standard of living. Most evidently this tribal community is still living with a societal stigma of criminality. It takes almost everything from their lives. The school going children of the Kharia Sabar have often dropped out from the high schools due to this stigma and non-acceptance from the non-tribal students. The term 'Birhor' comes from two words, 'Bir' means jungle and 'hor' means man. It signifies that the Birhors are the man of jungle. They have two social divisions one is Uthulu Birhor and another is Jagghi Birhor. The Uthulu Birhor tribe is still living in isolation and in the deep forested uplands. They have simple forest-based economy. They still practice barter system as a part of their economy. They sometimes comedown to the local markets and exchange ropes made with Chihorlata, a local creeping plant with food items. The Jagghi Birhor tribes are settled in the foothills. They maintain a close relationship with the non-tribal population. This leads to the cultural assimilation and transition. Hindu norms and cultural practices are slowly penetrated into their culture and beliefs. For instance, now a day the Jagghi Birhors use vermilion in their marriage system and they often take part in Hindu festivals. This reflects certain level of transitions to their lives. The impacts of the ecological degradation and the continuous cultural assimilation also affect the Santhal tribal community. They are living in both the uni-ethnic and the multi-ethnic villages. In the uni-ethnic villages, the effects of transition are relatively less than that of the multi-ethnic villages. Overall, the Santhal tribe is still maintaining their social structure, culture, economy and polity by their own natural laws and norms. So, after considering the present situations of these tribes it is believed that they are living in three worlds. The tribes who are still living in their traditional habitat with the age-old socio-economic condition fall in the first world. The Santhals of the uni-ethnic villages and The Uthulu Birhors can be considered under this category. The second world belongs to the tribal communities living in the mixed culture such as the Jagghi Birhors and the Santhals of the multi-ethnic villages. Here the influence of the cultural assimilation is gradually penetrating into their different norms and the practices. The Kharia Sabarsare belongs to the third world which is characterized by extreme poverty, lowest standard of living, ill health, high illiteracy and social reproach. After consideration of the present situation of the tribes the study comes to the realization that only education in real terms can revive the tribal communities from the distress condition. Only education can properly develop the self-reliance and can bring the prosperity to any society and the tribes are no exception.

23.2 Issues and Challenges of Tribal Development

The tribes always prefer to live in isolation. The tribal society, their culture and economy are comprehensively interlinked with the elements of ecology. Until the introduction of the forces of transition and transformation, the simple lifestyle with no notion of savings has been the predominant nature of the tribe. The consequences of the economic growth and the development of the tribal life can be observed through the present socio-ecological settings of the tribe. They have been uprooted from their natural environment and compelled to live with the uncertainty, deprivation and extreme economic backwardness. The tribes are now in a transition stage and this stage is not by their choice but by force. This transition leads to the change in their traditional habitat in particular. They are now living in the multi-ethnic, multi-lingual and mixed geographical space. It is inferred that the interaction between the tribal and the non-tribal societies may be viewed on the basis of the complementarity and reciprocity of the inter-human actions. Gouldner (1967) wrote in this regard that—'… complementarity connotes that one's rights are another's obligations, and vice versa. Reciprocity however connotes that each party has rights and duties'. In case of the interaction between the tribes and the non-tribes, the tribes always have found themselves at the receiving end. The complementary relationship affects the life and livelihood of the tribal communities. The tribes are now engaged as labourers in agricultural, industrial and construction sectors. This relationship also affects the tribal identity in the society. The tribes have now been identified by the materialistic terms such as 'poverty-stricken', 'backward', 'social exclusion', 'deprived', 'reservation', and 'illiterate', etc. Destructive modes of livelihood lead to moments of profound ecological and socio-cultural losses. Thus, the sense of one's self and the world is traumatically disrupted.

Ecology is the source of life and livelihood for the tribal society since decades. The introduction of complex monetary system and the consequent ecological degradation have destroyed the habitat and the sources of subsistence of the tribes. The factors like globalization, modernization, liberalization, mobilization of economy, privatization, politicization, etc. have affected badly the present state of the tribal society. The interaction with the outsiders has become a cause of disturbance to the otherwise self-contained tribal social world. With the opening up of the tribal regions, the tribal communities experienced some major impacts especially the destruction of their pristine ecological habitat and others like influx of non-tribal peasantry, transformation of simple forest based and subsistence economy to market-based economy leading to deplorable result in the tribal society. Identical thoughts shared by Roy Burman (2003) also says that "… the social ecology of the tribal is affected by large-scale immigration of non-tribal population as a sequel to opening up of tribal areas for commercial exploitation of resources and ancillary activities". Unfortunately, due to the effect of globalization and resultant consequences, the persistency of being a tribe is facing tremendous challenges. The impact of materialistic development naturally brings in the sense of inequality and deprivation in the levels of consumption. This is known as 'primitive accumulation' where one side of the coin, rich becomes richer and poor becomes poorer. Jhunjhunwala (2012), former professor of Economics, Indian Institute of Management, Bangalore, elaborated the concept of primitive accumulation in the tribal life where he wrote that '…quest for economic growth, the government deprives the common man of consumption. This is known as "primitive accumulation" in economics. Imagine you are living in a tribal community. Every family has a plot of land of equal size and a standard of living that is almost uniform. Now, there is a proposal to establish a factory, community ownership of business has not been successful. And no individual in the tribal community has the money to invest in this enterprise. Strict observance of equality will never make it possible for the community to put up a factory and the people will remain ever so backward. In such a situation, the chief finds ways of imposing inequality on the people. He asks them to undertake free labour for the establishment of the factory. Or he persuades them to violently confront the neighbour and enslave others to provide free labour. The chief establishes the factory and becomes rich while the people become poor as they haven't been paid for the work done'.

This sense of inequality among the tribal communities is being motivated by the extremists against Government and all those who enjoy the fruits of development. Low education level and unsatisfactory policy implementation help the extremists to motivate the tribal even easily. Though, nowadays, the tribal people are aware of the fact and ask for peace and real development in their region.

On the contrary, on a positive note, the interaction between tribal and non-tribal societies always should be reciprocal. In this view it envisages everyone's interest and overall improvement. No doubt, these two paradoxical views create two genres one who is in the side of thinking tribe and their development (material) by his or her own choice and profit and the other who leaves all rights to the tribals and believes development must come from within. Again, on a positive note, it is evident that presently the sense of urgency is observed in different governmental plans. Now modern sensibility learns to measure the destructive consequences of the modern development processes. Like various social activists, most of the civilized modern genres are keen to keep the tribal within their own indigenous world. In this context Roy (1989), a renowned anthropologist, addressed the basic problems of government policies on tribal development in his introductory note that '… most of the plans have been a failure mostly due to the inadequate knowledge about the tribal, the specific needs of the ecology, economy and society and their development requisitions'. Further, the tribal development policies have mostly disregarded the enrichment of non-material cultural aspects of the tribes such as customs, norms, values, societal hierarchies, forest-based life, and livelihood. Sukai (2010) opposed the general tribal development activities and wrote '…underlying non-tribal nature of any development approach is a serious factor to reckon with. Non-tribal people dominate the government machinery in the tribal area and not only feel bitter about the various welfare programmes designed for tribals, but also get into an exploitative relationship with the local traders, contractors and police. This results in very few benefits actually reaching the tribal people. While tribals revere the natural resources as life sustaining forces, the non-tribals outlook is one of utilitarian and short-term commercial exploitation. This disregard for tribal-nature symbiosis is causing not only a threat to tribal survival but is also leading to depletion of resources in the tribal regions … The officials and employees of the government posted in tribal areas approach the tribals with all the prejudices of a non-tribal society. The tribals are considered uncouth savages, steeped in crude superstitious and rituals. Their culture and social system are looked down upon and it is with condensation that they relate to tribals when they visit the tribal villages'. On a similar ground Bhagat (1989) expressed that '… the tribals, despite their desire, may not get adequate opportunity for responding to these programmes owing to lack of necessary infrastructural facilities'. Questioning on the fruitfulness of reservation policy on tribals, Ahuja (2007) wrote '… little will be achieved by a debate on the pros and cons of reservations. It will only aggravate the problems and lead to fragmentation of the country. The power elite, the government, political parties and people have to delve deeper into the very reasons why reservation has seemingly become necessary and what needs to be done to eliminate this pernicious practice'.

Over administrating from government concern goes against tribal development in real sense. First Prime Minister of India, late Pandit Jawaharlal Nehru (1958) gave a thoughtful Panchsheel, i.e. five fundamental principles for the tribal development which opposed over administration and addressed 'We should not over administer these areas or overwhelm them with multiplicity of schemes. We should rather work through, and not in rivalry to, their own social and cultural institutions'. Considering overall studies on tribal communities of the extended part of eastern Chotanagpur plateau, there are three broad issues against real tribal development that have been identified and these are the external issues, internal issues and the socio-economic issues. The external constraints here are mainly related to the initiation and implementation of the tribal development plans and policies from different government sources. It is mainly due to the misconception of the policy-makers towards the different socio-ecological identity of the tribes. The tribal socio-ecological elements are quite different for different tribes. For instance, the socio-cultural norms and practices of the Santhals are quite different than that of the Birhors. Interestingly within the Santhal tribal community there are twelve clans which are different from each other. Thus, before the initiation of the tribal development plans and polices, the policy-makers have to understand the surrounding ecology, social structure, lifestyle, economy and the natural laws of the tribes for the successful implementation of the plans.

The internal constraints denote some ill socio-cultural customary norms and traditions, superstitions prevalent within the tribal world such as the early marriage system among the Birhors, witchcraft among the Santhals and acute drinking of 'Handia' (a country beer) among the Kharia Sabars, etc. These issues can be the major hindrances to the tribal enrichment from within. The witchery often ends in diabolical murder or social boycott. Unfortunately, it is seen that despite being aware of such miserable consequences even educated tribes could not get rid of such superstitions and the curse. On the other hand, from the materialistic development point of view the mass illiteracy, malnutrition, and lack of awareness, etc. can also be some significant factors behind their backwardness.

The third issue is related to the present socio-economic condition of the tribes. The tribes now become settled close to the non-tribal territories which in turns transform their simple forest-based economy to the complex market-based economy. Now the tribes have been engaged as the agricultural labourer, industrial labourer, migrant worker, etc. On a comparative ground the economic sufficiency of the non-tribes is far better than that of the tribes. This leads to the state of the economic backwardness and the societal discrimination in the name of illiteracy, poverty, criminality, etc. This also has developed the sense of deprivation of consumption among the tribes. Presently they have been forced to live in the complex monetary world where they have to compete with the outsiders for subsistence. This process has ultimately given birth to the concept of primitive accumulation.

23.3 Strategies for Sustainable Tribal Development

The tribal development in real sense can be achieved only from within. This term 'within' means by own self and not by others. The tribal development is generally seen in the perspective of the materialistic world. The tribal development plans and policies always ignore the traditional habitat, ethos and beliefs of the tribes. A fruitful tribal development needs a sustainable developmental plan giving importance to the immaterial world which includes their traditional forest-based economy, the age-old tribal beliefs, customs and practices revolve around the ecology. The long-term sustainable tribal development should help to evolve the self-reliance which comes only by the education. The education is the only key to the real development, awareness and the empowerment in the tribal life.

There are increasing numbers of exponents in this field who have strongly advocated the education for the constructive development of the tribes. For instance, Gautam (2003) elaborated that 'education of ST children is considered important, not only because of the constitutional obligation but also as a crucial input for total development of the tribal communities'. The education becomes the long-term solution to overcome the economic backwardness. In this regard, Sujatha (1999) expressed that '... educational investment produces long-term effects, allowing one to eradicate the transmission of poverty from one generation to the next'. Abdulraheem (2011) in his contribution 'Education for the Economically and Socially Disadvantaged Groups in India: An assessment', identically wrote that '...education alone can be the solvation for poverty, and up-liftment of the socially discriminated ... Education not only helps them to promote their economic development but also helps to build their self-confidence and inner strength to face new challenges. Furthermore, Das and Bose (2010) gave importance to education not only as a means of expedite employment opportunities but also the generation of health consciousness and explained that '... better educational attainment will lead to more scope for employment and healthier dietary practices'.

Here the education means not only the formal education but also the education of the tribal people which can be embedded with their tradition, ecology, society, culture, economy and polity. In this educational system the prioritization should be given to their own languages. This simple step can raise an enthusiasm towards the educational system and can unite them for their betterment. The schools and the educational institutions should not be the place of discrimination. The 'National Policy of Education' (NPE) (1986) and the 'Programme of Action' (POA) (1992) specifically emphasized on indigenous education in the teaching and learning curriculum of the tribal education system. The important policies of NPE and POA have proposed the tribal education at the primary stage through their mother tongue, the encouragement of the tribal young generation to involve in teaching in the scheduled areas, etc. Further, a working group on Elementary and Adult Education for the 10th Five Year Plan (2002–07) has given priority on the improvement of the quality of education of the tribal children. In India, the deployment of community teachers was first experienced in the late 1970s and Assam first took the initiative to prepare separate teacher training modules and materials exclusively for Bodo tribal language in the year 1995.

Importance of the community participation is the need of the day. According to the Sustainable Development Goals (SDGs) of the United Nations, the development should be inclusive and equitable in nature. The seventeenth goal of it is about the partnerships for the goals. This promotes the participatory involvement of different sections of the society. Keeping this in mind the education for the tribes should include the community participation in the name of employment of the teachers from the tribal community. This will also help the better understanding of different lessons in their own languages at least in the primary level of education. There are several examples where the scholars have admitted this participatory involvement in the tribal education. Jha and Jhingran (2002) in their work 'Elementary Education for the Poorest and other Deprived Groups' rightly mentioned that '…increasing number of researchers strongly advocate the use of the mother tongue or home language as medium of instruction in early stages of education. This assumes greater significance in the context of education of tribal children because their mother tongue is often quite distinct from the prominent languages in the state or regional languages. ST children face problems wherever teachers do not speak their dialect at all. From the perspective of language, it is deliberate to have a local teacher from the same tribal community'. Vaidyanathan and Nair (2001) on a similar ground favoured community teacher in tribal education and elaborated that '…presence of tribal teachers, especially from the same community, has shown an improved school participation of ST children coming from the same community, it is believed that the teachers would understand and respect the culture and the ethos with much greater sensitivity'.

In connection with the introduction of mother tongue in tribal education, few scholars and researchers from different perspective envisaged to introduce other regional and global languages in the tribal education. Singh (2009) also mentioned that '…medium of instruction cannot only be the local dialect, because of practical constraints. It must however start with that. More importantly, class room transactions must be such that they show respect for child's language, identity and social background. Instilling this dignity within a child, can be done even in a multilingual setting, and be totally ignored even when the medium of instruction is the tribal language. You also have to account for people's (tribals) own aspirations of learning other languages'.

The sustainable tribal development through education also requires the evaluation of the cognitive qualities of the tribal children on the basis of their ecological and cultural contexts. It is required as the modern educational system of the dominant non-tribal population does not pay attention to the socio-ecological characteristics of the tribal domain. It is true that the tribal children are not culturally inferior and cognitively less competent than the children of the non-tribal society. So for the enrichment of the cognitive qualities among the tribal children the value-added education should include the customary laws and traditions of the tribal communities and all the aspects of the socio-ecological elements of the tribal world. The priority in education should be given to tribal women where literacy level is lamentably low. The impact of education of a girl has more lasting on the society than that of a boy. When the educated tribal girl becomes a mother, she will definitely arrange educational facilities for her children even after so many sacrifices and inconvenience for the family. It is evident that mainly few societal bigotries and the economic constrains are the obstacles in female education in tribal society. Patel (1991) in this context asserted that '...important reasons of failure of development activities ... is the prevalence of acute illiteracy and ignorance, combined with superstitions among the rural masses'. On the other hand, Rajasekaran (2008) pointed out the economic insufficiency as one of the drawbacks of girls' education where he wrote '...many tribal communities, parents give minimal importance to girl's education due to economic and social limitations, send them to school only intermittently ... Most frequently, girls, apart from taking part in agricultural activities and collection of forest products are engaged in sibling care. They are often forcibly pulled out from schools, and become child labour'. These age-old ill-societal orthodoxies are the internal hindrances of the real tribal development. No doubt, customs of early marriage in the Birhor tribal community may not be eliminated overnight. The fruits of sustainable tribal development and the educational advancement will be the measures of eradication of such internal issues among the tribes. The prioritization on the tribal female educational advancement has been favoured by so many scholars and researchers as well. For instance, Bagai and Nundy (2009) in their contribution 'Tribal Education A Fine Balance' advised that, '...empowerment of the tribal girl child is only possible through education. Low levels of educational attainment coupled with familial and social neglect has inhabited their growth potential. Strengthening of basic literacy and educational services (formal and non-formal) for the tribal girl child as well as orientation towards education through outreach activities is important. General education and awareness programs addressing the special development needs of the girl child (hygiene and safety), along with mobilization campaigns are needed to add fillip to the effort. This can be coupled with encouraging and facilitating for the involvement of community and NGOs in developing institutional mechanisms for tribal girl's education. Incentives for households to send girls to school, conditional cash transfers, scholarships and stipends, can be incorporated to generate motivation among tribal girls and their families about the value of girls' education ... Given the lack of structured marriage among tribals, and the early age for sexual interaction, health care needs of tribal girls become important. Educational initiatives can use gender-sensitive methods of teaching, and include discussions on sex education, information on reproduction, health, nutrition etc. ... Training of female teachers and exploring the benefits of single-sex classrooms is a good way of partially readdressing the imbalance'. Desai and Thakkar (2007) in their concluding remarks on 'Empowering Women Through Education' a chapter in their book (Chapter-III) 'Women in Indian Society' further revealed the prime need of female education in tribal empowerment as legitimate right and wrote '...the strategies of lowering the cost of girl's education will have to be given serious thought so that girls may not be the victims of gender discrimination. It is indeed a sad situation when a mother would like her daughter to go to school/college but is unable to send her because the child must help in the house work. In the nineteenth century, we pleaded for the education of women to make her a better partner for her husband; in the last century, education was for her empowerment, and today we are pleading for her right to education as a citizen'.

Much more attention on the socio-economic problems of Kharia Sabars and Particularly Vulnerable Tribal Groups (PTGs), the Birhors need to be paid through systematic, regular and purposive supervision and monitoring before inception of already mentioned congenial educational policies. Further, sincere and flexible approach towards tribal educational policy-making can change the economic structure of a tribal region. Special vocational education especially to tribal women in the form of craft like tailoring, cutting, embroidery, rope making and other non-traditional skills like computer education should imparted. Corroboration with the mentioned educational policy as a means of economic development of the tribal communities, Hoel and Jessen (1977) wrote that 'education ... an important infrastructure for the economic development of the tribals'. Furthermore, in connection with the parallel development of education and economy of the distress driven people and in the context of favouritism of non-formal education to generate job opportunity and self-confidence, Kakde (2008) simply explained that '... our demand should be to make the education system more meaningful and more relevant to the changing needs in society. It may be remembered in a way, that present unemployment is mostly false one. It is unemployment of unskilled people. It can be achieved by acquiring knowledge, skills and attitudes in the changing context'.

In fine, future plans and programmes will succeed only if the economic development and the spread of the education should go side by side. Planners should also consider 'different cultures different plans' for different strata of tribal society. The developmental processes are to be thought of as to their own material culture, ecology, spirits, ethos, etc., but what we are doing is that we are only trying to inject the extra-material culture on which they have deep aversion. So, they are not being intoxicated from the outside; they are allowed to participate in finding out their own destiny based on their social-ecological niches.

23.4 Conclusion

The Social ecology helps to understand the tribal world in a better way. The inter linkages among five broad elements of the Social ecology can explain the tribes and their present situation. Here with the help of the Social ecology, the study has found some issues and challenges within the tribes of the extended part of the Chotanagpur plateau. These issues are mainly the external, internal and socio-economic issues which have played detrimental role to achieve the sustainable tribal development in this region. The external issues are the external factors which are degrading the forest ecology and the economy of the tribes. The internal issues are the superstitions, ill-cultural beliefs, orthodox value systems of the tribes which have somehow restricted the tribes to access the sustainable rural and tribal developmental plans and policies. The socio-economic issues are related with the materialistic approach and action where tribes have been identified as backward, illiterate and in some cases criminal. The tribes of this region can be found in three different worlds. The Santhals are the most populated tribes with a well-organized social structure living both in pure Santhal villages and also in multi-ethnic villages. Interestingly they have maintained their own tradition in both the spaces. So, they are relatively in better position. The Birhors who are among the Particularly Vulnerable Tribal Group now settled close to non-tribal villages. The Jagghi Birhors have accepted the modern facilities and amenities like the measures of family planning, etc. The Uthulu Birhors are still living in the forested uplands and maintaining the forest-based simple life and livelihood. The barter system is still prevailing in their economy. The Kharia Sabars are the most affected tribes among all the tribes. They are still living with the stigma of ‘Criminal Tribe’. The identity acts like a curse to them and has forced to live them with the extreme poverty and social reproach.

The heterogeneous tribal conditions, issues and challenges make it difficult to think on the strategies for the tribal development of the region. The study finally finds some sustainable strategies and suggestions for long-term tribal development. The external issues can be solved by the introduction of the conflict negotiation approach where there will be a common sharing of the resources and universal political agendas between the dominating castes and the tribes upon which tribes may find their own place, voice and identity. This conflict negotiation resolution can eliminate caste conflict, political conflict and open the root towards the sustainable rural and tribal development. Effective decentralization by peoples’ participation in every sphere of life and livelihood may lead to sustainable rural as well as tribal development. The decentralization for sustainability may require the ‘participatory rural appraisal’, the ‘participatory action researches’, participatory technological development’, the ‘people’s participation in decision making’, etc. so that the developmental strategies can reflect the desires of the local tribal people. The people’s participation in the situation analysis can help to identify the helplessness and the priorities of the people. The superstitions, orthodox value system of the tribes cannot be eradicated so easily but the sustainable educational plans and policies again in terms of participatory approach and action can grow the rational thinking among the present generation of the tribes. The self-employment opportunities by imparting skill-based community education can help to divert the tribal youths from such ill-cultural beliefs. This may lead to the tribes into the processes of acceptance and utilization of the tribal developmental plans and policies. This also can develop the cognitive values more among the tribal youths and they will then find themselves in the general structure of the people of the nation. This will eradicate the societal barriers and place the tribes as a part of the society.

References

Abdulraheem A (2011) Education for the economically and socially disadvantaged groups in India: an assessment. Econ Aff 56(2):233, 241

Ahuja R (2007) Social problems in India. Rawat Publications, Jaipur, p 192

Bagai S, Nundy N (2009) Tribal education a fine balance. Dasra, Mumbai, pp 14, 29

Bhagat LN (1989) Role of education and value system in economic development of tribals: a case study of Oraon and Kharia tribes of Chotanagpur. Concept Publishing Company, New Delhi, p 67

Das S, Bose K (2010) Body mass index and chronic energy deficiency among Adult Santals of Purulia District, West Bengal, India. Int J Hum Sci 7(2):499

Desai N, Thakkar U (2007) Women in Indian society. National Book Trust, New Delhi, p 67

Gautam V (2003) Education of tribal children in India and the issues of medium of instruction: a Janshala experience. UN/Governmental Janhsala Programme, New Delhi, p 2

Gouldner AW (1967) The norm of reciprocity—a preliminary statement. Current perspectives in social psychology (Hollander EP, Hunt RG (eds), Oxford University Press). In: Das RK (ed) Social transmission and political orientation: the case of Midnapore tribals. Studies of tribes and tribals, vol 1, no 2. Kamla-Raj Enterprises, p 151

Guha R (2008) Social ecology. Oxford University Press, New Delhi, pp 1–18

Hoel PG, Jessen RJ (1977) Basic statistics for business and economics (Wiley, New Delhi). In: Roy K (ed) Education and health problems in tribal development a study of national integration. Concept Publishing Company, New Delhi, p 77

Jha J, Jhingran D (2002) Elementary education for the poorest and other deprival groups (Centre for Policy Research, New Delhi). In: Goutam V (ed) Education of tribal children in India and issue of medium of instruction: a Janshala experience. UN/Government Janshala Programme, New Delhi, 2003, p 4

Jhunjhunwala B (2012) Managing inequality—need to change the culture of the rich. Editorial, The Statesman, Kolkata, 28th January, p 6

Kakde S (2008) Globalisation and scheduled castes. In: Kakde J (ed) Development of scheduled caste and scheduled tribes in India. Cambridge Scholars Publishing, New Castle, p 17

Mukherjee RK (1942) Social ecology. Longmans, Green and Co., London

Nehru J (1958) Foreword. A philosophy for NEFA, V. Elwin, Shillong, adviser to the Governor of Assam. In: Vidyarthi LP, Rai BK (eds) The tribal culture of India. Concept Publishing Company, New Delhi, 1985, p 419

Patel S (1991) Tribal education in India. Mittal Publications, New Delhi, p 26

Rajasekaran G (2008) Tribal girls till the land in Bt. cotton fields (New India Press. com). In: Bagai S, Nundy N (eds) Tribal education a fine balance, 2009. Dasra, Mumbai, p 28

Roy K (1989) The framework and tribal problems. In: Roy K (ed) Education and health problems in tribal development a study of national integration. Concept Publishing Company, New Delhi, p 17

Roy Burman BK (2003) Indigenous and tribal peoples in world system perspective. Studies of tribes and tribals, vol 1, no 1. Kamla-Raj Enterprises, Delhi, p 22

Singh P (2009) Qualification. In: Bagai S, Nundy N (eds) Tribal education a fine balance. Dasra, Mumbai, 2009, p 13

Sujatha K (1999) Education of India schedule tribes: a study of community schools in the district of Vishakhapatnam, Andhra Pradesh. International Institute for Educational Planning/UNESCO, Paris, p 11

Sukai TB (2010) Extremism and tribal society in India. Soc Action A Q Rev Soc Trends 60(3):268

Vaidyanathan A, Nair PRG (2001) Elementary education in rural India: a grassroots view. In: Goutam V (ed) Education of tribal children in India and issue of medium of instruction: a Janshala experience. UN/Government Janshala Programme, New Delhi, 2003, p 4

Chapter 24
Development of Tribal Livelihood in Manbazar-II Block of Purulia District, West Bengal, India

Sumanta Kumar Baskey and Narayan Chandra Jana

Abstract Development is a multi-dimensional phenomenon. It includes some variables like the level of economic growth, level of education, level of health services, degree of modernization, status of women, level of nutrition, quality of housing, distribution of goods and services and access to communication, etc. Development of tribal livelihood is not uniform among the five tribes in Manbazar-II block because they have primitive traits, geographical isolation, and distinct culture, economic backwardness, and as well as of their limited engagement on different functional activities. It has been examined with mentioned variables regarding some selected developmental indices. Disparity has been found within them. Overall development will be increased through improvement of common minimum needs and their awareness. And also true development requires Government action especially for women. Role of social development such as literacy (particularly female literacy) in promoting basic capabilities emerges as the prerequisite to overall development.

Keywords Tribal livelihood · Workforce participation · Traditional economic activity · Quality of life · Human development

24.1 Introduction

Tribal economy is based on labour force on different sector, both agriculture and non-agriculture. In the study area maximum number of tribal people are engaged in as a labour on agriculture, coal mine, brick field, house construction, road construction, different factories, etc.

According to Dalton (1982) "an economy is a set of institutionalized activities, which combine natural resources, human labour, and technology to acquire, produce

S. K. Baskey (✉)
Department of Geography, Vivekananda College, East Udayrajpur, Kolkata, West Bengal Madhyamgram, India

N. C. Jana
Department of Geography, The University of Burdwan Golapbag, Burdwan, India

N. C. Jana et al. (eds.), *Livelihood Enhancement Through Agriculture, Tourism and Health*, Advances in Geographical and Environmental Sciences,
https://doi.org/10.1007/978-981-16-7310-8_24

and distribute material goods and specialist services in a structured and repetitive fashion". Dalton (1971) also described the tribal economy by giving some characteristics. According to him, there is three interrelated feature that characterize tribal economy. This is given below:

- It is a structural arrangement and enforced rules for the acquisition and production of material goods and services.
- In the process of acquisition and production of goods and services, division of human labour, use of natural resources and application of technology (tools and knowledge) are involved.
- In the distribution process, superficial devices and practices such as market place and device for measuring some types of transaction are involved.

Vidyarthi and Rai (1976a, b, c) have indentified nine structural features that characterized the tribal economies in India. They are as follows.

- Forest-based Economy
- Domestic mode of production
- Simple Technology
- Absence of profit Motive in Economic Dealings
- Community: A Unit of Economic Cooperation
- Gift and Ceremonial Exchange
- Periodical Markets
- Interdependence.

Dalton (1971) held that the factors such as low-level technology, small size of the economy and its relative isolation from outer world contribute to mutual dependence for people sharing many social relationships. In economic interactions, each tribal village community is considered as a cooperative unit. Economic anthropologist Dalton (1971) holds that the tribal mode of transaction is that of reciprocity, i.e. material gift and counter gift giving induced by social obligations of kinship. Vidyarthi and Rai (1976a, b, c) observed that the economic functional interdependence is similar to the *Jajmani* system, found among the Hindu caste groups in most of the regions of the country. Under the *jajmani* system each caste group, within a village, is expected to give certain standardized services to the people of other caste. The family head served an individual known as the *jajman,* while the man who performs as *kamin* of *Jajman*. Economic interdependence among the tribes has been observed in different tribal zones of the country in variety of ways.

According to Haimendorf (c1982 1982), there are no people in India poorer in material possessions than the Jungle *chenchu* tribes. Bows and arrows, knives and axes, digging sticks, etc. constitute their entire belongings. Most of them are unambitious, simple, honest, innocent, ignorant and lead a food gathering and pastoral life. The tribal people suffer from malnutrition and under nutrition.

Marketing plays an essential role in developed as well as developing economies by creating demand for different goods and services. Particularly in a developing economy like India, marketing is considered as a prospective instrument for creating effectual demand and supply. According to Mazur (1968), Marketing is the delivery of standard of living, i.e. marketing delivers a good standard of living in a society. Since, tribal economy represents a primitive society development of marketing practices in tribal areas will go a long way in improving the standard of living as well as life styles of these under developed people.

Manbazar-II block is a totally plateau area. It is the south-eastern part of Chhotanagpur plateau. The ruggedness rocky surface area of this region is mainly hilly and undulating. But low surface area is situated beside the main river of Kumari and Totko and their tributaries. The topographical slope decreases towards the east. Western part is the high elevation area. Two rivers (Kumari and Totko) are flowing through this block from west to east direction. Although different type of lands (*Tanr, Baid, Kanali, Shole or Bohal*) are situated within the physiographic divisions of this region. Wide range of forest cover belong to this region. The high undulating area consists of maximum forest cover. Tribal habitations are close to this hilly and forestry area. The soil characteristics and its properties depend on topographic elevation and forest cover. It is upstream area of Kangshabati Reservoir. The habitation area, agricultural activity, economic activity and some domestic needs of tribes of this block depend on all the physical characteristics of the study area. Five tribes (Bhumij, Mahali, Orang, Sabar and Santal) are present in the study area. Development of tribal livelihood is not uniform among the five tribes in Manbazar-II block because they have primitive traits, geographical isolation, distinct culture, and economic backwardness as well as of their limited engagement on different functional activities. Their engagement in different activities (agriculture, forest based, educational, transport and communication, medical facilities, commercial and credit, etc.) indicate status of the tribals livelihood amenities, educational status, income status, quality of life of this area.

24.2 Materials and Methods

24.2.1 Study Area

The study area lies in the Manbazar-II Block of Purulia District in the eastern plateau of India. The latitudinal extension of this area is 22° 50′ 00″ N to 23° 3′ 10″ N and 86° 30′ 55″ E to 86° 42′ 55″ E. The total geographical area of the plateau is 29528 km^2. Seven Gram Panchayats and 132 mouzas are there in this area. Two rivers, Kasai and Kumari, flow through the northern and southern part of this region, respectively. It is the upstream of the Kangshabati dam (Fig. 24.1).

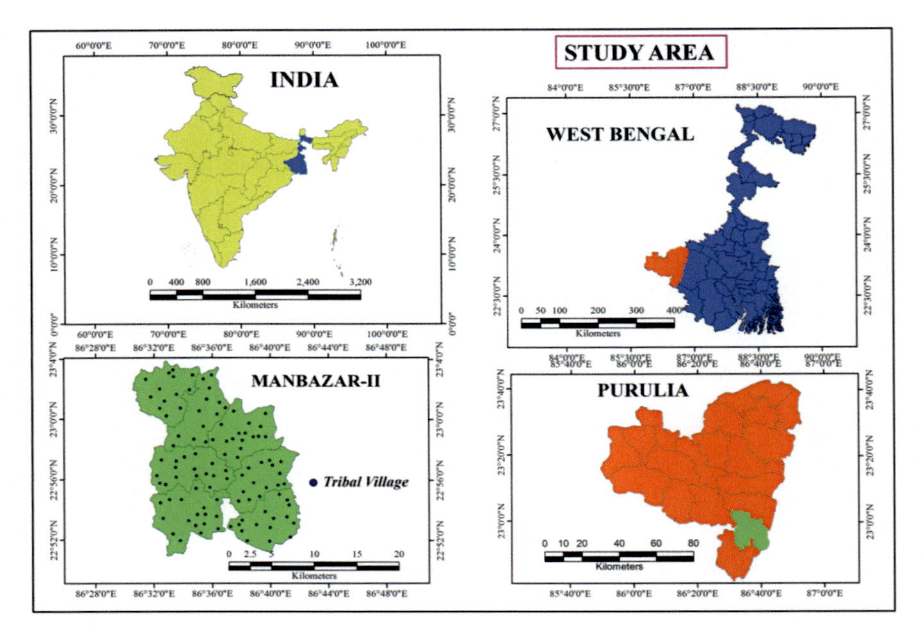

Fig. 24.1 Location map of the study area

24.2.2 Selection of the Study Area

In Manbazar-II Block, more than 55% populations are tribal. They live isolated throughout the area. Poor communication, low literacy rate, low economic conditions contribute to backwardness in the hilly forestry tribal region. Proper planning, educational development, rich communication can develop this region. Keeping in view the above situation the area has been selected for in-depth study.

24.2.3 Objectives of the Study

1. To study the livelihood amenities of the tribals
2. To study the functional activities of the tribals
3. To measure the status of tribal development.

24.2.4 Database

The entire study is based on primary survey. The primary data is collected at the village level for micro-level study from 2013 to 2019.

Primary Database

Extensive fieldwork was conducted to collect primary information regarding household amenities, workforce participation, agricultural activities, traditional economic activities, forest based economic activities, income status, different functional activities, etc.

24.2.5 Methodology

The entire work is divided into three main phases—Pre-field, Field and Post-field.

- In the time span of pre-field study extensive library work is done to review various literatures regarding the context of the study. Different offices like the Manbazar-II Block and all Gram Panchayats, Census of India are also consulted, along with the Anthropological Survey of India.
- Field study includes the primary survey in different mouzas with interview schedule.
- During the post-field study the analysis of the data, collected from the questionnaire survey, is made to recommend suitable measures for the recognized problems. For analyzing the data, the following are used:

  a. *Tools used*—interview schedule, thematic maps, charts, cartograms, tables, etc.
  b. *Parameters used*–living pattern, employment, backwardness, development indices, etc.
  c. *Analysis procedure*—statistical and cartographic techniques and representations are made with help of MS-excel, ArcGIS software.

24.3 Results and Discussion

24.3.1 Workforce Participation

The Indian Census defines a worker as a "person whose main activity is participation in any economically productive work by his physical or mental activity. Work involves not only actual work but effective supervision and direction of work" (District Census Handbook 1971). In the study area, the percentage of working population accounts for 65.87 of Bhumij, 64.99 of Mahali, 65.74 of Orang, 63.01 of Sabar and 65.64 of Santal tribe. The highest percentage of the total workers population belongs to Bhumij tribe and it is followed by Orang, Santal, Mahali and Sabar, respectively. And the highest percentage of non-working population among the tribes is found among Sabar tribe (36.99%). Child labour is present in the study area. The percentage of

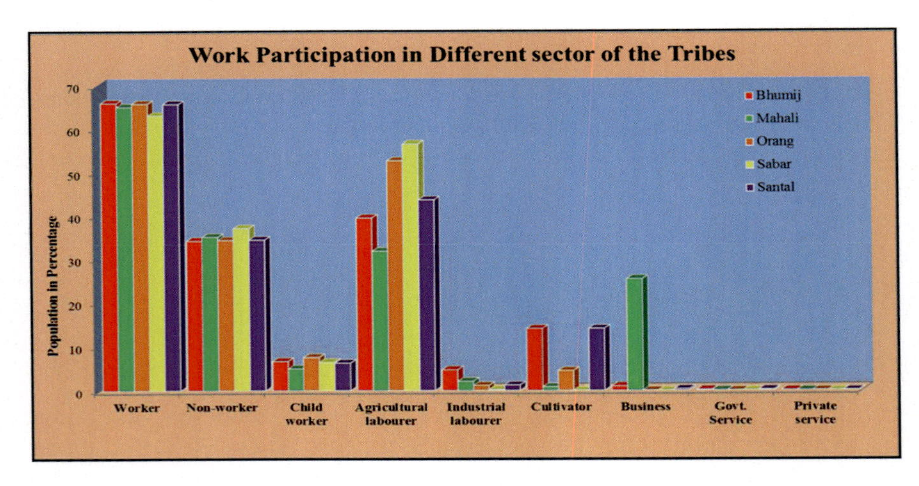

Fig. 24.2 Work participation in different sector of the tribes. *Source* Primary survey by authors, 2013–2019

child labour has been found to be 6.63% in Bhumij, 4.93% in Mahali, 7.52% in Orang, 6.38% in Sabar, and 6.24% in Santal tribe (Fig. 24.2).

The sources of livelihood among the tribals are found in agricultural and industrial labour, cultivator, business, fishing, mining, Government and Private Services, etc. The occupation of agricultural labour is an important means of livelihood among the tribals of the study area. A maximum number of tribal populations are engaged in agricultural labour through seasonal migration. Among the tribes, the percentage of agricultural labour of Bhumij is 39.41, Mahali is 31.75, Orang is 52.65, Sabar is 56.46 and Santal is 43.63. Among the tribes, the percentage of industrial labour of Bhumij is 4.68, Mahali is 2.09, Orang is 1.11, Sabar is 0.17 and Santal is 1.23. Among the tribes, the percentage of cultivator of Bhumij is 14, Mahali is 0.81, Orang is 4.46, Sabar is 0 and Santal is 14. Among the tribes, the percentage of business activity of Bhumij is 0.98, Mahali is 25.30, Orang is 0, Sabar is 0 and Santal is 0.33. Among the tribes, the percentage of fishing activity of Bhumij is 12.45, Mahali is 5.25, Orang is 11.50, Sabar is 35.74 and Santal is 18.25. Among the tribes, the percentage of mining activity of Bhumij is 2, Mahali is 0, Orang is 3, Sabar is 3 and Santal is 8. Among the tribes, the percentage of Government Services of Bhumij is 0.14, Mahali is 0.04, Orang is 0, Sabar is 0 and Santal is 0.16. Among the tribes, the percentage of Private Services of Bhumij is 0.04, Mahali is 0.01, Orang is 0, Sabar is 0 and Santal is 0.03 (Fig. 24.2).

24.3.2 Agricultural Activity

Two types of agricultural activity, i.e. *Kharif* and *Rabi* have been seen among the tribals in the study area. During the *Kharif* season five major crops (*Paddy, Babui. Arhor, Maize*) are cultivated. For the *paddy* cultivation they have used *Baid, Kanali* and *Shole* land. *Tanr* land has been used for *Babui* cultivation. For *Arhor* cultivation they use *Tanr* and *Baid* land. And *Baid* land has been used for *Maize* cultivation. The crops are sold in local market and to *Mahajans*. As a result, they do not get actual market value or price of different crops. If the actual market value of *paddy* crop is Rs. 850/packet, they get Rs. 750/packet. If the actual market value of *Babui* crop is Rs. 35/kg and Rs. 55/kg, they get Rs. 30/kg and Rs. 50/kg, respectively. If the actual market value of *Arhor* crop is Rs. 40/kg, they get Rs. 30/kg. If the actual market value of *Maize* crop is Rs. 4/piece, they get Rs. 2/piece (Table 24.1). So it

Table 24.1 Traditional economic activity

Agricultural activity	Season	Crops	Types of land/land used (*Tanr, Baid, Kanali, Sole*)	Market place	Market value	
					Obtain value	Actual value
Kharif	June–July to Nov–Dec	Paddy	Baid, Kanali, Shole	Local market, *Mahajans*	Rs. 750/packet	Rs. 850/packet
		Babui	Tanr		Rs. 30/kg and Rs. 50/Kg	Rs. 35/kg and Rs. 55/Kg
		Arhor	Tanr, Baid		Rs. 30/kg	Rs. 40/kg
		Maize	Baid		Rs. 2/piece	Rs. 4/piece
Rabi	Dec–Jan to April–May	Paddy	Shole		Rs. 850/packet	Rs. 950/packet
		Tomato	Baid, Kanali, Shole		Rs. 3/kg	Rs. 7/kg
		Till	Baid, Kanali		Rs. 800/packet	Rs. 1000/packet
		Linseed	Baid, Kanali		Rs. 700/packet	Rs. 900/packet
		Potato	Baid, Kanali, Shole		Rs. 150/packet	Rs. 200/packet
		Vegetables	Baid, Kanali, Shole		Rs. 4/ to 6/	

Source Primary survey by authors, 2013–2019

can be concluded that due to poor conscious they have been suffering huge loss in every crop.

During the *Rabi* season six major crops (*Paddy, Tomato, Till, Linseed, Potato and Vegetables*) are cultivated. For the *paddy* cultivation they use *Baid, Kanali* and *Shole* land. *Baid, Kanali* and *Shole* land has been used for tomato cultivation. For till and linseed cultivation they use *Baid* and *Kanali* land. *Baid, Kanali* and *Shole* land are used for potato cultivation. And for some vegetables and fruit (cauliflower, cabbage, brinjal, ridge gourd, bitter gourd, broad bean, calabash, small gourd, lady finger pumkin, watermelon, etc.) cultivation they use *Baid, Kanali* and *Shole* land. The *Rabi* crops are sold in local market and to *Mahajans.* And as a result they do not get actual market value of different crops. If the actual market value of *paddy* crop is Rs. 950/packet, they get Rs. 850/packet. If the actual market value of *tomato* is Rs. 7/kg they get Rs. 3/kg. If the actual market value of *till* is Rs. 1000/packet, they get Rs. 800/packet. If the actual market value of *linseed* is Rs. 900/packet, they get Rs. 700/packet. If the actual market value of *potato* is Rs. 200/packet, they get Rs. 150/packet. And for the vegetables there is a difference between actual and obtained value is more or less Rs. 4–6 (Table 24.1). So it can be concluded that due to poor conscious they have been suffering huge loss in every crop. On accounts of this ignorance and the *Mahajans* reap a huge profit out of their back-breaking toil.

24.3.3 Traditional Economic Activity

24.3.3.1 *Babui* Products and Economic Support

It is a one type of grass. Seeds are used for its plantation. June–July is the main time for the seed germination of *Babui* and October–November is its harvesting time. High and slope land (In plateau area known as ***Tanr*** land) is suitable for cultivation of *Babui* because water stagnation in its root is unfavorable for growth. It is 1.00–1.50 m length. **Santals** have been cultivating from their time of ancestor because Santals are the maximum owner of high land (*Tanr* land).

After harvesting *Babui*, it keeps in open sunlight area for one or two days to dry it because it is most useable than wet. From this *Babui* grass various types of products are made by tribals (Mainly Santal and Sabar) such as rope, cap, flower vase, flower pot, tray, fruit basket, bag, different ornaments and pots, etc. Santals have been making rope from *Babui* grass. It is sold to the *Mahajans* or in local market. The market value of this rope is Rs. 30/Kg. Very few numbers of Sabar people are engaged to make cap, flower vase, flower pot, tray, fruit basket, bag, different ornaments and pots, etc. from *Babui* grass. Businessmen (Non-tribal men) give tender to the Sabar people to make those items. Sabars get some remuneration from the businessmen. Table 24.2 shows the details of products, market, and market value.

Table 24.2 *Babui* and its different products made by tribals

Babui products	Market place	Market value	Remarks
Rope	Local market (Boro, Bandawan, Kuilapal) and Jharkhand, Bihar	Rs. 30/kg in normal bundle, Rs. 50/kg in special bundle	Santals are engaged in rope making and sell the product in local market. *Mahajans* and other non-tribal businessmen send this product to Jharkhand and Bihar
Cap	Fairs (Local, districts, State) and international market	50, 100,150	Sabars are engaged to make these products. They get tender from *Mahajans*
Tray		40 ,70, 100	
Flower vase		40, 80	
Flower pot		40, 80	
Fruit basket		40, 80	
Bag		50, 100, 150	
Ornaments		20, 30, 50	
Pots		20, 30, 40, 50	

Source Primary survey by authors, 2013–2019

24.3.3.2 *Bamboo* Products and Economic Support

Different baskets are made by Mahali people. Little amount of bamboo collects from jungle area but maximum bamboo buys from local area. Local market, local mart, is the main selling area. Sometimes they get order from nearer village during marriage ceremony and festivals. The market value of these products varies in sizes. Small basket is Rs. 20/, medium basket is Rs. 50/, big basket is Rs. 80/. Table 24.3 shows the details of bamboo products.

Table 24.3 Bamboo product made by tribals

Bamboo product	Market place	Market value	Remarks
Basket	Local market, mart, special order from people	Small-20/, Medium-50/Big-80/	Only Mahali tribes are engaged for these products making
Winnow		Small-20/, Medium-50/Big-80/	
Small basket		Small-20/, Medium-30/Big-50/	

Source Primary survey by authors, 2013–2019

24.3.3.3 *Sal Plates* and Economic Support

Santal and Sabar are making sal plate from *sal* leaf. They collect *sal* leaf from forest (12 miles), Kalajharna forest, Bandwan forest and local forest also. Raw leaf is used to make plate. After drying the plate in sunlight they sell it in local market and *Mahajans*. The market value of this plate is Rs. 1/plate. Table 24.4 shows the product details of *sal* plate.

24.3.3.4 *Cane* Products and Economic Support

Only Mahali tribe has been making basket from *cane*. They collect *cane* from 12 miles jungle. Local market, local fair, is the marketing place of this product. Market value of small basket is Rs. 20/, medium basket is Rs. 50/, big basket is Rs. 80/. The following table (Table 24.5) shows the details of cane products.

24.3.3.5 *Mahul* Products and Economic Support

Mahul is a flower of *Mahul* tree. Tribals collects *mahul* flower during April month from forest. After collecting this flower, they dry it in open sunlight. This dry *mahul* is sold at local market or *Mahajans*. The market value of this dry *mahul* is Rs. 25/Kg. This *mahul* flower is used for making liquid drug. Tribals do not sell it in market; they sell it in their locality because liquid drug making is anti-law activity. But mahajans are making liquid drug and selling it in local market, nearer state of Jharkhand and Bihar. *Mahul* is not only used for liquid drug making but also used for tea making. The dust of dry *mahul* flower is mixed with tea. The following table (Table 24.6) shows the details of *Mahul* products.

Table 24.4 Product of sal leaf

Product of sal leaf	Market place	Market value	Remarks
Sal plate	Local market, order from *Mahajans*	Rs. 1/plate	Only Sabar and Santal are engaged for this plate making

Source Primary survey by authors, 2013–2019

Table 24.5 Product of *cane*

Cane product	Market place	Market value (in Rs)	Remarks
Basket	Local market, mart, special order from people	Small-20/, Medium-50/, Big-80/	Only Mahali tribes are engaged for these products making

Source Primary survey by authors, 2013–2019

Table 24.6 Product of mahul flower

Mahul product	Market place	Market value	Remarks
Mahul flower	Local market, mahajans	Rs. 25/Kg	Orang, Sabar and Santal are engaged this activity
Liquid drug	Own locality	Rs. 10/glass	
Mahul dust	District market, Jharkhand, Bihar	–	*Mahajans* sells it

Source Primary survey by authors, 2013–2019

Table 24.7 Product of katchra

Katchra product	Market place	Market value	Remarks
Oil	Local market		Orang, Sabar and Santal are engaged this activity

Source Primary survey by authors, 2013–2019

24.3.3.6 *Katchra* Products and Economic Support

Katchra is the fruit of *Mahul* tree. The time of *katchra* fruit collection is the last week of May and first week of June. Oil is being prepared from the seeds of *katchra*. Tribals use this oil for cooking and also sell it in local market. Table 24.7 shows the details of *katchra* product.

24.3.3.7 Kandu Fruit and Economic Support

Kandu fruit is the fruit of *Kandu* tree. The time of *kandu* fruit collection is last week of April. Santal, Orang, Sabar collect it and sell it to *Mahajans* or local market. The market value of *kandu* fruit is Rs. 1/piece. Table 24.8 shows the details of *kandu* fruit.

So keeping in view of above discussion, as the tribals are very close to natural forest, some forest-based domestic and economic activity has been happening among the tribals. Bhumij tribe does not collect any forest product for economic purpose. Mahali tribe collects *Cane* from 12 miles forest area for making baskets. And they sell all products to local hut (Kuilapal and Boro), Bandwan market and local fair. And the market value varies in size like (Small-Rs. 15/ to 20/, medium-Rs. 30/ to 40/ and big-Rs. 60/ to 70/). Orang tribe collects *Mahul* flower and *katchra* from forest and

Table 24.8 Uses of kandu fruit

Product	Market place	Market value	Remarks
Kandu fruit	Local market, *Mahajans*	Rs. 1/piece	Santal, Orang, Sabar are engaged this activity

Source Primary survey by authors, 2013–2019

Table 24.9 Forest based economy

Tribes	Items collecting from forest area	Market place	Market value	Remarks
Bhumij	–	–	–	Do not collect any materials from forest for economic purpose
Mahali	*Cane collection for making baskets*	Local mart (Kuilapal, Boro); Bandwan market; local fair	Small-15/-20/Medium-30/-40/Big-60/-70/	Mahali of Buribandh G.P only has been collecting *Cane* from 12 miles Forest
Orang	*Mahul flower, Katchra*	–	1 kg-25/	*Mahajan* parches *Mahul flower*
Sabar	*Sal leaf, Wood*	Local market, some local households	Sal plate-50/ to 100/ per bundleWood-6/ to 7/ per bundle	*Mahajan* parches *sal plate*
Santal	*Sal leaf, Mahul flower, Katchra, Kandu fruit, Peal fruit, Stone*	Local market, Kuilapal, Bandwan	Sal plate-Rs. 50/ to 100/ per bundle Mahul flower- 25/kg Kandu-1/piece Stone-30/ per tin	*Mahajan* parches *sal plate and Mahul flower;* businessmen and other people come to parches *stone*

Source Primary survey by authors, 2013–2019

they sell these to the *Mahajans* by Rs. 25/kg. Sabar tribe collects sal leaf and wood from forest. They sell these in local market and local household and also *Mahajans*. The market value of sal plate is Rs. 50/ to 100/ per bundle and wood is Rs. 6/ or 7/bundle. Santal collects *Sal* leaf, *Mahul* flower, *Katchra, Kandu* fruit, Stone from forest. They sell these in local market, *Mahajans*, Kuilapal, Bandwan. The market value of sal plate is Rs. 50/ to 100/ per bundle, *Mahul* flower and *katchra is* Rs. 25/kg, and *Kandu*-1/piece (Table 24.9).

24.3.4 Income Status

Income index has been calculated through the gross income to maximum and minimum income which has been developed by the United Nation Development Programme (UNDP, http://hdr.undp.org/en/content/income-index). The maximum income has been considered to be Rs. 75,000 and minimum income is Rs. 100. The formula is given below.

Income index of the tribals (II):

$$II = \frac{In(GNIpc) - In(100)}{In(75000) - in(100)}$$

Source: UNDP, http://hdr.undp.org/en/content/income-index

Income index is not good among the tribes of the study area. And it is also not uniform among the tribes. Bhumij has the better condition among the other tribes because of their highest amount of land in comparison to other tribes. The income index of Bhumij is 0.39. Second-highest income index belongs to Santal tribe. The income index of Santal is 0.36. And the income index of Mahali, Orang and Sabar is 0.34, 0.25 and 0.18, respectively. Sabar has the lowest income index among the tribes because of their poor land ownership and little engagement on economic activity (Table 24.10 and Fig. 24.3).

There is a little bit difference of income index between male and female of the study area. Because tribal female works with male in different sectors. The income index of Bhumij male and female is 0.22 and 0.17, respectively. The income index of Mahali male and female is 0.18 and 0.16, respectively. The income index of Orang male and female is 0.16 and 0.09, respectively. The income index of Sabar male and female is 0.10 and 0.08, respectively. The income index of Santal male and female is 0.21 and 0.15, respectively (Fig. 24.3).

24.3.5 Functional Groups, Functions and Their Weightage

To measure the relative importance of different functions, functional weightage have been used for better explanation by R. S. Singh in 1988 and 1989. It has been calculated by selecting some functional groups, e.g. educational facilities, transport and communication, medical facilities, commercial and credit, etc. Functional weightage and concentration of settlement are inversely related.

Functional Weightage (W):

$$W = \left[\frac{N}{F}\right]$$

Source: R. S. Singh, 1988 & 1989.

where,

W = Functional Weightage

N = Total number of population

F = Number of population having in a particular function.

Table 24.10 Income index of the tribals

Tribes	Income index							
	Barijagda	Dighi	Kumari	Barajhoragora	Ankro-barakadam	Bargoria-jamtoria	Buribandh	Manbazar-II block
Bhumij	0.36	0.38	0.43	0.38	0.40	0.41	0.39	0.39
Mahali	0.34	0.34	0.32	0.34	0.36	0.36	0.31	0.34
Orang	–	–	0.25	–	–	–	–	0.25
Sabar	0.17	0.17	0.18	0.17	0.20	0.20	0.18	0.18
Santal	0.36	0.36	0.37	0.36	0.35	0.38	0.36	0.36

Source Primary survey and computed by authors, 2013–2019

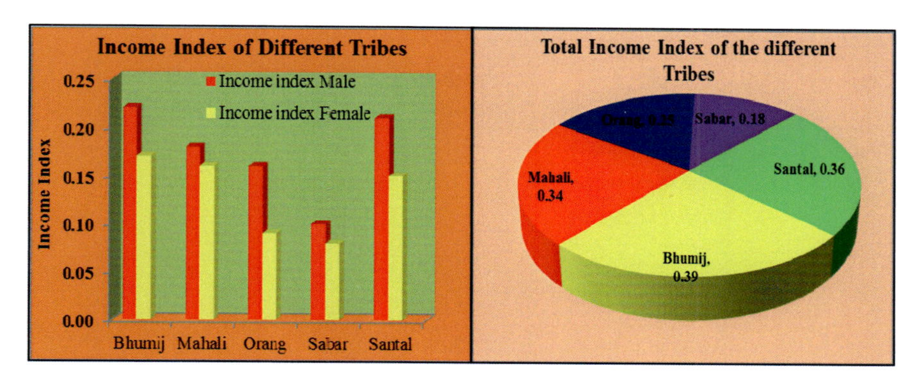

Fig. 24.3 Work participation in different sector of the tribes. *Source* Primary survey by authors, 2013–2019

The selected functions under the functional groups of educational facilities are Primary School, Secondary School, Higher Secondary School, College, University, Training College and Engineering College, etc. Among the tribes of the study area, there is a variation in having the functions of educational facilities. In the study area, maximum number of tribes has the functions of Primary School. The functional weightage in Primary School of Bhumij, Mahali, Orang, Sabar and Santal is 5.79, 7.38, 4.51, 7.18 and 5.56, respectively. That means maximum number of Orang and minimum number of Mahali and Sabar tribes is educated in primary level. The functional weightage in Secondary School of Bhumij, Mahali, Orang, Sabar and Santal is 8.85, 11.02, 9.32, 9.79 and 9.35, respectively. That means maximum number of Bhumij and minimum number of Mahali tribes is educated in secondary level. The functional weightage in Higher Secondary School of Bhumij, Mahali, Orang, Sabar and Santal is 14.96, 15.42, 32.11, 20.52 and 14.51, respectively. That means maximum number of Santal and minimum number of Orang tribes is educated in Higher Secondary level. The functional weightage in College of Bhumij, Mahali, Orang, Sabar and Santal is 41.39, 19.45, 72.25, 26.83 and 36.45, respectively. That means maximum number of Mahali and minimum number of Orang tribes is educated in College level. The functional weightage in University of Bhumij, Mahali, Orang, Sabar and Santal is 241.96, 379.23, 0, 895.37 and 209.48, respectively. That means maximum number of Santal and minimum number of Sabar tribes is educated in university level. The functional weightage in Training College of Bhumij, Mahali, Orang, Sabar and Santal is 599.14, 439.10, 0, 2387.67 and 483.96, respectively. That means maximum number of Mahali and minimum number of Sabar tribes is educated in Training College level. The functional weightage in Engineering College of Bhumij, Mahali, Orang, Sabar and Santal is to 1143.82, 1191.86, 0, 0 and 825.59, respectively. That means maximum number of Santal and minimum number of Mahali tribes is educated in Engineering College level (Table 24.11 and Fig. 24.4).

Table 24.11 Functional groups, functions and their weightage

Functional group	Functions	Weightage				
		Bhumij	Mahali	Orang	Sabar	Santal
Educational facilities	Primary school	5.79	7.38	4.51	7.18	5.56
	Secondary school	8.85	11.02	9.32	9.79	9.35
	Higher secondary school	14.96	15.42	32.11	20.52	14.51
	College	41.39	19.45	72.25	26.83	36.45
	University	241.96	379.23	0	895.37	209.48
	Training college	599.14	439.1	0	2387.67	483.96
	Engineering college	1143.82	1191.86	0	0	825.59
Medical facilities	P.H.C	1.11	1.27	1.14	1.27	1.07
	Quack physician	1.47	1.59	1.84	1.53	1.27
	Traditional healer	1.58	2.56	2.39	1.4	1.72
	Hospital	1.08	1.06	1.44	1.11	1.14
	Nursing home	103.98	379.23	144.5	0	127.59
	Animal husbandry	2.32	3.25	2.53	2.09	2.05
Transport and communication	N.H and A.H	18.05	20.2	26.27	57.76	17.26
	S.H	1.32	1.55	2.33	5.67	1.65
	Railway	34.28	38.27	41.28	73.84	35.01
	Post office	21.36	26.65	24.08	596.92	21.53
	Internet	47.12	85.13	57.8	895.37	115.99
	Mobile	2.16	2.15	2.67	2.98	2.16
Commercial and credit	Primary Co-operative societies	26.32	30.78	32.11	795.89	26.94
	Banks	1.44	1.48	2.12	2.31	1.44
	Others (LICI)	3.67	15.56	18.06	7.01	3.85

Source Primary survey and data computed by authors, 2013–2019

The selected functions under the functional groups of medical facilities are Primary Health Centre, Quack Physician, Traditional healer, Hospital, Nursing home and Animal husbandry, etc. Among the tribes of the study area, there is a variation in having the functions of health facilities. The functional weightage of Primary Health Centre of Bhumij, Mahali, Orang, Sabar and Santal belongs to 1.11, 1.27, 1.14, 1.27 and 1.07, respectively. The functional weightage of Quack physician of Bhumij, Mahali, Orang, Sabar and Santal is 1.47, 1.59, 1.84, 1.53 and 1.27, respectively. The functional weightage of Traditional healer of Bhumij, Mahali, Orang, Sabar and Santal is 1.58, 2.56, 2.39, 1.40 and 1.72, respectively. The functional weightage of Hospital of Bhumij, Mahali, Orang, Sabar and Santal is 1.08, 1.06, 1.44, 1.11 and 1.14, respectively. That means maximum number of tribes are taking health facilities in Primary health centre, Quack Physician, Traditional healer, and Hospital.

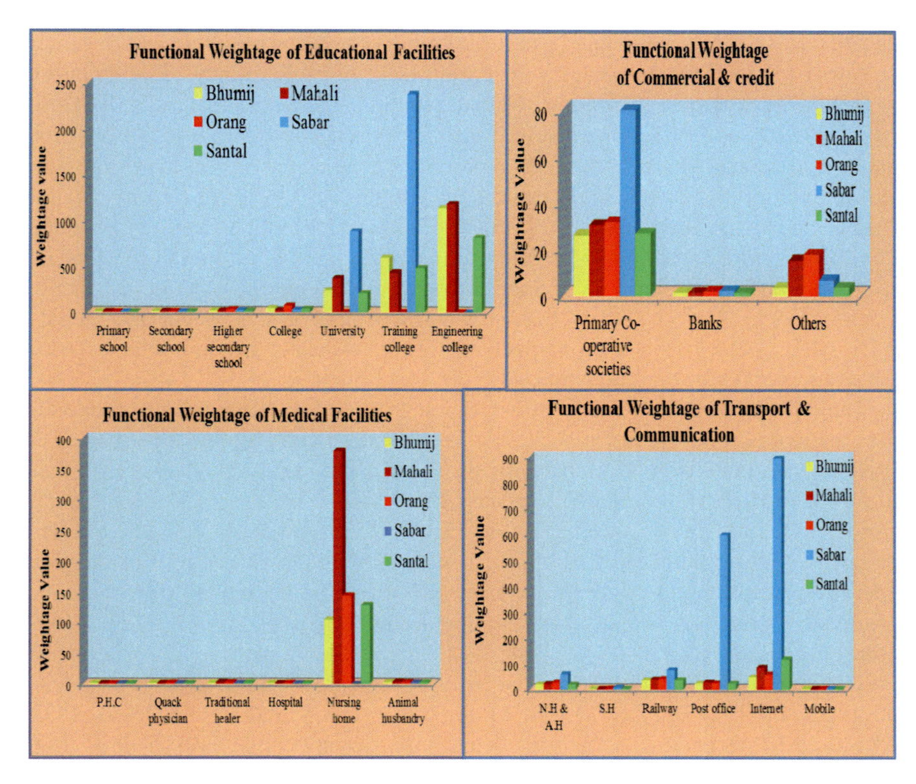

Fig. 24.4 Functional weightage of different functional groups. *Source* Primary survey by authors, 2013–2019

The functional weightage of Nursing home of Bhumij, Mahali, Orang, Sabar and Santal is 103.98, 379.23, 144.50, 0 and 127.59, respectively. That means maximum number of Bhumij tribes are taking health facilities in Nursing home. The functional weightage of Animal husbandry of Bhumij, Mahali, Orang, Sabar and Santal is 2.32, 3.25, 2.53, 2.09 and 2.05, respectively. That means maximum number of tribes are going to health centre for their domestic animals (Table 24.11 and Fig. 24.4).

The selected functions under the functional groups of transport and communication facilities are national highway, additional highway, sub-highway, railway, post office, mobile and Internet, etc. Among the tribes of the study area, there is a variation in having the functions of transport and communication facilities. The functional weightage of national highway, additional highway of Bhumij, Mahali, Orang, Sabar and Santal is 18.05, 20.20, 26.27, 57.76 and 17.26, respectively. The functional weightage of sub-highway of Bhumij, Mahali, Orang, Sabar and Santal is to 1.32, 1.55, 2.33, 5.67 and 1.65, respectively. That means maximum number of tribes are going to different places through sub-highway in short distance.

The functional weightage of railway of Bhumij, Mahali, Orang, Sabar and Santal belongs to 34.28, 38.27, 41.28, 73.84 and 35.01, respectively. That means maximum number of Bhumij tribes is going to different places through railway and minimum number of Sabar tribe is going to different places through railway. The functional weightage of post office of Bhumij, Mahali, Orang, Sabar and Santal is 21.36, 26.65, 24.08, 596.92 and 21.53, respectively. That means maximum number of Bhumij tribes are having the function of post office and minimum number of Sabar tribe are having the function of post office. The functional weightage of Internet of Bhumij, Mahali, Orang, Sabar and Santal is 47.12, 85.13, 57.80, 895.37 and 115.99, respectively. That means maximum number of Bhumij tribes are having the function of Internet facility and minimum number of Sabar tribe having the function of Internet facility. The functional weightage of mobile of Bhumij, Mahali, Orang, Sabar and Santal is 2.16, 2.15, 2.67, 2.98 and 2.16, respectively. That means maximum number of tribes are using mobile for communication (Table 24.11 and Fig. 24.4).

The selected functions under the functional groups of commercial and credit facilities are primary co-operative societies, banks and others. Among the tribes of the study area, there is a variation in having the functions of commercial and credit facilities. The functional weightage of primary co-operative societies of Bhumij, Mahali, Orang, Sabar and Santal is 26.32, 30.78, 32.11, 795.89 and 26.94, respectively. That means maximum number of Bhumij tribes are having the function of primary co-operative societies and minimum number of Sabar tribe are having the function of primary co-operative societies. The functional weightage of banks of Bhumij, Mahali, Orang, Sabar and Santal is 1.44, 1.48, 2.12, 2.31 and 1.44, respectively. That means maximum number of tribes using the banking facilities. The functional weightage of others (LICI) of Bhumij, Mahali, Orang, Sabar and Santal is 3.67, 15.56, 18.06, 7.01 and 3.85, respectively. That means maximum number of Bhumij tribes has done Life insurance and minimum number of Orang tribe has done Life insurance (Table 24.11 and Fig. 24.4).

24.3.6 Quality of Life

It is developed by the United Nation Development Programme (UNDP, Human Development Report, https://en.wikipedia.org/wiki/Physical_Quality_of_Life_Index). Quality of life index is related to condition of household, education, health and economy. It has been calculated with the different indicators of individual dimension. The formula of quality of life is given below.

$$QLI = \left[\frac{Sum\ of\ index\ of\ all\ indicators}{Number\ of\ indicators}\right]$$

Source: UNDP, Human Development Report, https://en.wikipedia.org/wiki/Physical_Quality_of_Life_Index.

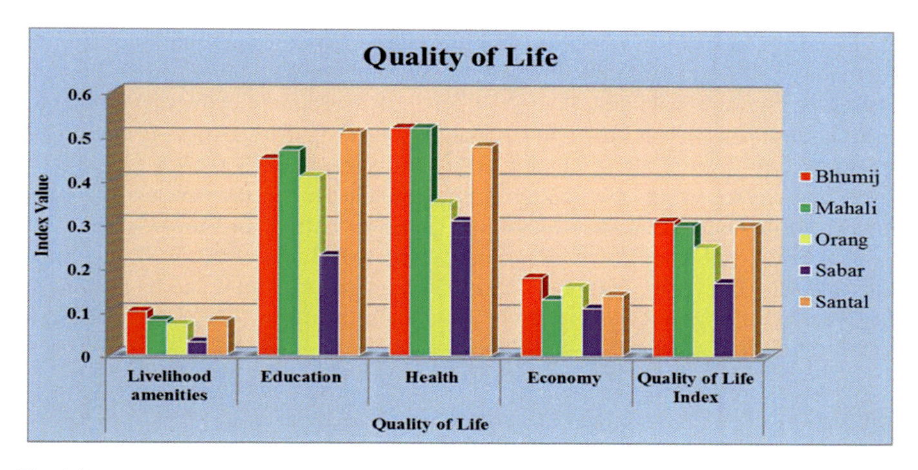

Fig. 24.5 Quality of life index. *Source* Primary survey by authors, 2013–2019

To understand the quality of life among the tribes, quality of life index has been calculated through some indicators. Household having safe drinking water, drainage facility, toilet, bathroom, separate kitchen and electricity and people do not sleep on floor have been considered for better assessment of quality of livelihood amenities among the tribes. Among the tribes, the quality of livelihood amenities of Bhumij, Mahali, Orang, Sabar and Santal is 0.10, 0.08, 0.07, 0.03 and 0.08, respectively. That means the quality of livelihood amenities index among the tribes is not good (Fig. 24.5).

To study educational quality among the tribes, educational quality of life has been analyzed through number of children going to school daily, number of children wanting to take higher education, number of children getting guide from father and mother. Among the tribes, the quality of education index of Bhumij, Mahali, Orang, Sabar and Santal is 0.45, 0.47, 0.41, 0.23 and 0.51 respectively. That means the quality of education of Santal is good among the tribes. The quality of education of Bhumij, Mahali and Orang is more or less uniform but Sabar has very low among the tribes (Fig. 24.5).

To study health quality among the tribes, health quality of life has been analyzed through household do not take traditional medicine, husband take care during his wife's pregnancy, children get vaccine in proper time, people do not take alcohol and people do not take tobacco. Among the tribes, the quality of health index of Bhumij, Mahali, Orang, Sabar and Santal is 0.52, 0.52, 0.35, 0.31 and 0.48, respectively. That means the quality of health of Bhumij and Mahali is good among the tribes but Sabar has bad quality of health among the tribes (Fig. 24.5).

To study employment quality among the tribes, employment quality of life has been analyzed through number of APL family, household having bank account, household having mediclaim, household having Government or Private employee, and household that do not have child labour. Among the tribes, the quality of employment index of Bhumij, Mahali, Orang, Sabar and Santal is 0.18, 0.16, 0.11 and 0.14, respectively. That means the quality of employment among the tribes is not good. Bhumij has the better condition among the other tribes (Fig. 24.5).

The overall quality of life index among the tribe of Bhumij, Mahali, Orang, Sabar and Santal is 0.31, 0.30, 0.25, 0.17 and 0.30, respectively. So it can be easily said that Bhumij belongs to good quality of life among the others. The quality of life of Sabar tribe is low among the other (Fig. 24.5).

24.3.7 Human Development

It was developed by the United Nation Development Programme (UNDP) in 1990. Three dimensions (health, knowledge and standard of living) and four indicators (life expectancy at birth, mean years of schooling, expected years of schooling, per capita income) have been taken for the calculation of HDI (http://hdr.undp.org/en/content/human-development-index-hdi). This is mentioned in Table 24.12. The formula of human development index is given below.

$$\mathrm{HDI} = \sqrt{\mathrm{LEI.\ EI.\ II}}$$

Among the tribes of the study area, the life expectancy index, education index and income index of Bhumij tribe are 0.56, 0.65 and 0.39, respectively. And the human development index is 0.37. Among the tribes of the study area, the life expectancy index, education index and income index of Mahali tribe are 0.49, 0.61 and 0.34, respectively. And the human development index is 0.32. Among the tribes of the study area, the life expectancy index, education index and income index of Orang tribe are 0.50, 0.53 and 0.0.25, respectively. And the human development index is

Table 24.12 Human development indicators

Dimension	Indicators	Values		Dimension indices
		maximum	Minimum	
Health	Life expectancy at birth	20 years	85 years	Life expectancy index (LEI)
Knowledge	Mean years of schooling	0	15 years	Education Index (EI)
	Expected years of schooling	0	18 years	
Standard of living	Per capita income	Rs. 100	Rs. 75,000	Income Index (II)

Table 24.13 Human Development Index (HDI)

Tribes	LEI	EI	II	HDI
Bhumij	0.56	0.65	0.39	0.37
Mahali	0.49	0.61	0.34	0.32
Orang	0.50	0.53	0.25	0.26
Sabar	0.46	0.49	0.18	0.20
Santal	0.51	0.64	0.36	0.34

Source Primary survey and data computed by authors, 2013–2019

0.26. Among the tribes of the study area, the life expectancy index, education index and income index of Sabar tribe are 0.46, 0.49 and 0.36, respectively. And the human development index is 0.20. Among the tribes of the study area, the life expectancy index, education index and income index of Santal tribe are 0.51, 0.64 and 0.36, respectively. And the human development index is 0.34 (Table 24.13, Fig. 24.6 and 24.7).

So it can be concluded that although human develop index of tribals is not good but not in uniform among the tribes. The human development index of Bhumij tribe is high among the other tribes. The second-highest human development index belongs to Santal tribe, third belongs to Mahali tribe and forth and fifth belongs to Orang and Sabar, respectively (Table 24.13, Figs. 24.6 and 24.7).

24.4 Conclusion

So, it may be concluded that there is no remarkable development among the tribals at the desired level. But mentally and to some extent economically they are gradually developing in comparison to their previous generations. The development is not uniform among the tribals due to variation in education, economic activity, awareness as well as perception of livelihoods. Bhumij are the most developing among the tribals because of their huge amount of land, entrepreneurship mentality, awareness of education and health, engagement in different economic activities, which ensure high capital and increasing income.

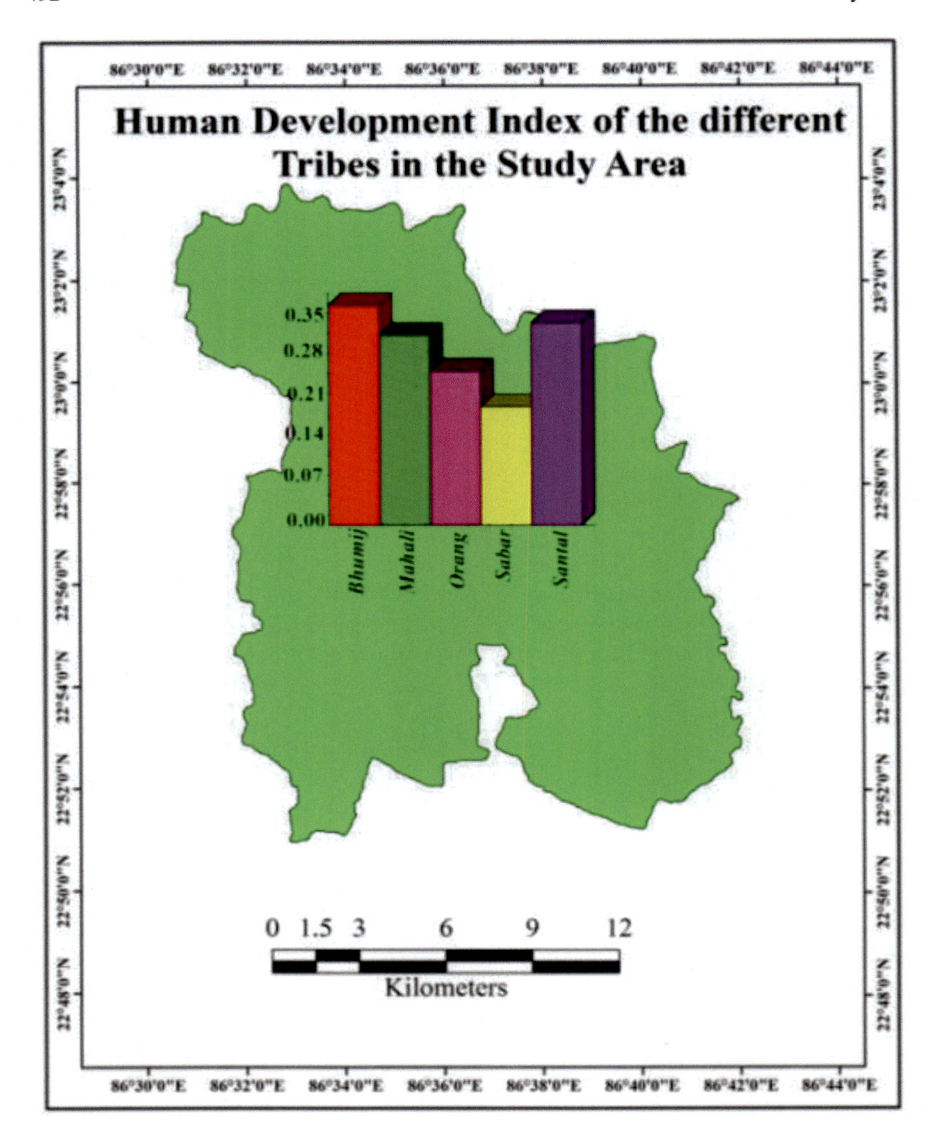

Fig. 24.6 Human development index. *Source* Primary survey by authors, 2013–2019

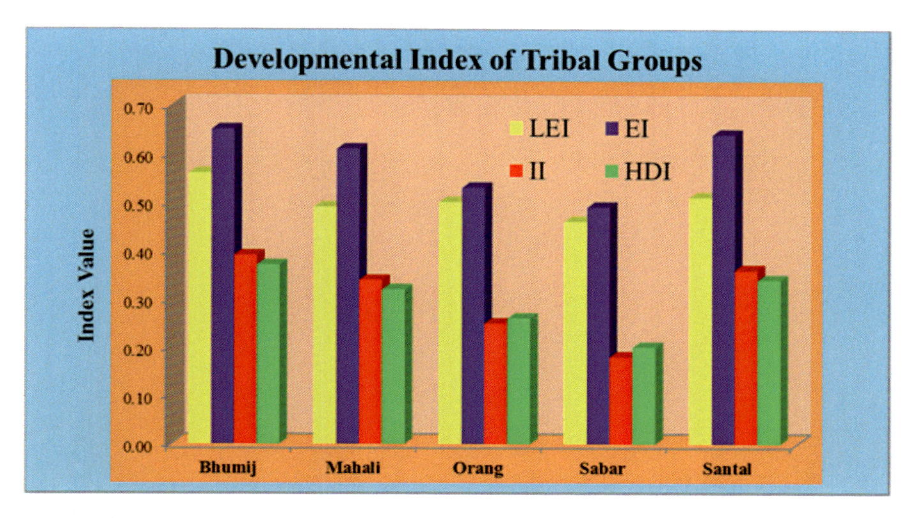

Fig. 24.7 Development index. *Source* Primary survey by authors, 2013–2019

References

Dalton G (1971) Economic anthropology and development. Essays on tribal and peasant economies. Basic Books, New York, p 386

Dalton G (1982) Barter. J Econ Issues 16:182

District Census Handbook, 1971, Census of India, New Delhi

http://hdr.undp.org/en/content/human-development-index-hdi

http://hdr.undp.org/en/content/income-index

https://en.wikipedia.org/wiki/Physical_Quality_of_Life_Index

Mazur A (1968) A nonrational approach to theories of conflict and coalitions. Social problems Research Group, First Published June 1, Sunnyvale, California, p 179

Singh RS (1988 & 1989) IMF policies towards less developed countries

Vidyarthi LP, Ray BK (1976a) The Tribal culture of India. Concept Publishing Company, New Delhi, pp 114–121

Vidyarthi LP, Ray BK (1976b) The tribal culture of India. Concept Publishing Company, New Delhi, pp 189–194

Vidyarthi LP, Ray BK (1976c) The tribal culture of India. Concept Publishing Company, New Delhi, pp 27–42

Von Fürer-Haimendorf C (c1982 1982). Tribes of India: the struggle for survival. University of California Press, Berkeley. http://ark.cdlib.org/ark:/13030/ft8r29p2r8/

Chapter 25
The Fourth Paradigm in Geographical Sciences

Sandeep Kundu

Abstract Science has evolved from being empirical (1000 years ago) and theoretical (100 years ago) to computational (few decades ago) and now, in the twenty-first century, it has entered into a new paradigm. This new paradigm, commonly referred to as the 'Fourth Paradigm', is based on data-driven science and is influencing the way we derive scientific insights. High volumes of data are now being generated at varying speeds and are spread over different geographies, which are stored digitally in a variety of formats. These constitute 'Big Data', on which scientific analytics are now being performed. Artificial intelligence and machine learning algorithms constitute the core of present-day analytics and are providing insights far advanced from the traditional approaches (i.e. empirical, theoretical and computational). Industry and businesses are leading the way forward in leveraging on this new paradigm which has tremendous potential for Geographical Sciences. Burning issues in health, population, migration, public policy, society, sales and marketing, climate and environment and, energy and sustainability are now utilizing this paradigm. This article discusses the fundamentals of data science with linkages to geographical science. It summarizes few key applications of 'The Fourth Paradigm' in Geographical sciences.

Keywords Geographical science · Data science · Pattern recognition · Artificial intelligence · Machine learning

25.1 Introduction

The world today is looking at optimizing resources to meet the demands of a growing population in an age where human activities are more digitally interconnected than ever. With globalization setting up an intricate web of supply-chain networks spread across distant geographies, we are increasingly witnessing domino effects of disasters. Conventional paradigms (both empirical and theoretical) are unable to address these cascading effects of extreme events. Insights from factual

S. Kundu (✉)
National University of Singapore, 21 Lower Kent Ridge Road, Singapore, Singapore

N. C. Jana et al. (eds.), *Livelihood Enhancement Through Agriculture, Tourism and Health*, Advances in Geographical and Environmental Sciences,
https://doi.org/10.1007/978-981-16-7310-8_25

data collected over years are now of great value to decision-makers as compared to fictitious computational simulations. Scientific disciplines are now transforming to adopt data analytics as the key decision driver to keep pace with today's development. Data availability has changed the way we perceive scientific problems and solutions. It has now occupied a central role in data-enabled science (Fig. 25.1) forming the foundation of this 'Fourth Paradigm'.

Data is now the new gold which can be mined from the vast archives we have built to date and continue to accrue ever since we stepped into the digital world. Today, data has become 'Big Data' and we are developing systems to handle its velocity, volume, variety and veracity. With science having progressed into a new era which is data-driven, key focus now has shifted towards bettering capture, storage, transformation, analysis and visualization of scientific information, without which informed decisions are not possible. Society today demands swift response to a variety of emerging problems. Therefore, new data-driven applications are constantly being developed to solve them. Over the last decade, big data has become a topic of global interest, and is increasingly attracting the attention of academia, industry, government and other organizations (Li et al. 2016).

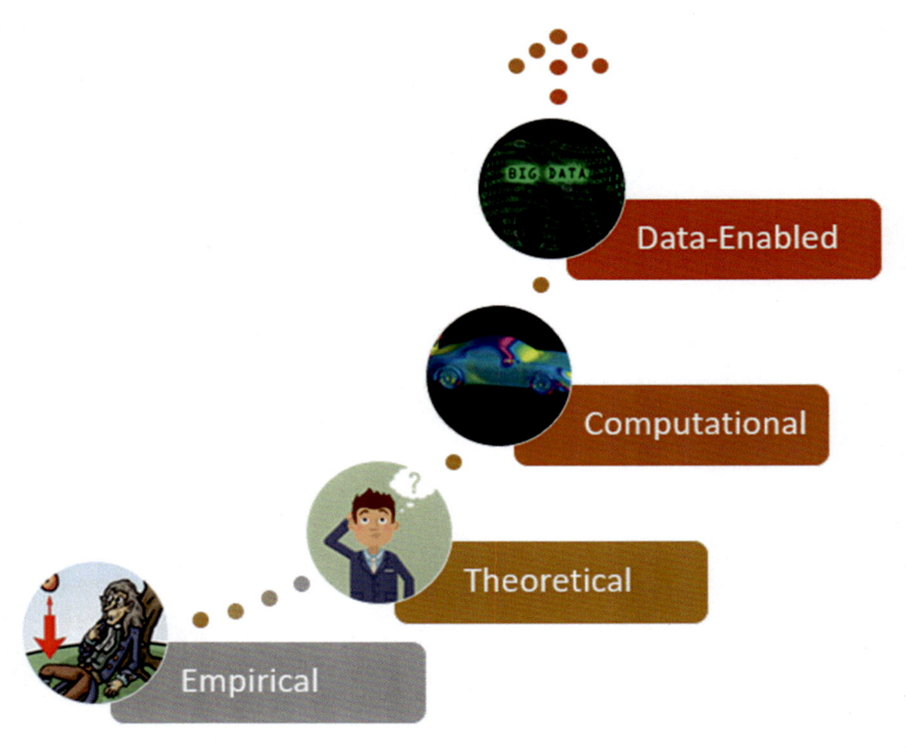

Fig. 25.1 The four paradigms of science

Geography, being the science of place and space, interconnects both natural and social sciences. Understanding contemporary urban and environmental problems requires an appreciation of the interdependence between human activities and the natural and cultural environment (Koutsopoulos 2011). Data science can establish this interdependence with impunity and therefore the 'Fourth Paradigm' has a huge scope in solving the major challenges our planet is confronted today (Boon and van Baalen 2019). This is especially important as most data is geographic in nature (Dempsey 2012), and geographic information science plays an important role in many scientific disciplines through its added dimensions like, spatial analyzes and geographic visualization (VoPham et al. 2018).

Experimental science was followed by theoretical science (e.g. Kepler's Laws, Newton's Laws of Motion, Maxwell's equations, etc.). But soon, problems became too complicated to be solved analytically, requiring computer simulations. These simulations supported solutions in the last half of the past millennium and in the process generated huge datasets. People now look for deriving information from current and historical data using complex algorithms and visualization methods on computers. The world of science has since changed and the new model is to capture data through devices and instruments which are then processed by software transforming the data into information or knowledge. Scientists, who only recently got to see big data in the pipeline, are expanding their horizon through techniques and technologies for data-intensive science. This can be differentiated from computational science and is termed as *the fourth paradigm.*

25.2 Geographical Sciences and Data

Geography's relevance to science and society originates from a distinctive set of perspectives through the lenses of geographers around the world which has three prime dimensions (Fig. 25.2). These are:

- Geographical perspectives of Place—Scale integration and interdependence
- Domains of synthesis—Human, Social and Environmental dynamics
- Spatial representation—from visual to cognitive.

25.2.1 Fundamental Principles

Principles of data science and geography must draw a common meeting point to be able to work together. The connection between geographical science and data science, despite being omnipresent, is little understood. Geographers and data scientists operate on the below two fundamental principles which brings their perspectives together. These principles are

- Tobler's principle

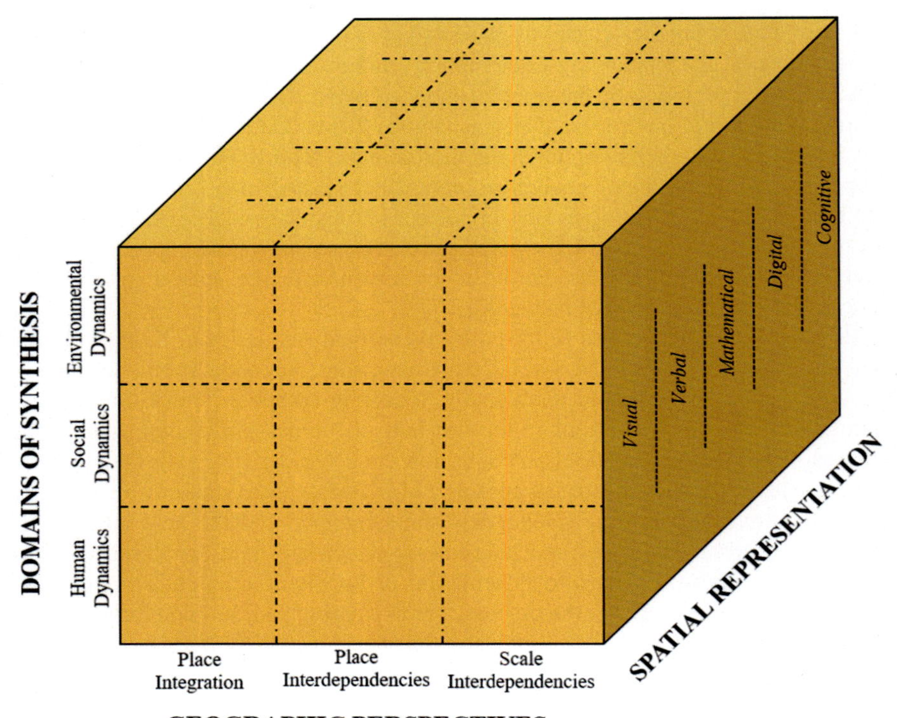

Fig. 25.2 Dimensions of geographical science adapted from NRC (1997)

- Bonferroni's principle.

Tobler's principle relates an instance of data with another based on the concept of Euclidian distance. Data instances close to each other are more related than the instances which are afar. Bonferroni's principle, on the other hand, states that patterns can be found and mathematically defined even from random datasets. 'Patterns' and 'Distance' are used as connectors to relate instances of random data. Data Science applications in geography, therefore, needs to utilize the concept of both distance and patterns to derive insights and solutions.

25.2.2 Data Levels and Mathematical Operations

Data is the basis of reasoning happenings around us. It is the variable which can assume different levels of measurements, namely Nominal, Ordinal, Interval or Ratio. These levels of measurement (Fig. 25.3) are fundamental to the type of mathematical operation that can be performed on data (Table 25.1).

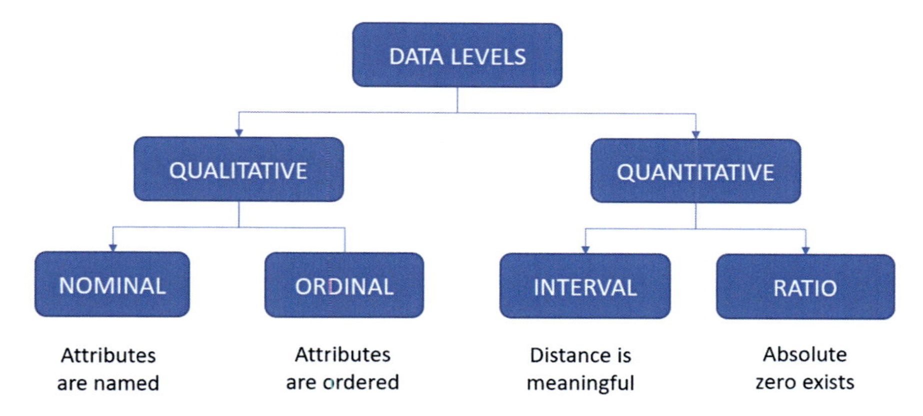

Fig. 25.3 Levels of data

Table 25.1 Data levels and mathematical operations

Type of operator	Nominal	Ordinal	Interval	Ratio
Mode, count, frequency	Yes	Yes	Yes	Yes
Median, min, maz, range	No	Yes	Yes	Yes
Mean, variance, standard deviation	No	No	Yes	Yes

The nominal level of measurement is used for classifying data. Nominal data can consist of words, letters and symbols. An example is the use of letters 'M' and 'F' for storing gender information. Ordinal level of measurement assumes an ordered relationship. Hence, at this level, data can be ordered in increasing or decreasing manner. An example is the score of students which can be used to rank or grade them. Interval data can be used to classify data and also to order and rank it. For example, a scale for pain between 1 and 10 is used by hospital emergencies to rank the criticality of a patient. Temperature, too, is a good example of interval data. The next level of data is ratio data where in addition to having equal intervals data can also accomodate no value or zero. Distance and weights are good examples of ratio data. Table 25.1 summarizes the mathematical operators that can be applied on different levels of data. Generally speaking, nominal and categorical data values are 'qualitative data' as they cannot be subjected to many computations (Huisman and Rolf 2001).

25.3 Data Science Value Chain

The fourth paradigm in geographical science adopts a value-chain pathway to generate knowledge from data. This value chain realizes the information flow in

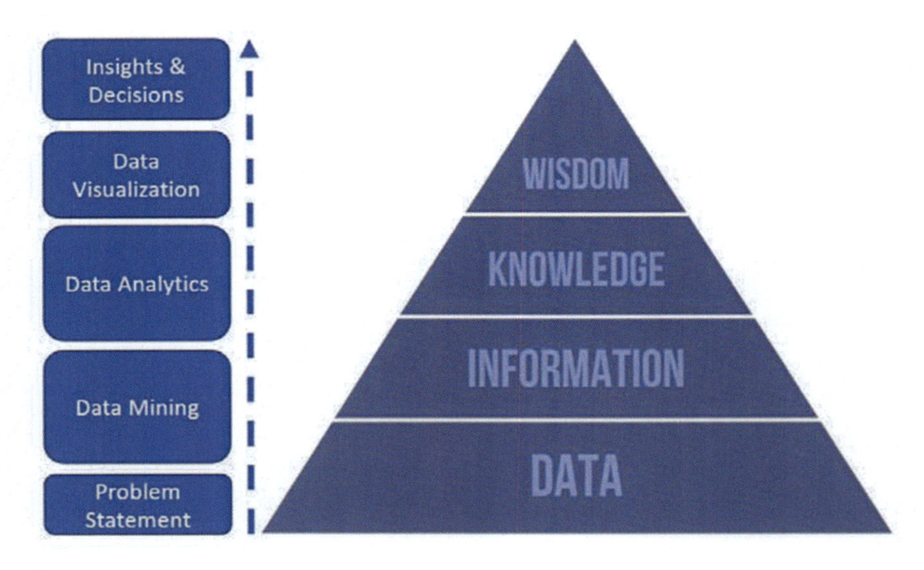

Fig. 25.4 DIKW model and data science value-chain elements

the data analytical system through a series of steps. These steps generate value and insights from data thereby enabling the analysis of spatial big data to be used as a decision support tool. Essentially, the data science value chain essentially transforms a dump of raw data into information that is used to generate knowledge and wisdom following the DIKW Pyramid (Rowley 2007). Data science value chain involves generic steps like Data Mining, Data Analytics, Visualization and eventually Insights for making decisions (Fig. 25.4).

25.3.1 Data Mining

Data mining is essentially a process of collecting and collating large datasets, exploring and finding patterns and using the right model to address the problem statement. This is done mostly in stages as the initial data comes in diverse and disparate forms from multiple agencies. Each agency collects and stores data for their own business objectives which could be different to the research problem statement. Hence, data collected from multiple agencies needs to be explored on how they can be transformed into a common and usable format. For example, if we want to find the part of a city that has the highest crime and why, then we must source various datasets like crime incidents from the police, census data from the social welfare department and administrative boundaries from the planning department. These three datasets have different structures and are stored is different formats.

Data exploration or exploratory data analysis is to find patterns on datasets which address its quality and usability. It is common to have many gaps and errors in the data which can influence the analysis and results. At the data exploration stage, the aim is to detect these and derive subsets from the whole data population, that are helpful in answering the questions at hand. A data analyst is required to visually explore data for its content and characteristics. This include size, volume, completeness, correctness, integrity and relationship between the data elements (Idreos et al. 2015). Once analyzed, data can be transformed (converted) to a common and usable format amenable for integration into a modeling system. Data transformation ranges from being simple to extremely complex depending on the requirement. Data transformation is executed through a mix of manual and automated steps. Tools and technologies for data transformation vary widely and are based on the format, structure, complexity, and volume of the data at hand (Morcos et al. 2015).

Data Modeling is the semi-automatic analysis to extract interesting patterns through clustering, association, classification and correlation-regression algorithms. Visualization like graphs, plots and maps are used to explore results. Spatial indexing may also be required for geographical data for enhancing search and query. Patterns detected at this stage provide a summary of the input data and paves the way for advanced analysis. This involves machine learning and artificial intelligence. Advanced analysis supports predictive and prescriptive analyses for decision support systems (Eom and Lee 1990).

25.3.2 Data Analytics

Data analytics is examining data with an aim to extract information. It can be considered as advanced data exploration. Data analytics can be both quantitative and qualitative. The former involves analysis of numerical data variables that can be treated statistically. The qualitative approach is interpretive and requires domain expertise. It is based on non-numerical data like text, images, audio and video, natural languages, themes and people's points of view. The outcomes from various analytics is commonly categorised into four types (Fig. 25.5).

Descriptive analytics are used more at the data mining stage and is conducted to infer on the health of data through visualization of patterns.

Diagnostic analytics is more investigative as it is used to detect the root cause of data patterns or data anomalies. This type of analytics detects patterns which deviate from expected trends and is used by businesses mostly to check the health of their business operations.

Predictive analytics attempts to portray future scenarios from trends so that it helps one to be prepared to face them ahead of time. This involves the use of Machine Learning (ML). A good example is predicting climate from observed historical climate data. Predictive analysis helps us at mitigating extreme events like heat waves, storm surges and floods.

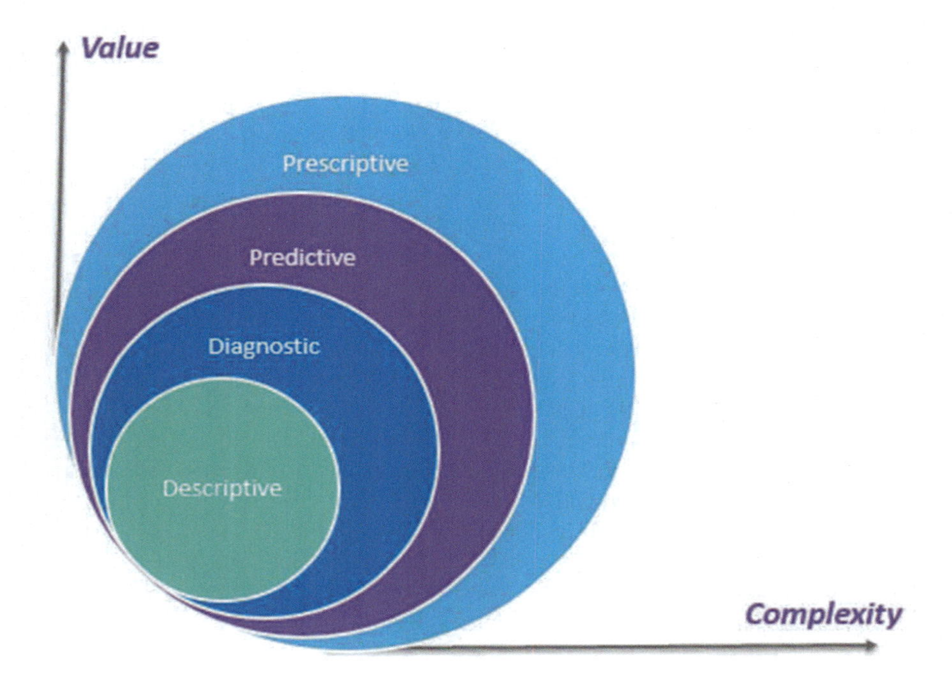

Fig. 25.5 Types of data analytics

Prescriptive analysis is a step which is further advanced. It helps in prescribing action plans based on a predictive scenario. This is more artificial intelligence (AI) driven where a decision based on a ML predictive outcome is implemented. Prescriptive analysis helps in developing self-driven cars where the AI drives the decision based on which a driverless car can navigate on roads avoiding accidents.

25.3.3 Data Visualization, Insights and Decision Support

Data visualization is the graphical representation of information or data and is used at all stages of the data science value-chain. Data visualization elements like charts, graphs and maps present a useful currency for understanding trends, outliers and patterns in data. This is the basis of business intelligence (BI) dashboards which most businesses and governments use today.

Dashboards are information visualization tools where trends and analyses are shown to portray key performance indicators (KPI) metrics to monitor operational health. Underneath these dashboards are several tiers of information technology (IT) tools. These connect to databases and perform several tasks in the data science value chain and produce information in form of tables, line charts, bar charts and gauges.

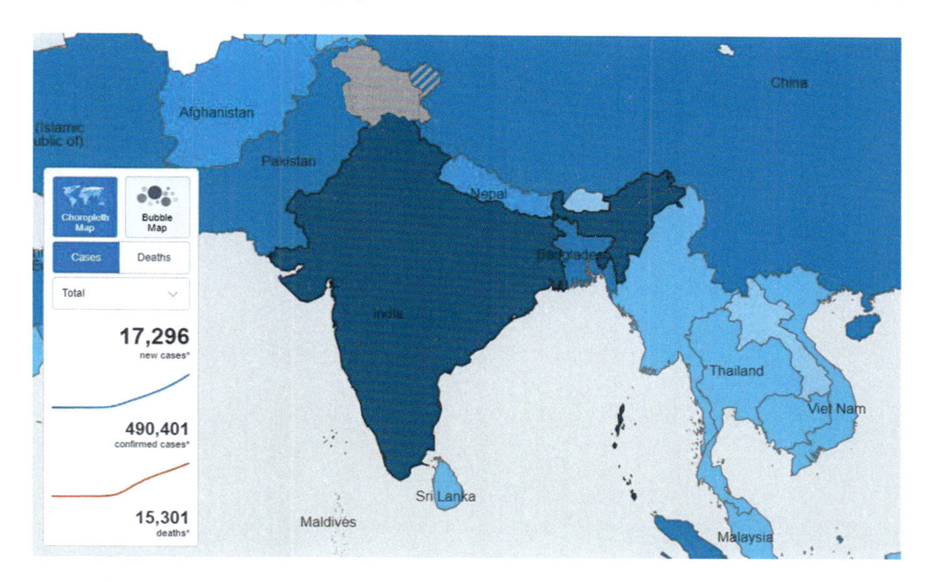

Fig. 25.6 World Health Organization—Covid dashboard (https://Covid19.Who.Int/Region/Searo/Country/In, retrieved on 27June 2020)

Dashboard is the most efficient way to track, monitor and analyze performance. Environmental and weather dashboards are realtime and support a variety of businesses. An example of geographical science application is the World Health Organization (WHO) COVID-19 dashboard that reflects the spread of the pandemic across the globe (Fig. 25.6).

25.4 Methods and Applications

The data science value chain is a pathway where appropriate methods are deployed for geographical applications. Algorithms are aplenty, but their appropriateness is justified by use cases. Below are the groups of methods that are mostly applied for geographical inferences.

25.4.1 Pattern Recognition

Finding patterns in data is the first step and regression analysis is a very good means to identify patterns. A group of random variables are studied in pairs to detect patterns between them. Regression analysis is based on several mathematical formula.

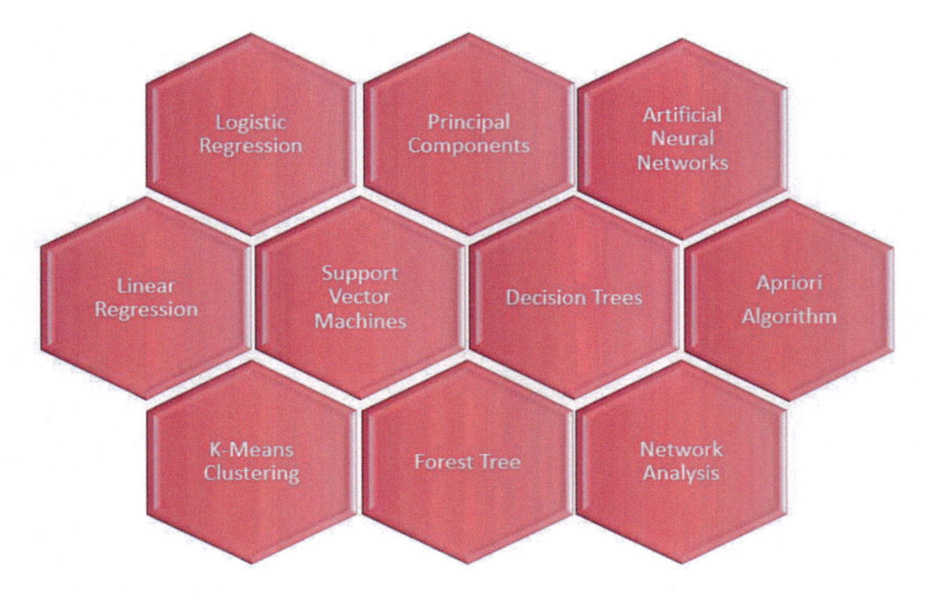

Fig. 25.7 Pattern recognition and classification methods

The simplest formula is Linear regression, in which the relationship between two variables X and Y is denoted by a linear equation $Y = a + bX + c$, where a, b and c are constants. Other regression methods are polynomial (binomial, quadratic, etc.). Advanced regression methods like Logistic regression are also used where appropriate. Regression methods were used to establish the relationship between carbon emissions and temperature rise. Regression is also used to reduce the number of variables used for analysis. Dependent variables from a set of random variables can be identified and the redundant ones can be discarded. Principal component analysis is an example where regression is used for not only identifying dependent variables but also to calculate new axioms (variables) for discriminating a thematic class from a set of random variables. Object recognition from satellite images uses principal component analysis. Other methods for pattern recognition and discrimination (classification) are K-means, Support Vector machines, Decision trees, Forest Tree classifiers, etc. Advanced methods, that use a combination of one or more methods to identify patterns or eliminate redundant data, are Apriori algorithms, Network analysis and Artificial Neural Networks (Fig. 25.7).

25.4.2 Correlation and Causation

While regression describes the numerical relationship of an independent variable with a dependent variable, it is seldom used for causation. Correlation is used for causation and is a statistical measure to determine the co-relationship between two random variables.

Causation from correlation is often implicit and depends on other observational evidences (or variables). To elaborate further, on a sunny day, melting of ice-cream correlates well with one having a sunburn. But sunburns do not cause the melting of the ice-cream and vice versa. In this case, a good correlation of ice-cream melting and sunburn instances does not essentially find the prime cause. The cause, in this case, lies in a third variable, which is sunny hot weather. The Granger causality test (Granger 1969) and several other convergent cross mapping methods have been proposed to test causality inferences from correlation.

25.4.3 Prediction

Predictive modeling is useful for forecasting. It is a process that uses models for statistical inferences on data to predict outcomes. In businesses, sales and revenue forecasts are common. In geographical science, prediction is key to understand migration, population dynamics, growth, resource modeling and on anything that is linked to generate temporal trends peeking into futuristic scenarios.

Machine learning is often used based on statistical techniques to allow a computer to construct predictive models. This is why machine learning and predictive analysis are often used as synonyms today. A geospatial example of machine learning techniques on prediction are spaghetti models (Viega and Ferreira 2010), which are commonly used for predicting storm tracks (Fig. 25.8).

25.5 Social Media and Public Policy

Blogs and social networks have recently become a valuable resource for mining sentiments of people. People's sentiments on customer relationship management can be gathered through public opinion tracking and text filtering from social media platforms (Mostafa 2013). Geographical origin of users, who use these platforms, provide a spatial connation that can be linked to other spatial data for indexing. Social media text mining is now being increasingly used by businesses and governments alike to market their products and policies. In developing countries, where data levels are rudimentary, social media-based sentiment analysis is mostly being used by governments for civic engagement and policy processes (Oginni and Moitui 2015).

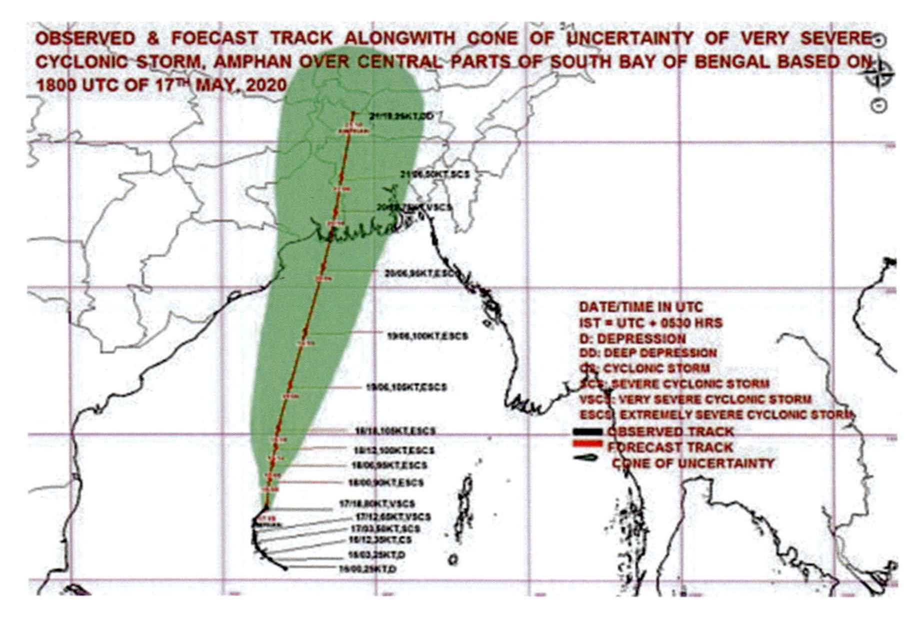

Fig. 25.8 Storm track of cyclone amphan (*source* Rediff News)

25.6 Future Directions

As we see an explosion of data in today's digital age, data analytics will continue to influence the way we infer and plan our present and future. The Fourth paradigm or Data Science, therefore, shall influence the way we live and operate. Geographical perspectives shall be critical to decision making, hence Geographers should embrace data science as an essential part of their educational curriculum and also adopt it in their research. This shall support insightful decisions and shall render the discipline more effective and acceptable in today's society. There is no doubt that data, spatial data and spatial big data shall be the basis of scientific discourse for decades to come.

References

Boon M, van Baalen SJ (2019) Epistemology for interdisciplinary research: Shifting philosophical paradigms of science. Eur J Philos Sci 9(1):1–28. https://doi.org/10.1007/s13194-018-0242-4

Dempsey C (2012) Where is the Phrase "80% of Data is Geographic" From? http://www.gislounge.com/80-percent-data-is-geographic/ (retrieved 26 June 2020)

Eom HB, Lee SM (1990) Decision support systems applications research: A bibliography (1971–1988). Eur J Oper Res 46(3):333–342. https://doi.org/10.1016/0377-2217(90)90008-Y

Granger C (1969) Investigating causal relations by econometric models and cross-spectral methods. Econometrica 37(3):424–438. https://doi.org/10.2307/1912791

Huisman O, Rolf A de By (eds) (2001) Principles of geographic information systems, ITC educational textbook series;1, Netherlands.
Idreos S, Papaemmanouil O, Chaudhuri S (2015). Overview of data exploration techniques. https://doi.org/10.1145/2723372.2731084
Koutsopoulos K (2011) "Changing paradigms of geography."
Li S, Dragicevic S, Castro FA, Sester M, Winter S, Coltekin A, Cheng T (2016) Geospatial big data handling theory and methods: A review and research challenges. ISPRS J Photogramm Remote Sens 115:119–133. https://doi.org/10.1016/j.isprsjprs.2015.10.012
Morcos J, Abedjan Z, Ilyas IF, Ouzzani M, Papotti P, Stonebraker M (2015) DataXFormer: An interactive data transformation tool. Paper presented at the 2015 ACM SIGMOD International conference on management of data, pp 883–888. doi:https://doi.org/10.1145/2723372.2735366
Mostafa MM (2013) More than words: Social networks' text mining for consumer brand sentiments. Expert Syst Appl 40(10):4241–4251. https://doi.org/10.1016/j.eswa.2013.01.019
NRC, Rediscovering Geography Committee. (1997) Rediscovering geography: New relevance for science and society. National Academy Press, Washington, D.C
Oginni SO, Moitui JN (2015) Social media and public policy process in africa: Enhanced policy process in digital age. Consilience 14:158–172. https://doi.org/10.7916/D85Q4VS0
Rowley J (2007) The wisdom hierarchy: representations of the DIKW hierarchy. J Inf Commun Sci. 33(2):163–180. https://doi.org/10.1177/0165551506070706
Veiga GM, Ferreira DR (2010) Understanding spaghetti models with sequence clustering for ProM. 43:92–103. https://doi.org/10.1007/978-3-642-12186-9_10
VoPham T, Hart JE, Laden F, Chiang Y (2018) Emerging trends in geospatial artificial intelligence (geoAI): Potential applications for environmental epidemiology. Environ Health 17(1):40–46. https://doi.org/10.1186/s12940-018-0386-x

Chapter 26
Establishing Relationships of Cellular Communication Coverage Provided by Governmental and Non-governmental Companies as a Function of Digital Elevation, Population Density, and Transport Infrastructure in Jodhpur District, Rajasthan

Aswathy Puthukkulam, Sanjay Gaur, T. R. Vinod, and Anand Plappally

Abstract With recent unplanned development and steep human population density increase in Jodhpur District during 2010–2019, spectral congestion can be an impending problem. The article emphasis lies in the analysis of coverage calculated based on Okamura–Hata model with respect to several distinct parameters. Jodhpur city, lying on the Vindhyan porous plateau (with high water management potential), has the highest density of mobile towers in the district. Mobile towers installed by government communication companies are mostly across the rustic champaign. These are distributed with relatively uniform density. Most of the district is characterized by negative normalized difference vegetation indices and low slopes confirming desert (Thar Desert) climate. Therefore, population density, railway routes, road infrastructures and townships are major parameters, which define cellular tower installations. Governmental company-based tower installations account for less than 20% of the non-governmental cellular tower installations. Coverage of the governmental towers is much higher than those of non-government cellular tower installations in terms of land surface area. WebApp Builder in ArcGIS is utilized to present this scenario which hereat provides an opportunity to perform design, development and planning of future cellular tower installations.

Keywords Desert · Antenna · Coverage · Wireless · App · Tower

A. Puthukkulam (✉) · S. Gaur
Jodhpur Institute of Engineering and Technology, Rajasthan Technical University, Kota, Rajasthan, India

T. R. Vinod
Center for Environment and Development, Thiruvanathapuram, Kerala, India

A. Plappally
IIT Jodhpur, Jodhpur, Rajasthan, India

N. C. Jana et al. (eds.), *Livelihood Enhancement Through Agriculture, Tourism and Health*, Advances in Geographical and Environmental Sciences,
https://doi.org/10.1007/978-981-16-7310-8_26

26.1 Introduction

Choropleth of arid regions of Jodhpur District at the edges of Thar Desert in India becomes the baseline data over which mobile towers of the region gets overlaid for our study (Graser and Peterson 2016). These thematic maps will provide a platform to decipher the hidden relationships of the wireless infrastructure expanse in this region against numerical attributes defining the statistics of the region and its present development. In order to provide better clarity on multi-attribute effects on the relations bivariate choropleth methodology will be enumerated wherever required in this document (Graser and Peterson 2016).

This document distinctly discusses pastoral and urban development of mobile towers bringing in location-specific-related infrastructure development by government (Bharat Sanchar Nigam Ltd) as well as private companies. The distinction of the extent of coverage in rural and urban locations is also another area which is discussed in relations with their installation density, location as well as the height of the towers (King 2013).

Fibre optic cable is the future of communication even if technology is 3G, 4G, 5G, or 6G (King 2013; Flagship 6G 2019). Correlating imagery and sensing, with proper positioning (with high mobility and elevation) will open new interdisciplinary applications in future (Flagship 6G 2019).

The fibre optic cable will not be only at the base of the cell tower but at its height during this rise of the small cell spectrum technology (Flagship 6G 2019; Hamdy 2018; Benham 2012). Further to meet wireless traffic demand in urban areas the observation can be that operators (both government and private communication network companies) opt for cell tower densification to increase network capacity (Hamdy 2018). Therefore, local operators will depend on height of wireless tower, and small cell architecture, which may play an important role in future (Flagship 6G 2019). Therefore, the implication of the work in this paper can be on the design of enhanced signal integrity, capacity improvement, footprint and lower energy expenditure.

Geo-spatial modelling of electronic devices in a communication network and associating rules with the device features using commercial data models, practical consumer behaviour can be captured in these devices (Benham 2012; McGregor 2016). For example, a fibre optic cable equipment can be created that would prevent it connect to a Cu splice thus improving design performance (ArcUser 2001). This means that location is getting linked to information of a location. Such a thought also opens up a large number of subjects, which look to geographical information systems (GIS) for assistance and support. For example, the anthropological alterations of the Earth's surface change the way the land was used (Dickenson 1995).

Land management applications require local soil properties and information of its character (Zhu 2001). For example, desert soils across the world showcase soil erosion due to wind, drought, sparse vegetation and anthropogenic activities such as mining (Lambin and Geist 2001). The change is also closely related to the ecosystem services, agricultural character, the climate, location-specific biodiversity, ecology and hydrology (Amuti and Luo 2014). Therefore, land cover and use information are required for spatial planning, for example, from campestral to urban planning for region-specific development (Eiden 2018).

Geographical imagery devices do not record these natural and anthropogenic land activities directly (Anderson et al. 1983). The sensing devices collect a response based on characteristics such as forest cover or anthropogenic construction on the land surface. The analysis of these images uses correlation between points, textures, color contrast, two dimensional shapes, latitude/longitudinal, frequency of light waves and time-based rotations to extract information about the type of land utility (Anderson et al. 1983).

Soils forms a major raster parameter or attribute while remotely studying desert or arid ecosystems (Hill 1994). The optical properties of soils are biased and dependent upon the overlay of soil mineral material composition spectra, shadows, moisture levels and surface types (Leblon et al. 1996; Vodacek 2018). For example, Landsat 4 images of Pali, Rajasthan show highest reflectance in salty soils with encrustation than those compared to sodic soils, salty soils without encrustations and lands which became saline due to salty irrigation water (Kalra and Joshi 1994).

The authors have used Landsat images from Turkey acquired from two distinct ages separated by more than a quarter century. Reis further studied land use land cover changes by using pixel to pixel comparison of the imagery. Further studies according to the topographic structure confirmed that the land cover changes are maximum at low slope values (Reis 2008).

Shalaby and Tateishi 2007 performed similar studies using Landsat images acquired in 1987 and 2001 for the north-western coast of Egypt. They were worried that encroachment of agricultural lands may influence land degradation and desertification (Shalaby and Tateishi 2007). They said that grassland converted to cropland can be considered a positive change. The wind and water erosion during the time was responsible for loss of the fertile top layer in different parts of north western Egypt and will influence desertification (Shalaby and Tateishi 2007).

For starting any GIS-based project, it is important to identify the location (coordinates) and identify the problems. Corresponding to the acquired information and analysis, maps can be drawn which may depict natural surface resources, underground resources or manmade structures (Bădescu et al. 2009). In this article, we are plotting manmade structures (mobile towers) overlaid on natural and underground resources.

Energy reflected from the surface cannot independently characterize the green cover, therefore multiple spectral information taken together should provide the vegetation indices. Vegetation Indices (VIs) thus calculated corresponds proportionally to the behaviour and form of vegetation. Indices showcase enhanced reflected signal from measured spectral responses such as red (0.6–0.7 m) and near infrared (0.7–1.1 m) combination. It describes the relative density of vegetation and its health for each pixel. Thus, it is a dimensionless number related to the amount of vegetation in a given remotely captured image pixel.

There are several variations and types of vegetation indices, which were propounded by researchers across the world from time to time (Jensen 2007).

They are Simple ratio (SR), Normalized difference vegetation index (NDVI—in this case), Kauth-Thomas Transformation, Normalized Difference Moisture or Water Index(NDMI OR NDWI), Perpendicular vegetation Index (PVI), Leaf Relative Water Content Index (LWCI), Soil adjusted vegetation Index (SAVI), Atmospherically Resistant Vegetative Index (ARVI), Soil and Atmospherically Resistant Vegetative Index (SARVI), Enhanced Vegetation Index (EVI), New Vegetation Index (NVI), Aerosol Free Vegetation Index (AFVI), Triangular Vegetation Index(TVI), Reduced Simple Ratio (RSR), Ratio TCARI, Normalized Difference Built-up Index, Red-Edge Position, and Visible Atmospherically Resistant Index (VARI) (Jensen 2007). In the present case, Normalized Difference Vegetation Index (NDVI) is used. Rouse et al. (1974) introduced normalized Difference Vegetation Index (NDVI) (Rouse et al. 1974).

Mohamed et al. predicts that Okumura Hata model is suitable for low slope urban environment by adding correction factor to it (Mohamed 2018). Therefore, this article will also discretize between the rural and urban environment and installations there in.

Girma et al. confirms that Hata model and its extensions were found to be efficient in modelling low aspect regions of northern India (Solomon et al. 2018; Chaurasia 2006). Therefore, in this article coding will be performed using Okumura Hata extension for the calculation of coverage.

The Okumura Hata model is suitable for transmitter antenna height 30–200 m, 1–20 km distance from the transmitter and frequency range of 100–1500 MHz. In this article, Okumura Hata model is considered since all these conditions are satisfied for our case study (ITU 2002; Popoola and Oseni 2014). Correlation factor has been added to it according to the type of terrain, whether small city or large city, urban or rural (Pahlavan and Krishnamurthy 2013).

The study aims at developing inter-relationships with coverage and cell densification corresponding to the different statistical attributes which defines the developmental indices of the Jodhpur District as well as its geological nature. The major objectives of the current work are.

- To expound the correlation between population density and tower density.
- To develop a web-based app to locate the nearly wireless communication towers.
- To effectively calculate the pastoral and urban coverage provided by the government company in Jodhpur District.

26.2 Methodology

26.2.1 Setting

The study is set in the Jodhpur District of Rajasthan, India. Rajasthan is situated in Western India sharing its borders with Pakistan. Thar Desert is shared by both India and Pakistan. Most of the Thar Desert lies in India. Rajasthan has state of Gujarat

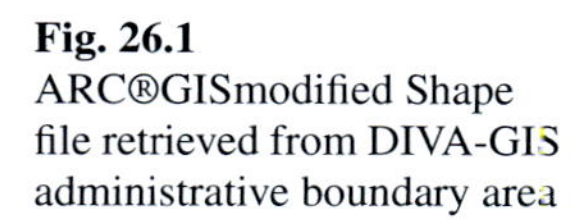

Fig. 26.1 ARC®GISmodified Shape file retrieved from DIVA-GIS administrative boundary area

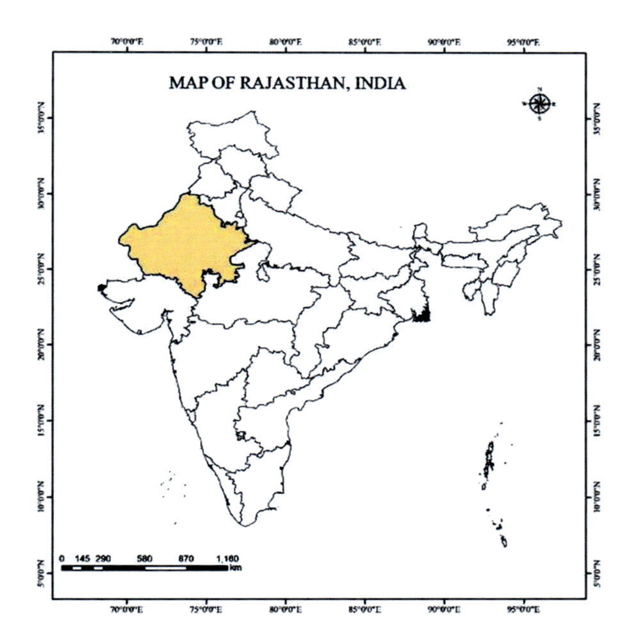

on its southern side, and Madhya Pradesh on its west. On the north, Rajasthan has Punjab and Haryana as neighbours. The shape file of Rajasthan state is depicted in Fig. 26.1.

ArcGIS Desktop package comprises several inbuilt applications, like ArcMap, ArcCatalog, ArcToolbox, ArcScene, ArcGlobe and ArcGIS Pro. Arcmap is an application used to create maps and perform analysis on geospatial data collected from various sources. It can handle multiple layers which can be managed by table of contents pane. Arc toolbox is the tools for analyzing and manipulating geospatial data.

A shapefile stores vector data. The data is in the form of either point, line or polygon. Roadways map is an example of shapefile with line feature. Using GIS software (ARC®GIS or QGIS) shapefile can be analyzed. Shapefile can be prepared in three different forms, namely with extensions.shp,.dbf and.shx. Here.shp file stores the point, line or polygon feature values,.dbf file will store the feature values in a database and.shx is an index file that combines the other two files together.

The shapefile of the study area has been prepared in Arcmap 10.2 by taking the Indian administrative boundary files from https://www.diva-gis.org/Data. After unzipping the file downloaded, the file has to be opened in ArcMap.

From the attribute table, the study area (Jodhpur District) has been selected. And using export data option, the shapefile layer of Jodhpur was created. For preparing the map, legend, title, grid, north arrow and scale has been added. Once the map is prepared, using export map option under the file menu, a jpeg file has been created. Jodhpur is defined at distinct latitude and longitude coordinates, respectively, forming a quadrilateral. These are 27° 37′ 59″ N, 71° 45′ 37″ E (top left corner), 27° 37′ 59″

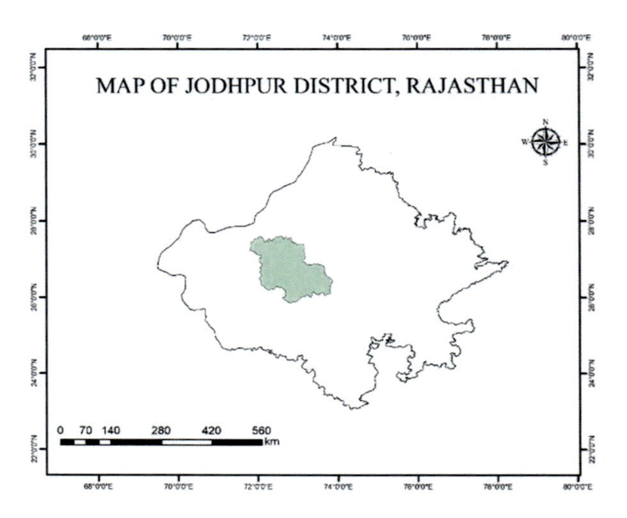

Fig. 26.2 ARC®GIS modified Shape file of Rajasthan with study area highlighted

N, 73° 56′10″E (top right corner), 25° 50′ 34″ N 73° 56′10″E (lower left corner) and 25° 50′ 34″ N 71° 45′ 37″ E (lower right corner), respectively.

Rajasthan as illustrated in Fig. 26.2 is characterized by the Vindhyan Era hills, and plateaus in 40% of its geography (Salunkhe et al. 2018). Further, it has loamy soils with semi-arid climates near the Thar Desert.

Jodhpur District as illustrated in Fig. 26.2 covers an area of 22,850 km^2. Further, 22,594.18 km^2 is categorized under pastoral area and 255.82 km^2 is said to be urban according to Census of India, 2011 (COI 2011). The district is bordered by Bikaner District on the north, Nagaur District on the east, Jaisalmer District on the west and Pali and Barmer districts in the south. Earlier Jodhpur District had seven blocks or tehsils Phalodi, Osian, Bhopalgarh, Jodhpur, Luni, Shergarh and Bilara but recent governments merged Luni and Jodhpur to generate a new Jodhpur block (COI 2011).

The district experiences arid or semi-arid climate with average annual precipitation in the range 21.6 to 46 cm. Almost 80% of this annual precipitation is received during the south-west monsoon. Thus, rainfall is experienced from first week of July to the middle of September. Several parts of the district lie in the Thar Desert. Therefore, the area experiences extreme summer and extreme winters. Maximum temperature gradients during the day tonight transit is experienced in the months of May and June. The yearly average of the temperature ranges from 26 to 28 °C. Aravalli remnants (Malani Igneous Suite), rocky plateaus or pediments, alluvial pileups, sandy plains with water filled depressions and sand dunes dot Jodhpur District.

Water filled porous depressions and sand dunes are found in the north-western and western parts of this district. The slope of Jodhpur District terrain is towards the west. The Luni River with its alluvial plains and its tributary Jojri present elaborate drainage networks in east and south-east locations of the district. The Vindhyan Sandstone and Limestone found in Jodhpur are porous is nature and considered effective with their water tapping capacities yielding in the range 5–25 cubic meter per hour (Salunkhe et al. 2018). Considering this geological Jodhpur City was established on the water harvesting porous Chonka-Daijar plateau (Agarwal and Narain 2008).

The satellite images are taken from USGS Earth explorer. The satellite images are acquired during the month of April 2019. The date of acquisition of images is 1, 9, 16 and 23 April 2019.

The satellite images (Geotiff format) of study area during the study period were downloaded. For the study area selected five images were mosaiced using Arctool box. From the downloaded zipped folder band 5 and band 4 was selected and mosaiced using Mosaic to new raster tool in Data Management toolset. NDVI calculation was performed using raster calculator option in MapAlgebra.

For Landsat 8 image,

$$\mathrm{NDVI} = (\mathrm{Band5} - \mathrm{Band4})/(\mathrm{Band5} + \mathrm{Band4}) \quad (26.1)$$

NDVI values range from -1 to $+\,1$. For barren, rocky or no vegetation regions, the values will be negative. Region with sparse vegetation will be having values around positive 0.1. Values above this indicate area is having average or high-density vegetation. Figure 26.3 shows the NDVI map of Jodhpur District.

According to the result of NDVI, the maximum value for Jodhpur is around 0.5 and minimum value is -0.3. The negative value indicates that Jodhpur District is having barren, rocky or sandy areas. The positive value indicates moderate vegetation in this area. It can be seen that dominant color in the map is reddish/ yellow indicating the area is not having high vegetation.

From data of Census handbook 2011, population densities of 6 tehsils have been calculated. After finding population density, these values have been mapped to shapefile of Jodhpur district using attribute table option of ArcGIS. Under properties

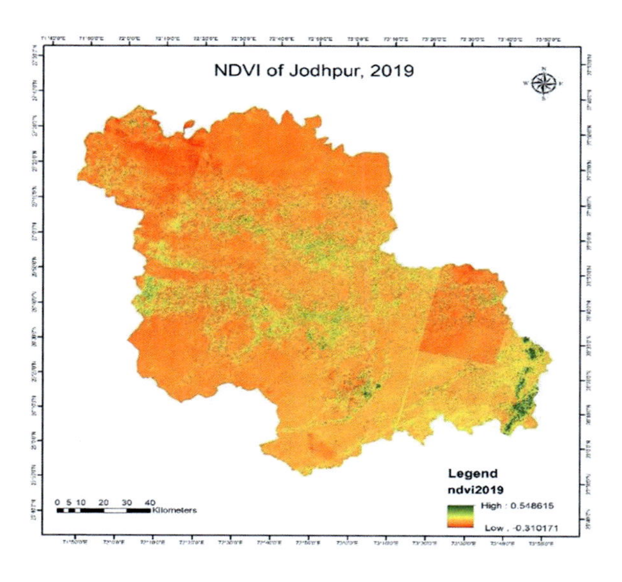

Fig. 26.3 NDVI Map of Jodhpur, Rajasthan generated using Arcmap

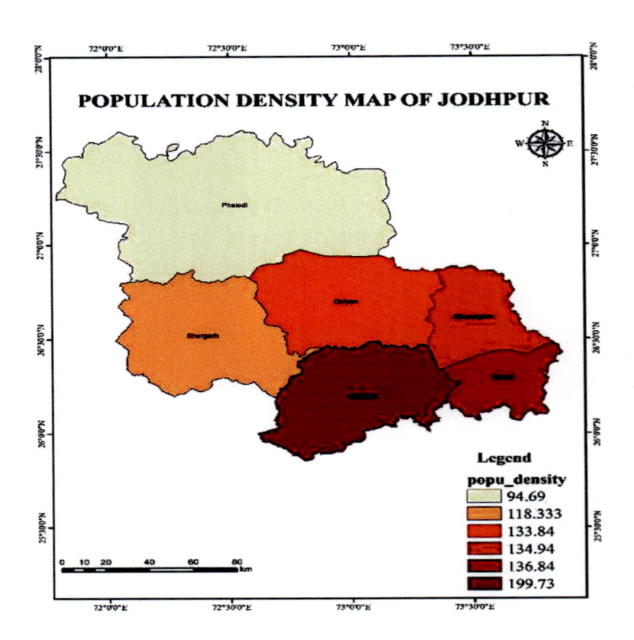

Fig. 26.4 The choropleth of population density of Jodhpur district (Census 2011)

window, using symbology option color coding has been given to population density mapping.

From Fig. 26.4 it is evident that Jodhpur city is having the highest population density whereas Phalodi is having the least population density.

In Jodhpur District Bharat Sanchar Nigam Limited (BSNL) is having 341towers.The data received from BSNL elaborates the site name, their respective location and antenna heights for a set sample of the data set of 341 towers used in the analysis and calculations in this article.

From Fig. 26.5, it is clear that the rural as well as urban antenna heights showcase much deviation from the actual range with some prominent outliers towards the extreme values (Cells 2018; Molisch 2011; Mathur 2013; Haan 1977). This observation hints at importance to be provided to the tails of the distribution of antenna heights in a region.

For the present study, 2249 towers operated by private communication companies were plotted in Fig. 26.6 and spatially referred (Cells 2018). This data has been retrieved from archives of radiocells.org.

Government agency (the BSNL) has spread the communication towers to the rural areas across the Jodhpur District to provide services. The major duty of the government towards providing services to its people irrespective of location is clear from Figs. 26.6 and 26.7 with intensification of antenna or wireless tower installation near to the city as well as almost uniform spacing of wireless towers across pastoral or rural Jodhpur. The private communication players are cautiously expanding to the specific quarters keeping track of scope of development in and around Jodhpur City. This behaviour is clear from Figs. 26.6 and 26.7.

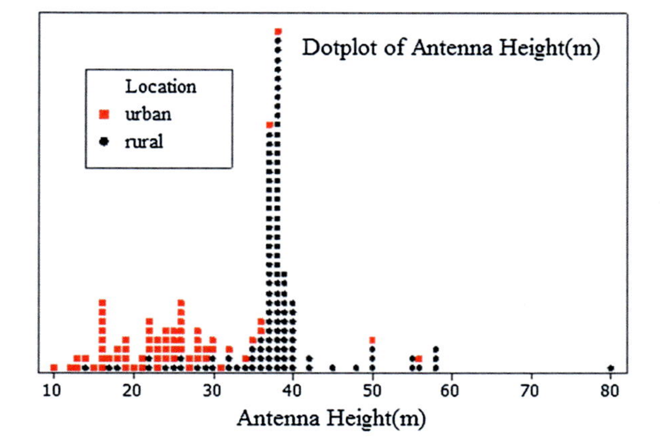

Fig. 26.5 The dot plot for the heights of the campestral and urban towers installed by BSNL in the District of Jodhpur

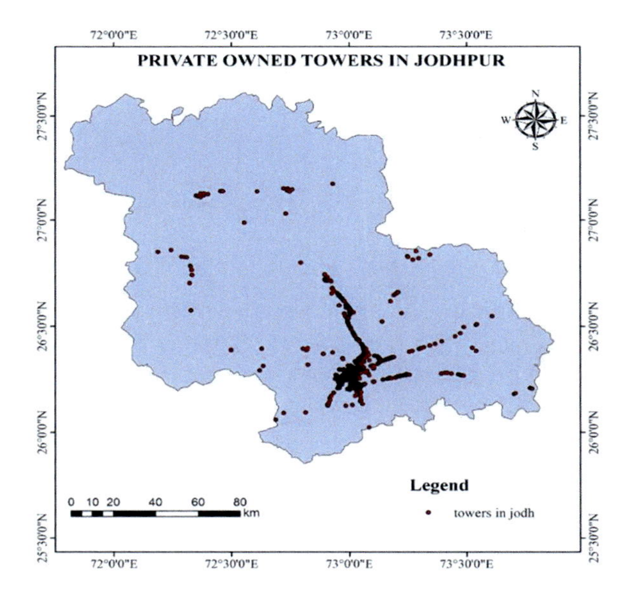

Fig. 26.6 Private company tower location overlaid on shapefile of Jodhpur, Rajasthan

For this study the data set corresponding to the government antenna is considered as a case and analyzed. ArcGIS Online is a web application published by ESRI for sharing geographic information. Using ArcGIS online, creating story maps, web apps, native apps and perform/share surveys becomes much easier. Using ArcGIS online in this study a web app has been created showcasing the location and details of mobile towers in Jodhpur District.

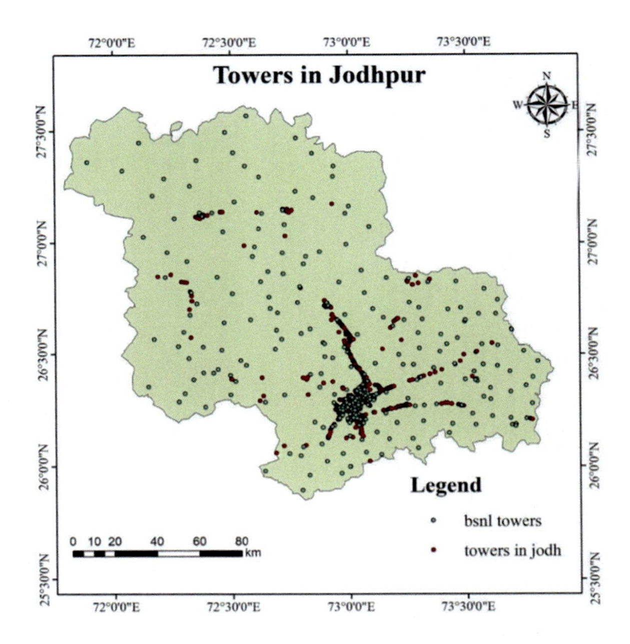

Fig. 26.7 Overlay of government (BSNL) and private wireless mobile towers on shape file of Jodhpur, Rajasthan

26.2.2 Okumura Hata Equations

A mobile network consists of base station (BS) with antennas mounted on masts, poles or towers. In urban areas, antennas are mounted on top of buildings or any built-up areas. For good signal reception, BS should be close to the users. In rural areas the number of BS can be less since there are less scatters or obstruction between BS and the mobile user. Since the number of built up and users are more in urban areas, the number of BS required will be high.

Mobile Station (MS) is the mobile phone of the user. Every cell has its own Base Transceiver Station (BTS) at its centre. When a person makes a call, radio frequency signals are sent and received by the mobile phone set. The antenna of the nearest base station picks up this signal and is sent to the Base Station Controller Centre (BSC). BSC controls the working of all base stations. From the BSC the signal is transferred to Mobile Switching Centre (MSC). The exchange connects the call to the receiver. MSC s is different for different areas.

A cell can be considered like a basic geographic unit in mobile communication. In cellular network, area is covered in regular shaped cells. The cells can be of circular, hexagonal, square or some other shapes. Since hexagonal shape satisfies the conditions like non-overlap and maximum area it is preferred over other geometrical shapes. Coverage range of a base station is also called a cell.

Determining Coverage Radius

$$RSSI = EIRP - Pathloss \tag{26.2}$$

where RSSI is Received Signal Strength Intensity,

EIRP is Effective Isotropic Radiated Power.

The Okumura-Hata model is a model for predicting signal propagation. It was initially designed for city areas by Okumura. Hata's model gives values of pathloss at given distances between a base station and a mobile device. It depends on various factors such as transmit frequency, antenna height and the type of location (Mathur 2013).

The model can be expressed as

$$L_{50}(db) = L_f + A_{mu}\big(f,d\big) - G(h_{te}) - G(h_{re}) - G_{Area} \quad (26.3)$$

where L_{50} = 50th percentile value of path loss between transmitter and receiver in dB

L_F	is free space propagation loss in dB
A_{mu}	is the median attenuation relative to free space
$G(h_{te})$	is mobile antenna height gain factor in dB
$G(h_{re})$	is transmitter antenna height gain factor in dB
G_{Area}	is gain due to the type of environment

The pathloss is written as

$$PL = A + Blog(d) + C \quad (26.4)$$

where A, B and C are factors that depend on transmission frequency and antenna height (Molisch 2011).

$$A = 69.55 + 26.16log(f_c) - 13.82log(h_b) - a(h_m) \quad (26.5)$$

$$B = 44.9 - 6.55log(h_b) \quad (26.6)$$

where f_c is given in MHz and d in km.

The factor $a(h_m)$ and the factor C depend on the environment as shown below:

For small and medium-size cities,

$$a(h_m) = (1.1log(f_c) - 0.7)h_m - (1.56log(f_c) - 0.8) \quad (26.7)$$

$$C = 0 \quad (26.8)$$

For metropolitan areas,

$$a(h_m) = 8.29\big(log(1.54h_m)^2\big) - 1.1\text{forf} \leq 200\,\text{MHz} \quad (26.9)$$

$$3.2(log(11.75h_m))^2 - 4.97 \text{ for f} \geq 400\,\text{MHz} \quad (26.10)$$

$$C = 0 \quad (26.11)$$

For suburban environments

$$C = -2[log(f_c/28)]^2 - 5.4 \quad (26.12)$$

For rural area

$$C = -4.78[log(f_c)]^2 + 18.33log(f_c) - 40.98 \quad (26.13)$$

The function $a(h_m)$ in suburban and rural areas is the same as for urban (small- and medium-sized cities) areas (Pahlavan and Krishnamurthy 2013).

26.3 Results and Discussions

Density of the towers in Jodhpur block of Jodhpur District is highest compared to the other blocks as shown in Fig. 26.8. Jodhpur City is located at Jodhpur block in the district of Jodhpur. Due to the availability of water in the Chonka-Daijar plateau on which the city is located serves as a hub for residential and industrial development (Agarwal and Narain 2008). From the Fig. 26.9, moderately efficient river basins covering Osian, Bilara, Bhopalgarh and Jodhpur showcase high intensity of development of communication network structures to support development services in these locations. This supports the fact that Luni River cuts the rocky limestone terrain and its tributary, Jojri which presently (2001–2019) carries municipal and industrial waste (textile, wooden artefacts and mining units in Jodhpur suburbs) drains into the eastern and south-eastern parts of the district (Haan 1977).

Thus, population density is also high due to these unstructured and unplanned industrial developments which increased the labour force and their requirement of wireless connectivity. This also paves the way for the development of services along the major rail route as illustrated in Fig. 26.10 joining Jodhpur City with other campestral parts of Jodhpur District or other neighbouring districts.

It is important to note that a web-app is developed to provide people with a hands-on access to find the nearest tower neighbouring the roads they travel in Jodhpur District.

Using the data collected from radiocells.org and BSNL, Jodhpur, a web map and web application has been developed. For this ArcGIS online products have been utilized. It was under the secure international 4 weeks online training program by ESRI (Shramek 2019).

ArcGIS online helps to make interactive maps where all the information will be stored in cloud. It helps to make maps, analyze the data and share the data with people

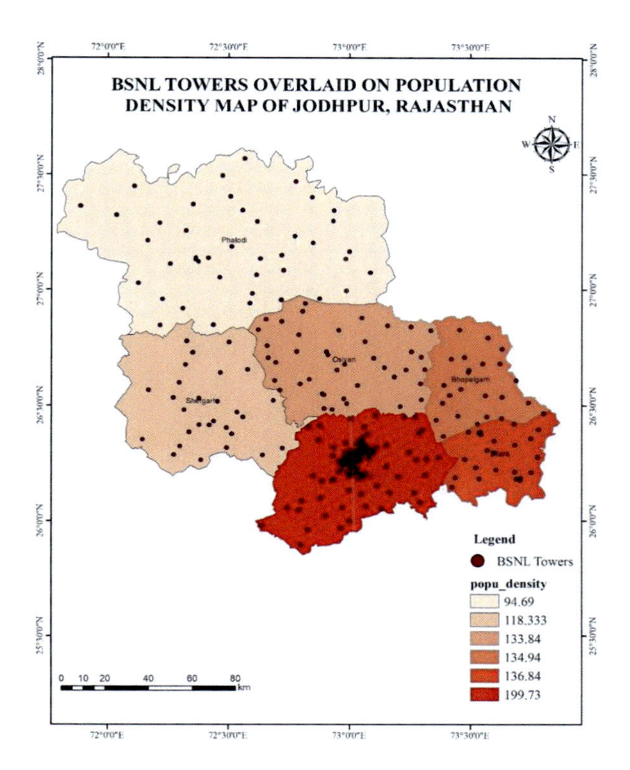

Fig. 26.8 Government mobile tower locations overlaid on population density choropleth of Jodhpur District, Rajasthan

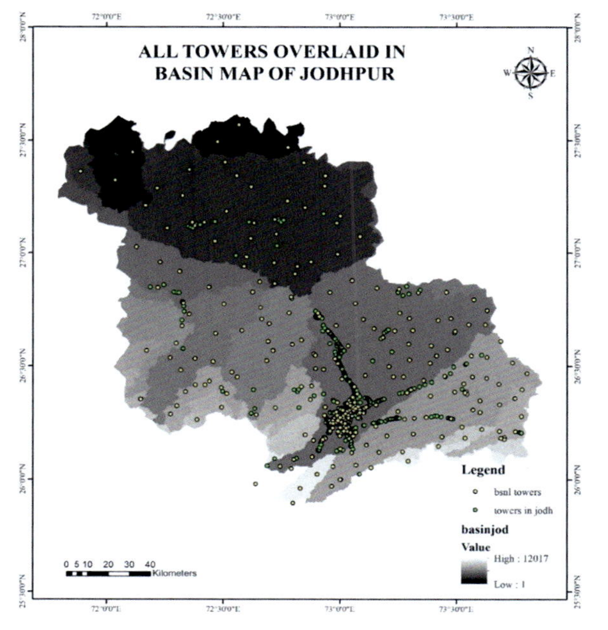

Fig. 26.9 Government- and Private-owned company mobile towers overlaid on river basin expanse map of Jodhpur district, Rajasthan

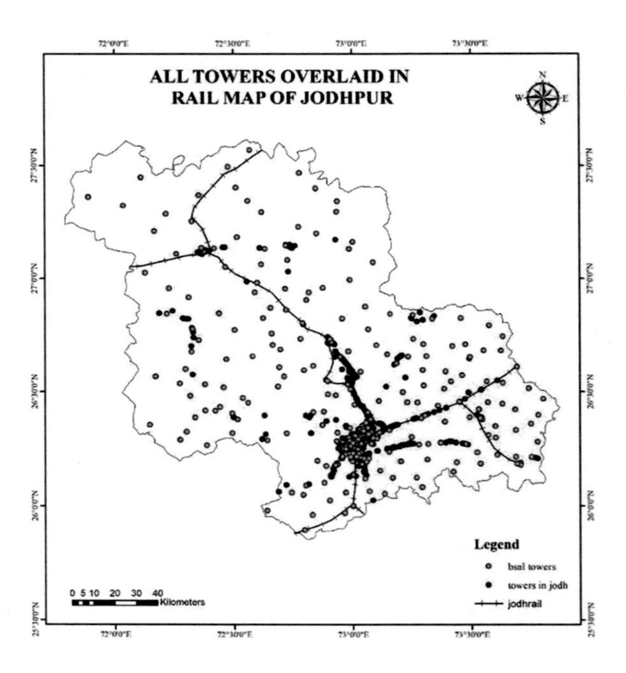

Fig. 26.10 Overlay of Mobile towers on the Rail Map of Jodhpur, Rajasthan

around the world. For making web map, story map or web app, several steps has to be followed. The first and foremost is having a login credentials to access ArcGIS online products.

For making a web map, Map option in ArcGIS.com is used. The data collected can be mapped in map Viewer by using option "Add from File" or by dragging the file into the map browser. The collected data should be in Comma Separated Value format (csv). After that required base map can be selected. In this work Open Street Map base map layer has been opted. Under the content option in the map browser, the points can be mapped in different format and style. The attribute table of the.csv file can be viewed and modified. Under the configure pop up option the author can manage how the pop will be displayed as it is shown in Fig. 26.11 (Shramek 2019).

The map should be saved after setting the pop-up option, base map layers and setting the legend of the map. The map can be shared in private or public mode.

Figure 26.12 shows the mobile tower locations of Jodhpur District. The red dots indicate the BSNL towers in the area and grey dot indicate other towers.

Before making a web app, web layers and web map has to be made. From the Mapviewer browser, using the share option clicks CreateNewApp (Web AppBuilder). WebApp Builder for ArcGIS window will open. Under theme option, desired theme for your app can be selected as in Fig. 26.13.

The widgets required for your web app can be chosen under the widget options. The widgets chosen for the app created here are home, search, layers, base map gallery and legend. Figure 26.14 displays the widget option in Web app builder for creating a web app.

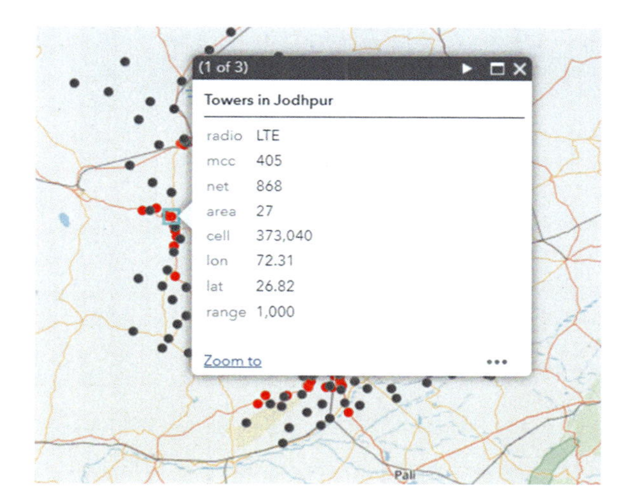

Fig. 26.11 Map showing the pop up for each location

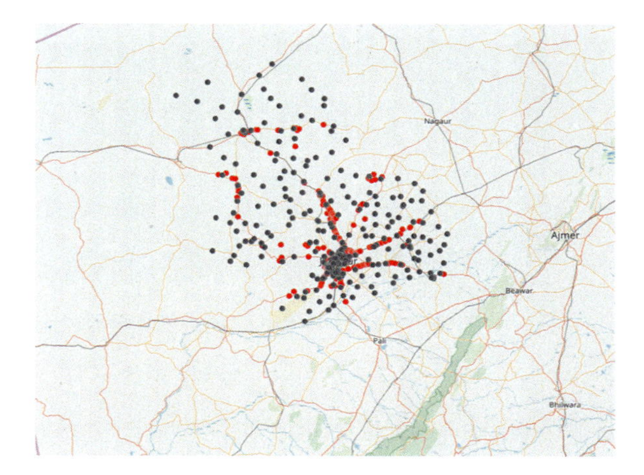

Fig. 26.12 Web map of mobile towers in Jodhpur with open street map base map layer

Using web app builder, a web app has been made with details of mobile towers like latitude, longitude, cell site and antenna height. Figure 26.15 is a web app made using ArcGIS online with pop up and desired widget (Shramek 2019).

Web app made is mobile friendly and gives information about towers in Jodhpur District, both government owned and privately owned companies. It is having various widgets like home screen, search option to locate any area, layer option to view multiple layers and legend (Shramek 2019).

Using Okumura Hata model, coverage distance d for 341 towers was calculated using Matlab code. After finding the coverage of 341 towers, the average coverage of rural and urban towers has been calculated. After finding the coverage distance for rural and urban fields with the help of code, it can be plotted on an Arcmap using

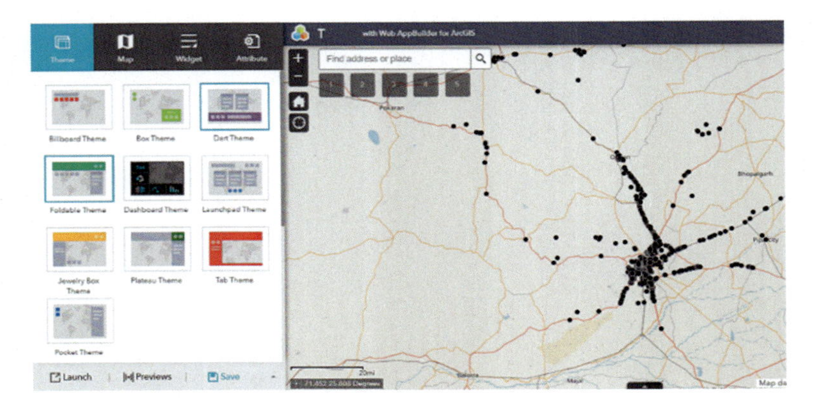

Fig. 26.13 Themes of Web App builder for ArcGIS (Shramek 2019)

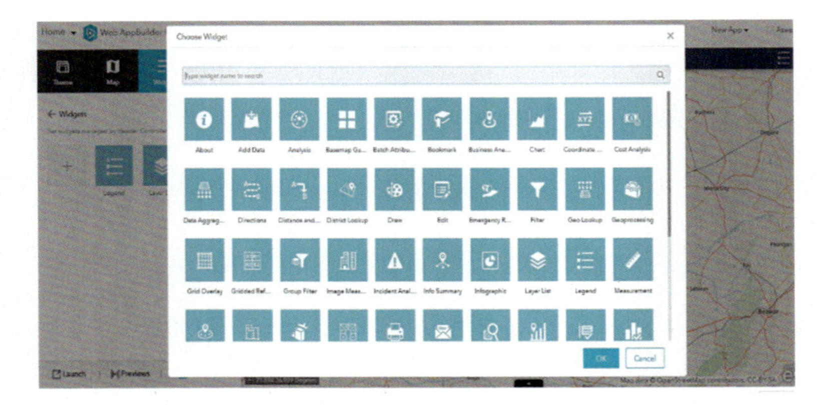

Fig. 26.14 Widget option for Web App builder for ArcGIS (Shramek 2019)

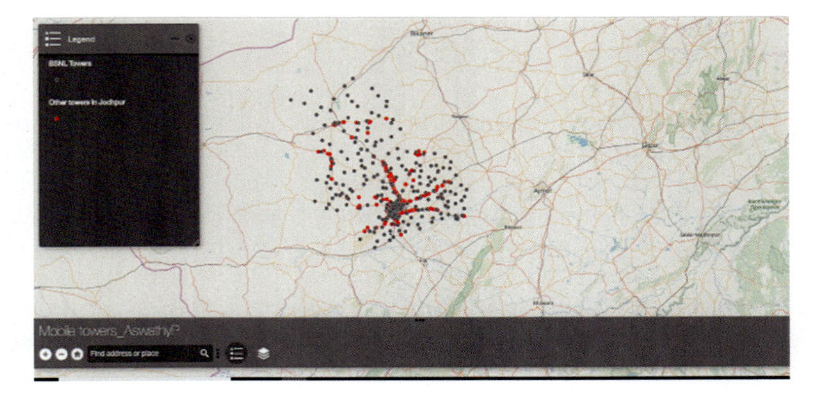

Fig. 26.15 Web app showing mobile tower location of Jodhpur and its details

Table 26.1 Average coverage distance of towers in rural and urban area, Bsnl, Jodhpur

S. No	Area classification	Average coverage (km, radial)
1	Urban	1.02388
2	Pastoral/Rural	7.20118

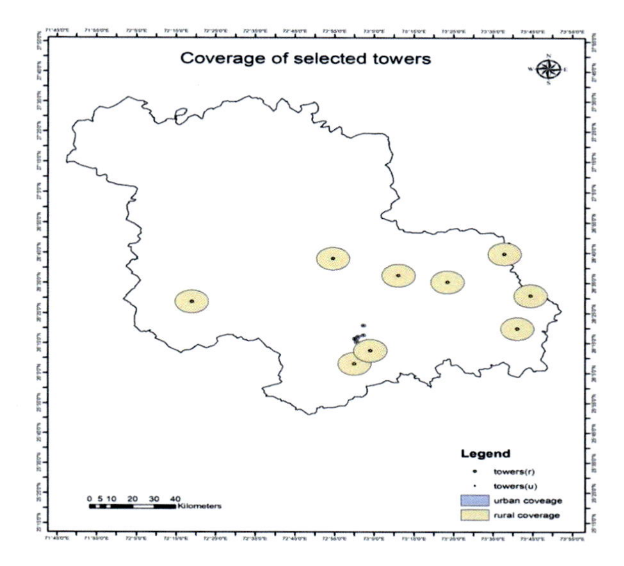

Fig. 26.16 Coverage of a sample 15 selected towers of BSNL using average coverage values

buffer tool in ArcGIS. The result of average coverage has been displayed in Table 26.1.

The calculated average coverage is used in Fig. 26.16 to showcase visual difference between the extent of coverage provided by urban and rural BSNL towers, respectively. It is important to note that pastoral and urban coverages represented for all the 341 BSNL towers in Jodhpur District are shown in Fig. 26.17.

Government agency (the BSNL) has spread the communication towers to the rural areas across the Jodhpur District to provide services. The major duty of the government towards providing services to its people irrespective of location is clear from Fig. 26.17. It is a common knowledge that wireless coverage improves with increase of antenna height or the tower height. The confirmation of this thought is illustrated with Fig. 26.18.

From Figs. 26.19 and 26.20, the urban towers are found to have very distinct and small coverage radii when compared with taller rural towers with large coverage. This also signifies proliferation of government towers across the rural areas. It also points towards myriads of services that may differ with coverage differences with that of the urban towers.

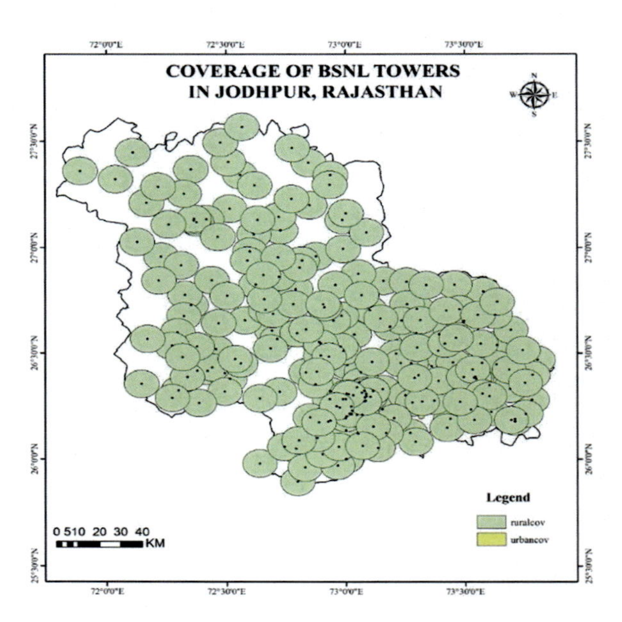

Fig. 26.17 The overall average rural coverage of towers mapped across Jodhpur, Rajasthan

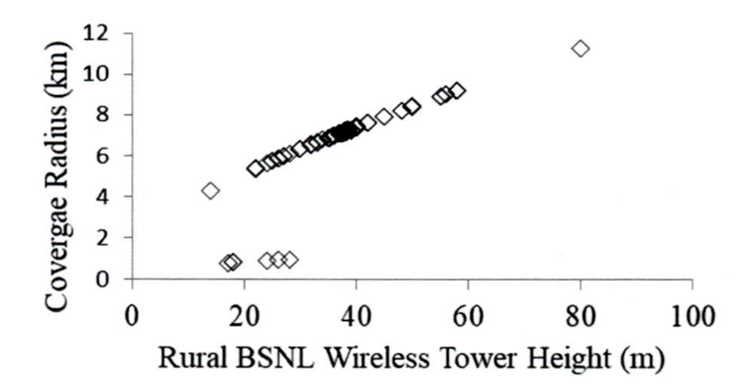

Fig. 26.18 The coverage radius as a function of tower heights from campestral Jodhpur District

26.4 Conclusion

Foreseeing the future growth of small cell technologies and excessive use of services such as proliferation of Internet of things, artificial intelligence and smart management systems for large administrative bodies as well as industrial operations, planning of communication networks become pertinent. In this regard, present assessment of services available through installed towers is important. This enables to understand the development of services across arid environments of Jodhpur District. The major conclusions of study are as follows:

- With increase in density of population it is observed that there is an increase in density of tower installations. For example, Jodhpur block having the maximum density of population is dotted with maximum number of towers.

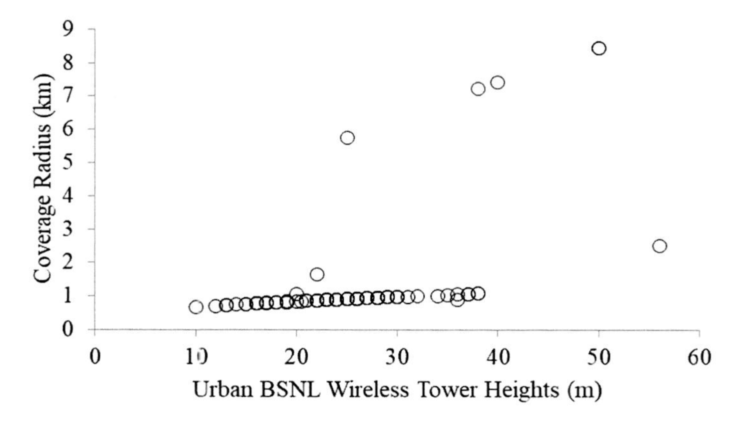

Fig. 26.19 The coverage radius as a function of tower heights from urban Jodhpur District

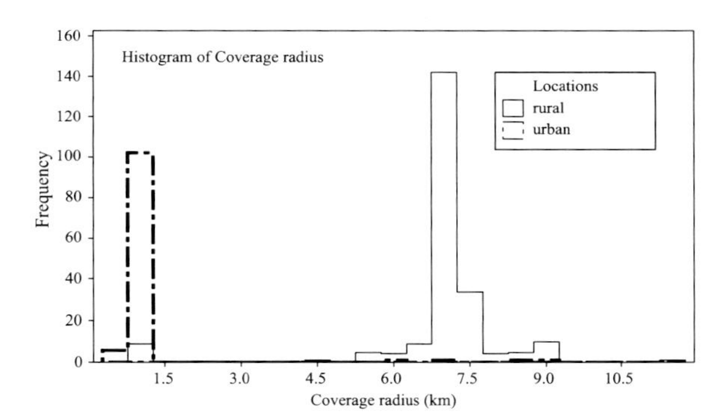

Fig. 26.20 The histogram showing the coverage radius trends of the pastoral and urban wireless communication towers (BSNL) across Jodhpur District

- A web app has been developed that provides a hands-on visualization of tower locations with an interface as street map.
- The antenna heights in the urban areas are comparatively much smaller than those at campestral locations.
- Using the NDVI study, the landscape was classified as an arid landscape.
- Even though many private towers amounting to 2246 have been studied in this document, it is important to note that most of these towers are located at the urban locations as well as on the major transport routes connecting locations of development. The rural locations are almost devoid of private communication tower installations in Jodhpur District.
- The government communication companies ensure last mile connectivity to the remote rustic corners, but private firms cautiously expand to those locations (main roads, rail and populated areas) where they observe availability of large user density.
- The maximum density of communication towers is observed to be in locations with high groundwater tapping capacities in Jodhpur District.

Acknowledgements I would like to thank Mr. P.S. Prasood and other staffs of Center for Environment and Development, Trivandrum for encouraging throughout. I would like to thank the BSNL Jodhpur, Rajasthan for providing the information about its infrastructure and technical information of their installations.

References

Agarwal A, Narain S (2008) Dying wisdom: rise, fall and potential of India's traditional water harvesting systems. New Delhi

Amuti T, Luo G (2014) Analysis of land cover change and its driving forces. Solid earth-solid earth discuss 6:1907–1947

Anderson JR, Hardy EE, Roach JT, Witmer RE (1983) A land use and land cover classification system for use with remote sensor data. Washington

ArcUser (2001) GIS use in telecommunications growing, October

Bădescu G, Ovidiu S, Nicolae AB, HreniucNP, Iulius EK, Radulescu ATBR (2009) The efficient use of the GIS technology in creating strategies for regional development and environment protection. In: 2nd International conference on environmental and geological science and engineering. WSEAS.

Benham C (2012) A GIS based decision and suitability model: solving the tower location problem in support of electric power smart grid initiatives. Northwest Missouri State University

Cells R (2018) Radio Cells. https://radiocells.org/downloads/raw_data. Accessed 20 Aug 2012

Chaurasia VK (2006) Planning optimum location for wireless tower in GIS environment. IIT Rourkee

COI (2011) District census handbook, Jodhpur, Census of India (COI) 2011, Rajasthan, Series-09

Dickenson R (1995) Land processes in climate models. Remote Sens Environ 51:27–38

Eiden G (2018) Land-Cover and Land-Use mapping in land use, land cover and soil sciences. Encycl. Life Support Syst.

Flagship 6G (2019) 6G Flagship, Key drivers and research challenges for ubiquitous wireless intelligence. In: 6G research visions series. First 6G Wireless Summit, University of Oulu

Graser A, Peterson GN (2016) QGIS Map design. Locate Press

Haan CT (1977) Statistical methods in hydrology, 3rd edn. Iowa State University Press, Ames, Iowa

Hamdy MN (2018) Small cells big challenges a practical solutions guide. CommScope, Inc., WP-112815-EN (07/18)

Hill J (1994) Spectral properties of soils and the use of optical remote sensing systems for soil erosion mapping. Chemistry of aquatic systems: local and Global perspectives. Springer, Netherlands, Dordrecht, pp 497–526

ITU (2002) ITU publications: terrestrial land mobile radio wave propagation in the VHF/UHF bands. Radio communication Bureau, Geneva

Jensen JR (2007) Remote Sensing of the Environment: An Earth Resource Perspective, 2nd edn. Pearson

Kalra N, Joshi D (1994) Spectral reflectance characteristics of salt-affected arid soils of Rajasthan. J Indian Soc Remote Sens 22:183–193

King SC (2013) Fiber management solutions for the cell tower. Commun. Markets Div.

Lambin EF, Geist HJ (2001) Global LU/land-cover changes—What have we learned so far? IGBP Glob Chang Newslet 46:27–30

Leblon B, Gallant L, Granberg H (1996) Effects of shadowing types on ground-measured visible and near-infrared shadow reflectances. Remote Sens Environ 58:322–328

Mathur SM (2013) Physical geology of India, First. National book trust, India, New Delhi

McGregor P (2016) A spatial analysis of cellular tower placement along cities and highways to determine optimal tower placement criteria using geographic information science (GIS). Saint Mary's University of Minnesota University Central Services Press, Winona, MN. Pap Resour Anal 19:10
Mohamed I (2018) Path-loss estimation for wireless cellular networks using Okumura/Hata model. Sci J Circuits, Syst Signal Process 7:20
Molisch A (2011) Wireless Communications, 2nd edn. John Wiley & Sons Ltd, West Sussex
Pahlavan P, Krishnamurthy K (2013) Principles of wireless access and localization. Wiley, First
Popoola SI, Oseni OF (2014) Empirical path loss models for GSM network deployment in Makurdi, Nigeria. Int Ref J Eng Sci 3:85–94
Reis S (2008) Analyzing land use/land cover changes using remote sensing and GIS in Rize, North-East Turkey. Sensors 8:6188–6202
Rouse JW, Haas RH, Schell JA (1974) Monitoring vegetations in the great plains with ERTS. In: Third Earth resource technology, Satellite 1 Sym-posium, Greenbelt, NASA. SP-351, pp 3010–3017
Salunkhe Y, Bera S, Rao A, Venkataraman S, Raj V, Murthy U (2018) Evaluation of indicators for desertification risk assessment in part of Thar desert region of Rajasthan using geospatial techniques. Earth Syst Sci 127:1–24
Shalaby A, Tateishi R (2007) Remote sensing and GIS for mapping and monitoring land cover and land-use changes in the Northwestern coastal zone of Egypt. Appl Geogr 27:28–41
Shramek J (2019) Do it yourself Geo Apps. In: Training. https://www.esri.com/training/. Accessed 30 Oct 2019
Solomon TG, Dominic B, Konditi OC (2018) A novel radio wave propagation modeling method using system identification technique over wireless links in East Africa. Int J Antennas Propag doi:https://doi.org/10.1155/2018/2162570
Vodacek A (2018) 1051-553 Special topics: Environmental applications of remote sensing
Zhu J-K (2001) Plant salt tolerance. Trends Plant Sci 6:66–71